# Differential and Integral Equations

# Differential and Integral Equations

Edited by
Michael Greco

Willford Press
www.willfordpress.com

Published by Willford Press,
118-35 Queens Blvd., Suite 400,
Forest Hills, NY 11375, USA



ISBN: 978-1-68285-571-3

**Cataloging-in-Publication Data**

Differential and integral equations / edited by Michael Greco.
p. cm.
Includes bibliographical references and index.
ISBN 978-1-68285-571-3
1. Differential equations. 2. Integral equations. 3. Mathematics. I. Greco, Michael.
QA371 .D54 2019
515.35--dc21

For information on all Willford Press publications
visit our website at www.willfordpress.com

Willford Press

# Contents

# Preface

Differential and integral equations are studied in calculus. They are at the core of complex mathematical calculations and are applied in physics, computational mathematics, economics, etc. Differential and integral equations are closely related to each other and each can be formulated in terms of the other. The techniques used in differential and integral equations are frequently used in advanced classical mathematics, electromagnetism, general relativity and other branches of physics. The topics included in this book on differential and integral equations are of the utmost significance and bound to provide incredible insights to readers. This book further elucidates the advanced techniques and their applications in a multidisciplinary manner. The extensive content of this book will provide mathematicians, physicists and other experts with a thorough understanding of the subject.

Various studies have approached the subject by analyzing it with a single perspective, but the present book provides diverse methodologies and techniques to address this field. This book contains theories and applications needed for understanding the subject from different perspectives. The aim is to keep the readers informed about the progresses in the field; therefore, the contributions were carefully examined to compile novel researches by specialists from across the globe.

Indeed, the job of the editor is the most crucial and challenging in compiling all chapters into a single book. In the end, I would extend my sincere thanks to the chapter authors for their profound work. I am also thankful for the support provided by my family and colleagues during the compilation of this book.

**Editor**

# 1

# $\sigma$-Approximately module amenable Banach algebras

**Maryam Momeni · Taher Yazdanpanah**

**Abstract** In this paper, we define the notion of sigma-approximate module amenability of Banach algebras and give some properties about this notion. Also for Banach $\mathfrak{A}$-bimodule $\mathcal{A}$, and $\mathcal{J}$, the closed ideal of $\mathcal{A}$ generated by elements of form $(\alpha \cdot a)b - a(b \cdot \alpha)$, $(a, b \in \mathcal{A}, \quad \alpha \in \mathfrak{A})$, we considered some corollaries about $\widehat{\sigma}$-approximate amenability of $\frac{\mathcal{A}}{\mathcal{J}}$ as a Banach $\mathcal{A}$-bimodule, where $\widehat{\sigma} : \frac{\mathcal{A}}{\mathcal{J}} \to \frac{\mathcal{A}}{\mathcal{J}}$ by $\widehat{\sigma}(a + \mathcal{J}) = \sigma(a) + \mathcal{J}$ has a dense range.

**Keywords** $\sigma$-Approximate module amenable · Banach algebras · Banach module · $\sigma$-Module derivation

**Mathematics Subject Classification (2010)** 46H25

## Introduction

The concept of module amenability for Banach algebras was introduced by Amini [1]. Let $\mathfrak{A}$ and $\mathcal{A}$ be Banach algebras such that $\mathcal{A}$ is a Banach $\mathfrak{A}$-bimodule with the following compatible actions:

$$\alpha \cdot (ab) = (\alpha \cdot a)b \quad \text{and} \quad (ab) \cdot \alpha = a(b \cdot \alpha),$$

for all $a, b \in \mathcal{A}$, $\alpha \in \mathfrak{A}$. Let $\mathcal{X}$ be a Banach $\mathcal{A}$-bimodule and a Banach $\mathfrak{A}$-bimodule with compatibility of actions:

$$\alpha \cdot (a \cdot x) = (\alpha \cdot a) \cdot x \quad \text{and} \quad a \cdot (x \cdot \alpha) = (a \cdot x) \cdot \alpha,$$

for all $a \in \mathcal{A}$, $\alpha \in \mathfrak{A}$, $x \in \mathcal{X}$, and the same for the other side actions. Then, we say that $\mathcal{X}$ is a Banach $\mathcal{A}$-$\mathfrak{A}$-bimodule. If moreover, $\alpha \cdot x = x \cdot \alpha$, $(\alpha \in \mathfrak{A}, x \in \mathcal{X})$, then $\mathcal{X}$ is called a commutative $\mathcal{A}$-$\mathfrak{A}$-bimodule. Note that, when $\mathcal{A}$ acts on itself by algebra multiplication from both sides, it is not in general a Banach $\mathcal{A}$-$\mathfrak{A}$-bimodule because $\mathcal{A}$ does not satisfy $a \cdot (\alpha \cdot b) = (a \cdot \alpha) \cdot b$, $(\alpha \in \mathfrak{A}, \quad a, b \in \mathcal{A})$ [1].

If $\mathcal{A}$ is a commutative $\mathfrak{A}$-bimodule and acts on itself by algebra multiplication from both sides, then it is also a Banach $\mathcal{A}$-$\mathfrak{A}$-bimodule. Also, if $\mathcal{A}$ is a commutative Banach algebra, then it is a commutative $\mathcal{A}$-$\mathfrak{A}$-bimodule.

Now suppose that $\mathcal{X}$ be an $\mathcal{A}$-$\mathfrak{A}$-bimodule, then a continuous map $T : \mathcal{A} \to \mathcal{X}$ is called an $\mathfrak{A}$-bimodule map, if $T(a \pm b) = T(a) \pm T(b)$ and $T(\alpha \cdot a) = \alpha \cdot T(a)$ and $T(a \cdot \alpha) = T(a) \cdot \alpha$, for each $\alpha \in \mathfrak{A}$, $\mathfrak{a}, \mathfrak{b} \in \mathcal{A}$. The space of all $\mathfrak{A}$-bimodule maps $T : \mathcal{A} \to \mathcal{X}$ such that $T(ab) = T(a)T(b)$, $(a, b \in \mathcal{A})$, is denoted by $\mathrm{Hom}_{\mathfrak{A}}(\mathcal{A}, \mathcal{X})$. Also we denote $\mathrm{Hom}_{\mathfrak{A}}(\mathcal{A}, \mathcal{A})$, by $\mathrm{Hom}_{\mathfrak{A}}(\mathcal{A})$.

Let $\mathcal{A}$ and $\mathfrak{A}$ be as above and $\mathcal{X}$ be a Banach $\mathcal{A}$-$\mathfrak{A}$-bimodule. A bounded $\mathfrak{A}$-bimodule map $D : \mathcal{A} \to \mathcal{X}$ is called a module derivation if

$$D(ab) = D(a) \cdot b + a \cdot D(b), \quad (a, b \in \mathcal{A}).$$

$D$ is not necessary linear, but its boundedness implies its norm continuity, because it preserves subtraction. When $\mathcal{X}$ is commutative $\mathcal{A}$-$\mathfrak{A}$-bimodule, each $x \in \mathcal{X}$ defines a module derivation

$$\delta_x(a) = a \cdot x - x \cdot a, \quad (a \in \mathcal{A}),$$

which is called an inner module derivations.

Let $\mathcal{A}$ be a Banach $\mathfrak{A}$-bimodule and $\sigma \in \mathrm{Hom}_{\mathfrak{A}}(\mathcal{A})$. A $\sigma$-module derivation from $\mathcal{A}$ into a Banach $\mathcal{A}$-bimodule $\mathcal{X}$

M. Momeni (✉)
Department of Mathematics, Ahvaz Branch, Islamic Azad University (IAU), Ahvaz, Iran
e-mail: srb.maryam@gmail.com

T. Yazdanpanah
Department of Mathematics, Persian Gulf University, 75169 Bushehr, Iran
e-mail: yazdanpanah@pgu.ac.ir

is a bounded $\mathfrak{A}$-bimodule map $D : \mathcal{A} \longrightarrow \mathcal{X}$ satisfying

$$D(ab) = \sigma(a) \cdot D(b) + D(a) \cdot \sigma(b), \quad (a, b \in \mathcal{A}).$$

When $\mathcal{X}$ is commutative $\mathcal{A}$-$\mathfrak{A}$-bimodule, each $x \in \mathcal{X}$ defines a $\sigma$-module derivation

$$\delta_x^\sigma : \mathcal{A} \longrightarrow \mathcal{X}, \quad \delta_x^\sigma(a) = \sigma(a) \cdot x - x \cdot \sigma(a), \quad (a \in \mathcal{A}),$$

which is called a $\sigma$-inner module derivation.

## $\sigma$-Approximate module amenability

We start this section with definition of sigma-approximate module amenability, then we consider some hereditary properties of this concept.

**Definition 1** Let $\mathcal{A}$ be a Banach $\mathfrak{A}$-bimodule and $\sigma \in Hom_{\mathfrak{A}}(\mathcal{A})$. We say that $\mathcal{A}$ is a $\sigma$-approximately module amenable ($\sigma$-(AMA)), if for each commutative Banach $\mathcal{A}$-$\mathfrak{A}$-bimodule, $\mathcal{X}$, every $\sigma$-module derivation $D : \mathcal{A} \longrightarrow \mathcal{X}^*$ is $\sigma$-approximately inner, i.e, there is a net $(x_i)_{i \in \mathfrak{I}} \in \mathcal{X}^*$ such that $D(a) = \lim_i \delta_{x_i}^\sigma(a) = \lim_i \sigma(a)x_i - x_i\sigma(a)$, $(a \in \mathcal{A})$. Also we say that $\mathcal{A}$ is a $\sigma$-approximately module contractible ($\sigma - (AMC)$), if for each commutative Banach $\mathcal{A}$-$\mathfrak{A}$-bimodule, $\mathcal{X}$, every $\sigma$-module derivation $D : \mathcal{A} \longrightarrow \mathcal{X}$ is $\sigma$-approximately inner.

The two following results is the $\sigma$-approximate version of [1, Proposition 2.1] and [5], respectively.

**Proposition 2** *Let $\mathcal{A}$ be a Banach $\mathfrak{A}$-bimodule and $\sigma \in Hom_{\mathfrak{A}}(\mathcal{A})$. Suppose that $\mathfrak{A}$ has a bounded approximate identity and $\mathcal{A}$ is $\sigma$-approximately amenable. Then $\mathcal{A}$ is $\sigma$-(AMA).*

*Proof* Let $\mathcal{X}$ be a commutative $\mathcal{A}$-$\mathfrak{A}$-bimodule and $D : \mathcal{A} \longrightarrow \mathcal{X}^*$ be a $\sigma$-module derivation. By [1, Proposition 2.1], $D$ is a $\sigma$-derivation, i.e, $D$ is $\mathbb{C}$-linear. Now since $\mathcal{A}$ is $\sigma$-approximately amenable, $\mathcal{A}$ is $\sigma$-(AMA). □

**Proposition 3** *Let $\mathcal{A}$ be an essential left Banach $\mathfrak{A}$-bimodule and $\sigma \in Hom_{\mathfrak{A}}(\mathcal{A})$. If $\mathcal{A}$ is $\sigma$-approximately amenable, then $\mathcal{A}$ is $\sigma$-(AMA).*

*Proof* Let $\mathcal{X}$ be a commutative $\mathcal{A}$-$\mathfrak{A}$-bimodule and $D : \mathcal{A} \longrightarrow \mathcal{X}^*$ be a $\sigma$-module derivation. Since $\mathcal{A}$ is an essential left Banach $\mathfrak{A}$-bimodule, $D$ is $\mathbb{C}$-linear [5]. Now since $\mathcal{A}$ is $\sigma$-approximately amenable, $D$ is $\sigma$-approximately inner and thus $\mathcal{A}$ is $\sigma$-(AMA). □

**Proposition 4** *Let $\mathcal{A}$ be a Banach $\mathfrak{A}$-bimodule and $\sigma \in Hom_{\mathfrak{A}}(\mathcal{A})$. If $\mathcal{A}$ is $\sigma$-(AMA), then $\mathcal{A}$ is $(\lambda \circ \sigma, \mu \circ \sigma)$-(AMA), for each $\lambda, \mu \in \mathrm{Hom}_{\mathfrak{A}}(\mathcal{A})$.*

*Proof* Let $\mathcal{X}$ be a commutative $\mathcal{A}$-$\mathfrak{A}$-bimodule and $D : \mathcal{A} \longrightarrow \mathcal{X}^*$ be a $(\lambda \circ \sigma, \mu \circ \sigma)$-module derivation. Then $\mathcal{X}$ is an $\mathcal{A}$-module derivation with the following module actions:

$$a * x = \lambda(a) \cdot x \quad \text{and} \quad x * a = x \cdot \mu(a), \quad (a \in \mathcal{A},\, x \in \mathcal{X}).$$

It is easy to see that $\mathcal{X}$ is a commutative $\mathcal{A}$-$\mathfrak{A}$-bimodule with this product. We have

$$\begin{aligned} D(ab) &= (\lambda \circ \sigma)(a) \cdot D(b) + D(a) \cdot (\mu \circ \sigma)(b) \\ &= \sigma(a) * D(b) + D(a) * \sigma(b), \quad (a, b \in \mathcal{A}). \end{aligned}$$

Thus, $D$ is a $\sigma$-module derivation. So there exists a net $(x_i) \in \mathcal{X}^*$ such that $D(a) = \lim_i \delta_{x_i}^\sigma(a)$, $(a \in \mathcal{A})$. So we have

$$\begin{aligned} D(a) &= \lim_i(\sigma(a) * x_i - x_i * \sigma(a)) \\ &= \lim_i((\lambda \circ \sigma)(a) \cdot x_i - x_i \cdot (\mu \circ \sigma)(a)), \quad (a \in \mathcal{A}). \end{aligned}$$

Which shows that $D$ is $(\lambda \circ \sigma, \mu \circ \sigma)$-approximately inner. Thus, $\mathcal{A}$ is $(\lambda \circ \sigma, \mu \circ \sigma)$-(AMA). □

**Corollary 5** *Let $\mathcal{A}$ be a Banach $\mathfrak{A}$-bimodule. If $\mathcal{A}$ is (AMA), then $\mathcal{A}$ is $(\lambda, \mu)$- (AMA), for each $\lambda, \mu \in \mathrm{Hom}_{\mathfrak{A}}(\mathcal{A})$.*

**Proposition 6** *Let $\mathcal{A}$ be a Banach $\mathfrak{A}$-bimodule and $\sigma \in \mathrm{Hom}_{\mathfrak{A}}(\mathcal{A})$. Suppose that $\sigma$ is an idempotent epimorphism and $\mathcal{A}$ is $\sigma$-(AMA). Then, $\mathcal{A}$ is (AMA).*

*Proof* Let $\mathcal{X}$ be a commutative $\mathcal{A}$-$\mathfrak{A}$-bimodule and $D : \mathcal{A} \longrightarrow \mathcal{X}^*$ be a module derivation. So $\widetilde{D} = D \circ \sigma$ is a $\sigma$-module derivation, because, for each $a, b \in \mathcal{A}$ and $\alpha \in \mathfrak{A}$ we have

$$\begin{aligned} \widetilde{D}(ab) &= D \circ \sigma(ab) = D(\sigma(a)\sigma(b)) \\ &= \sigma(a)(D \circ \sigma)(b) + (D \circ \sigma)(a)\sigma(b), \end{aligned}$$

and

$$\widetilde{D}(\alpha a) = D(\sigma(\alpha a)) = D(\alpha\sigma(a)) = \alpha D(\sigma(a)).$$

Since $\mathcal{A}$ is $\sigma$-(AMA), there exists a net $(x_i)_{i \in \mathfrak{I}} \in \mathcal{X}^*$ such that $\widetilde{D}(a) = \lim_i(\sigma(a)x_i - x_i\sigma(a))$, $(a \in \mathcal{A})$. Now for each $b \in \mathcal{A}$, there exists $a \in \mathcal{A}$ such that $b = \sigma(a)$. Therefore,

$$\begin{aligned} D(b) = D(\sigma(a)) = \widetilde{D}(a) &= \lim_i(\sigma(a)x_i - x_i\sigma(a)) \\ &= \lim_i(bx_i - x_ib), \quad (b \in \mathcal{A}). \end{aligned}$$

So $D$ is approximately inner and $\mathcal{A}$ is (AMA). □

**Proposition 7** *Let $\mathcal{A}$ and $\mathcal{B}$ be Banach $\mathfrak{A}$-bimodules and $\sigma \in Hom_{\mathfrak{A}}(\mathcal{A})$ and $\tau \in \mathrm{Hom}_{\mathfrak{A}}(\mathcal{B})$. Suppose that $\varphi \in \mathrm{Hom}_{\mathfrak{A}}(\mathcal{A}, \mathcal{B})$ be a surjective map such that $\varphi \circ \sigma = \tau \circ \varphi$. If $\mathcal{A}$ is $\sigma$-(AMA), then $\mathcal{B}$ is $\tau$-(AMA).*

*Proof* Let $\mathcal{X}$ be a commutative Banach $\mathcal{B}$-$\mathfrak{A}$-bimodule and $\mathcal{D} : \mathcal{B} \to \mathcal{X}^*$ be a $\tau$-module derivation. Then, $(\mathcal{X}, *)$ can be considered as a Banach $\mathcal{A}$-$\mathfrak{A}$-bimodule by the following module actions:

$$a * x = \varphi(a) \cdot x \quad \text{and} \quad x * a = x \cdot \varphi(a), \quad (a \in \mathcal{A}, x \in X).$$

Therefore, $\widetilde{D} = D \circ \varphi : \mathcal{A} \to (\mathcal{X}^*, *)$ is a $\sigma$-module derivation, because

$$\begin{aligned}\widetilde{D}(ab) &= D(\varphi(a)\varphi(b)) \\ &= D(\varphi(a))\tau(\varphi(b)) + \tau(\varphi(a))D(\varphi(b)) \\ &= \widetilde{D}(a)\varphi(\sigma(b)) + \varphi(\sigma(a))\widetilde{D}(b) \\ &= \widetilde{D}(a) * \sigma(b) + \sigma(a) * \widetilde{D}(b), \quad (a, b \in \mathcal{A}).\end{aligned}$$

Since $\mathcal{A}$ is $\sigma$-(AMA), there exists a net $(x_i)_{i\in\mathfrak{I}} \in \mathcal{X}^*$ such that $\widetilde{D} = \lim_i \delta^{\sigma}_{x_i}$. So we have

$$\begin{aligned}\widetilde{D}(a) &= \lim_i \sigma(a) * x_i - x_i * \sigma(a) \\ &= \lim_\alpha \varphi(\sigma(a)) \cdot x_i - x_i \cdot \varphi(\sigma(a)) \\ &= \lim_\alpha \tau(\varphi(a)) \cdot x_i - x_i \cdot \tau(\varphi(a)), \quad (a \in \mathcal{A}).\end{aligned}$$

Since $\varphi$ is a surjective map, so $D(b) = \lim_i \tau(b) \cdot x_i - x_i \cdot \tau(b)$, $(b \in \mathcal{B})$. Hence, $\mathcal{B}$ is $\tau$-(AMA). □

**Proposition 8** *Suppose that $\mathcal{A}$ and $\mathcal{B}$ are Banach $\mathfrak{A}$-modules and $\varphi \in Hom_{\mathfrak{A}}(\mathcal{A}, \mathcal{B})$ be a surjective map. If $\mathcal{A}$ is (AMA), then $\mathcal{B}$ is $\sigma$-(AMA), for each $\sigma \in \mathrm{Hom}_{\mathfrak{A}}(\mathcal{B})$.*

*Proof* Let $\mathcal{X}$ be a Banach $\mathcal{B}$-$\mathfrak{A}$-bimodule and $\sigma \in \mathrm{Hom}_{\mathfrak{A}}(\mathcal{B})$. Then $(\mathcal{X}, *)$ is a Banach $\mathcal{A}$-$\mathfrak{A}$-bimodule with the following module actions:

$$a * x = \sigma(\varphi(a)) \cdot x \quad \text{and} \quad x * a = x \cdot \sigma(\varphi(a)), \quad (a \in \mathcal{A}, x \in \mathcal{X}).$$

Now, let $D : \mathcal{B} \to \mathcal{X}^*$ be a $\sigma$-module derivation. So $\widetilde{D} = D \circ \varphi : \mathcal{A} \to (\mathcal{X}^*, *)$ is a module derivation, because for each $\alpha \in \mathfrak{A}$ and $a, b \in \mathcal{A}$, we have

$$\widetilde{D}(\alpha a) = D(\varphi(\alpha a)) = D(\alpha\varphi(a)) = \alpha D(\varphi(a)),$$

and

$$\begin{aligned}\widetilde{D}(ab) &= D(\varphi(ab)) \\ &= D(\varphi(a))\sigma(\varphi(b)) + \sigma(\varphi(a))D(\varphi(b)) \\ &= \widetilde{D}(a) * b + a * \widetilde{D}(b).\end{aligned}$$

So there exists a net $(x_i)_{i\in\mathfrak{I}} \in X^*$ such that $\widetilde{D} = \lim_i \delta_{x_i}$ and we have

$$\begin{aligned}\widetilde{D}(a) &= \lim_i \delta_{x_i}(a) \\ &= \lim_i (a * x_i - x_i * a) \\ &= \lim_i \sigma(\varphi(a)) \cdot x_i - x_i \cdot \sigma(\varphi(a)), \quad (a \in \mathcal{A}).\end{aligned}$$

Since $\varphi$ is surjective, for each $b \in \mathcal{B}$, there exists $a \in \mathcal{A}$, such that $b = \varphi(a)$. So for each $b \in \mathcal{B}$ we have

$$\begin{aligned}D(b) = D(\varphi(a)) = \widetilde{D}(a) &= \lim_i \sigma(\varphi(a)) \cdot x_i - x_i \cdot \sigma(\varphi(a)) \\ &= \lim_i \sigma(b) \cdot x_i - x_i \cdot \sigma(b).\end{aligned}$$

Which shows that $D$ is $\sigma$-approximately inner. Thus, $\mathcal{B}$ is $\sigma$-(AMA). □

Let $\mathcal{A}$ be a Banach $\mathfrak{A}$-bimodule with compatible actions and $\mathcal{J}$ be the closed ideal of $\mathcal{A}$ generated by elements of form $(\alpha \cdot a)b - a(b \cdot \alpha)$, for all $a, b \in \mathcal{A}$ and $\alpha \in \mathfrak{A}$. Then, the quotient Banach algebra $\frac{\mathcal{A}}{\mathcal{J}}$ is Banach $\mathcal{A}$-bimodule with compatible actions [2]. The following Lemma is proved in [3].

**Lemma 9** Let $\mathcal{A}$ be a Banach $\mathfrak{A}$-bimodule and $\mathfrak{A}$ has a bounded approximate identity for $\mathcal{A}$. Suppose that $\sigma \in Hom_{\mathfrak{A}}(\mathcal{A})$ such that $\sigma(\mathcal{J}) \subseteq \mathcal{J}$. Then $\widehat{\sigma} : \frac{\mathcal{A}}{\mathcal{J}} \to \frac{\mathcal{A}}{\mathcal{J}}$ by $\widehat{\sigma}(a + \mathcal{J}) = \sigma(a) + \mathcal{J}$ is $\mathbb{C}$-linear.

**Proposition 10** *Let $\mathcal{A}$ be a Banach $\mathfrak{A}$-bimodule and $\mathfrak{A}$ has a bounded approximate identity for $\mathcal{A}$. Let $\sigma$ be as in above lemma. If $\frac{\mathcal{A}}{\mathcal{J}}$ is $\widehat{\sigma}$-approximately amenable, then $\mathcal{A}$ is* $\sigma$-(AMA).

*Proof* Let $\mathcal{X}$ be a commutative Banach $\mathcal{A}$-$\mathfrak{A}$-bimodule. It is easy to see that $\mathcal{J} \cdot \mathcal{X} = \mathcal{X} \cdot \mathcal{J} = 0$. So $\mathcal{X}$ is a Banach $\frac{\mathcal{A}}{\mathcal{J}}$-bimodule with the following module actions;

$$(a + \mathcal{J}) \cdot x = ax \quad \text{and} \quad x \cdot (a + \mathcal{J}) = xa, \quad (a \in \mathcal{A}, x \in \mathcal{X}).$$

Suppose that $D : \mathcal{A} \to \mathcal{X}^*$ be a $\sigma$-module derivation. Define $\widehat{D} : \frac{\mathcal{A}}{\mathcal{J}} \to \mathcal{X}^*$ by $\widehat{D}(a + \mathcal{J}) = D(a)$, $(a \in \mathcal{A})$. $\widehat{D}$ is well defined [3, Proposition 2.6] and it is $\mathbb{C}$-linear [1, Proposition 2.1]. Also, it is easy to see that $\widehat{D}(ab + \mathcal{J}) = \widehat{D}(a + \mathcal{J})\ \widehat{\sigma}(b + \mathcal{J}) + \widehat{\sigma}(a + \mathcal{J})\widehat{D}(b + \mathcal{J})$. Moreover according to the above Lemma, $\widehat{\sigma}$ is $\mathbb{C}$-linear. Therefore, $\widehat{D}$ is $\widehat{\sigma}$-derivation. Thus, there exists a net $(x_i)_{i\in\mathfrak{I}} \in X^*$ such that $\widehat{D} = \lim_i \delta^{\widehat{\sigma}}_{x_i}$ and we have

$$\begin{aligned}D(a) = \widehat{D}(a + \mathcal{J}) &= \lim_i (\widehat{\sigma}(a) \cdot x_i - x_i \cdot \widehat{\sigma}(a)) \\ &= \lim_i (\sigma(a) + \mathcal{J}) \cdot x_i - x_i \cdot (\sigma(a) + \mathcal{J}) \\ &= \lim_i \sigma(a)x_i - x_i\sigma(a), \quad (a \in \mathcal{A}).\end{aligned}$$

Which shows that $D$ is $\sigma$-approximately inner and therefore $\mathcal{A}$ is $\sigma$-(AMA). □

In [4], section 3, we stated some properties of $\sigma$-approximate contractibility when $\mathcal{A}$ has an identity and considered some corollaries when $\sigma(\mathcal{A})$ is dense in $\mathcal{A}$. In proof of the following proposition we use those results. Recall that, the Banach algebra $\mathfrak{A}$ acts trivially on $\mathcal{A}$ from left if for each $\alpha \in \mathfrak{A}$ and $a \in \mathcal{A}$, $\alpha \cdot a = f(\alpha)a$, where $f$ is a continuous linear functional on $\mathfrak{A}$.

**Proposition 11** *Let $\mathcal{A}$ be a Banach $\mathfrak{A}$-bimodule with trivial left actions and $\sigma$ be as in above lemma. Suppose that $\mathcal{A}$ is $\sigma$-(AMA). If $\frac{A}{\mathcal{J}}$ has an identity and $\overline{\widehat{\sigma}(\frac{A}{\mathcal{J}})} = \frac{A}{\mathcal{J}}$, then $\frac{A}{\mathcal{J}}$ is $\widehat{\sigma}$-approximately amenable.*

*Proof* By [4, Corollary 3.3.], we can assume that $\mathcal{X}$ is a $\sigma$-unital Banach $\frac{A}{\mathcal{J}}$-bimodule. Let $e + \mathcal{J}$ be the identity in $\frac{A}{\mathcal{J}}$. So $\widehat{\sigma}(e + \mathcal{J})$ is a unit for $\widehat{\sigma}(\frac{A}{\mathcal{J}})$. Thus by density of $\widehat{\sigma}(\frac{A}{\mathcal{J}})$ in $\frac{A}{\mathcal{J}}$, we see that $\widehat{\sigma}(e + \mathcal{J}) = e + \mathcal{J}$. Now let $\widehat{D} : \frac{A}{\mathcal{J}} \to \mathcal{X}^*$ be a $\widehat{\sigma}$-derivation. By [4, Lemma 3.7], $\widehat{D}(e + \mathcal{J}) = 0$. Now similar to [3, Proposition 2.7], we can see $\mathcal{X}$ as a commutative Banach $\mathcal{A}$-$\mathfrak{A}$-bimodule and $D = \widehat{D} \circ \pi : \mathcal{A} \to \mathcal{X}^*$ is a $\sigma$-module derivation, where $\pi : \mathcal{A} \to \frac{A}{\mathcal{J}}$ is the natural $\mathfrak{A}$-module map. Since $\mathcal{A}$ is $\sigma$-(AMA), there exists a net $(x_i)_{i\in\mathfrak{I}} \in X^*$ such that $D = \lim_i \widehat{\delta^{\sigma}_{x_i}}$ and we have,

$$\begin{aligned}\widehat{D}(a + \mathcal{J}) &= \lim_i (\sigma(a) + \mathcal{J}) \cdot x_i - x_i \cdot (\sigma(a) + \mathcal{J}) \\ &= \lim_i \widehat{\sigma}(a + \mathcal{J}) \cdot x_i - x_i \cdot \widehat{\sigma}(a + \mathcal{J}), \quad (a \in \mathcal{A}).\end{aligned}$$

Which means that $\widehat{D}$ is $\widehat{\sigma}$-approximately inner and $\widehat{\sigma}$-approximately amenable. □

Let $\mathfrak{A}$ be a non-unital Banach algebra. Then, $\mathfrak{A}^{\#} = \mathfrak{A} \oplus \mathbb{C}$, the unitization of $\mathfrak{A}$, is a unital Banach algebra which contains $\mathfrak{A}$ as a closed ideal. Let $\mathcal{A}$ be a Banach algebra and a Banach $\mathfrak{A}$-bimodule with compatible actions. Then, $\mathcal{A}$ is a Banach algebra and a Banach $\mathfrak{A}^{\#}$-bimodule with the following actions:

$$(\alpha, \lambda)a = \alpha a + \lambda a \quad \text{and} \quad a(\alpha, \lambda) = a\alpha + a\lambda, \\ (\alpha \in \mathfrak{A}, \lambda \in \mathbb{C}, \mathfrak{a} \in \mathcal{A}).$$

Let $\mathcal{A}$ be a Banach algebra and a Banach $\mathfrak{A}$-bimodule with compatible actions and let $\mathcal{A}^{\#} = \mathcal{A} \oplus \mathfrak{A}^{\#}$. Then $(\mathcal{A}^{\#}, \cdot)$ is a Banach algebra, where the multiplication $\cdot$ is defined by $(a, \mathfrak{u}) \cdot (\mathfrak{b}, \mathfrak{v}) = (\mathfrak{ab} + \mathfrak{av} + \mathfrak{ub}, \mathfrak{uv})$, $(a, b \in \mathcal{A}, \mathfrak{u}, \mathfrak{v} \in \mathfrak{A})$. Also $\mathcal{A}^{\#}$ is a Banach $\mathfrak{A}^{\#}$-bimodule with the following module actions:

$$(a, \mathfrak{u}) \cdot \mathfrak{v} = (a \cdot \mathfrak{v}, \mathfrak{uv}) \quad \text{and} \quad \mathfrak{v} \cdot (a, \mathfrak{u}) = (\mathfrak{v} \cdot a, \mathfrak{vu}) \\ (a \in \mathcal{A}, \mathfrak{u}, \mathfrak{v} \in \mathfrak{A}^{\#}).$$

So $\mathcal{A}^{\#}$ is a unital Banach $\mathfrak{A}^{\#}$-bimodule with compatible actions.

A similar result of [5, Theorem 3.1], for approximate module amenability, is as follows:

**Proposition 12** *Let $\mathcal{A}$ be a Banach $\mathfrak{A}$-bimodule, $\sigma \in \mathrm{Hom}_{\mathfrak{A}}(\mathcal{A})$. Then $\widehat{\sigma}(a, \mathfrak{u}) = \sigma(\mathfrak{a}) \oplus \mathfrak{u}$, $(a \in \mathcal{A}, \mathfrak{u} \in \mathfrak{A}^{\#})$ is in $\mathrm{Hom}^{\#}_{\mathfrak{A}}(\mathcal{A}^{\#})$ and the following are equivalent;*

(i) $\mathcal{A}$ is $\sigma$-(AMA) as an $\mathfrak{A}^{\#}$-bimodule.
(ii) $\mathcal{A}^{\#}$ is $\widehat{\sigma}$-(AMA) as an $\mathfrak{A}^{\#}$-bimodule.

*Proof* It is easy to see that $\widehat{\sigma} \in \mathrm{Hom}^{\#}_{\mathfrak{A}}(\mathcal{A}^{\#})$.

$i \Rightarrow 2$. Let $\mathcal{X}$ be a commutative Banach $\mathcal{A}^{\#}$–$\mathfrak{A}^{\#}$-bimodule and $\widehat{D} : \mathcal{A}^{\#} \to \mathcal{X}^*$ be a $\widehat{\sigma}$-module derivation. By [4, Lemma 3.1], $\widehat{D}(1) = 0$. So $D = \widehat{D}\,|_{\mathcal{A}}: \mathcal{A} \to \mathcal{X}^*$ is a $\sigma$-module derivation. Thus by the hypothesis, there exists a net $(x_i)_{i\in\mathfrak{I}} \in X^*$ such that $D = \lim_i \delta^{\sigma}_{x_i}$. Note that $\mathcal{X}$ is a commutative Banach $\mathcal{A}$-$\mathfrak{A}^{\#}$-module and $\widehat{D}(a, 0) = \lim_i \widehat{\sigma}(a, 0)x_i - x_i\widehat{\sigma}(a, 0)$, $(a \in \mathcal{A})$. Also we have

$$\begin{aligned}\widehat{D}(a, \mathfrak{u}) &= \widehat{D}((a, 0) + (0, \mathfrak{u})) = \widehat{D}(a, 0) + \widehat{D}(0, \mathfrak{u}) \\ &= \widehat{D}(a, 0), \quad ((a, \mathfrak{u}) \in \mathcal{A}^{\#}).\end{aligned}$$

Thus, $\widehat{D}$ is $\widehat{\sigma}$-approximately inner and therefore $\mathcal{A}^{\#}$ is $\widehat{\sigma}$-(AMA).

$ii \Rightarrow i$. Let $\mathcal{X}$ be a commutative Banach $\mathcal{A}$-$\mathfrak{A}^{\#}$-bimodule and $D : \mathcal{A} \to \mathcal{X}^*$ be a $\sigma$-module derivation. Define $\widehat{D} : \mathcal{A}^{\#} \to \mathcal{X}^*$ by $\widehat{D}(a, \mathfrak{u}) = D(a)$, $((a, \mathfrak{u}) \in \mathcal{A})$. Thus $\widehat{D}$ is $\widehat{\sigma}$-$\mathfrak{A}^{\#}$-module derivation, because,

$$\begin{aligned}\widehat{D}((a, \mathfrak{u})(b, \mathfrak{v})) &= \widehat{D}((ab + a\mathfrak{v} + \mathfrak{u}b), \mathfrak{uv}) \\ &= D(ab + a\mathfrak{v} + \mathfrak{u}b) \\ &= D(ab) + +D(a)\mathfrak{v} + \mathfrak{u}D(b) \\ &= \sigma(a)D(b) + D(a)\sigma(b) + D(a)\mathfrak{v} + \mathfrak{u}D(b) \\ &= (\sigma(a) + \mathfrak{u})D(b) + D(a)(\sigma(b) + \mathfrak{v}) \\ &= \widehat{\sigma}(a, \mathfrak{u})D(b) + D(a)\widehat{\sigma}(b, \mathfrak{v}) \\ &= \widehat{\sigma}(a, \mathfrak{u})\widehat{D}(b, \mathfrak{v}) + \widehat{D}(a, \mathfrak{u})\widehat{\sigma}(b, \mathfrak{v}), \\ &\quad (a, b \in \mathcal{A}, \mathfrak{u}, \mathfrak{v} \in \mathfrak{A}^{\#}).\end{aligned}$$

and by (ii) is a module $\widehat{D}$-approximately inner. Therefore, $D$ is module approximately inner. So $\mathcal{A}$ is $\sigma$-*(AMA)*. □

## References

1. Amini, M.: Module amenability for semigroup algebras. Semigroup Forum **69**, 243–254 (2004)
2. Amini, M., Bodaghi, A.: Module amenability and weak module amenability for second dual of Banach algebras. Chamchuri J. Math. **2**, 57–71 (2010)
3. Bodaghi, A.: Module $(\varphi, \psi)$-amenability of Banach algebras, Archivum mathemeticum (BRNO). Tomus **46**, 227–235 (2010)
4. Momeni, M., Yazdanpanah, T., Mardanbeigi, M.R.: $\sigma$-Approximately contractible Banach algebras. Abstr. Appl. Anal. **2012**, Article ID 653140 (2012)
5. Pourmahmood-Aghababa, H., Bodaghi, A.: Module approximate amenability of Banach algebras. Bull. Iran. Math. Soc.

2

# Extension of some theorems to find solution of nonlinear integral equation and homotopy perturbation method to solve it

**Mohsen Rabbani**[1] · **Reza Arab**[1]

**Abstract** In this paper, the concept of contraction via the measure of non-compactness on a Banach space is investigated by generalizing some results which have been previously discussed in literatures. Furthermore, to validity of the theorems and homotopy perturbation method (HPM), as a technical solution, they are applied on some nonlinear singular integral equations.

**Keywords** Measure of non-compactness · Integral equation · Homotopy perturbation

## Introduction and auxiliary facts

Integral equation is an essential branch of sciences that it has applications in engineering sciences, physical sciences, etc. Measures of non-compactness used for existence of solution fractional integral equations [5], singular Volterra integral equations discussed in [2] and also in [3, 6] Darbo fixed point theorem was created by measures of non-compactness. But we consider solvability of the nonlinear problem with fractional order in the following form:

$$u(s) = g(s) + h(s, u(s)) \int_0^s \frac{(s^m - \xi^m)^{\alpha-1}}{\Gamma(\alpha)} m\xi^{m-1} k(f(s,\xi))u(\xi)\, d\xi, \quad s \in [0,1], \quad 0 < \alpha \leq 1, \quad m > 0, \tag{1.1}$$

where $\Gamma(\alpha) = \int_0^\infty t^{\alpha-1} e^{-t} dt$ and $h(s, u)$ is generated by the superposition operator of $H$ such that $(Hu)(s) = h(s, u(s))$, where $u = u(s)$ defined on [0, 1] in [4]. We prove the existence of some non-decreasing solutions for Eq. (1.1) in $C[0, 1]$ (set of all continuous functions on [0, 1]). In the following for ability and validity of the proposed method, we solve an example of Eq. (1.1) by homotopy perturbation method.

In this section, we suppose $A \neq \emptyset$ and $A \subseteq E$, where $(E, \|\cdot\|)$ is a real Banach space. Also $\mathfrak{M}_E \neq \emptyset$ is a family of bounded subsets of $E$ and $\mathfrak{N}_E$ a subfamily consisting of all relatively compact sets.

**Definition 1** [6] A mapping $\eta : \mathfrak{M}_E \to \mathbb{R}^+$ is a measure of non-compactness in $E$ if it satisfies the following conditions:

($1^0$) The family $\ker\eta = \{A \in \mathfrak{M}_E : \eta(A) = 0\}$ is nonempty and $\ker\eta \subset \mathfrak{N}_E$,
($2^0$) $A \subset B \Rightarrow \eta(A) \leq \eta(B)$,
($3^0$) $\eta(\bar{A}) = \eta(A)$,
($4^0$) $\eta(\text{Conv}A) = \eta(A)$,
($5^0$) $\eta(\lambda A + (1-\lambda)B) \leq \lambda\eta(A) + (1-\lambda)\eta(B)$ for $\lambda \in [0, 1]$,
($6^0$) If $\{A_n\}$ be a sequence of closed sets from $m_E$ such that $A_{n+1} \subset A_n$ for $n \in \mathbb{N}$ and if $\lim_{n\to\infty} \eta(A_n) = 0$, then the set $A_\infty = \bigcap_{n=1}^\infty A_n$ is nonempty.

✉ Mohsen Rabbani
mrabbani@iausari.ac.ir

Reza Arab
mathreza.arab@iausari.ac.ir

[1] Department of Mathematics, Sari Branch, Islamic Azad University, Sari, Iran

## Extension of Darbo fixed point theorem

For obtaining the generalization of Darbo fixed point theorem (see [1]), we present a new kind of contraction. So throughout this paper we assume that functions of $G, \Theta, \phi : [0, +\infty) \to [0, +\infty)$ satisfy in these conditions:

(a) $G \in C[0,+\infty)$ and $G(0)=0<G(s), \forall s>0$;
(b) $\phi(s)<\Theta(s), \forall s>0$ and $\phi(0)=\Theta(0)=0$;
(c) $\phi(s), \Theta(s) \in C[0,+\infty)$;
(d) $\Theta$ is increasing.

Also, let $\mathbb{G}=\{G : G \text{ satisfy (a)}\}$ and $\Psi=\{(\Theta,\phi) : \Theta$ and $\phi$ satisfy (b), (c) and (d)$\}$.

Now, we illustrate the generalized $(\Theta, G, \phi)$-contractive mappings via the measure of non-compactness by the following definition and theorem.

**Definition 2** Let $v \neq \emptyset$, subset of a Banach space $E$ and $\tau : v \to v$ be a mapping. We say that $\tau$ is a generalized $(\Theta, G, \phi)$-contractive mapping if for any $0<a<b<\infty$ there exist $0<\rho(a,b)<1$, $G \in \mathbb{G}$ and $(\Theta,\phi) \in \Psi$ which for all $A \subseteq v$ and $\eta$ (arbitrary measure of non-compactness), then

$$a \le G(\eta(A)) \le b \Longrightarrow \Theta(G(\eta(\tau A))) \le \rho(a,b)\phi(G(\eta(A))). \tag{2.1}$$

**Theorem 1** *Let $v \neq \emptyset$, bounded, closed, convex and subset of a Banach space $E$ and $\tau : v \to v$ be a generalized $(\Theta, G, \phi)$-contractive continuous mapping. Then $\tau$ has at least one fixed point in $v$.*

*Proof* Let $v_0 = v$, we construct a sequence $\{v_n\}$ where $v_{n+1} = \mathrm{Conv}(\tau v_n)$, for $n \ge 0$. $\tau v_0 = \tau v \subseteq v = v_0$, $v_1 = \mathrm{Conv}(\tau v_0) \subseteq v = v_0$, therefore by continuing this process, we have

$$v_0 \supseteq v_1 \supseteq \cdots \supseteq v_n \supseteq v_{n+1} \supseteq \cdots$$

If $\exists N \in \mathbb{N}$; $G(\eta(v_N))=0$, i.e., $\eta(v_N)=0$, then $v_N$ is relatively compact. On the other hand, since $\tau(v_N) \subseteq \mathrm{Conv}(\tau v_N) = v_{N+1} \subseteq v_N$ so, $\tau$ is compact. Thus from Shauder Theorem (see [1]) we conclude that $\tau$ has a fixed point. Otherwise we suppose,

$$G(\eta(v_n)) > 0, \quad \forall n \ge 1. \tag{2.2}$$

If

$$G(\eta(v_{n_0})) < G(\eta(v_{n_0+1})), \tag{2.3}$$

for some $n_0 \in \mathbb{N}$, according to (2.2) and (2.3), we can get

$$0<a := G(\eta(v_{n_0})) \le G(\eta(v_{n_0})) < G(\eta(v_{n_0+1})) := b.$$

By considering $\tau$ and Definition 2, there exists $0<\rho(a,b)<1$ such that

$$\begin{aligned}\Theta(G(\eta(v_{n_0+1}))) &= \Theta(G(\eta(\mathrm{conv}(\tau v_{n_0})))) = \Theta(G(\eta(\tau v_{n_0})))\\ &\le \rho(a,b)\phi(G(\eta(v_{n_0})))\\ &< \rho(a,b)\Theta(G(\eta(v_{n_0})))\\ &< \rho(a,b)\Theta(G(\eta(v_{n_0+1}))),\end{aligned}$$

which implies that $\rho(a,b) > 1$, and this is a contradiction. So, we can write,

$$G(\eta(v_{n+1})) \le G(\eta(v_n)),$$

for all $n \in \mathbb{N}$, that is, the sequence $\{G(\eta(v_n))\}$ is nonincreasing and nonnegative, we infer that

$$\lim_{n\to\infty} G(\eta(v_n)) = \delta. \tag{2.4}$$

Now, if $\delta > 0$, then

$$0<a := \delta \le G(\eta(v_n)) \le G(\eta(v_0)) =: b, \quad \text{for all } n \ge 0.$$

By considering $\tau$ and Definition 2, there exists $0<\rho(a,b)<1$ such that

$$\begin{aligned}\Theta(G(\eta(v_{n+1}))) &= \Theta(G(\eta(\mathrm{conv}(\tau v_n)))) = \Theta(G(\eta(\tau v_n)))\\ &\le \rho(a,b)\phi(G(\eta(v_n)))\\ &< \rho(a,b)\Theta(G(\eta(v_n))),\end{aligned} \tag{2.5}$$

from (2.4) and continuity of the $\Theta$ and $\phi$ in (2.5), we get

$$\Theta(\delta) = \lim_{n\to\infty} \Theta(G(\eta(v_{n+1}))) = \lim_{n\to\infty} \phi(G(\eta(v_n))) = \phi(\delta),$$

from (b) it is concluded that $\delta = 0$ and this is a contradiction. So in the above process $\delta = 0$ and

$$\lim_{n\to\infty} G(\eta(v_n)) = 0.$$

It follows that

$$\lim_{n\to\infty} \eta(v_n) = 0.$$

From $v_n \supseteq v_{n+1}$ and $\tau v_n \subseteq v_n$ for $n \in \mathbb{N}$, as a result of $(6^0)$, we can write

$$v_\infty = \bigcap_{n=1}^{\infty} v_n \neq \emptyset,$$

is a convex closed set, invariant under $\tau$ and belongs to $Ker\eta$. The proof is completed by Shauder Theorem (see [1]). □

We consider in the following a result of Theorem 1.

**Theorem 2** *Let $v \neq \emptyset$, bounded, closed, convex and subset of a Banach space $E$ and $\tau : v \to v$ be continuous function and $\eta$ be a measure of non-compactness, also $\exists \lambda, 0<\lambda<1, G \in \mathbb{G}$ and $(\Theta,\phi) \in \Psi$ such that*

$$\forall A \subseteq v, \quad \Theta(G(\eta(\tau A))) \le \lambda\phi(G(\eta(A))),$$

*then $\tau$ has at least one fixed point in $v$.*

**Corollary 1** *Let $v, \tau$ and $\eta$ be as mentioned in Theorem 2 and also $\exists \lambda, 0<\lambda<1$ such that $\forall A \subseteq v$, $G(\eta(\tau A)) \le \lambda G(\eta(A))$, then $\tau$ has at least one fixed point in $v$.*

*Proof* Put in $\Theta(s) = s$ and $\phi(s) = \lambda s$ for each $s \in [0, +\infty)$ and apply Theorem 2. □

*Remark 1* Taking $G = I$ in Corollary 1, we obtain the Darbo fixed point theorem.

**Corollary 2** *Let $v, \tau$ and $\eta$ be as mentioned in Theorem 2 and also $\exists \lambda, 0 < \lambda < 1$ and $G \in \mathbb{G}$ such that $\forall A \subseteq v$,*

$$\int_0^{G(\eta(\tau A))} f(\xi)\mathrm{d}\xi \leq \lambda \int_0^{G(\eta(A))} f(\xi)\mathrm{d}\xi,$$

*where $f : [0, \infty) \to [0, \infty)$ is a Lebesgue-integrable, summable and nonnegative function also for each $\epsilon > 0$, $\int_0^\epsilon f(\xi)\mathrm{d}\xi > 0$. Then $\tau$ has at least one fixed point in $v$.*

*Proof* Let $\Theta(s) = \int_0^s f(\xi)\mathrm{d}\xi$ and $\phi(s) = \lambda\Theta(s)$ for each $s \in [0, +\infty)$ and apply Theorem 2. □

**Corollary 3** *Let $v, \tau$ and $\eta$ be as mentioned in Theorem 2 and we suppose that for any $0 < a < b < \infty$, there exists $0 < \rho(a, b) < 1$ and $(\Theta, \phi) \in \Psi$ such that for all $A \subseteq v$,*

$$a \leq \eta(A) + \phi(\eta(A)) \leq b \Longrightarrow \Theta(\eta(\tau A) + \phi(\eta(\tau A))) \leq \rho(a, b)\phi[\eta(A) + \phi(\eta(A))],$$

*where $\phi : \mathbb{R}^+ \longrightarrow \mathbb{R}^+$ is continuous function with $\phi(0) = 0$ and $\phi(s) > 0$ for all $s > 0$. Then $\tau$ has at least one fixed point in $v$.*

*Proof* Let $G(s) = s + \phi(s)$ for each $s \in [0, +\infty)$, and apply Definition 2 and Theorem 1. □

*Remark 2* Theorem 3.1 of [5] is special case of Corollary 3.

An immediate consequence of Corollary 3 is the following form.

**Corollary 4** *Let $v, \tau$ and $\eta$ be as mentioned in Theorem 2 and also $\exists \lambda, 0 < \lambda < 1$ such that for any nonempty $A \subseteq v$,*

$$\eta(\tau A) + \phi(\eta(\tau A)) \leq \lambda[\eta(A) + \phi(\eta(A))],$$

*where $\phi : \mathbb{R}^+ \longrightarrow \mathbb{R}^+$ is continuous function with $\phi(0) = 0$ and $\phi(s) > 0$ for all $s > 0$. Then $\tau$ has at least one fixed point in $v$.*

*Remark 3*

(i) Theorem 3.2 of [5] is a special case of Corollary 4.
(ii) Darbo fixed point theorem is concluded from Corollary 4 by taking $\phi \equiv 0$.

**Corollary 5** *Let $v, \tau, G, \phi, \Theta$ and $\eta$ be as mentioned in Theorem 2 and also for all $A \subseteq v$,*

$$G(\eta(\tau A)) < \alpha(G(\eta(A)))G(\eta(A)),$$

*where $\alpha : \mathbb{R}^+ \longrightarrow [0, 1)$ and $\alpha(s_n) \to 1 \Longrightarrow s_n \to 0$, then $\tau$ has at least one fixed point in $v$.*

*Proof* Let $\Theta(s) = s$ and $\phi(s) = \alpha(s)s$ for each $s \in [0, +\infty)$ and apply Theorem 2. □

## Application

We apply Theorem 1 and the above discussion for existence of solution nonlinear integral equations. Consider $C[0, 1]$ as a Banach space with the following norm:

$$||u|| = \max\{|u(s)| : s \geq 0\},$$

and suppose that $A \neq \varnothing$ be a bounded subset of $C[0, 1]$. For $u \in A$ and $\epsilon \geq 0$, we put in,

$$\Omega(u, \epsilon) : = \sup\{|u(s) - u(\xi)| : s, \xi \in [0, 1], |s - \xi| \leq \epsilon\},$$
$$\Omega(A, \epsilon) : = \sup\{\Omega(u, \epsilon) : u \in A\}, \Omega_0(A) := \lim_{\epsilon \to 0} \Omega(A, \epsilon),$$
$$J(u) : = \sup\{|u(\xi) - u(s)| - [u(\xi) - u(s)] : s, \xi \in [0, 1], s \leq \xi\},$$
$$J(A) : = \sup\{J(u) : u \in A\}.$$

Thus it is easy that, all of the functions belonging to $A$ are non-decreasing on [0, 1] if and only if $J(A) = 0$. In the following we define $\eta$ on $\mathfrak{M}_C[0, 1]$ by

$$\eta(A) := \Omega_0(A) + J(A).$$

According to [7], it is straight forward to show that the function of $\eta$ is a measure of non-compactness on $C[0, 1]$.

Now, we investigate Eq. (1.1) by conditions as follows:

$(b_1)$ $g : [0, 1] \to \mathbb{R}^+$ is a continuous, non-decreasing and nonnegative function on [0, 1];
$(b_2)$ $h : [0, 1] \times \mathbb{R} \to \mathbb{R}$ is continuous function in $s$ and $u$ such that $h([0, 1] \times \mathbb{R}^+) \subseteq \mathbb{R}^+$ and there exists a continuous and non-decreasing function $\phi : \mathbb{R}^+ \to \mathbb{R}^+$ with $\phi(0) = 0$ and for each $s > 0$, $\phi(s) < s$ such that

$$|h(s, u) - h(s, z)| \leq \phi(|u - z|), \quad \forall s \in [0, 1], \quad \forall u, \quad z \in \mathbb{R}, \tag{3.1}$$

also $\phi$ is superadditive ($\phi(s) + \phi(\xi) \leq \phi(s + \xi)$ for all $s, \xi \in \mathbb{R}^+$);
$(b_3)$ In Eq. (1.1) the operator $H$ satisfies any nonnegative function as $u$ in the condition of $J(Hu) \leq \phi(J(u))$, where $\phi$ is introduced in $(b_2)$;
$(b_4)$ $f : [0, 1] \times [0, 1] \to \mathbb{R}$ is continuous and also it is non-decreasing in terms of variables $s$ and $\xi$, separately;
$(b_5)$ $k : \mathrm{Imf} \to \mathbb{R}^+$ is a continuous and non-decreasing function on the compact set Imf;
$(b_6)$ With assumptions $M_1 = \max\{|g(s)| : s \in [0, 1]\}$ and $M_2 = \max\{|h(s, 0)| : s \in [0, 1]\}$, inequality

$$M_1\Gamma(\alpha + 1) + (\phi(r) + M_2)||k||r \leq \Gamma(\alpha + 1)r, \tag{3.2}$$

has a positive solution as $r_0$, where $\lambda = \dfrac{||k||r_0}{\Gamma(\alpha + 1)} < 1$.

Theorem 3 *Under conditions* $(b_1)$–$(b_6)$, *Eq.* (1.1) *has at least one non-decreasing solution as* $u = u(\xi) \in C[0,1]$.

*Proof* We define operators $G$ and $\tau$ on $C[0, 1]$ by the formulas

$$(Gu)(s) = \int_0^s \frac{(s^m - \xi^m)^{\alpha-1}}{\Gamma(\alpha)} m\xi^{m-1}k(f(s,\xi))u(\xi)\,\mathrm{d}\xi,$$
$$(\tau u)(s) = g(s) + h(s, u(\xi))(Gu)(s).$$

Firstly, we prove that $G$ is self-map on $C[0, 1]$. Suppose $\epsilon > 0$ is given and let $u \in C[0,1]$ and $s_1, s_2 \in [0,1]$ (without loss of generality) let $s_2 \geq s_1$ and $|s_2 - s_1| \leq \epsilon$. Then,

$$\begin{aligned}
|(Gu)(s_2) - (Gu)(s_1)| &= \left|\int_0^{s_2} \frac{(s_2^m - \xi^m)^{\alpha-1}}{\Gamma(\alpha)} m\xi^{m-1}k(f(s_2,\xi))u(\xi)\mathrm{d}\xi \right.\\
&\left. - \int_0^{s_1} \frac{(s_1^m - \xi^m)^{\alpha-1}}{\Gamma(\alpha)} m\xi^{m-1}k(f(s_1,\xi))u(\xi)\mathrm{d}\xi\right|\\
&\leq \left|\int_0^{s_2} \frac{(s_2^m - \xi^m)^{\alpha-1}}{\Gamma(\alpha)} m\xi^{m-1}k(f(s_2,\xi))u(\xi)\mathrm{d}\xi \right.\\
&\left. - \int_0^{s_2} \frac{(s_2^m - \xi^m)^{\alpha-1}}{\Gamma(\alpha)} m\xi^{m-1}k(f(s_1,\xi))u(\xi)\mathrm{d}\xi\right|\\
&+ \left|\int_0^{s_2} \frac{(s_2^m - \xi^m)^{\alpha-1}}{\Gamma(\alpha)} m\xi^{m-1}k(f(s_1,\xi))u(\xi)\mathrm{d}\xi \right.\\
&\left. - \int_0^{s_1} \frac{(s_2^m - \xi^m)^{\alpha-1}}{\Gamma(\alpha)} m\xi^{m-1}k(f(s_1,\xi))u(\xi)\mathrm{d}\xi\right|\\
&+ \left|\int_0^{s_1} \frac{(s_2^m - \xi^m)^{\alpha-1}}{\Gamma(\alpha)} m\xi^{m-1}k(f(s_1,\xi))u(\xi)\mathrm{d}\xi \right.\\
&\left. - \int_0^{s_1} \frac{(s_1^m - \xi^m)^{\alpha-1}}{\Gamma(\alpha)} m\xi^{m-1}k(f(s_1,\xi))u(\xi)\mathrm{d}\xi\right|\\
&\leq \int_0^{s_2} \frac{(s_2^m - \xi^m)^{\alpha-1}}{\Gamma(\alpha)} m\xi^{m-1}|k(f(s_2,\xi)) - k(f(s_1,\xi))||u(\xi)|\mathrm{d}\xi\\
&+ \int_{s_1}^{s_2} \frac{(s_2^m - \xi^m)^{\alpha-1}}{\Gamma(\alpha)} m\xi^{m-1}|k(f(s_1,\xi))||u(\xi)|\mathrm{d}\xi\\
&+ \int_0^{s_1} \frac{|(s_2^m - \xi^m)^{\alpha-1} - (s_1^m - \xi^m)^{\alpha-1}|}{\Gamma(\alpha)} m\xi^{m-1}|k(f(s_1,\xi))||u(\xi)|\mathrm{d}\xi.
\end{aligned}$$

Therefore, if we put

$$\Omega_{\mathrm{kof}}(\epsilon,.) = \sup\{|k(f(s,\xi)) - k(f(s',\xi))| : s, s', \xi \in [0,1]\text{and}|s - s'| \leq \epsilon\},$$

then we have

$$\begin{aligned}
&|(Gu)(s_2) - (Gu)(s_1)|\\
&\leq \frac{||u||\Omega_{kof}(\epsilon,.)}{\Gamma(\alpha)} \int_0^{s_2} (s_2^m - \xi^m)^{\alpha-1} m\xi^{m-1}\mathrm{d}\xi\\
&+ \frac{||u||||k||}{\Gamma(\alpha)} \int_{s_1}^{s_2} (s_2^m - \xi^m)^{\alpha-1} m\xi^{m-1}\mathrm{d}\xi\\
&+ \frac{||u||||k||}{\Gamma(\alpha)} \int_0^{s_1} [(s_1^m - \xi^m)^{\alpha-1} - (s_2^m - \xi^m)^{\alpha-1}] m\xi^{m-1}\mathrm{d}\xi\\
&\leq \frac{||u||\Omega_{\mathrm{kof}}(\epsilon,.)}{\Gamma(\alpha)} \frac{s_2^{m\alpha}}{\alpha} + \frac{||u||||k||}{\Gamma(\alpha)} \frac{(s_2^m - s_1^m)^\alpha}{\alpha}\\
&+ \frac{||u||||k||}{\Gamma(\alpha)}\left[\frac{(s_2^m - s_1^m)^\alpha}{\alpha} + \frac{s_1^{m\alpha}}{\alpha} - \frac{s_2^{m\alpha}}{\alpha}\right]\\
&\leq \frac{||u||\Omega_{\mathrm{kof}}(\epsilon,.)}{\Gamma(\alpha+1)} + \frac{2||u||||k||}{\Gamma(\alpha+1)}(s_2^m - s_1^m)^\alpha.
\end{aligned}$$

Obviously, from the uniform continuity of the function *kof* on the set $[0,1] \times [0,1]$ we can get $\Omega_{\mathrm{kof}}(\epsilon,.) \to 0$ as $\epsilon \to 0$. Thus $Gu \in C[0,1]$, and consequently, $\tau u \in C[0,1]$. Also, we have

$$\begin{aligned}
|(Gu)(s)| &\leq \int_0^s \frac{(s^m - \xi^m)^{\alpha-1}}{\Gamma(\alpha)} m\xi^{m-1}|k(f(s,\xi))||u(\xi)|\mathrm{d}\xi\\
&\leq \frac{||k||||u||}{\Gamma(\alpha)} \int_0^s (s^m - \xi^m)^{\alpha-1} m\xi^{m-1}\mathrm{d}\xi \leq \frac{||k||||u||}{\Gamma(\alpha+1)}
\end{aligned} \tag{3.3}$$

for all $s \in [0,1]$. Therefore,

$$\begin{aligned}
|(\tau u)(s)| &\leq |g(s)| + |h(s,u)||Gu(\xi)|\\
&\leq M_1 + [|h(s,u) - h(s,0)| + |h(s,0)|]\frac{||k||||u||}{\Gamma(\alpha+1)}\\
&\leq M_1 + (\phi(||u||) + M_2)\frac{||k||||u||}{\Gamma(\alpha+1)}.
\end{aligned}$$

Hence,

$$||\tau u|| \leq M_1 + (\phi(||u||) + M_2)\frac{||k||||u||}{\Gamma(\alpha+1)}.$$

Thus, if $||u|| \leq r_0$ we conclude the following estimation by assumption $(b_6)$

$$||\tau u|| \leq M_1 + (\phi(r_0) + M_2)\frac{||k||r_0}{\Gamma(\alpha+1)} \leq r_0.$$

Consequently, the operator $\tau$ maps the ball $B_{r_0} \subset C[0,1]$ into itself. To prove continuity of $\tau$ on $B_{r_0}$, let $\{u_n\}$ be a sequence in $B_{r_0}$ such that $u_n \to u$. We have to show that $\tau u_n \to \tau u$. In fact, $\forall s \in [0,1]$, we have

$$\begin{aligned}
|(Gu_n)(s) - (Gu)(s)| &= \left|\int_0^s \frac{(s^m - \xi^m)^{\alpha-1}}{\Gamma(\alpha)} m\xi^{m-1}k(f(s,\xi))u_n(\xi)\mathrm{d}\xi \right.\\
&\left. - \int_0^s \frac{(s^m - \xi^m)^{\alpha-1}}{\Gamma(\alpha)} m\xi^{m-1}k(f(s,\xi))u(\xi)\mathrm{d}\xi\right|\\
&\leq \int_0^s \frac{(s^m - \xi^m)^{\alpha-1}}{\Gamma(\alpha)} m\xi^{m-1}|k(f(s,\xi))||u_n(\xi) - u(\xi)|\mathrm{d}\xi,
\end{aligned}$$

thus

$$||Gu_n - Gu|| \leq \frac{||k||}{\Gamma(\alpha+1)}||u_n - u||.$$

As,

$$\begin{aligned}
&|(\tau u_n)(s) - (\tau u)(s)| = |h(s,u_n(s))(Gu_n)(s) - h(s,u(s))(Gu)(s)|\\
&\leq |h(s,u_n(s))(Gu_n)(s) - h(s,u(s))(Gu_n)(s)|\\
&+ |h(s,u(s))(Gu_n)(s) - h(s,u(s))(Gu)(s)|\\
&\leq |h(s,u_n(s)) - h(s,u(s))||(Gu_n)(s)| + |h(s,u(s))||(Gu_n)(s) - (Gu)(s)|\\
&\leq \phi(|u_n(s) - u(s)|)\int_0^s \frac{(s^m - \xi^m)^{\alpha-1}}{\Gamma(\alpha)} m\xi^{m-1}|k(f(s,\xi))||u_n(\xi)|\mathrm{d}\xi\\
&+ (\phi(|u(s)|) + M_2)\int_0^s \frac{(s^m - \xi^m)^{\alpha-1}}{\Gamma(\alpha)} m\xi^{m-1}|k(f(s,\xi))||u_n(\xi) - u(\xi)|\mathrm{d}\xi.
\end{aligned}$$

It follows that

$$||\tau u_n - \tau u|| \leq \phi(||u_n - u||)\frac{||k||}{\Gamma(\alpha+1)}||u_n|| + (\phi(||u||) + M_2)\frac{||k||}{\Gamma(\alpha+1)}||u_n - u||.$$

So $\tau$ is continuous on $B_{r_0}$. we introduce,

$$B_{r_0}^{\sim} = \{u \in B_{r_0} : u(s) \geq 0, \quad \text{for } s \in [0,1]\} \subseteq B_{r_0}.$$

Obviously $B_{r_0}^{\sim} \neq \varnothing$ is bounded, closed and convex. By assumptions $(b_1)$, $(b_2)$ and $(b_5)$, if $u(s) \geq 0$ then $(\tau u)(s) \geq 0$ for all $s \in [0,1]$. Thus $\tau$ projects $B_{r_0}^{\sim}$ into itself. Moreover $\tau$ is continuous on $B_{r_0}^{\sim}$. Let $A \neq \varnothing$ be a subset of $B_{r_0}^{\sim}$, also $\epsilon > 0$ and

$$s_1, s_2 \in [0,1]; |s_2 - s_1| \leq \epsilon.$$

For simplicity, we suppose that $s_2 \geq s_1$. So we get

$$\begin{aligned}
&|(\tau u)(s_2) - (\tau u)(s_1)| \\
&= |g(s_2) + h(s_2, u(s_2))(Gu)(s_2) - g(s_1) - h(s_1, u(s_1))(Gu)(s_1)| \\
&\leq |g(s_2) - g(s_1)| + |h(s_2, u(s_2))(Gu)(s_2) - h(s_1, u(s_2))(Gu)(s_2)| \\
&+ |h(s_1, u(s_2))(Gu)(s_2) - h(s_1, u(s_1))(Gu)(s_2)| \\
&+ |h(s_1, u(s_1))(Gu)(s_2) - h(s_1, u(s_1))(Gu)(s_1)| \\
&\leq |g(s_2) - g(s_1)| + |h(s_2, u(s_2)) - h(s_1, u(s_2))||(Gu)(s_2)| \\
&+ |h(s_1, u(s_2)) - h(s_1, u(s_1))||(Gu)(s_2)| \\
&+ |h(s_1, u(s_1))||(Gu)(s_2) - (Gu)(s_1)| \\
&\leq \Omega(g,\epsilon) + \rho_{r_0}(h,\epsilon)\frac{||u||||k||}{\Gamma(\alpha+1)} + \phi(|u(s_2) - u(s_1)|)\frac{||u||||k||}{\Gamma(\alpha+1)} \\
&+ (\phi(||u||) + M_2)\left[\frac{||u||\Omega_{\text{kof}}(\epsilon,.)}{\Gamma(\alpha+1)} + \frac{2||u||||k||}{\Gamma(\alpha+1)}(s_2^m - s_1^m)^\alpha\right],
\end{aligned}$$

where we denoted

$$\rho_{r_0}(h,\epsilon) = \sup\{|h(s,u) - h(s',u)| : s, s' \in [0,1], u \in [0, r_0], |s - s'| \leq \epsilon\}.$$

According to mean value theorem ($|s_2^m - s_1^m|^\alpha \leq m^\alpha|s_2 - s_1|^\alpha$) in the last inequality, we conclude that,

$$\begin{aligned}
&|(\tau u)(s_2) - (\tau u)(s_1)| \\
&\leq \Omega(g,\epsilon) + \rho_{r_0}(h,\epsilon)\frac{||u||||k||}{\Gamma(\alpha+1)} + \phi(|u(s_2) - u(s_1)|)\frac{||u||||k||}{\Gamma(\alpha+1)} \\
&+ (\phi(||u||) + M_2)\left[\frac{||u||\Omega_{\text{kof}}(\epsilon,.)}{\Gamma(\alpha+1)} + \frac{2||u||||k||}{\Gamma(\alpha+1)}(m\epsilon)^\alpha\right].
\end{aligned}$$

Hence,

$$\begin{aligned}
\Omega(\tau u,\epsilon) &\leq \Omega(g,\epsilon) + \rho_{r_0}(h,\epsilon)\frac{r_0||k||}{\Gamma(\alpha+1)} + \phi(\Omega(u,\epsilon))\frac{r_0||k||}{\Gamma(\alpha+1)} \\
&+ (\phi(r_0) + M_2)\left[\frac{r_0\Omega_{\text{kof}}(\epsilon,.)}{\Gamma(\alpha+1)} + \frac{2r_0||k||}{\Gamma(\alpha+1)}(m\epsilon)^\alpha\right].
\end{aligned}$$

By computing supremum on $A$, we can write

$$\begin{aligned}
\Omega(\tau A,\epsilon) &\leq \Omega(g,\epsilon) + \rho_{r_0}(h,\epsilon)\frac{r_0||k||}{\Gamma(\alpha+1)} + \phi(\Omega(A,\epsilon))\frac{r_0||k||}{\Gamma(\alpha+1)} \\
&+ (\phi(r_0) + M_2)\left[\frac{r_0\Omega_{\text{kof}}(\epsilon,.)}{\Gamma(\alpha+1)} + \frac{2r_0||k||}{\Gamma(\alpha+1)}(m\epsilon)^\alpha\right].
\end{aligned}$$

Since $g$ is continuous on [0, 1] and also, $h$ and kof are uniform continuous on $[0,1] \times [0, r_0]$ and $[0,1] \times [0,1]$, respectively, so when $\epsilon \to 0$ then $\Omega(g,\epsilon) \to 0$, $\rho_{r_0}(h,\epsilon) \to 0$, $\Omega_{\text{kof}}(\epsilon,.) \to 0$ and also in the following, we have

$$\Omega_0(\tau A) \leq \frac{r_0||k||}{\Gamma(\alpha+1)}\phi(\Omega_0(A)). \tag{3.4}$$

Suppose $u \in A$ and $s_1, s_2 \in [0,1]$ such that $s_1 < s_2$, thus

$$\begin{aligned}
&|(\tau u)(s_2) - (\tau u)(s_1)| - [(\tau u)(s_2) - (\tau u)(s_1)] \\
&= |g(s_2) + h(s_2,u(s_2))(Gu)(s_2) - g(s_1) - h(s_1,u(s_1))(Gu)(s_1)| \\
&- [g(s_2) + h(s_2,u(s_2))(Gu)(s_2) - g(s_1) - h(s_1,u(s_1))(Gu)(s_1)] \\
&\leq \{|g(s_2) - g(s_1)| - [g(s_2) - g(s_1)]\} + |h(s_2,u(s_2))(Gu)(s_2) \\
&- h(s_1,u(s_1))(Gu)(s_2)| + |h(s_1,u(s_1))(Gu)(s_2) - h(s_1,u(s_1))(Gu)(s_1)| \\
&- \{[h(s_2,u(s_2))(Gu)(s_2) - h(s_1,u(s_1))(Gu)(s_2)] \\
&+ [h(s_1,u(s_1))(Gu)(s_2) - h(s_1,u(s_1))(Gu)(s_1)] \\
&\leq \{|h(s_2,u(s_2)) - h(s_1,u(s_1))| - [h(s_2,u(s_2)) - h(s_1,u(s_1))]\}(Gu)(s_2) \\
&+ h(s_1,u(s_1))\{|(Gu)(s_2) - (Gu)(s_1)| - [(Gu)(s_2) - (Gu)(s_1)]\} \\
&\leq J(Hu)\frac{r_0||k||}{\Gamma(\alpha+1)}.
\end{aligned}$$

Also we conclude that,

$$J(\tau u) \leq \phi(J(u))\frac{r_0||k||}{\Gamma(\alpha+1)},$$

and consequently,

$$J(\tau A) \leq \frac{r_0||k||}{\Gamma(\alpha+1)}\phi(J(A)). \tag{3.5}$$

From (3.4) and (3.5) and the definition of $\eta$, we get

$$\begin{aligned}
\eta(\tau A) &= \Omega_0(\tau A) + J(\tau A) \leq \frac{r_0||k||}{\Gamma(\alpha+1)}\phi(\Omega_0(A)) + \frac{r_0||k||}{\Gamma(\alpha+1)}\phi(J(A)) \\
&\leq \frac{r_0||k||}{\Gamma(\alpha+1)}(\phi(\Omega_0(A)) + \phi(J(A))) \leq \frac{r_0||k||}{\Gamma(\alpha+1)}(\phi(\Omega_0(A) + J(A))) \\
&\leq \lambda\phi(\eta(A)).
\end{aligned}$$

By the above inequality and because $\frac{r_0||k||}{\Gamma(\alpha+1)} < 1$, with applying Theorem 1 for in the case of $G(s) = \Theta(s) = 1$, we complete the proof. Also, such a solution is non-decreasing in Remark 3 and the definition of $\mu$, was given in Sect. 2. □

**Corollary 6** *Let the conditions of Theorem 3 be satisfied, then some of the integral equations with fractional order have at least one solution in C[0, 1], such as in the case of (i, ii, iii):*

(i) *for* $m = 1$,

$$u(s) = g(s) + h(s, u(s)) \int_0^s \frac{(s-\xi)^{\alpha-1}}{\Gamma(\alpha)} k(f(s,\xi))u(\xi)\,d\xi,$$

(ii) *for* $m = 1$ *and* $h(s, u(s)) = 1$,

$$u(s) = g(s) + \int_0^s \frac{(s-\xi)^{\alpha-1}}{\Gamma(\alpha)} k(f(s,\xi))u(\xi)\,d\xi,$$

(iii) *for* $m = 1, k = I, h(s, u(s)) = 1$ *and* $g(s) = 0$,

$$u(s) = \int_0^s \frac{(s-\xi)^{\alpha-1}}{\Gamma(\alpha)} f(s,\xi)u(\xi)\,d\xi.$$

Now, we consider an example by applying Theorem 3.

*Example 4* Suppose, integral equation with singular kernel and fractional order is given in the following form,

$$u(s) = \frac{1}{5}s^3 + \frac{2su(s)}{5(1+s)} \int_0^s \frac{2\xi}{\Gamma(\frac{1}{2})\sqrt{s^2-\xi^2}} \left[\frac{1}{8}(s+\xi) + \frac{1}{4}\right] u(\xi)\,d\xi, \tag{3.6}$$

where $s \in [0,1]$. In this example, we have $g(s) = \frac{1}{5}s^3$ and this function satisfies assumption $(b1)$ and $M_1 = \frac{1}{5}$. Here $f(s,\xi) = \frac{1}{4}\sqrt{s+\xi}$ and this function satisfies assumption $(b_4)$. Let $k : [0, \frac{\sqrt{2}}{4}] \to \mathbb{R}^+$ be given by $k(z) = 2z^2 + \frac{1}{4}$, then $k$ satisfying assumption $(b_5)$ with $||k|| = \frac{1}{2}$. Moreover, the function $h(s,u) = \dfrac{2su}{5(1+s)}$ satisfies hypothesis $(b_2)$ with assumption $\phi(s) = \dfrac{1}{5}s$,

$$|h(s,u) - h(s,z)| \le \frac{1}{5}|u - z| = \phi(|u - z|), \quad \forall u, z \in \mathbb{R},\ s \in [0,1],$$

also $h$ satisfies in $(b_3)$. In fact, by choosing an arbitrary nonnegative function $u \in C[0,1]$ and $s_1, s_2 \in [0,1]$ ($s_1 \le s_2$), we can write

$$\begin{aligned}
&|(Hu)(s_2) - (Hu)(s_1)| - [(Hu)(s_2) - (Hu)(s_1)] \\
&= |h(s_2, u(s_2)) - h(s_1, u(s_1))| - [h(s_2, u(s_2)) - h(s_1, u(s_1))] \\
&= \left|\frac{2s_2u(s_2)}{5(1+s_2)} - \frac{2s_1u(s_1)}{5(1+s_1)}\right| - \left[\frac{2s_2u(s_2)}{5(1+s_2)} - \frac{2s_1u(s_1)}{5(1+s_1)}\right] \\
&\le \left|\frac{2s_2u(s_2)}{5(1+s_2)} - \frac{2s_2u(s_1)}{5(1+s_2)}\right| + \left|\frac{2s_2u(s_1)}{5(1+s_2)} - \frac{2s_1u(s_1)}{5(1+s_1)}\right| \\
&\quad - \left[\frac{2s_2u(s_2)}{5(1+s_2)} - \frac{2s_2u(s_1)}{5(1+s_2)} + \frac{2s_2u(s_1)}{5(1+s_2)} - \frac{2s_1u(s_1)}{5(1+s_1)}\right] \\
&\le \frac{2s_2}{5(1+s_2)}|u(s_2) - u(s_1)| + \left|\frac{2s_2}{5(1+s_2)} - \frac{2s_1}{5(1+s_1)}\right| u(s_1) \\
&\quad - \frac{2s_2}{5(1+s_2)}[u(s_2) - u(s_1)] - \left[\frac{2s_2}{5(1+s_2)} - \frac{2s_1}{5(1+s_1)}\right] u(s_1) \\
&\le \frac{2s_2}{5(1+s_2)}\{|u(s_2) - u(s_1)| - [u(s_2) - u(s_1)]\} \\
&\le \frac{2s_2}{5(1+s_2)}J(u) \le \frac{1}{5}J(u) = \phi(J(u)).
\end{aligned}$$

According to the example, (3.2) converts to this form,

$$\frac{1}{5}\Gamma\left(\frac{3}{2}\right) + \frac{1}{10}r^2 \le \Gamma\left(\frac{3}{2}\right)r,$$

and $r_0 = 1$ is as a positive solution of it. Also,

$$\lambda = \frac{||k||r_0}{\Gamma\left(\frac{3}{2}\right)} = \frac{1}{\sqrt{\pi}} < 1.$$

Thus, Theorem 3 guarantees that Eq. (3.6) has a non-decreasing solution.

## Homotopy perturbation method (HPM) for solving functional I.E.

In this section, we solve functional integral Eq. (3.6) by using (HPM). In [10], perturbation method which depends on a small parameter can be led to imprecise solution by choosing unsuitable small parameter. But homotopy perturbation method introduced in [9], by an important concept of topology it can convert a nonlinear problem to a finite number of linear problems without dependence to the small parameter, this independence is very important. For introducing homotopy perturbation method according to the above-mentioned references, we consider the nonlinear problem:

$$\begin{aligned}
&M(u) - g(s) = 0, \quad s \in D \\
&\Lambda\left(u, \frac{\partial u}{\partial n}\right) = 0, \quad n \in \Upsilon,
\end{aligned} \tag{4.1}$$

where $M$ and $\Lambda$ are differential and boundary operators, respectively, also $g(s)$ is a known analytic function and $\Upsilon$ is the boundary of the domain $D$. we assume operator $M$ is divided into linear and nonlinear operators such as $\ell$ and $\aleph$. So, we can write Eq. (4.1) to this form,

$$\ell(u) + \aleph(u) - g(s) = 0. \tag{4.2}$$

Homotopy perturbation $H(v,p)$ can be written as follows [9]:

$$\begin{aligned}
&H : D \times [0,1] \longrightarrow \Re, \\
&H(v,p) = (1-p)[\ell(v) - \ell(v_0)] + p[M(v) - g(s)] = 0,
\end{aligned} \tag{4.3}$$

where $p$ is an embedding parameter, $v$ is an approximation of $u$ and $v_0$ is an initial approximation of $u$. Of course some kinds of modifications of homotopy perturbation method can be seen in [8, 11]. We solve nonlinear integral Eq. (3.6) by Eq. (4.3). Let us consider Eq. (3.6) to the following form,

$$u(s) = \frac{1}{5}s^3 + u(s)\int_0^s \frac{4s\xi\left(\frac{1}{8}(s+\xi)+\frac{1}{4}\right)}{5\sqrt{\pi}(1+s)\sqrt{s^2-\xi^2}}u(\xi)\,\mathrm{d}\xi, \tag{4.4}$$

the general form of Eq. (4.4) is as follows:

$$u(s) - u(s)\int_0^s k(s,\xi)u(\xi)\,\mathrm{d}\xi = \frac{1}{5}s^3, \tag{4.5}$$

according to the nonlinear Eq. (4.1), we can write,

$$M(u(s)) = g(s); \;\; g(s) = \frac{1}{5}s^3. \tag{4.6}$$

In the homotopy perturbation (4.3) we approximate solution of Eq. (4.5) in terms of power series of $p$,

$$v = v_0 + pv_1 + p^2v_2 + \cdots = \sum_{i=0}^{\infty} p^i v_i. \tag{4.7}$$

Also in Eq. (4.3), we choose linear and nonlinear operators to these forms,

$$\ell(v) = v, M(v) = v(s) - v(s)\int_0^s k(s,\xi)v(\xi)\,\mathrm{d}\xi, g(s) = \frac{1}{5}s^3.$$

So, we can write,

$$\begin{aligned} H(v,p) &= (1-p)(v-v_0) \\ &+ p\left[v(s) - v(s)\int_0^s k(s,\xi)v(\xi)\mathrm{d}\xi - \frac{1}{5}s^3\right] = 0, \end{aligned} \tag{4.8}$$

by substituting Eq. (4.7) in the homotopy formula Eq. (4.8), we have

$$\begin{aligned} pv_1 + p^2v_2 + \cdots + pv_0 - pv_0(s)\int_0^s k(s,\xi)v_0(\xi)\,\mathrm{d}\xi \\ - p^2v_0(s)\int_0^s k(s,\xi)v_1(\xi)\,\mathrm{d}\xi \\ - p^2v_1(s)\int_0^s k(s,\xi)v_0(\xi)\,\mathrm{d}\xi + \cdots - p\frac{1}{5}s^3 = 0, \end{aligned} \tag{4.9}$$

with ordering the above relations in terms of $p$ powers, we have

$$p^1 : \left(v_1 + v_0 - v_0(s)\int_0^s k\left(s,\xi\right)v_0(\xi)\,\mathrm{d}\xi - \frac{1}{5}s^3\right),$$

$$p^2 : \left(v_2 - v_0(s)\int_0^s k(s,\xi)v_1(\xi)\,\mathrm{d}\xi - v_1(s)\int_0^s k(s,\xi)v_0(\xi)\,\mathrm{d}\xi\right),$$

$$p^3 : \cdots$$

By considering to Eq. (4.9), we put in the coefficients of $p$ powers equal to zero and by suitable choosing initial guess $v_0(s)$, we obtain

**Table 1** Absolute errors for Eq. (4.4) by HPM

| $t$ | Absolute errors |
|---|---|
| 0.0 | 0.0 |
| 0.1 | $5.9\times10^{-18}$ |
| 0.2 | $4.8\times10^{-14}$ |
| 0.3 | $9.3\times10^{-12}$ |
| 0.4 | $3.9\times10^{-10}$ |
| 0.5 | $7.0\times10^{-9}$ |
| 0.6 | $7.5\times10^{-8}$ |
| 0.7 | $5.5\times10^{-7}$ |
| 0.8 | $3.1\times10^{-6}$ |
| 0.9 | $1.4\times10^{-5}$ |
| 1.0 | $5.7\times10^{-5}$ |

$$\begin{aligned} v_0(s) &= \frac{1}{5}s^3, \\ v_1(s) &= v_0(s)\int_0^s k(s,\xi)v_0(\xi)\,\mathrm{d}\xi, \\ v_2(s) &= v_0(s)\int_0^s k(s,\xi)v_1(\xi)\,\mathrm{d}\xi - v_1(s)\int_0^s k(s,\xi)v_0(\xi)\,\mathrm{d}\xi, \end{aligned} \tag{4.10}$$

where $k(s,\xi)$ is given by Eq. (4.4), therefore,

$$v_1(s) = \frac{4s^4}{125\sqrt{\pi}(1+s)}\int_0^s \frac{\frac{1}{8}(s+\xi)+\frac{1}{4}}{\sqrt{s^2-\xi^2}}\xi^4\,\mathrm{d}\xi = \frac{s^8(128s+45\pi(2+s))}{(240)(250)\sqrt{\pi}(1+s)}.$$

By taking two terms of Eq. (4.7) into account, we can approximate the solution of Eq. (4.4) as follows:

$$v(s) = \frac{1}{5}s^3 + \frac{s^8(128s+45\pi(2+s))}{240\times 250\sqrt{\pi}(1+s)}. \tag{4.11}$$

By substituting (4.11) in Eq. (4.4) and comparing both sides of it, we reach absolute errors in points (see Table 1).

## Conclusion

In this paper, we try to introduce a mixed plan of pure and applied mathematics, where measure of non-compactness on a Banach space is used for the generalization of Darbo fixed point theorem for existence of solution singular integral equations with fractional order. Also, by homotopy perturbation method we obtain an approximation of a solution with high accuracy.

## References


1. Agarwal, R.P. , O'Regan, D.: Fixed Point Theory and Applications. Cambridge University Press, Cambridge (2004)
2. Aghajani, A., Banaś, J., Jalilian, Y.: Existence of solution for a class of nonlinear Volterra singular integral equations. Comput. Math. Appl. **62**, 1215–1227 (2011)

3. Aghajani, A.: Banaś, J., Sabzali, N.: Some generalizations of Darbo fixed point theorem and applications. Bull. Belg. Math. Soc. Simon Stevin **20**, 345–358 (2013)
4. Appell, J., Zabrejko, P.: Nonlinear superposition operators. In: Cambridge Tracts in Mathematics, vol. 95. Cambridge University Press, Cambridge (1990)
5. Arab, R.: Some generalizations of Darbo fixed point theorem and its application. Miskolc Math. Notes (in press)
6. Banaś, J., Goebel, K.: Measures of non-compactness in Banach spaces. In: Lecture Notes in Pure and Applied Mathematics, vol. 60. Dekker, New York (1980)
7. Banaś, J., Olszowy, L.: Measures of non-compactness related to monotonicity. Comment. Math. **41**, 13–23 (2001)
8. Glayeri, A., Rabbani, M.: New technique in semi-analytic method for solving non-linear differential equations. Math. Sci. **4**, 395–404 (2011)
9. He, J.: A new approach to non-linear partial differential equations. Commun. Non Linear Sci. Numer. Simul. **2**(4), 230–235 (1997)
10. Nayfeh, A., Mook, H.D.T.: Non-Linear Oscillations. Wiley, New York (1979)
11. Rabbani, M.: Modified homotopy method to solve non-linear integral equations. Int. J. Nonlinear Anal. Appl. **6**, 133–136 (2015)

# 3

# Graph convergence and generalized Yosida approximation operator with an application

**Rais Ahmad[1] · Mohd. Ishtyak[1] · Mijanur Rahaman[1] · Iqbal Ahmad[1]**

**Abstract** In this paper, we introduce a Yosida inclusion problem as well as a generalized Yosida approximation operator. Using the graph convergence of $H(\cdot,\cdot)$-accretive operator and resolvent operator convergence discussed in Li and Huang (Appl Math Comput 217:9053–9061, 2011), we establish the convergence for generalized Yosida approximation operator. As an application, we solve a Yosida inclusion problem in $q$-uniformly smooth Banach spaces. An example is constructed, and through MATLAB programming, we show some graphics for the convergence of generalized Yosida approximation operator.

**Keywords** Graph convergence · Resolvent operator · Smooth Banach space · Yosida approximation operator · Yosida inclusion

**Mathematics Subject Classification** 47H09 · 49J40 · 65K15

✉ Mijanur Rahaman
mrahman96@yahoo.com

Rais Ahmad
raisain_123@rediffmail.com

Mohd. Ishtyak
ishtyakalig@gmail.com

Iqbal Ahmad
iqbalahmad120@gmail.com

1 Department of Mathematics, Aligarh Muslim University, Aligarh 202002, India

## Introduction

A reasonable attention has been shown by many researchers for the study of variational inclusions (inequalities) and their generalized forms, which occupies a leading and significant role to connect research between analysis, geometry, biology, elasticity, optimization, image processing, biomedical and mathematical sciences, etc. A broad range of problems with which we encounter in physics, economics, management sciences, and operations research can be formulated as an inclusion problem $0 \in T(x)$, for a given set-valued mapping $T$ on a Hilbert space $H$. Thus, the problem of finding a zero of $T$, i.e., a point $x \in H$, such that $0 \in T(x)$ is a fundamental problem in many areas of applied sciences.

On the other hand, it is well known that monotone operators on Hilbert spaces can be regularized into single-valued Lipschitzian monotone operators via a process known as the Yosida approximation. This Yosida approximation operators are instrumental to approximate the solutions of general variational inclusion problems using non-expansive resolvent operators. Recently, many authors [2, 3, 5, 6, 8–10] have applied Yosida approximation operators and their generalized forms to solve some variational inclusion problems. Zou and Huang [14], Ahmad et al. [1] introduced and studied the graph convergence of $H(\cdot,\cdot)$-accretive operators and $H(\cdot,\cdot)$-co-accretive operators, respectively, for solving variational inclusion problems and their system. For more details, we refer to [4, 11, 12, 15].

This paper deals with the introduction of a generalized Yosida approximation operator with some of its properties. Under the concept of graph convergence of $H(\cdot,\cdot)$-accretive operators, we prove the convergence of generalized Yosida approximation operator. Finally, we solve a Yosida inclusion problem in $q$-uniformly Banach spaces. A MATLAB programming related to graph convergence of generalized Yosida approximation operator is discussed with a consolidated example. Our results are applicable and new in this direction and refinement of results of Li and Huang [7].

## Preliminaries

Let $X$ be a real Banach Space with its dual space $X^*$. We denote the duality pairing between $X$ and $X^*$ by $\langle\cdot,\cdot\rangle$, and $2^X$ is the family of all nonempty subsets of $X$.

The generalized duality mapping $F_q : X \to 2^{X^*}$ is defined by

$$F_q(x) = \left\{f^* \in X^* : \langle x, f^*\rangle = \|x\|^q, \|f^*\| = \|x\|^{q-1}\right\}, \quad \forall x \in X,$$

where $q > 1$ is a constant. For $q = 2$, $F_q$ coincides with the normalized duality mapping. If $X$ is a Hilbert space, $F_2$ becomes the identity mapping on $X$. It is to be noted that if $X$ is uniformly smooth, then $F_q$ is single-valued. Throughout the paper, we assume that $X$ is a real Banach space and $F_q$ is single-valued.

The function $\rho_X : [0,\infty) \to [0,\infty)$ is called modulus of smoothness of $X$, such that

$$\rho_X(t) = \left\{\frac{\|x+y\| + \|x-y\|}{2} - 1 : \|x\| \le 1, \|y\| \le t\right\}.$$

A Banach space $X$ is called

1. uniformly smooth if $\lim_{t\to 0} \frac{\rho_X(t)}{t} = 0$;
2. $q$-uniformly smooth if there exists a constant $c > 0$, such that

$$\rho_X(t) \le ct^q, \quad q > 1.$$

While encountered with the characteristic inequalities, Xu [13] proved the following important Lemma in $q$-uniformly smooth Banach spaces.

**Lemma 1** *Let $X$ be a real uniformly smooth Banach space. Then, $X$ is $q$-uniformly smooth if and only if there exists a constant $c_q > 0$, such that for all $x, y \in X$,*

$$\|x+y\|^q \le \|x\|^q + q\langle y, F_q(x)\rangle + c_q\|y\|^q.$$

The following definitions and concepts are essential to achieve the aim of this paper.

**Definition 1** [14] Let $A, B : X \to X$ and $H : X \times X \to X$ be the single-valued mappings.

1. $A$ is said to be accretive, if

$$\langle A(x) - A(y), F_q(x-y)\rangle \ge 0, \quad \forall x, \quad y \in X;$$

2. $A$ is said to be strictly accretive, if $A$ is accretive and

$$\langle A(x) - A(y), F_q(x-y)\rangle = 0, \quad \text{if and only if } x = y;$$

3. $A$ is said to be $\delta_A$-strongly accretive, if there exists a constant $\delta_A > 0$, such that

$$\langle Ax - Ay, F_q(x-y)\rangle \ge \delta_A\|x-y\|^q;$$

4. $A$ is said to be $\gamma_A$-Lipschitz continuous, if there exists a constant $\gamma_A > 0$, such that

$$\|Ax - Ay\| \le \gamma_A\|x-y\|, \quad \forall x, \quad y \in X;$$

5. $H(A,\cdot)$ is said to be $\alpha$-strongly accretive with respect to $A$, if there exists a constant $\alpha > 0$, such that

$$\langle H(Ax,\cdot) - H(Ay,\cdot), F_q(x-y)\rangle \ge \alpha\|x-y\|^q, \quad \forall x, \quad y \in X;$$

6. $H(\cdot,B)$ is said to be $\beta$-relaxed accretive with respect to $B$, if there exists a constant $\beta > 0$, such that

$$\langle H(\cdot,Bx) - H(\cdot,By), F_q(x-y)\rangle \ge -\beta\|x-y\|^q, \quad \forall x, \quad y \in X;$$

7. $H(A,\cdot)$ is said to be $\sigma$-Lipschitz continuous with respect to $A$, if there exists a constant $\sigma > 0$, such that

$$\|H(Ax,\cdot) - H(Ay,\cdot)\| \le \sigma\|x-y\|, \quad \forall x, \quad y \in X.$$

Similarly, we can define the Lipschitz continuity of $H$ with respect to $B$.

**Definition 2** [14] Let $H : X \to X$ be a single-valued mapping and $M : X \to 2^X$ be a set-valued mapping. The mapping $M$ is said to be

1. accretive, if

$$\langle u - v, F_q(x-y)\rangle \ge 0, \quad \forall x, y \in X, \quad u \in M(x), v \in M(y);$$

2. $m$-accretive, if $M$ is accretive and $(I + \lambda M)(X) = X$, for all $\lambda > 0$, where $I$ is the identity operator on $X$;
3. $H$-accretive, if $M$ is accretive and $(H + \lambda M)(X) = X$, for all $\lambda > 0$.

**Definition 3** [14] Let $A, B : X \to X$, $H : X \times X \to X$ be the single-valued mappings and $M : X \to 2^X$ be a set-valued mapping. The mapping $M$ is said to be $H(\cdot,\cdot)$-accretive with respect to $A$ and $B$, if $M$ is accretive and $[H(A,B) + \lambda M](X) = X$, for every $\lambda > 0$.

**Lemma 2** [14] *Let $H(A, B)$ be $\alpha$-strongly accretive with respect to $A$, $\beta$-relaxed accretive with respect to $B$ and $\alpha > \beta$. Let $M$ be an $H(\cdot,\cdot)$-accretive operator with respect to $A$ and $B$. Then, the operator $[H(A,B) + \lambda M]^{-1}$ is single-*

*valued and is called the resolvent operator, i.e.,* $R^{H(\cdot,\cdot)}_{M,\lambda} : X \to X$, *such that*

$$R^{H(\cdot,\cdot)}_{M,\lambda}(u) = [H(A,B) + \lambda M]^{-1}(u), \quad \forall u \in X, \quad \lambda > 0. \tag{1}$$

Furthermore, the resolvent operator defined by Eq. (1) is $\frac{1}{(\alpha-\beta)}$-Lipschitz continuous.

**Lemma 3** [3] *Let* $\{a_n\}$ *and* $\{b_n\}$ *be two non-negative real sequences satisfying*

$$a_{n+1} \le k a_n + b_n,$$

*with* $0<k<1$ *and* $b_n \to 0$. *Then,* $\lim_{n\to\infty} a_n = 0$.

## Generalized Yosida approximation operator and its convergence

We define the generalized Yosida approximation operator using the resolvent operator defined by Eq. (1), that is

$$R^{H(\cdot,\cdot)}_{M,\lambda}(u) = [H(A,B) + \lambda M]^{-1}(u), \quad \forall x \in X, \quad \lambda > 0.$$

**Definition 4** The generalized Yosida approximation operator denoted by $J^{H(\cdot,\cdot)}_{M,\lambda}$ is defined as

$$J^{H(\cdot,\cdot)}_{M,\lambda}(u) = \frac{1}{\lambda}\left[I - R^{H(\cdot,\cdot)}_{M,\lambda}\right](u), \quad \forall u \in X \quad \text{and} \quad \lambda > 0, \tag{2}$$

where $I$ is the identity mapping on $X$.

**Lemma 4** *The generalized Yosida approximation operator defined by* Eq. (2) *is*

1. $\theta_1$*-Lipschitz continuous, where* $\theta_1 = \frac{[\alpha-\beta+1]}{\lambda(\alpha-\beta)}, \alpha > \beta$.
2. $\theta_2$*-strongly monotone, where* $\theta_2 = \frac{[(\alpha-\beta)-1]}{\lambda(\alpha-\beta)}, \alpha > \beta$.

*Proof*

1. Let $u, v \in X$ and $\lambda > 0$. Using Lemma 2, we have

$$\begin{aligned}
&\left\|J^{H(\cdot,\cdot)}_{M,\lambda}(u) - J^{H(\cdot,\cdot)}_{M,\lambda}(v)\right\| \\
&= \frac{1}{\lambda}\left\|[I(u) - R^{H(\cdot,\cdot)}_{M,\lambda}(u)] - [I(v) - R^{H(\cdot,\cdot)}_{M,\lambda}(v)]\right\| \\
&\le \frac{1}{\lambda}\left[\|u-v\| + \left\|R^{H(\cdot,\cdot)}_{M,\lambda}(u) - R^{H(\cdot,\cdot)}_{M,\lambda}(v)\right\|\right] \\
&\le \frac{1}{\lambda}[\|u-v\| + \frac{1}{[\alpha-\beta]}\|u-v\|] \\
&= \frac{1}{\lambda}\left[\frac{\alpha-\beta+1}{\alpha-\beta}\right]\|u-v\|,
\end{aligned}$$

i.e.,

$$\left\|J^{H(\cdot,\cdot)}_{M,\lambda}(u) - J^{H(\cdot,\cdot)}_{M,\lambda}(v)\right\| \le \theta_1\|u-v\|, \tag{3}$$

where $\theta_1 = \frac{[\alpha-\beta+1]}{\lambda(\alpha-\beta)}, \alpha > \beta$.

2. For any $u, v \in X$, and $\lambda > 0$ and using Lemma 2, we have

$$\begin{aligned}
&\left\langle J^{H(\cdot,\cdot)}_{M,\lambda}(u) - J^{H(\cdot,\cdot)}_{M,\lambda}(v), F_q(u-v)\right\rangle \\
&= \frac{1}{\lambda}\left\langle I(u) - R^{H(\cdot,\cdot)}_{M,\lambda}(u) - [I(v) \right. \\
&\quad \left. - R^{H(\cdot,\cdot)}_{M,\lambda}(v)], F_q(u-v)\right\rangle \\
&= \frac{1}{\lambda}\left[\left\langle u-v, F_q(u-v)\right\rangle \right. \\
&\quad \left. - \left\langle R^{H(\cdot,\cdot)}_{M,\lambda}(u) - R^{H(\cdot,\cdot)}_{M,\lambda}(v), F_q(u-v)\right\rangle\right] \\
&\ge \frac{1}{\lambda}\left[\|u-v\|^q - \left\|R^{H(\cdot,\cdot)}_{M,\lambda}(u) - R^{H(\cdot,\cdot)}_{M,\lambda}(v)\right\| \right. \\
&\quad \left. \|u-v\|^{q-1}\right] \\
&\ge \frac{1}{\lambda}\left[\|u-v\|^q - \frac{1}{[\alpha-\beta]}\|u-v\|\|u-v\|^{q-1}\right] \\
&= \frac{1}{\lambda}\left[\|u-v\|^q - \frac{1}{[\alpha-\beta]}\|u-v\|^q\right] \\
&= \frac{[(\alpha-\beta)-1]}{\lambda(\alpha-\beta)}\|u-v\|^q.
\end{aligned}$$

i.e.,

$$\left\langle J^{H(\cdot,\cdot)}_{M,\lambda}(u) - J^{H(\cdot,\cdot)}_{M,\lambda}(v), \quad F_q(u-v)\right\rangle \ge \theta_2\|u-v\|^q, \quad \forall u, v \in X, \lambda > 0$$

and $\theta_2 = \frac{[(\alpha-\beta)-1]}{\lambda(\alpha-\beta)}, \alpha > \beta$.

□

*Note 1* It is interesting to note that resolvent operator defined by Eq. (1) and generalized Yosida approximation operator defined by Eq. (2) are connected by the following relation:

$$\lambda J^{H(\cdot,\cdot)}_{M,\lambda}(x) \in [\lambda M + H(A,B) - I]\left(R^{H(\cdot,\cdot)}_{M,\lambda}(x)\right).$$

Let $M : X \to 2^X$ be a set-valued mapping. The graph of the mapping $M$ is defined by

$$graph(M) = \{(x,y) \in X \times Y : y \in M(x)\}$$

**Definition 5** [7] Let $A, B : X \to X$ and $H : X \times X \to X$ be the single-valued mappings. Let $M_n, M : X \to 2^X$ be $H(\cdot,\cdot)$-accretive operators for $n = 0, 1, 2, \ldots$. The sequence $\{M_n\}$ is said to be graph convergence to $M$, denoted by $M_n \xrightarrow{G} M$, if for every $(x,y) \in graph(M)$, there exists a sequence $(x_n, y_n) \in graph(M_n)$, such that

$$x_n \to x, \quad y_n \to y \quad as\ n \to \infty.$$

**Theorem 1** [7] *Let $M_n, M : X \to 2^X$ be $H(\cdot,\cdot)$-accretive operators for $n = 0, 1, 2, \ldots$. Assume that $H : X \times X \to X$ is a single-valued mapping, such that*

1. *H(A, B) is $\alpha$-strongly accretive with respect to A and $\beta$-relaxed accretive with respect to B, $\alpha > \beta$;*
2. *H(A, B) is $\gamma_1$-Lipschitz continuous with respect to A and $\gamma_2$-Lipschitz continuous with respect to B.*

*Then, $M_n \xrightarrow{G} M$ if and only if*

$$R^{H(\cdot,\cdot)}_{M_n,\lambda}(u) \to R^{H(\cdot,\cdot)}_{M,\lambda}(u), \quad \forall u \in X, \quad \lambda > 0,$$

*where $R^{H(\cdot,\cdot)}_{M_n,\lambda} = [H(A,B) + \lambda M_n]^{-1}$ and $R^{H(\cdot,\cdot)}_{M,\lambda} = [H(A,B) + \lambda M]^{-1}$.*

Now, we prove the convergence of generalized Yosida approximation operator in the light of graph convergence of $H(\cdot,\cdot)$-accretive operator without using the convergence of resolvent operator defined by Eq. (1).

**Theorem 2** *Let $M_n, M : X \to 2^X$ be $H(\cdot,\cdot)$-accretive operators for $n = 0, 1, 2, \ldots$, and $H : X \times X \to X$ be a single-valued mapping, such that conditions (1) and (2) of Theorem 1 hold.*

*Then $M_n \xrightarrow{G} M$ if and only if*

$$J^{H(\cdot,\cdot)}_{M_n,\lambda}(x) \to J^{H(\cdot,\cdot)}_{M,\lambda}(x), \quad \forall x \in X, \quad \lambda > 0,$$

*where*

$$J^{H(\cdot,\cdot)}_{M_n,\lambda}(x) = \frac{1}{\lambda}\left[I - R^{H(\cdot,\cdot)}_{M_n,\lambda}\right](x),$$

$$J^{H(\cdot,\cdot)}_{M,\lambda}(x) = \frac{1}{\lambda}\left[I - R^{H(\cdot,\cdot)}_{M,\lambda}\right](x), \quad \forall x \in X,$$

*and $R^{H(\cdot,\cdot)}_{M_n,\lambda}$ and $R^{H(\cdot,\cdot)}_{M,\lambda}$ are defined in Theorem 1.*

*Proof* Necessary part: Suppose that $M_n \xrightarrow{G} M$. For any given $x \in X$, let

$$z_n = J^{H(\cdot,\cdot)}_{M_n,\lambda}(x) \quad \text{and} \quad z = J^{H(\cdot,\cdot)}_{M,\lambda}(x).$$

Then,

$$z = J^{H(\cdot,\cdot)}_{M,\lambda}(x) = \frac{1}{\lambda}\left[I - R^{H(\cdot,\cdot)}_{M,\lambda}\right](x),$$

implies that

$$(x - \lambda z) = R^{H(\cdot,\cdot)}_{M,\lambda}(x) = [H(A,B) + \lambda M]^{-1}(x),$$

i.e.,

$$H(A,B)(x - \lambda z) + \lambda M(x - \lambda z) = x.$$

It follows that

$$\frac{1}{\lambda}[x - H(A,B)(x - \lambda z)] \in M(x - \lambda z).$$

That is

$$\left(x - \lambda z, \frac{1}{\lambda}[x - H(A,B)(x - \lambda z)]\right) \in \text{graph}(M).$$

By Definition 4, there exists a sequence $(w_n, y_n) \in \text{graph}(M_n)$, such that

$$w_n \to (x - \lambda z), \quad y_n \to \frac{1}{\lambda}[x - H(A,B)(x - \lambda z)]. \tag{4}$$

Since $y_n \in M_n(w_n)$, we have

$$H(Aw_n, Bw_n) + \lambda y_n \in [H(A,B) + \lambda M_n](w_n),$$

and so,

$$\begin{aligned} w_n &= [H(A,B) + \lambda M_n]^{-1}[H(Aw_n, Bw_n) + \lambda y_n], \\ &= R^{H(\cdot,\cdot)}_{M_n,\lambda}[H(Aw_n, Bw_n) + \lambda y_n], \\ &= \left[I - \lambda J^{H(\cdot,\cdot)}_{M_n,\lambda}\right][H(Aw_n, Bw_n) + \lambda y_n], \end{aligned}$$

which implies that

$$\frac{1}{\lambda}w_n = \frac{1}{\lambda}H(Aw_n, Bw_n) + y_n - J^{H(\cdot,\cdot)}_{M_n,\lambda}[H(Aw_n, Bw_n) + \lambda y_n]. \tag{5}$$

Using (1) of Lemma 4 and Eq. (5), we have

$$\begin{aligned} &\|z_n - z\| \\ &= \left\|J^{H(\cdot,\cdot)}_{M_n,\lambda}(x) - z\right\| \\ &= \left\|J^{H(\cdot,\cdot)}_{M_n,\lambda}(x) + \frac{1}{\lambda}w_n - \frac{1}{\lambda}w_n - z\right\| \\ &= \Big\|J^{H(\cdot,\cdot)}_{M_n,\lambda}(x) + \frac{1}{\lambda}H(Aw_n, Bw_n) + y_n - J^{H(\cdot,\cdot)}_{M_n,\lambda} \\ &\qquad [H(Aw_n, Bw_n) + \lambda y_n] - \frac{1}{\lambda}w_n - z\Big\| \\ &\le \left\|J^{H(\cdot,\cdot)}_{M_n,\lambda}(x) - J^{H(\cdot,\cdot)}_{M_n,\lambda}[H(Aw_n, Bw_n) + \lambda y_n]\right\| \\ &\quad + \left\|\frac{1}{\lambda}H(Aw_n, Bw_n) + y_n - \frac{1}{\lambda}w_n - z\right\| \\ &\le \theta_1\|x - H(Aw_n, Bw_n) - \lambda y_n\| \\ &\quad + \left\|\frac{1}{\lambda}H(Aw_n, Bw_n) + y_n - \frac{1}{\lambda}x\right\| + \left\|\frac{1}{\lambda}w_n - \frac{1}{\lambda}x + z\right\| \\ &= \left(\theta_1 - \frac{1}{\lambda}\right)\|x - H(Aw_n, Bw_n) - \lambda y_n\| + \frac{1}{\lambda}\|w_n - x + \lambda z\| \\ &= \left(\theta_1 - \frac{1}{\lambda}\right)\|x - H(Aw_n, Bw_n) + H(A,B)(x - \lambda z) \\ &\quad - H(A,B)(x - \lambda z) - \lambda y_n\| + \frac{1}{\lambda}\|w_n - x + \lambda z\| \\ &\le \left(\theta_1 - \frac{1}{\lambda}\right)\|x - H(A,B)(x - \lambda z) - \lambda y_n\| \\ &\quad + \left(\theta_1 - \frac{1}{\lambda}\right)\|H(A,B)(x - \lambda z) - H(Aw_n, Bw_n)\| \\ &\quad + \frac{1}{\lambda}\|w_n - x + \lambda z\|. \end{aligned} \tag{6}$$

Since $H$ is $\gamma_1$-Lipschitz continuous with respect to $A$ and $\gamma_2$-Lipschitz continuous with respect to $B$, we have

$$\begin{aligned}
&\left\|H(A,B)(x-\lambda z)-H(A,B)w_n\right\| \\
&\quad=\left\|H(A(x-\lambda z),B(x-\lambda z))-H(A(x-\lambda z),Bw_n)\right. \\
&\qquad\left.+H(A(x-\lambda z),Bw_n)-H(Aw_n,Bw_n)\right\| \\
&\quad\le\|H(A(x-\lambda z),B(x-\lambda z))-H(A(x-\lambda z),Bw_n)\| \\
&\qquad+\|H(A(x-\lambda z),Bw_n)-H(Aw_n,Bw_n)\| \\
&\quad\le\gamma_2\|x-\lambda z-w_n\|+\gamma_1\|x-\lambda z-w_n\| \\
&\quad=(\gamma_1+\gamma_2)\|x-\lambda z-w_n\|. \qquad (7)
\end{aligned}$$

Using Eqs. (7), (6) becomes

$$\begin{aligned}
\|z_n-z\|\le&\left(\theta_1-\frac{1}{\lambda}\right)\|x-H(A,B)(x-\lambda z)-\lambda y_n\| \\
&+\left[\left(\theta_1-\frac{1}{\lambda}\right)(\gamma_1+\gamma_2)+\frac{1}{\lambda}\right]\|w_n-x+\lambda z\|.
\end{aligned}$$

By Eq. (4), we have

$$w_n\to(x-\lambda z),\quad y_n\to\frac{1}{\lambda}[x-H(A,B)(x-\lambda z)],$$

i.e.,

$$\|w_n-x+\lambda z\|\to 0,\quad \frac{1}{\lambda}\|x-H(A,B)(x-\lambda z)-\lambda y_n\|\to 0,$$

and so

$$\|z_n-z\|\to 0,\quad \text{as } n\to\infty,$$

i.e.,

$$J^{H(\cdot,\cdot)}_{M_n,\lambda}(x)\to J^{H(\cdot,\cdot)}_{M,\lambda}(x).$$

**Sufficient Part:** Suppose that

$$J^{H(\cdot,\cdot)}_{M_n,\lambda}(x)\to J^{H(\cdot,\cdot)}_{M,\lambda}(x),\quad \forall x\in X,\quad \lambda>0.$$

For any $(x,y)\in graph(M)$, we have $y\in M(x)$, and hence

$$H(Ax,Bx)+\lambda y\in[H(A,B)+\lambda M](x).$$

Therefore,

$$x=\left[I-\lambda J^{H(\cdot,\cdot)}_{M,\lambda}\right](H(Ax,Bx)+\lambda y).$$

Let $x_n=\left[I-\lambda J^{H(\cdot,\cdot)}_{M_n,\lambda}\right](H(Ax,Bx)+\lambda y)$. This implies that

$$\frac{1}{\lambda}[H(Ax,Bx)-H(Ax_n,Bx_n)+\lambda y]\in M_n(x_n).$$

Let $y'_n=\frac{1}{\lambda}[H(Ax,Bx)-H(Ax_n,Bx_n)+\lambda y]$ and using the same arguments as for Eq. (7), we have

$$\begin{aligned}
\|y'_n-y\|&=\left\|\frac{1}{\lambda}[H(Ax,Bx)-H(Ax_n,Bx_n)+\lambda y]-y\right\| \\
&=\frac{1}{\lambda}\|H(Ax,Bx)-H(Ax_n,Bx_n)\| \\
&=\frac{1}{\lambda}\|H(Ax,Bx)-H(Ax_n,Bx) \\
&\quad+H(Ax_n,Bx)-H(Ax_n,Bx_n)\| \qquad (8)\\
&\le\frac{1}{\lambda}\|H(Ax,Bx)-H(Ax_n,Bx)\| \\
&\quad+\frac{1}{\lambda}\|H(Ax_n,Bx)-H(Ax_n,Bx_n)\| \\
&\le\left(\frac{\gamma_1+\gamma_2}{\lambda}\right)\|x_n-x\|.
\end{aligned}$$

Using above arguments, we have

$$\begin{aligned}
\|x_n-x\|&=\left\|\left(I-\lambda J^{H(\cdot,\cdot)}_{M_n,\lambda}\right)[H(Ax,Bx)+\lambda y]-\left(I-\lambda J^{H(\cdot,\cdot)}_{M,\lambda}\right)\right. \\
&\quad\left.[H(Ax,Bx)+\lambda y]\right\| \\
&=\left\|\left[\left(I-\lambda J^{H(\cdot,\cdot)}_{M_n,\lambda}\right)-\left(I-\lambda J^{H(\cdot,\cdot)}_{M,\lambda}\right)\right][H(Ax,Bx)+\lambda y]\right\|. \qquad (9)
\end{aligned}$$

Since $J^{H(\cdot,\cdot)}_{M_n,\lambda}(x)\to J^{H(\cdot,\cdot)}_{M,\lambda}(x)$, we have from (9) that

$$\|x_n-x\|\to 0\quad \text{as } n\to\infty.$$

Thus, from (8), it follows that $y'_n\to y$ as $n\to\infty$, i.e.,

$$M_n\xrightarrow{G}M.$$

This completes the proof. □

Combining Theorems 1 and 2, we have the following remark.

*Remark 1* The convergence of the resolvent operator $R^{H(\cdot,\cdot)}_{M_n,\lambda}(x)\to R^{H(\cdot,\cdot)}_{M,\lambda}(x)$, and the convergence of the generalized Yosida approximation operator $J^{H(\cdot,\cdot)}_{M_n,\lambda}(x)\to J^{H(\cdot,\cdot)}_{M,\lambda}(x)$ are equivalent if and only if the operator $M_n\xrightarrow{G}M$.

*Proof* Suppose that $M_n\xrightarrow{G}M$ and $R^{H(\cdot,\cdot)}_{M_n,\lambda}(x)\to R^{H(\cdot,\cdot)}_{M,\lambda}(x)$. Then

$$\begin{aligned}
&R^{H(\cdot,\cdot)}_{M_n,\lambda}(x)\to R^{H(\cdot,\cdot)}_{M,\lambda}(x),\quad \forall x\in X \\
&\quad\Rightarrow\left[I-R^{H(\cdot,\cdot)}_{M_n,\lambda}\right](x)\to\left[I-R^{H(\cdot,\cdot)}_{M,\lambda}\right](x) \\
&\quad\Rightarrow\frac{1}{\lambda}\left[I-R^{H(\cdot,\cdot)}_{M_n,\lambda}\right](x)\to\frac{1}{\lambda}\left[I-R^{H(\cdot,\cdot)}_{M,\lambda}\right](x) \\
&\quad\Rightarrow J^{H(\cdot,\cdot)}_{M_n,\lambda}(x)\to J^{H(\cdot,\cdot)}_{M,\lambda}(x),\quad \forall x\in X.
\end{aligned}$$

On similar way, we can show that $J^{H(\cdot,\cdot)}_{M_n,\lambda}(x)\to J^{H(\cdot,\cdot)}_{M,\lambda}(x)$ implies that $R^{H(\cdot,\cdot)}_{M_n,\lambda}(x)\to R^{H(\cdot,\cdot)}_{M,\lambda}(x)$. □

We construct the following consolidated example which shows that the mapping $M$ is $H(\cdot,\cdot)$-accretive with respect to $A$ and $B$, $M_n \xrightarrow{G} M$ and $J^{H(\cdot,\cdot)}_{M_n,\lambda} \to J^{H(\cdot,\cdot)}_{M,\lambda}$. Through MATLAB programming, we show some graphics for the convergence of generalized Yosida approximation operator.

*Example 1* Let $X = \mathbb{R}$; $A, B : \mathbb{R} \to \mathbb{R}$ and $H : \mathbb{R} \times \mathbb{R} \to \mathbb{R}$ be the mappings defined by

$$A(x) = \frac{x^3}{8}, \quad B(x) = \frac{x}{2},$$

and

$$H(A(x), B(x)) = A(x) - B(x), \quad \forall x \in \mathbb{R},$$

with the condition $x^2 + y^2 + xy \geq 1$. Suppose $M_n, M : \mathbb{R} \to 2^{\mathbb{R}}$ are the set-valued mappings defined by

$$M_n(x) = \frac{x}{2} + \frac{1}{n^2},$$

and

$$M(x) = \frac{x}{2}.$$

Then, for any fixed $u \in \mathbb{R}$, we have

$$\begin{aligned}\langle H(Ax, u) - H(Ay, u), x - y\rangle &= \langle Ax - Ay, x - y\rangle \\ &= \frac{1}{8}(x-y)^2(x^2 + y^2 + xy) \\ &\geq \frac{1}{8}(x-y)^2 = \frac{1}{8}\|x - y\|^2.\end{aligned}$$

Hence, $H(A, B)$ is $\frac{1}{8}$-strongly accretive with respect to $A$. In addition

$$\begin{aligned}&\langle H(u, Bx) - H(u, By), x - y\rangle = -\langle Bx - By, x - y\rangle \\ &= -\frac{1}{2}(x-y)^2 \geq -\frac{3}{2}(x-y)^2.\end{aligned}$$

Hence, $H(A, B)$ is $\frac{3}{2}$-relaxed accretive with respect to $B$.

One can easily verify that for $\lambda = 1$,

$$[H(A, B) + \lambda M]^{-1}(\mathbb{R}) = \mathbb{R}.$$

Hence, $M$ is $H(\cdot,\cdot)$-accretive with respect to $A$ and $B$.

Now, we show that $M_n \xrightarrow{G} M$. For any $(x, y) \in \text{graph}(M)$, there exists a sequence $(x_n, y_n) \in \text{graph}(M_n)$, where let

$$x_n = \left(1 + \frac{1}{n}\right)x,$$

and

$$y_n = M_n(x_n) = \frac{x_n}{2} + \frac{1}{n^2}, \quad \forall n \in \mathbb{N}.$$

Since

$$\lim_n x_n = \lim_n\left[\left(1 + \frac{1}{n}\right)x\right] = x,$$

we have,

$$x_n \to x \quad \text{as } n \to \infty.$$

In addition, by definition of graph, it follows that

$$\lim_n y_n = \lim_n\left(\frac{x_n}{2} + \frac{1}{n^2}\right) = \frac{1}{2}x = M(x) = y.$$

It follows that $y_n \to y$ as $n \to \infty$ and hence, $M_n \xrightarrow{G} M$.

Furthermore, we show that $J^{H(\cdot,\cdot)}_{M_n,\lambda} \to J^{H(\cdot,\cdot)}_{M,\lambda}$ as $M_n \xrightarrow{G} M$.

Let for $\lambda = 1$, the resolvent operators are given by

$$R^{H(\cdot,\cdot)}_{M_n,\lambda}(x) = [H(A, B) + \lambda M_n]^{-1}(x) = 2\sqrt[3]{\left(x - \frac{1}{n^2}\right)},$$

and

$$R^{H(\cdot,\cdot)}_{M,\lambda}(x) = [H(A, B) + \lambda M]^{-1}(x) = 2\sqrt[3]{x},$$

and the generalized Yosida approximation operators are given by

$$J^{H(\cdot,\cdot)}_{M_n,\lambda}(x) = \frac{1}{\lambda}\left[I - R^{H(\cdot,\cdot)}_{M_n,\lambda}\right](x) = \left[x - 2\sqrt[3]{\left(x - \frac{1}{n^2}\right)}\right],$$

and

$$J^{H(\cdot,\cdot)}_{M,\lambda}(x) = \frac{1}{\lambda}\left[I - R^{H(\cdot,\cdot)}_{M,\lambda}\right](x) = \left(x - 2\sqrt[3]{x}\right).$$

We evaluate

$$\left\|J^{H(\cdot,\cdot)}_{M_n,\lambda} - J^{H(\cdot,\cdot)}_{M,\lambda}\right\| = \left\|\left[x - 2\sqrt[3]{\left(x - \frac{1}{n^2}\right)}\right] - \left(x - 2\sqrt[3]{x}\right)\right\|,$$

which shows that

$$\left\|J^{H(\cdot,\cdot)}_{M_n,\lambda} - J^{H(\cdot,\cdot)}_{M,\lambda}\right\| \to 0 \quad \text{as } n \to \infty,$$

i.e.,

$$J^{H(\cdot,\cdot)}_{M_n,\lambda} \to J^{H(\cdot,\cdot)}_{M,\lambda} \quad \text{as } M_n \xrightarrow{G} M.$$

Using the above example, the convergence of generalized Yosida approximation operator $J^{H(\cdot,\cdot)}_{M_n,\lambda}$ to $J^{H(\cdot,\cdot)}_{M,\lambda}$ is illustrated in the following figure for $n = 1, 2, 5, 15$.

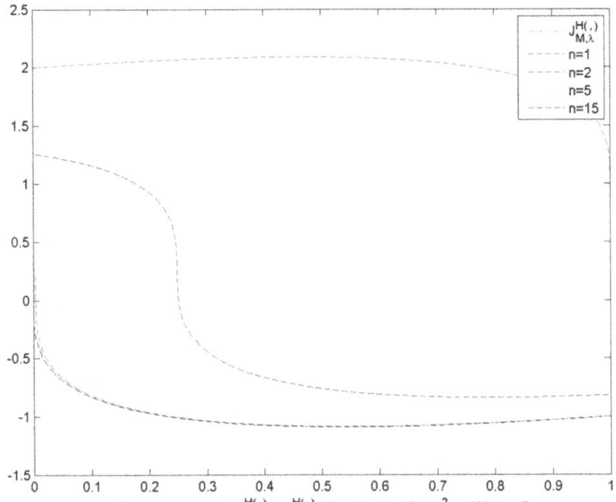

The convergence of $J^{H(.,.)}_{M_n,\lambda}$ to $J^{H(.,.)}_{M,\lambda}$ taking $M_n(x)=x/2+1/n^2$ and $M(x)=x/2$

## A Yosida inclusion problem and existence of solution

First, we state a Yosida inclusion problem and its equivalence with a fixed point problem.

Let $X$ be $q$-uniformly smooth Banach space and let $M : X \to 2^X$ be $H(\cdot,\cdot)$-accretive operator. We consider the following problem.

Find $x \in X$, such that

$$0 \in J^{H(\cdot,\cdot)}_{M,\lambda}(x) + M(x), \quad \forall x \in X, \quad \lambda > 0, \tag{10}$$

where $J^{H(\cdot,\cdot)}_{M,\lambda}$ is the generalized Yosida approximation operator defined by Eq. (2). Problem (10) is called Yosida inclusion problem.

The fixed point formulation of the problem Eq. (10) is as follows:

$$x = R^{H(\cdot,\cdot)}_{M,\lambda}\left[H(A,B)x - \lambda J^{H(\cdot,\cdot)}_{M,\lambda}(x)\right], \forall x \in X,\ \lambda > 0. \tag{11}$$

Using the definition of the resolvent operator $R^{H(\cdot,\cdot)}_{M,\lambda}$ defined by Eq. (1), one can easily obtain the equivalence of Eqs. (10) and (11).

Based on Eq. (11), we construct the following iterative algorithm for solving Yosida inclusion problem Eq. (10).

**Algorithm 1** For any $x_0 \in X$, compute the sequence $\{x_n\} \subset X$ by the following scheme:

$$x_{n+1} = R^{H(\cdot,\cdot)}_{M_n,\lambda}\left[H(A,B)x_n - \lambda J^{H(\cdot,\cdot)}_{M_n,\lambda}(x_n)\right], \quad \text{where } \lambda > 0, \quad n = 0,1,2,\ldots. \tag{12}$$

If $J^{H(\cdot,\cdot)}_{M_n,\lambda} = T$, where $T : X \to X$ is a mapping, then the Yosida inclusion problem (10) and Algorithm 1 reduces to the variational inclusion problem (10) and Algorithm 1 of Li and Huang [7], respectively, and note that for suitable choice of operators in the formulation of (12), one can obtain many existing problems and algorithms in literature.

**Theorem 3** *Let $X$ be a $q$-uniformly smooth Banach space and $A, B : X \to X$ be the single-valued mappings. Let $H : X \times X \to X$ be a single-valued mapping and $M_n, M : X \to 2^X$ be the $H(\cdot,\cdot)$-accretive operators, such that $M_n \xrightarrow{G} M$. Assume that*

1. *$H(A, B)$ is $\alpha$-strongly accretive with respect to $A$ and $\beta$-relaxed accretive with respect to $B$ and $\alpha > \beta$;*
2. *$H(A, B)$ is $\gamma_1$-Lipschitz continuous with respect to $A$ and $\gamma_2$-Lipschitz continuous with respect to $B$;*
3. $(\alpha - \beta) \geq \sqrt[q]{1 + c_q(\gamma_1 + \gamma_2)^q - q(\alpha - \beta)} + \sqrt[q]{1 - q\lambda\theta_2 + c_q\lambda^q\theta_1}$;
4. $(\alpha - \beta) \geq [\gamma_1 + \gamma_2 + \lambda\theta_1]$.

*where $\theta_1 = \frac{[(\alpha-\beta)+1]}{\lambda(\alpha-\beta)}$, $\theta_2 = \frac{[(\alpha-\beta)-1]}{\lambda(\alpha-\beta)}$, $\alpha > \beta$ and $c_q$ is same as in Lemma 2.1. Then, the Yosida inclusion problem (10) has a unique solution and the iterative sequence $\{x_n\}$ generated by Algorithm 1 converges strongly to $x$.*

*Proof* Let the mapping $F : X \to X$ be defined by

$$F(x) = R^{H(\cdot,\cdot)}_{M,\lambda}\left[H(A,B)x - \lambda J^{H(\cdot,\cdot)}_{M,\lambda}(x)\right], \quad \forall x \in X, \quad \lambda > 0.$$

For any $x, y \in X$ and using Lemma 2, we have

$$\begin{aligned}
\|F(x) - F(y)\| &= \Big\|R^{H(\cdot,\cdot)}_{M,\lambda}\left[H(A,B)x - \lambda J^{H(\cdot,\cdot)}_{M,\lambda}(x)\right] \\
&\quad - R^{H(\cdot,\cdot)}_{M,\lambda}\left[H(A,B)y - \lambda J^{H(\cdot,\cdot)}_{M,\lambda}(y)\right]\Big\| \\
&\leq \frac{1}{(\alpha-\beta)}\left\|H(A,B)x - \lambda J^{H(\cdot,\cdot)}_{M,\lambda}(x) - H(A,B)y + \lambda J^{H(\cdot,\cdot)}_{M,\lambda}(y)\right\| \\
&= \frac{1}{(\alpha-\beta)}\Big\|H(A,B)x - H(A,B)y - (x-y) \\
&\quad + (x-y) - \lambda J^{H(\cdot,\cdot)}_{M,\lambda}(x) + \lambda J^{H(\cdot,\cdot)}_{M,\lambda}(y)\Big\| \\
&\leq \frac{1}{(\alpha-\beta)}\|H(A,B)x - H(A,B)y - (x-y)\| \\
&\quad + \frac{1}{(\alpha-\beta)}\left\|(x-y) - \lambda\left(J^{H(\cdot,\cdot)}_{M,\lambda}(x) - J^{H(\cdot,\cdot)}_{M,\lambda}(y)\right)\right\|.
\end{aligned} \tag{13}$$

Using the same arguments as used in Li and Huang [7], we have

$$\|H(A,B)x - H(A,B)y - (x-y)\|^q \leq \left[1 + c_q(\gamma_1+\gamma_2)^q - q(\alpha-\beta)\right]\|x-y\|^q,$$

and hence

$$\|H(A,B)x - H(A,B)y - (x-y)\| \leq \sqrt[q]{\left[1 + c_q(\gamma_1+\gamma_2)^q - q(\alpha-\beta)\right]}\,\|x-y\|. \tag{14}$$

Using (1) and (2) of Lemma 4, we obtain

$$\left\|(x-y)-\lambda\left(J_{M,\lambda}^{H(\cdot,\cdot)}(x)-J_{M,\lambda}^{H(\cdot,\cdot)}(y)\right)\right\|^q$$
$$\leq \|x-y\|^q - q\lambda\left\langle J_{M,\lambda}^{H(\cdot,\cdot)}(x)-J_{M,\lambda}^{H(\cdot,\cdot)}(y), F_q(x-y)\right\rangle$$
$$+ c_q\lambda^q\left\|J_{M,\lambda}^{H(\cdot,\cdot)}(x)-J_{M,\lambda}^{H(\cdot,\cdot)}(y)\right\|$$
$$\leq \|x-y\|^q - q\lambda\theta_2\|x-y\|^q + c_q\lambda^q\theta_1\|x-y\|^q$$
$$= \left(1-q\lambda\theta_2+c_q\lambda^q\theta_1\right)\|x-y\|^q,$$

i.e.,

$$\left\|(x-y)-\lambda\left(J_{M,\lambda}^{H(\cdot,\cdot)}(x)-J_{M,\lambda}^{H(\cdot,\cdot)}(y)\right)\right\| \leq \sqrt[q]{1-q\lambda\theta_2+c_q\lambda^q\theta_1}\|x-y\|^q. \tag{15}$$

Using Eqs. (14), (15), (13) becomes

$$\|F(x)-F(y)\| \leq \frac{1}{\alpha-\beta}\left[\sqrt[q]{1+c_q(\gamma_1+\gamma_2)^q-q(\alpha-\beta)} + \sqrt[q]{1-q\lambda\theta_2+c_q\lambda^q\theta_1}\right]\|x-y\|,$$

i.e.,

$$\|F(x)-F(y)\| \leq k\|x-y\|, \tag{16}$$

where

$$k = \frac{1}{\alpha-\beta}\left[\sqrt[q]{1+c_q(\gamma_1+\gamma_2)^q-q(\alpha-\beta)} + \sqrt[q]{1-q\lambda\theta_2+c_q\lambda^q\theta_1}\right].$$

By condition (3), it follows that $0<k<1$ and so (16) implies that the mapping $F$ has a unique fixed point $x \in X$. Thus, $x$ is a unique solution of Yosida inclusion problem (10).

Next, we show that the sequence $\{x_n\}$ generated by the Algorithm 1 strongly converges to $x$.

Using Eqs. (11) and (12), we obtain

$$\begin{aligned}
\|x_{n+1}-x\| &= \left\|R_{M_n,\lambda}^{H(\cdot,\cdot)}\left[H(A,B)x_n-\lambda J_{M_n,\lambda}^{H(\cdot,\cdot)}(x_n)\right] - R_{M,\lambda}^{H(\cdot,\cdot)}\left[H(A,B)x-\lambda J_{M,\lambda}^{H(\cdot,\cdot)}(x)\right]\right\| \\
&= \left\|R_{M_n,\lambda}^{H(\cdot,\cdot)}\left[H(Ax_n,Bx_n)-\lambda J_{M_n,\lambda}^{H(\cdot,\cdot)}(x_n)\right] - R_{M,\lambda}^{H(\cdot,\cdot)}\left[H(Ax_n,Bx_n)-\lambda J_{M_n,\lambda}^{H(\cdot,\cdot)}(x_n)\right]\right. \\
&\quad \left.+ R_{M,\lambda}^{H(\cdot,\cdot)}\left[H(Ax_n,Bx_n)-\lambda J_{M_n,\lambda}^{H(\cdot,\cdot)}(x_n)\right] - R_{M,\lambda}^{H(\cdot,\cdot)}\left[H(Ax,Bx)-\lambda J_{M,\lambda}^{H(\cdot,\cdot)}(x)\right]\right\| \\
&\leq \left\|R_{M_n,\lambda}^{H(\cdot,\cdot)}\left[H(Ax_n,Bx_n)-\lambda J_{M_n,\lambda}^{H(\cdot,\cdot)}(x_n)\right] - R_{M,\lambda}^{H(\cdot,\cdot)}\left[H(Ax_n,Bx_n)-\lambda J_{M_n,\lambda}^{H(\cdot,\cdot)}(x_n)\right]\right\| \\
&\quad + \left\|R_{M,\lambda}^{H(\cdot,\cdot)}\left[H(Ax_n,Bx_n)-\lambda J_{M_n,\lambda}^{H(\cdot,\cdot)}(x_n)\right] - R_{M,\lambda}^{H(\cdot,\cdot)}\left[H(Ax,Bx)-\lambda J_{M,\lambda}^{H(\cdot,\cdot)}(x)\right]\right\| \\
&\leq b_n + \frac{1}{\alpha-\beta}\left\|H(Ax_n,Bx_n)-\lambda J_{M_n,\lambda}^{H(\cdot,\cdot)}(x_n) - \left[H(Ax,Bx)-\lambda J_{M,\lambda}^{H(\cdot,\cdot)}(x)\right]\right\|,
\end{aligned} \tag{17}$$

where

$$b_n = \left\|R_{M_n,\lambda}^{H(\cdot,\cdot)}\left[H(Ax_n,Bx_n)-\lambda J_{M_n,\lambda}^{H(\cdot,\cdot)}(x_n)\right] - R_{M,\lambda}^{H(\cdot,\cdot)}\left[H(Ax_n,Bx_n)-\lambda J_{M_n,\lambda}^{H(\cdot,\cdot)}(x_n)\right]\right\|.$$

Using Lipschitz continuity of $H(A, B)$ in both the arguments and Lipschitz continuity of generalized Yosida approximation operator, we obtain

$$\begin{aligned}
&\left\|H(Ax_n,Bx_n)-H(Ax,Bx)-\lambda\left[J_{M_n,\lambda}^{H(\cdot,\cdot)}(x_n)-J_{M,\lambda}^{H(\cdot,\cdot)}(x)\right]\right\| \\
&\leq \left\|H(Ax_n,Bx_n)-H(Ax_n,Bx)+H(Ax_n,Bx)-H(Ax,Bx)\right. \\
&\quad \left.-\lambda\left[J_{M_n,\lambda}^{H(\cdot,\cdot)}(x_n)-J_{M,\lambda}^{H(\cdot,\cdot)}(x)\right]-J_{M,\lambda}^{H(\cdot,\cdot)}(x_n)+J_{M,\lambda}^{H(\cdot,\cdot)}(x_n)\right\| \\
&\leq \|H(Ax_n,Bx_n)-H(Ax_n,Bx)\| + \|H(Ax_n,Bx)-H(Ax,Bx)\| \\
&\quad + \lambda\left\|J_{M_n,\lambda}^{H(\cdot,\cdot)}(x_n)-J_{M,\lambda}^{H(\cdot,\cdot)}(x_n)\right\| + \lambda\left\|J_{M,\lambda}^{H(\cdot,\cdot)}(x_n)-J_{M,\lambda}^{H(\cdot,\cdot)}(x)\right\| \\
&\leq \gamma_2\|x_n-x\| + \gamma_1\|x_n-x\| + \lambda c_n + \lambda\theta_1\|x_n-x\|,
\end{aligned} \tag{18}$$

where $c_n = \left\|J_{M_n,\lambda}^{H(\cdot,\cdot)}(x_n)-J_{M,\lambda}^{H(\cdot,\cdot)}(x_n)\right\|$.

Using Eq. (18), (17) becomes

$$\|x_{n+1}-x\| \leq b_n + \frac{1}{\alpha-\beta}[\gamma_1+\gamma_2+\lambda\theta_1]\|x_n-x\| + \lambda c_n,$$

where $\theta_1 = \frac{[\alpha-\beta+1]}{\lambda(\alpha-\beta)}$.

By Theorems 1 and 2, we have

$$R_{M_n,\lambda}^{H(\cdot,\cdot)}\left[H(Ax_n,Bx_n)-\lambda J_{M_n,\lambda}^{H(\cdot,\cdot)}(x_n)\right] \to R_{M,\lambda}^{H(\cdot,\cdot)}\left[H(Ax_n,Bx_n)-\lambda J_{M_n,\lambda}^{H(\cdot,\cdot)}(x_n)\right]$$

and hence,

$$J_{M_n,\lambda}^{H(\cdot,\cdot)}(x_n) \to J_{M,\lambda}^{H(\cdot,\cdot)}(x_n).$$

Thus, $b_n \to 0$ and $c_n \to 0$ as $n \to \infty$. It follows that

$$\|x_{n+1}-x\| \leq P(\theta)\|x_n-x\| + d_n,$$

where $d_n = b_n + \lambda c_n$, and $P(\theta) = \frac{1}{(\alpha-\beta)}[\gamma_1+\gamma_2+\lambda\theta_1]$. By condition (4), we have $0<P(\theta)<1$ and $d_n \to 0$ as $b_n, c_n \to 0 (n \to \infty)$. By Lemma 3, we have

$$\|x_{n+1}-x\| \to 0.$$

This completes the proof. □

*Remark 2* If we take $J_{M,\lambda}^{H(\cdot,\cdot)} = T$, where $T: X \to X$ is a mapping and deleting condition (4) from Theorem 3, we can obtain Theorem 4.1 of Li and Huang [7].

## References

1. Ahmad, R., Akram, M., Dilshad, M.: Graph Convergence for the $H(\cdot,\cdot)$-co-accretive mapping with an application. Bull. Malays. Math. Sci. Soc. **38**(4), 1481–1506 (2015)
2. Agrawal, R.P., Verma, R.U.: Generalized system of $(A, \eta)$-maximal relaxed monotone variational inclusion problems based on

generalized hybrid algorithms. Commun. Nonlinear Sci. Numer. Simul. **15**(2), 238–251 (2010)
3. Cao, H.W.: Yosida approximation equations technique for system of generalized set-valued variational inclusions, J. Inequal. Appl. **2013**, 455 (2013)
4. Fang, Y.P., Huang, N.J.: $H$-monotone operator and resolvent operator technique for variational inclusions. Appl. Math. Comput. **145**, 795–803 (2003)
5. Ikehata, R., Okozawa, N.: Yosida approximation and Nonlinear Hybrid Equation. Nonlinear Anal. **15**(5), 479–495 (1990)
6. Iwamiya, T., Okochi, H.: Monotonicity, resolvents and Yosida approximations of operators on Hilbert manifolds. Nonlinear Anal. **54**(2), 205–214 (2003)
7. Li, X., Huang, N.J.: Graph convergence for the $H(\cdot,\cdot)$-accretive operator in Banach spaces with an application. Appl. Math. Comput. **217**, 9053–9061 (2011)
8. Lan, H.Y.: Generalized Yosida approximations based on relativly $A$-maximal $m$-relaxed monotonicity frameworks. Abstr. Appl. Anal. **2013** (2013) (**article ID 157190**)
9. Moudafi, A.: A Duality algorithm for solving general variational inclusions. Adv. Model. Optim. **13**(2), 213–220 (2011)
10. Penot, J.-P., Ratsimahalo, R.: On the Yosida approximaton of operators. Proc. R. Soc. Edinb. Math. **131A**, 945–966 (2001)
11. Verma, R.U.: General system of $A$-monotone nonlinear variational inclusion problems with applications. J. Optim. Theory Appl. **131**(1), 151–157 (2006)
12. Verma, R.U.: General nonlinear variational inclusion problems involving $A$-monotone mapping. Appl. Math. Lett. **19**, 960–963 (2006)
13. Xu, H.K.: Inequalities in Bancah Spaces with applications. Nonlinear Anal. **16**(12), 1127–1138 (1991)
14. Zou, Y.Z., Huang, N.J.: $H(\cdot,\cdot)$-accretive operator with an application for solving varaitional inclusions in Bancah spaces. Appl. Math. Comput. **204**, 809–816 (2008)
15. Zou, Y.Z., Huang, N.J.: A new system of Variational Inclusions involving $(A,\eta)$-accretive operator in Bancah spaces. Appl. Math. Comput. **212**, 135–144 (2009)

4

# Operational matrices of Chebyshev polynomials for solving singular Volterra integral equations

Monireh Nosrati Sahlan[1] · Hadi Feyzollahzadeh[1]

**Abstract** An effective technique based on fractional calculus in the sense of Riemann–Liouville has been developed for solving weakly singular Volterra integral equations of the first and second kinds. For this purpose, orthogonal Chebyshev polynomials are applied. Properties and some operational matrices of these polynomials are first presented and then the unknown functions of the integral equations are represented by these polynomials in the matrix form. These matrices are then used to reduce the singular integral equations to some linear algebraic system. For solving the obtained system, Galerkin method is utilized via Chebyshev polynomials as weighting functions. The method is computationally attractive, and the validity and accuracy of the presented method are demonstrated through illustrative examples. As shown in the numerical results, operational matrices, even for first kind integral equations, have relatively low condition numbers, and thus, the corresponding matrices are well posed. In addition, it is noteworthy that when the solution of equation is in power series form, the method evaluates the exact solution.

**Keywords** Chebyshev polynomials · Singular integral equations · Operational matrix · Fractional calculus · Galerkin method

**Mathematics Subject Classification** 41A50 · 26A33 · 65L60

✉ Monireh Nosrati Sahlan
nosrati@bonabu.ac.ir

[1] Department of Mathematics and Computer Science, Technical Faculty, University of Bonab, Bonab, Iran

## Introduction

The aim of this study is to present a high-order computational method for solving special cases of singular Volterra integral equations of the first and second kinds, namely Abel's integral equations, defined by

$$f(x) = \int_0^x |x-t|^{-\alpha} y(t)\mathrm{d}t, \tag{1}$$

and

$$y(x) = f(x) + \int_0^x |x-t|^{-\alpha} y(t)\mathrm{d}t, \tag{2}$$

$$0<\alpha<1, \quad 0\leq x\leq T,$$

where $f(x) \in C[0, T]$ is the known function and $y(x)$ is the unknown function that to be determined, and $T$ is a positive constant.

Abel's equation is one of the integral equations derived directly from a concrete problem of mechanics or physics (without passing through a differential equation). Historically, Abel's problem is the first one that led to the study of integral equations. The generalized Abel's integral equations on a finite segment appeared in the paper of Zeilon [15] for the first time.

A comprehensive reference on Abel-type equations, including an extensive list of applications, can be found in [8, 9].

The construction of high-order methods for the equations is, however, not an easy task because of the singularity in the weakly singular kernel. In fact, in this case, the solution $y$ is generally not differentiable at the endpoints of the interval [3], and due to this, to the best of the authors' knowledge, the best convergence rate ever achieved remains only at polynomial order. For example, if we set uniform meshes with $n+1$ grid points and apply the spline method of order $m$, then the convergence rate is only $O(n^{-2P})$ at most (see [4, 12]), and it cannot be improved by increasing $m$. One way of remedying this is to introduce graded meshes [13]. By doing so, the rate is improved to $O(n^{-m})$, which now depends on m, but still at polynomial order. Rashit Ishik [10] used Bernstein series solution for solving linear integro-differential equations with weakly singular kernels. In [5] and [6], wavelets method was applied for solution of nonlinear fractional integro-differential equations in a large interval and systems of nonlinear singular fractional Volterra integro-differential equations. Authors of [11] applied fractional calculus for solving Abel integral equations. The expansions approach for solving cauchy integral equation of the first kind is discussed in [14].

In this paper, we use the Chebyshev polynomials operational matrices via Galerkin method for solving weakly singular integral equations. Our method consists of reducing the given weakly singular integral equation to a set of algebraic system by expanding the unknown function by Chebyshev polynomials of the first kind. Galerkin method is utilized to solve the obtained system.

The structure of this paper is arranged as follows. The main problem and brief history of some presented methods are expressed in Sect. 1. In Sect. 2, we present some necessary definitions and mathematical preliminaries of the fractional calculus theory in the sense of Riemann–Liouville. Section 3 is devoted to introducing Chebyshev polynomials, properties and some operational matrices of these functions. In Sect. 4, Chebyshev polynomials are applied as testing and weighting functions of Galerkin method for efficient solution of Eq. 1. In Sect. 5, we report our numerical founds and compare with other methods in solving these integral equations, and Sect. 6 contains our conclusion.

## Some preliminaries in fractional calculus

In this section, we briefly present some definitions and results in fractional calculus for our subsequent discussion. The fractional calculus is the name for the theory of integrals and derivatives of arbitrary order, which unifies and generalizes the notions of integer-order differentiation and n-fold integration [7]. There are various definitions of fractional integration and differentiation, such as Grunwald–Letnikov, and Caputo and Riemann–Liouville's definitions. In this study, fractional calculus in the sense of Riemann–Liouville is considered.

**Definition 1** Let $f$ be a real function on $[a, b]$ and $0<\alpha<1$. Then, the left and right Riemann–Liouville fractional integral operators of order $\alpha$ for the function $f$ are defined, respectively, as

$$_aI_x^\alpha f(x)=\frac{1}{\Gamma(\alpha)}\int_a^x (x-t)^{\alpha-1}f(t)\mathrm{d}t,$$

$$_xI_b^\alpha f(x)=\frac{1}{\Gamma(\alpha)}\int_x^b (t-x)^{\alpha-1}f(t)\mathrm{d}t,$$

$$x\in[a,b],\quad \alpha>0.$$

**Definition 2** For $f\in C[a,b]$, the left and right Riemann–Liouville fractional derivatives are defined, respectively, as

$$_aD_x^\alpha f(x)=\frac{1}{\Gamma(1-\alpha)}\frac{\mathrm{d}}{\mathrm{d}t}\int_a^x (x-t)^{-\alpha}f(t)\mathrm{d}t,$$

$$_xD_b^\alpha f(x)=\frac{1}{\Gamma(1-\alpha)}\frac{\mathrm{d}}{\mathrm{d}t}\int_x^b (t-x)^{-\alpha}f(t)\mathrm{d}t.$$

In this study, the left Riemann–Liouville fractional integral operator is utilized to transform singular integral equation to some algebraic system. Therefore, for abbreviation, the mentioned operator is denoted by $I^\alpha$.

**Theorem 1** *The operator $I^\alpha$ (stand for left and right Riemann–Liouville fractional integral operator) satisfies the following properties*:

$$(1)\quad I^\alpha\left(\sum_{i=o}^n \mu_i f_i(x)\right)=\sum_{i=0}^n \mu_i I^\alpha f_i(x),$$

$$(2)\quad I^\alpha x^\beta=\frac{\Gamma(\beta+1)}{\Gamma(\beta+\alpha+1)}x^{\alpha+\beta},\quad \beta>-1.$$

*Proof* We prove the proposition for left Riemann–Liouville fractional integral operator and the proof for right Riemann–Liouville fractional integral operator can be done similarly. For the part (1), we have

$$\begin{aligned}I^\alpha\left(\sum_{i=o}^n \mu_i f_i(x)\right)&=\frac{1}{\Gamma(\alpha)}\int_a^x (x-t)^{\alpha-1}\left(\sum_{i=o}^n \mu_i f_i(t)\right)\mathrm{d}t\\&=\frac{1}{\Gamma(\alpha)}\sum_{i=o}^n \mu_i\left(\int_a^x (x-t)^{\alpha-1}f_i(t)\mathrm{d}t\right)\\&=\sum_{i=o}^n \mu_i\frac{1}{\Gamma(\alpha)}\left(\int_a^x (x-t)^{\alpha-1}f_i(t)\mathrm{d}t\right)\\&=\sum_{i=0}^n \mu_i I^\alpha f_i(x).\end{aligned}$$

In addition, for part (2), we have

$$I^\alpha x^\beta=\frac{1}{\Gamma(\alpha)}\int_a^x (x-t)^{\alpha-1}t^\beta\mathrm{d}t,$$

by changing the variable $t=rx$, we get

$$I^\alpha x^\beta=\frac{1}{\Gamma(\alpha)}\int_0^1 x^{\alpha-1}(1-r)^{\alpha-1}r^\beta x^\beta x\mathrm{d}r,$$

now, using the formulae of the beta function, we have

$$I^{\alpha}x^{\beta} = \frac{x^{\alpha+\beta}}{\Gamma(\alpha)}\int_0^1 (1-r)^{\alpha-1}r^{\beta}\mathrm{d}r = \frac{x^{\alpha+\beta}}{\Gamma(\alpha)}B(\alpha,\beta+1).$$

On the other hand, we know that the beta function can be written in terms of the Gamma function as follows:

$$B(a,b) = \frac{\Gamma(a)\Gamma(b)}{\Gamma(a+b)},$$

so we have

$$I^{\alpha}x^{\beta} = \frac{\Gamma(\beta+1)}{\Gamma(\alpha+\beta+1)}x^{\alpha+\beta}.$$

□

## Chebyshev polynomials

In this section, a brief summary of orthogonal Chebyshev polynomials is expressed.

**Definition 3** The $n$th degree of Chebyshev polynomials is defined by

$$T_n(t) = \cos(n\theta), \quad 0 \leq \theta \leq \pi,$$

where $t = \cos(\theta)$. The roots of Chebyshev polynomial of degree $n+1$ can be obtained by

$$t_i = \cos\left(\frac{(2i+1)\pi}{2n+1}\right), \quad i = 0,\ldots,n.$$

In addition, the following successive relation holds for Chebyshev polynomials:

$$T_{n+1}(x) = 2xT_n(x) - T_{n-1}(x),$$

where

$$T_0(x) = 1, \quad T_1(x) = x.$$

Chebyshev polynomials are orthogonal with respect to the weight function $w(x) = \frac{1}{\sqrt{1-x^2}}$ in the interval $[-1,1]$, that is

$$\int_{-1}^{1} \frac{T_n(x)T_m(x)}{\sqrt{1-x^2}}\mathrm{d}x = \begin{cases} 0\,; & m \neq n, \\ \frac{\pi}{2}\,; & m = n \neq 0 \\ \pi\,; & m = n = 0. \end{cases}$$

### Matrix form

We can represent Chebyshev polynomials in the matrix form. Put

$$\mathbf{T(x)} = (\mathbf{T_0(x)}, \mathbf{T_1(x)}, \ldots, \mathbf{T_n(x)})^{\mathbf{T}}, \quad \mathbf{X} = (\mathbf{1}, \mathbf{x}, \ldots, \mathbf{x^n})^{\mathbf{T}}, \tag{3}$$

then we can write

$$\mathbf{T(x)} = \mathbf{TX}, \tag{4}$$

where $T$ is a $(n+1)\times(n+1)$ matrix defined by

$$T = \begin{pmatrix} 1 & 0 & 0 & 0 & 0 & 0 & \ldots & 0 \\ 0 & 1 & 0 & 0 & 0 & 0 & \ldots & 0 \\ -1 & t_{21} & 2 & 0 & 0 & 0 & \ldots & 0 \\ 0 & t_{31} & t_{32} & 2^2 & 0 & 0 & \ldots & 0 \\ 1 & t_{41} & t_{42} & t_{43} & 2^3 & 0 & \ldots & 0 \\ \vdots & & & \vdots & & & \ddots & \\ \cos(\frac{n\pi}{2}) & t_{n1} & t_{n2} & t_{n3} & t_{n4} & \cdots & & 2^{n-1} \end{pmatrix}, \tag{5}$$

and the first element of each row is $t_{i0} = \cos(\frac{i\pi}{2})$, $i = 0,\ldots,n$. In addition, other elements are defined by $t_{i,j} = sign(t_{i-1,j-1})(2|t_{i-1,j-1}| + |t_{i-2,j}|$. The Chebyshev polynomials are defined in the $[-1,1]$, but the integration interval of Eqs. (1) and (2) is [0, $T$]. To transform the interval $[-1,1]$ to [0, $T$], we apply the $(n+1)\times(n+1)$ shift matrix $R$, which is defined by

$$R_{ij} = \begin{cases} \binom{i}{j}\gamma_1^{i-j}\gamma_2^{j}\,; & j = 0,1,\ldots,i, \quad i = 0,1,\ldots,n, \\ 0; & i<j, \end{cases} \tag{6}$$

and $\gamma_1 = -1, \gamma_2 = \frac{2}{T}$. Thus, the shifted Chebyshev polynomial matrix is written as $WX$, where $W = TR$.

### Function approximation

A function $y(x) \in L_2[0,T]$, can be expressed in terms of the shifted Chebyshev polynomials as [2]

$$y(x) = \sum_{j=0}^{\infty} c_j\varphi_j(x)\mathrm{d}x = C^T \cdot W \cdot X, \tag{7}$$

where $\varphi_j$ is the shifted Chebyshev polynomial of degree $j$. The coefficients $c_j$ are given by

$$c_0 = \frac{1}{\pi}\int_0^T \frac{f(x)\varphi_0(x)}{\sqrt{\frac{4x}{T} - \frac{4x^2}{T^2}}}\mathrm{d}x,$$

and

$$c_j = \frac{2}{\pi}\int_0^T \frac{f(x)\varphi_j(x)}{\sqrt{\frac{4x}{T} - \frac{4x^2}{T^2}}}\mathrm{d}x, \quad j = 1,2,\ldots.$$

### Operational matrix of fractional Integration

The fractional integral of Chebyshev polynomials can be defined by

$$I^{\alpha}(TX) = (A * T)X^{(\alpha)}, \tag{8}$$

where

$$X^{(\alpha)} = x^\alpha \cdot X = \left[x^\alpha\ x^{\alpha+1}\ \ldots\ x^{\alpha+n}\right]^T,$$

and $*$ is point by point product and $A$ is $(n+1) \times (n+1)$ lower triangular matrix defined by

$$A_{ij} = \begin{cases} \dfrac{\Gamma(j+1)}{\Gamma(\alpha+j+1)}, & i \geq j, \\ 0, & i < j. \end{cases}$$

In addition, for each function $g(x)$, approximated by shifted Chebyshev functions ($g(x) = D^T WX$), the fractional integral can be written as

$$I^\alpha(g(x)) = I^\alpha(D^T WX) = D^T(A * W)X^{(\alpha)}. \tag{9}$$

## Numerical implementation

In this section, the shifted Chebyshev polynomials are applied for solving singular integral Eqs. (1) and (2). For this purpose, initially, the singular integral equation is transformed to nonsingular integral equation, utilizing Riemann–Liouville calculus.

Putting $-\alpha = \beta - 1$ in the integral part of the Eqs. (1) and (2), we get

$$\int_0^x (x-t)^{-\alpha} y(t)\mathrm{d}t = \Gamma(\beta)\left(\frac{1}{\Gamma(\beta)}\int_0^x (x-t)^{\beta-1} y(t)\mathrm{d}t\right),$$

by definition of Riemann–Liouville fractional integral operator, the current relation can be rewritten as

$$\int_0^x (x-t)^{-\alpha} y(t)\mathrm{d}t = \Gamma(\beta) I^\beta y(x). \tag{10}$$

Now, the unknown function $y(x)$ is approximated by shifted Chebyshev polynomials as

$$y(x) \simeq y_n(x) = C^T WX, \tag{11}$$

where $C^T$ is the unknown vector, that to be determined. In the following, we describe the method in detail for the first and second kinds.

### The first kind

Consider the weakly singular Volterra integral equation of the first kind (1), according to Eq. (10), we have

$$f(x) = \Gamma(\beta) I^\beta y(x), \tag{12}$$

substituting Eqs. (11) and (8) in (12), we can get

$$f(x) = \Gamma(\beta) C^T (A * W) X^\beta. \tag{13}$$

Therefore, singular integral equation (1) is transformed to the above algebraic system. For solving this system, Galerkin procedure is utilized via shifted Chebyshev polynomials. Put $\Gamma(\beta)(A * W) = \Lambda$ and suppose $\varphi_i(x)$ be the shifted Chebyshev polynomial of degree $i$, which can be written as

$$\varphi_i(x) = W_i X, \quad i = 0, 1, \ldots, n,$$

where $W_i$ is the $i$th row of the matrix $W$. Multiplying Eq. (13) by $\varphi_i(x)$, we get

$$C^T \Lambda X^\beta W_i X = f(x) W_i X, \quad i = 0, 1, \ldots, n. \tag{14}$$

Putting

$$\tilde{X} = \begin{pmatrix} 1 & x & x^2 & \ldots & x^n \\ x & x^2 & x^3 & \ldots & x^{n+1} \\ \vdots & \vdots & \vdots & \ddots & \vdots \\ x^n & x^{n+1} & x^{n+2} & \ldots & x^{2n} \end{pmatrix}, \tag{15}$$

and

$$\tilde{X}^\beta = x^\beta \tilde{X},$$

we get

$$C^T \Lambda \tilde{X}^\beta W_i^T = f(x) W_i X, \quad i = 0, 1, \ldots, n. \tag{16}$$

by integrating current equations from 0 to $T$, we have

$$W_i P^\beta \Lambda^T C = W_i P_x, \tag{17}$$

where $P_x$ and $P^\beta$ are related integration operational matrix, defined by

$$P_x = \int_0^T X f(x)\mathrm{d}x, \quad (P_x)_{i1} = \int_0^T x^i f(x)\mathrm{d}x,$$

$$P^\beta = \int_0^T \tilde{X}^\beta \mathrm{d}x, \quad (P^\beta)_{ij} = \frac{T^{i+j+\beta+1}}{i+j+\beta+1},$$

$$i = 0, \ldots, n, \quad j = 0, \ldots, n.$$

Considering Eq. (17) for $i = 0, \ldots, n$, we get

$$WP^\beta \Lambda^T C = WP_x, \tag{18}$$

a system of $n+1$ equations and $n+1$ unknowns that can be solved easily.

### The second kind

Similarly, for the second kind singular integral equation (2), we have

$$y(x) = f(x) + \int_0^x (x-t)^{-\alpha} y(t)\mathrm{d}t = f(x) + \Gamma(\beta) I^\beta (C^T WX), \tag{19}$$

so Eq. (19) can be written as

$$C^T WX - \Gamma(\beta) C^T (A * W) X^\beta = f(x). \tag{20}$$

Multiplying Eq. (20) by $\varphi_i(x)$ and integrating from 0 to $T$, we can rewrite the current equation in the following form:

$$\left(WPW^T - WP^{\beta}\Lambda^T\right)C = WPf(x), \tag{21}$$

where

$$P = \int_0^T \tilde{X}\mathrm{d}x, \quad P_{ij} = \frac{T^{i+j+1}}{i+j+1}.$$

Eq. (21) is a system of $n+1$ equations and $n+1$ unknowns that can be solved easily.

## Illustrative examples

In this section, for showing the accuracy and efficiency of the described method, we present some examples. Moreover, the condition number of the operational matrices, defined by

$$cond(A) = \|A\|.\|A^{-1}\|, \tag{22}$$

are given in corresponding tables. In Examples 1 and 2 singular Volterra integral equation of the second kind, and in examples 3 and 4, singular Volterra integral equation of the first kind is solved. As we know, numerically solving the first kind integral equations is so difficult, because their operational matrices have large condition numbers and in the other words are bad-conditioned, while, as seen in Tables 2 and 3, the maximum condition number of the problem is 64. The abbreviation c.n.m in the tables denotes the condition number of the operational matrices.

*Example 1* Consider the following singular integral equation of the second kind:

$$y(x) + \int_0^x \frac{y(t)}{(x-t)^{1/2}}\mathrm{d}t = x^2 + \frac{16}{15}x^{5/2},$$

with the exact solution $y(x) = x^2$. The unknown coefficients for $c_i$ are obtained through the method explained in Sect. 4 for $N = 4$.

$$c_0 = \frac{3}{8}, \quad c_1 = \frac{1}{2} \quad c_2 = \frac{1}{8} \quad c_3 = 0 \quad c_4 = 0.$$

The function $y(x)$ is a polynomial of degree 2 and the least approximation level for Chebyshev polynomials, in this study, is $N = 4$. Therefore, the approximated solution through the presented method is the same as the exact solution, that is $y_4(x) = x^2$.

*Example 2* Consider the following singular integral equation:

$$y(x) - \int_0^x \frac{y(t)}{(x-t)^{1/2}}\mathrm{d}t = x^{3/2} - \frac{3}{8}\pi x^2,$$

**Table 1** Exact and approximate solutions of Example 2

| $x$ | $N = 4$ | $N = 8$ | $N = 12$ | Exact |
|---|---|---|---|---|
| 0.1 | 0.0326 | 0.0316 | 0.0316 | 0.0316 |
| 0.2 | 0.0896 | 0.0895 | 0.0894 | 0.0894 |
| 0.3 | 0.1638 | 0.1643 | 0.1643 | 0.1643 |
| 0.4 | 0.2526 | 0.2529 | 0.2530 | 0.2530 |
| 0.5 | 0.3536 | 0.3536 | 0.3536 | 0.3536 |
| 0.6 | 0.4651 | 0.4648 | 0.4648 | 0.4648 |
| 0.7 | 0.5859 | 0.5856 | 0.5857 | 0.5857 |
| 0.8 | 0.7154 | 0.7155 | 0.7155 | 0.7155 |
| 0.9 | 0.8535 | 0.8538 | 0.8538 | 0.8538 |
| 1 | 1.0006 | 1.0001 | 1.0000 | 1.0000 |
| *c.n.m** | 11.23 | 22.32 | 32.85 | |

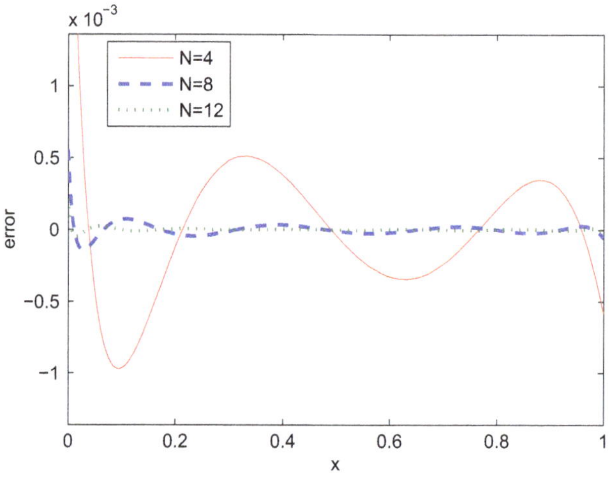


**Fig. 1** Graph of estimated solution of Example 2 for $N = 4, 8$ and 12

with the exact solution $y(x) = x^{\frac{3}{2}}$. The solution for $y(x)$ is obtained by the method in Sect. 4 for $N = 4, 8$ and 12. The unknown coefficients for $c_i$ are obtained through the method explained in Sect. 4 for $N = 4$:

$$c_0 = 0.4242, \quad c_1 = 0.5097 \quad c_2 = 0.0724$$
$$c_3 = -0.0076 \quad c_4 = 0.0018,$$

and the approximate solution for $N = 4$ is calculated as

$$y_4(x) = 0.2253x^4 + 0.6927x^3 + 1.2240x^2 + 0.2477x - 0.0038.$$

In Table 1, we have presented exact and approximated solutions of Example 2 in some arbitrary points. In addition, the last line of Table 1 shows the condition number of operational matrices. The errors of approximate solutions in the levels $N = 4, 8$, and 12 are shown in Fig. 1.

*Example 3* Consider the following singular integral equation of the first kind:

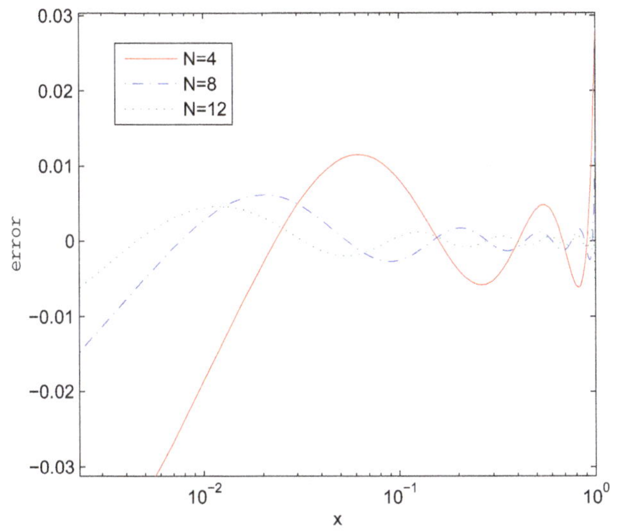


**Fig. 2** Graph of estimated solution of Example 3 for $N = 4, 8$ and 12

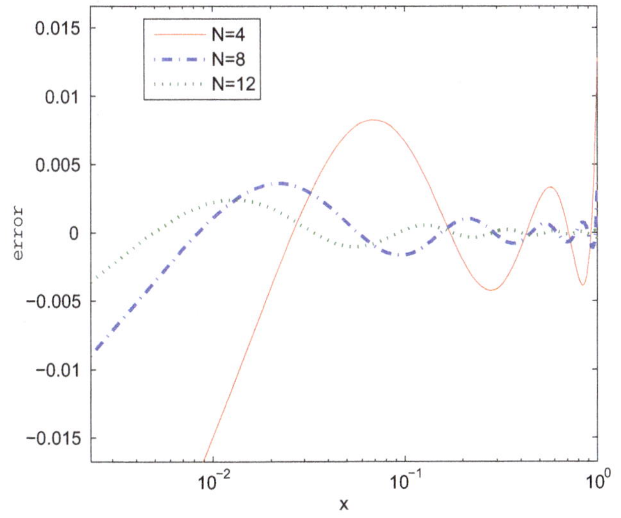


**Fig. 3** Graph of estimated solution of Example 4 for $N = 4, 8$ and 12

$$\int_0^x \frac{y(t)}{(x-t)^{1/3}}\mathrm{d}t = x^{7/6},$$

with the exact solution $y(x) = \frac{7\Gamma(1/6)}{18\Gamma(2/3)}\sqrt{\frac{x}{\pi}}$. The solution for $y(x)$ is obtained by the method in Sect. 4 for $N = 4, 8$, and 12. The unknown coefficients for $c_i$ are obtained through the method explained in Sect. 4 for $N = 4$:

$$c_0 = 0.5739, \quad c_1 = 0.3719 \quad c_2 = -0.0768$$
$$c_3 = 0.0218 \quad c_4 = -0.0172,$$

and the approximate solution for $N = 4$ is calculated as

$$y_4(x) = -2.1983x^4 + 5.0939x^3 - 4.4082x^2 + 2.2999x + 0.0862.$$

**Table 2** Exact and approximate solutions of Example 3

| $x$ | Approximate $N = 4$ | $N = 8$ | $N = 12$ | Method of [1] $N = 20$ | Exact |
|---|---|---|---|---|---|
| 0.1 | 0.2870 | 0.2878 | 0.2852 | 0.2848 | 0.2852 |
| 0.2 | 0.4071 | 0.4017 | 0.4033 | 0.4032 | 0.4033 |
| 0.3 | 0.4940 | 0.4992 | 0.4936 | 0.4944 | 0.4936 |
| 0.4 | 0.5706 | 0.5713 | 0.5703 | 0.5704 | 0.5704 |
| 0.5 | 0.6335 | 0.6366 | 0.6378 | 0.6374 | 0.6377 |
| 0.6 | 0.6946 | 0.6985 | 0.6986 | 0.6989 | 0.6986 |
| 0.7 | 0.7556 | 0.7558 | 0.7546 | 0.7547 | 0.7546 |
| 0.8 | 0.8126 | 0.8055 | 0.8066 | 0.8063 | 0.8067 |
| 0.9 | 0.8567 | 0.8565 | 0.8556 | 0.8551 | 0.8556 |
| 1 | 0.8736 | 0.8890 | 0.9018 | 0.8992 | 0.9019 |
| $c.n.m^*$ | 13.94 | 26.78 | 43.68 | | |

**Table 3** Exact and approximate solutions of Example 4

| $x$ | $N = 4$ | $N = 8$ | $N = 12$ | Exact |
|---|---|---|---|---|
| 0.1 | 0.2732 | 0.2815 | 0.2797 | 0.2799 |
| 0.2 | 0.4445 | 0.4463 | 0.4433 | 0.4443 |
| 0.3 | 0.5863 | 0.5823 | 0.5820 | 0.5822 |
| 0.4 | 0.7063 | 0.7058 | 0.7052 | 0.07052 |
| 0.5 | 0.8159 | 0.8178 | 0.8185 | 0.8183 |
| 0.6 | 0.9209 | 0.9239 | 0.9240 | 0.92418 |
| 0.7 | 1.0237 | 1.0248 | 1.0242 | 1.0241 |
| 0.8 | 1.1227 | 1.1191 | 1.1194 | 1.1195 |
| 0.9 | 1.2131 | 1.2110 | 1.2110 | 1.2109 |
| 1 | 1.2863 | 1.2951 | 1.2995 | 1.2990 |
| $c \cdot n \cdot m^*$ | 16.01 | 40.32 | 64.18 | |

In Table 2, we present exact and approximate solutions of Example 3 in some arbitrary points and compare them by the results of [1] for $n = 20$. In addition, the last line of Table 2 shows the condition number of operational matrices. Figure 2 illustrates the error of approximate solutions in the levels $N = 4, 8$, and 12.

*Example 4* Consider the following singular integral equation of the first kind:

$$\int_0^x \frac{y(t)}{(x-t)^{2/3}}\mathrm{d}t = \pi x,$$

with the exact solution $y(x) = \frac{3\sqrt{3}}{4}x^{2/3}$. The solution for $y(x)$ is obtained through the method in Sect. 4 for $N = 4, 8$, and 12. The unknown coefficients for $c_i$ are obtained through the method explained in Sect. 4 for $N = 4$.

$$c_0 = 07539, \quad c_1 = 0.5968 \quad c_2 = -0.0738$$
$$c_3 = 0.0212 \quad c_4 = -0.0118,$$

and the approximate solution for $N = 4$ is calculated as

$$y_4(x) = -1.5114x^4 + 3.7014x^3 - 3.4978x^2 + 2.5439x + 0.0502.$$

In Table 3, exact and approximated solutions of Example 4 in some arbitrary points are given. The condition number of operational matrices for $N = 4, 8$, and 12 are calculated by Eq. 22, and is shown in the last row of Table 3. Figure 3 is helpful in geometric understanding the errors of approximated solutions in the levels $N = 4, 8$, and 12.

## Conclusions

In this study, a numerical approach based on Chebyshev polynomials operational matrices was developed to approximate the solution of the weakly singular Volterra integral equations of the first and second kinds. Applying fractional derivative of these polynomials, we have transformed the singular integral equations to some linear algebraic system. The numerical results obtained support the validity and efficiency of the proposed method. It is noteworthy that when the solution of equation is in power series form, the method evaluates the exact solution, such as Example 1. In addition, as can be seen, the operational matrices of first kind integral equations have relatively low condition numbers. Thus, the corresponding matrices are well posed.

## References


1. Avazzadeh, Z., Shafiee, B., Loghmani, G.: Fractional calculus for solving Abel's integral equations using Chebyshev polynomials. Appl. Math. Sci. **5**(45), 2207–2216 (2011)
2. Doha, E.H., Bhrawy, A.H., Ezz-Eldien, S.S.: Efficient Chebyshev spectral methods for solving multi-term fractional orders differential equations. Appl. Math. Model. **35**, 5662–5672 (2011)
3. Graham, I.G.: Singularity expansions for the solutions of second kind Fredholm integral equations with weakly singular convolution kernels. J. Integr. Equ. **4**, 1–30 (1982)
4. Graham, I.G.: Galerkin methods for second kind integral equations with singularities. Math. Comput. **39**, 519–533 (1982)
5. Heydari, M.H., Hooshmandasl, M.R., Maalek Ghaini, F.M., Li, M.: Chebyshev wavelets method for solution of nonlinear fractional integro-differential equations in a large interval. Adv. Math. Phys., Article ID 482083 (2013)
6. Heydari, M.H., Hooshmandasl, M.R., Mohammadi, F., Cattani, C.: Wavelets method for solving systems of nonlinear singular fractional Volterra integro-differential equations. Commun. Nonlinear Sci. Numer. Simul. **19**(1), 37–48 (2014)
7. Kilbas, A.A., Srivastava, H.M., Trujillo, J.J.: Theory and applications of fractional differential equations. North-Holland Mathematics Studies, vol. 204. Elsevier (2006)
8. Nieto, J.J., Okrasinski, W.: Existence, uniqueness, and approximation of solutions to some nonlinear diffusion problems. J. Math. Anal. Appl. **210**(1), 231–240 (1997)
9. Okrasinski, W., Vila, S.: Determination of the interface position for some nonlinear diffusion problems. Appl. Math. Lett. **11**(4), 85–89 (1998)
10. Rasit Isik, O., Sezer, M., Guney, Z.: Bernstein series solution of a class of linear integro-differential equations with weakly singular kernel. Appl. Math. Comput. **217**(16), 7009–7020 (2011)
11. Saleh, M.H., Amer, S.M., Mohamed, DSh, Mahdy, A.E.: Fractional calculus for solving generalized Abels integral equations using Chebyshev polynomials. Int. J. Comput. Appl. **100**(8), 19–23 (2014)
12. Schneider, C.: Product integration for weakly singular integral equations. Math. Comput. **36**, 207–213 (1981)
13. Vainikko, G., Uba, P.: A piecewise polynomial approximation to the solution of an integral equation with weakly singular kernel. J. Austral. Math. Soc. Ser. B. **22**, 431–438 (1981)
14. Yaghobifar, M., Nik Long, N.M.A., Eshkuvatov, Z.K.: The expansions approach for solving cauchy integral equation of the first kind. Applied Mathematical Sciences. **4**(52), 2581–2586 (2010)
15. Zeilon, N.: Sur quelques points de la theorie de l'equation integrale d'Abel [On some points of the theory of integral equation of Abel type]. Arkiv. Mat. Astr. Fysik. **18**, 1–19 (1924)

# 5

# Analytical study for Soret, Hall, and Joule heating effects on natural convection flow saturated porous medium in a vertical channel

**K. Kaladhar[1] · Ch. RamReddy[2] · D. Srinivasacharya[2] · T. Pradeepa[2]**

**Abstract** This paper analyzes the laminar, incompressible natural convective transport inside vertical channel in an electrically conducting fluid saturated porous medium. In addition, this model incorporates the combined effects of Hall current, Soret and Joule heating. The non-linear governing equations with their corresponding boundary conditions are initially cast into a dimensionless form by using suitable transformations and Adomian decomposition method has been used to solve the system of equations. To explore the influence of various parameters on fluid flow properties, quantitative analysis is exhibited graphically and shown in the tabular form.

**Keywords** Natural convection · Soret effect · Hall effect · Joule heating parameter · ADM

## Introduction

Considerable attention has been paid to the immense investigation of natural convection flow saturated porous medium due to its diverse application in engineering and industrial process such as dispersion of fog, solar energy collecting devices, air conditioning of a room, material processing, cooling of molten metals, petroleum industries, moisture transport in thermal insulation etc. Different simulation procedures have been adopted by several researchers to analyze the natural convection flow saturated porous medium in vertical channel with various fluid models. The analysis of combined free and forced convection flow between two asymmetrically and symmetrically heated vertical parallel walls in a porous medium with viscous dissipation effect has been considered by Ingham et al. [1]. Paul et al. [2] presented an analytical solution for the free convection flow between vertical walls partially filled with porous matrix and a clear fluid placing an interface vertically. Umavathi [3] discussed the natural convection flow of immiscible fluids in a vertical channel filled with a porous medium taking into account of Darcy–Brinkman–Forchheimer equation model. The free convection flow between vertical walls in porous medium with radiation effect has been studied by Mishra et al. [4].

Soret effect (thermal-diffusion) refers to the differentiation of species. Soret effect is neglected in many cases related to the transfer of heat and mass due to that it is a smaller order of magnitude than the effects described by Fourier's and Fick's laws. The applications and early literature can be found in [5]. Srinivasacharya and Kaladhar [6] analyzed the free convection flow of a chemical reacting couple stress fluid in a vertical channel with the influence of Soret and Dufour effects. With the influence of magnetic field on natural convection flow of a power law fluid in a porous medium with stratification, Dufour and Soret effects has been reported by Srinivasacharya et al. [7] (also see the citations therein).

Several researchers combined the MHD flow problems with Hall effect owing that the Hall current effect cannot neglect when the magnetic field strength is strong. in view of applications, Tani [8] considered the steady motion viscous fluid in a channel with Hall effect. Srinivasacharya and Mekonnen [9–11] presented the Hall and ion-slip effects in non-Newtonian fluids. The unsteady free

✉ K. Kaladhar
kaladhar@nitpy.ac.in

1 Department of Mathematics, National Institute of Technology Puducherry, Karaikal 609605, India

2 Department of Mathematics, National Institute of Technology Warangal, Warangal, India

convection flow of an electrically conducting fluid through a porous medium in a vertical channel with effects of thermal radiation, thermal diffusion and Hall current has been reported by Manglesh and Gorla [12]. Garg et al. [13] examined the effect of Hall current viscoelastic fluid in a vertical porous channel filled with porous medium and oscillatory magneto-hydrodynamic convection.

In all erstwhile studies, the effect of Joule heating was ignored and works on Joule heating effects of participating MHD are fewer. Joule heating is the predominant heat mechanism for heat generation in integrated circuits and is an undesired effect. Chen [14] considered the radiative heat transfer and free convection flow past permeable stretching surfaces with Joule heating and viscous dissipation effects. Hossain and Gorla [15] scrutinized the effect of Joule heating on mixed convection boundary layer flow of an electrically conducting fluid past a vertical surface in the presence of a uniform transverse magnetic field fixed relative to the surface. The combined viscous and Joule heating effects on nonlinear convection flow on stagnation point through stretching or shrinking sheet in the presence of homogeneous and heterogeneous reactions has been studied by Nandkeolyar [16].

The present investigation carries out the electrically conducting natural convection in a vertical porous medium with the influence of thermal diffusion, Hall current and Joule heating effects. Inspite of the complex structure of the problem and to provide more accurate analysis to the technical and industrial applications, the final nonlinear system of equations are solved using Adomian decomposition method. Interesting features of the present setup has been given at the end of the study.

## Formulation of the problem

Consider a laminar, incompressible electrically conducting fluid saturated porous medium in a vertical channel. The plates are placed vertically upwards (in the direction of $x$-axis) and $y$-axis is perpendicular to the plates. The plates are extended infinitely in directions of $x$ and $z$ and is presented in Fig. 1. The effects of Joule heating and Hall currents produces due to the applied uniform magnetic field (applied in the direction of $y$-axis). The additional force in the $z$-direction is generated due to the effect of Hall current, which gives rise to a cross flow therefore the flow becomes three dimensional. In this present study, the magnetic Reynolds number is very small with that the induced magnetic field has been neglected in comparison with the applied magnetic field. Buoyancy forces causes the natural convection flow. With the above presumptions, the governing equations

**Fig. 1** Physical model and coordinate system

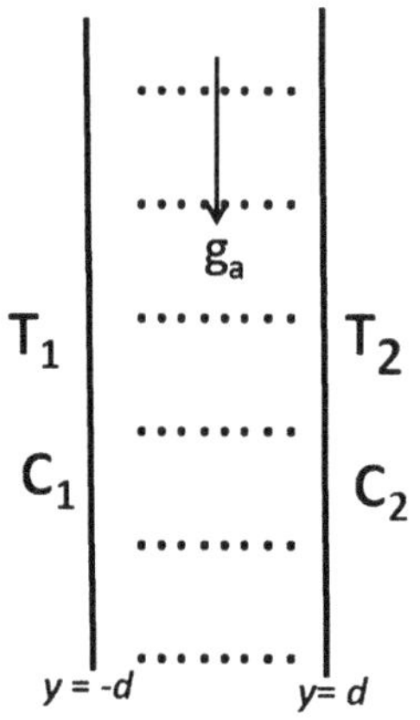


for an electrically conducting fluid saturated porous medium with thermal-diffusion are

$$\frac{\partial v}{\partial y} = 0 \Rightarrow v = v_0 = \text{constant} \quad (1)$$

$$\rho v \frac{\partial u}{\partial y} = \mu \frac{\partial^2 u}{\partial y^2} + \rho g_a \beta_T \left( (T - T_1) + \frac{\beta_C}{\beta_T}(C - C_1) \right) - \frac{\sigma B_0^2}{1 + \beta_h^2}(u + \beta_h w) - \frac{\varepsilon \mu}{K_f} u \quad (2)$$

$$\rho v \frac{\partial w}{\partial y} = \mu \frac{\partial^2 w}{\partial y^2} + \frac{\sigma B_0^2}{1 + \beta_h^2}(\beta_h u - w) - \frac{\varepsilon \mu}{K_f} w \quad (3)$$

$$\rho C_p v \frac{\partial T}{\partial y} = K_f \frac{\partial^2 T}{\partial y^2} + \mu \left[ \left( \frac{\partial u}{\partial y} \right)^2 + \left( \frac{\partial w}{\partial y} \right)^2 \right] + \frac{\sigma B_0^2}{1 + \beta_h^2}(u^2 + w^2) \quad (4)$$

$$v \frac{\partial C}{\partial y} = D \frac{\partial^2 C}{\partial y^2} + \frac{D K_f}{T_m} \frac{\partial^2 T}{\partial y^2} \quad (5)$$

with

$$u = w = 0 \quad \text{at y} = \pm \text{d} \quad (6a)$$

$$T = T_1, \quad C = C_1 \quad \text{at y} = -\text{d} \quad (6b)$$

$$T = T_2, \quad C = C_2 \quad \text{at y} = \text{d} \quad (6c)$$

where the velocity components in $x-, y-$ and $z-$ directions are taken as $u$, $v$, $w$ respectively, $g_a$ is the acceleration due to gravity, $\rho$ is the density, $C_p$ is the specific heat, $\mu$ is the coefficient of viscosity, $\varepsilon$ is the porosity, $\beta_h$ is the Hall parameter, $\beta_T$ is the coefficient of thermal expansion, $\beta_C$ is the coefficient of solutal expansion, $\sigma$ is the electric conductivity of the fluid, $B_0$ is the uniform magnetic field, $D$ is the mass diffusivity, $K_f$ is the coefficient of thermal conductivity and $T_m$ is the mean fluid temperature. The last terms of (4), (5) represents the Joule heating and thermal diffusions respectively.

Introducing the following dimensionless variables

$$\eta = \frac{y}{d}, \quad u = \frac{\nu \mathrm{Gr}}{d^2} f, \quad w = \frac{\nu \mathrm{Gr}}{d^2} g$$
$$\theta = \frac{T - T_1}{T_2 - T_1}, \quad \phi = \frac{C - C_1}{C_2 - C_1} \tag{7}$$

in Eqs. (2)–(5), we obtain the governing dimensionless equations as

$$f'' - \mathrm{Re} f' + \theta + N\phi - \frac{\mathrm{Ha}^2}{1+\beta_h^2}(f + \beta_h g) - \frac{\epsilon}{\mathrm{Da}} f = 0 \tag{8}$$

$$g'' - \mathrm{Re} g' + \frac{\mathrm{Ha}^2}{1+\beta_h^2}(\beta_h f - g) - \frac{\epsilon}{\mathrm{Da}} g = 0 \tag{9}$$

$$\theta'' - \mathrm{RePr}\theta' + \mathrm{BrGr}^2\left[(f')^2 + (g')^2\right] + \frac{J}{1+\beta_h^2}\mathrm{Gr}^2(f^2 + g^2) = 0 \tag{10}$$

$$\phi'' - \mathrm{ReSc}\phi' + \mathrm{ScSr}\theta'' = 0 \tag{11}$$

with

$$f = g = \theta = \phi = 0 \quad \text{at} \quad \eta = -1$$
$$f = g = 0,\ \theta = \phi = 1 \quad \text{at} \quad \eta = 1 \tag{12}$$

where the primes denote differentiation with respect to $\eta$, $\mathrm{Ha} = B_0 d\sqrt{\frac{\sigma}{\mu}}$ is the Hartmann parameter, $N = \frac{\beta_C(C_2 - C_1)}{\beta_T(T_2 - T_1)}$ is the buoyancy parameter, $\mathrm{Re} = \frac{\rho v_0 d}{\mu}$ is the Reynolds number, $\mathrm{Gr} = \frac{g_a \beta_T (T_2 - T_1) d^3}{\nu^2}$ is the Grashof number, $\mathrm{Pr} = \frac{\mu C_p}{K_f}$ is the Prandtl number, $\mathrm{Br} = \frac{\mu \nu^2}{K_f d^2 (T_2 - T_1)}$ is the Brinkman parameter, $J = \mathrm{Ha}^2\mathrm{Br}$ is the Joule heating parameter and $S_r = \frac{DK_T(T_2 - T_1)}{\nu T_m (C_2 - C_1)}$ is the thermo-diffusion parameter.

The shearing stress, heat and mass fluxes at the wall surfaces can be obtained from

$$\tau_w = \mu \frac{\partial u}{\partial y}\bigg|_{y=\pm d}; \quad q_w = -K_f \left.\frac{\partial T}{\partial y}\right]_{y=\pm d}; \quad q_m = -D\left.\frac{\partial C}{\partial y}\right]_{y=\pm d}$$

The non-dimensional shear stress $C_f = \frac{\tau_w}{\rho u_0^2}$, the Nusselt number $\mathrm{Nu} = \frac{q_w d}{K_f(T_2 - T_1)}$ and the Sherwood number $\mathrm{Sh} = \frac{q_m d}{D(C_2 - C_1)}$ are given by

$$\mathrm{Re}C_{f1} = 2f'(-1); \mathrm{Re}C_{f2} = 2f'(1);$$
$$\mathrm{Nu}_{1,2} = -\theta'(\eta)|_{\eta=-1,1}; \quad \mathrm{Sh}_{1,2} = -\phi'(\eta)|_{\eta=-1,1}.$$

## The ADM solution

Consider the equation $Fu(t) = g(t)$, where $F$ represents a general nonlinear ordinary or partial differential operator including both linear and nonlinear terms. The main idea of the ADM is that decomposing the linear terms into $L + R$ and rewriting the equations into the following form

$$(L + R)u + Nu = g \tag{13}$$

where $L$ is easily invertible(usually the highest order derivative), $R$ is the remained of the linear operator and $Nu$ indicates the nonlinear terms.

By solving this equation for $Lu$, since $L$ is invertible, we can write

$$L^{-1}Lu = L^{-1}g + L^{-1}Ru + L^{-1}Nu \tag{14}$$

If $L$ is a second-order operator, $L^{-1}$ is a twofold indefinite integral. By solving Eq. (14), we have

$$u = A + Bt + L^{-1}g + L^{-1}Ru + L^{-1}Nu \tag{15}$$

where $A$ and $B$ are integration constants and can be obtained from the initial or boundary conditions. Adomian decomposition method assumes the solution $u$ that can be expanded into infinite series as

$$u = \sum_{n=0}^{\infty} u_n \tag{16}$$

Also, the nonlinear term $Nu$ will be written as

$$Nu = \sum_{n=0}^{\infty} A_n \tag{17}$$

where $A_n's$ are the special Adomian polynomials. By specified $A_n's$ , next component of $u$ can be determined.

$$u_{n+1} = L^{-1}\sum_{n=0}^{\infty} A_n \tag{18}$$

After getting the considerable accuracy, the solution is presented in Eq. (15). In Eq. (18), the Adomian polynomials can be generated by several means. Here we used the following recursive formulation

$$A_n = \frac{1}{n!}\left[\frac{d^n}{d\lambda^n}\left[N\left(\sum_{i=0}^{n}\lambda^i u_i\right)\right]\right]_{\lambda=0}, \quad n = 0, 1, 2, \ldots \tag{19}$$

The calculated solution is more realistic than those achieved by simplifying the model of the physical problem since it does not resort to assumption of weak nonlinearity or linearization.

According to Eq. (14), the governing Eqs. (8)–(11) must be written as following

$$L_1 f = \mathrm{Re} f' - \theta - N\phi + \frac{H_a^2}{1+\beta_h^2}(f + \beta_h g) + \frac{\epsilon}{\mathrm{Da}} f \tag{20}$$

$$L_2 g = \mathrm{Re}\, g' - \frac{H_a^2}{1+\beta_h^2}(\beta_h f - g) + \frac{\epsilon}{\mathrm{Da}} g \tag{21}$$

$$L_3\theta = \mathrm{Re}\,\mathrm{Pr}\,\theta' - \mathrm{Br}\,\mathrm{Gr}^2\left[(f')^2 + (g')^2\right] - \frac{J}{1+\beta_h^2}\mathrm{Gr}^2(f^2+g^2) \tag{22}$$

$$L_4\phi = \mathrm{Re}\,\mathrm{Sc}\,\phi' - \mathrm{Sc}\,\mathrm{Sr}\,\theta'' \tag{23}$$

where the differential operator $L_1$, $L_2$, $L_3$ and $L_4$ are given by $L_1 = L_2 = L_3 = L_4 = \frac{d^2}{d\eta^2}$. Assume the inverse of the operator $L_1^{-1}$, $L_2^{-1}$, $L_3^{-1}$ and $L_4^{-1}$ exists and it can be integrated from 0 to $\eta$. i.e. $L_1^{-1} = L_2^{-1} = L_3^{-1} = L_4^{-1} = \int_0^\eta \int_0^\eta (.) d\eta d\eta$.

After operating $L_1^{-1}, L_2^{-1}, L_3^{-1}, L_4^{-1}$ on Eqs. (20)–(23) and exerting boundary condition on it, we have

$$f(\eta) = f(0) + f'(0)\eta + L_1^{-1}(N_1 f) \tag{24}$$

$$g(\eta) = g(0) + g'(0)\eta + L_2^{-1}(N_2 g) \tag{25}$$

$$\theta(\eta) = \theta(0) + \theta'(0)\eta + L_3^{-1}(N_3\theta) \tag{26}$$

$$\phi(\eta) = \phi(0) + \phi'(0)\eta + L_4^{-1}(N_4\phi) \tag{27}$$

where

$$N_1 f = \mathrm{Re} f' - \theta - N\phi + \frac{H_a^2}{1+\beta_h^2}(f+\beta_h g) + \frac{\epsilon}{\mathrm{Da}} f;$$

$$N_2 g = \mathrm{Re}\, g' - \frac{H_a^2}{1+\beta_h^2}(\beta_h f - g) + \frac{\epsilon}{\mathrm{Da}} g$$

$$N_3\theta = \mathrm{Re}\,\mathrm{Pr}\,\theta' - \mathrm{Br}\,\mathrm{Gr}^2\left[(f')^2+(g')^2\right] - \frac{J}{1+\beta_h^2}\mathrm{Gr}^2(f^2+g^2);$$

$$N_4\phi = \mathrm{Re}\,\mathrm{Sc}\,\phi' - \mathrm{Sc}\,\mathrm{Sr}\,\theta''$$

The ADM introduced the following expression

$$f(\eta) = \sum_{m=0}^{\infty} f_m(\eta) = f_0(\eta) + L_1^{-1}(N_1 f) \tag{28}$$

$$g(\eta) = \sum_{m=0}^{\infty} g_m(\eta) = g_0(\eta) + L_2^{-1}(N_2 g) \tag{29}$$

$$\theta(\eta) = \sum_{m=0}^{\infty} \theta_m(\eta) = \theta_0(\eta) + L_3^{-1}(N_3\theta) \tag{30}$$

$$\phi(\eta) = \sum_{m=0}^{\infty} \phi_m(\eta) = \phi_0(\eta) + L_4^{-1}(N_4\phi) \tag{31}$$

To determine the components of $f_m(\eta), g_m(\eta), \theta_m(\eta)$ and $\phi_m(\eta)$, the initial values of $f_0(\eta), g_0(\eta), \theta_0(\eta)$ and $\phi_0(\eta)$ are defined by applying the boundary conditions

$$f_0(\eta) = a_1 + a_2\eta, \quad g_0(\eta) = a_3 + a_4\eta, \\ \theta_0(\eta) = a_5 + a_6\eta, \quad \phi_0(\eta) = a_7 + a_8\eta \tag{32}$$

and

$$f_1(\eta) = \left[\mathrm{Re}\,a_2 - a_5 - N a_7 + \frac{H_a^2}{1+\beta_h^2}(a_1+\beta_h a_3) + \epsilon\frac{a_1}{\mathrm{Da}}\right]\frac{\eta^2}{2} + \left[-a_6 - N a_8 + \frac{H_a^2}{1+\beta_h^2}(a_2+\beta_h a_4) + \epsilon\frac{a_2}{\mathrm{Da}}\right]\frac{\eta^3}{6} \tag{33}$$

$$g_1(\eta) = \left[\mathrm{Re}\,a_4 - \frac{H_a^2}{1+\beta_h^2}(\beta_h a_1 - a_3) + \epsilon\frac{a_3}{\mathrm{Da}}\right]\frac{\eta^2}{2} + \left[\epsilon\frac{a_4}{\mathrm{Da}} - \frac{H_a^2}{1+\beta_h^2}(\beta_h a_2 + a_4)\right]\frac{\eta^3}{6} \tag{34}$$

$$\theta_1(\eta) = -\frac{J}{1+\beta_h^2}\mathrm{Gr}^2(2a_1a_2 + 2a_3a_4)\frac{\eta^3}{6} \times \left[\mathrm{Re}\,Pr\,a_6 - \mathrm{Br}\,\mathrm{Gr}^2(a_2^2+a_4^2) - \frac{J}{1+\beta_h^2}\mathrm{Gr}^2(a_2^2+a_4^2)\right]\frac{\eta^2}{2} - \frac{J}{1+\beta_h^2}\mathrm{Gr}^2(a_2^2+a_4^2)\frac{\eta^4}{12} \tag{35}$$

$$\phi_1(\eta) = (\mathrm{Re}\,\mathrm{Sc}\,a_8)\frac{\eta^2}{2} \tag{36}$$

and $f_m(\eta), g_m(\eta), \theta_m(\eta)$ and $\phi_m(\eta)$ for $m \geq 2$ be determined in similar way.

Then using the above in the following series expansions

$$f(\eta) = \sum_{m=0}^{\infty} f_m(\eta), \; g(\eta) = \sum_{m=0}^{\infty} g_m(\eta) \\ \theta(\eta) = \sum_{m=0}^{\infty} \theta_m(\eta), \; \phi(\eta) = \sum_{m=0}^{\infty} \phi_m(\eta) \tag{37}$$

lead to following equations

$$f(\eta) = a_1 + a_2\eta + \left[\mathrm{Re}\,a_2 - a_5 - N a_7 + \frac{H_a^2}{1+\beta_h^2}(a_1+\beta_h a_3) + \epsilon\frac{a_1}{\mathrm{Da}}\right]\frac{\eta^2}{2} + \left[-a_6 - N a_8 + \frac{H_a^2}{1+\beta_h^2}(a_2+\beta_h a_4) + \epsilon\frac{a_2}{\mathrm{Da}}\right]\frac{\eta^3}{6} + \cdots \tag{38}$$

$$g(\eta) = a_3 + a_4\eta + \left[\mathrm{Re}\, a_4 - \frac{H_a^2}{1+\beta_h^2}(\beta_h a_1 - a_3) + \epsilon\,\frac{a_3}{\mathrm{Da}}\right]\frac{\eta^2}{2} + \left[\epsilon\,\frac{a_4}{\mathrm{Da}} - \frac{H_a^2}{1+\beta_h^2}(\beta_h a_2 + a_4)\right]\frac{\eta^3}{6} + \cdots \tag{39}$$

$$\begin{aligned}\theta(\eta) = {} & a_5 + a_6\eta - \frac{J}{1+\beta_h^2}\mathrm{Gr}^2(2a_1a_2 + 2a_3a_4)\frac{\eta^3}{6} \\ & + \left[\mathrm{Re}\,\mathrm{Pr}\, a_6 - \mathrm{Br}\,\mathrm{Gr}^2\left(a_2^2 + a_4^2\right)\right]\frac{\eta^2}{2} \\ & - \left[\frac{J}{1+\beta_h^2}\mathrm{Gr}^2\left(a_2^2 + a_4^2\right)\right]\frac{\eta^2}{2} \\ & - \frac{J}{1+\beta_h^2}\mathrm{Gr}^2\left(a_2^2 + a_4^2\right)\frac{\eta^4}{12} + \cdots\end{aligned} \tag{40}$$

$$\begin{aligned}\phi(\eta) = {} & a_7 + a_8\eta - [\mathrm{Re}\,\mathrm{Sc}\, a_8]\frac{\eta^2}{2} \\ & + \left[R^2\,\mathrm{Sc}^2\, a_8 + \mathrm{Sc}\,\mathrm{Sr}\frac{J}{1+\beta_h^2}\mathrm{Gr}^2(2a_1a_2 + 2a_3a_4)\right]\frac{\eta^3}{6} \\ & - \mathrm{Sc}\,\mathrm{Sr}\left(\mathrm{Re}\,\mathrm{Pr}, a_6 - \mathrm{Br}\,\mathrm{Gr}^2(a_2^2 + a_4^2)\right)\frac{\eta^2}{2} \\ & + \mathrm{Sc}\,\mathrm{Sr}\left(\frac{J}{1+\beta_h^2}\mathrm{Gr}^2\left(a_1^2 + a_3^2\right)\right)\frac{\eta^2}{2} \\ & + \mathrm{Sc}\,\mathrm{Sr}\frac{J}{1+\beta_h^2}\mathrm{Gr}^2\left(a_2^2 + a_4^2\right)\frac{\eta^4}{12}\cdots\end{aligned} \tag{41}$$

BY increasing the number of terms, the ADM gives the more accurate solution. For the complete solution of equations above $a_1, a_2, a_3, a_4, a_5, a_6, a_7$ and $a_8$ should be determined, with boundary conditions.

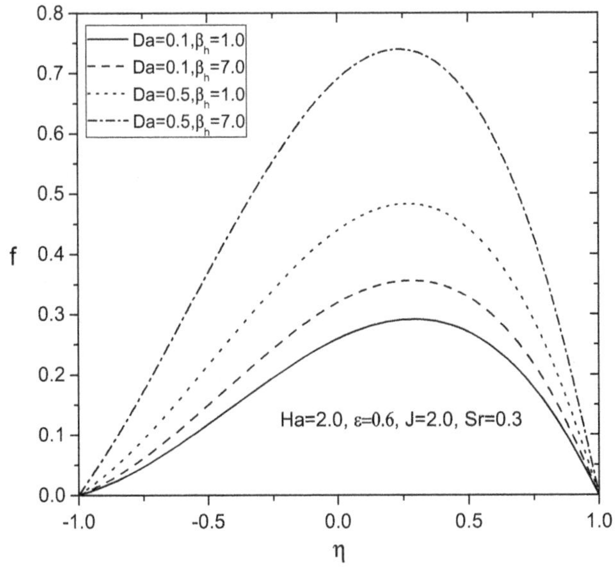


**Fig. 2** Effect of Da and $\beta_h$ on velocity profile

## Results and discussion

The velocities $(f(\eta), g(\eta))$, temperature $(\theta(\eta))$ and concentration $(\phi(\eta))$ profiles of are calculated and shown in Figs. 2, 3, 4, 5, 6, 7, 8, 9, 10, 11, 12 and 13 with different values of Ha, $\beta_h$, Sr, $J$ and $\epsilon$, Da. Computations were carried out by fixing the parameters

$$\mathrm{Re} = 2.0,\ N = 5,\ \mathrm{Pr} = 0.71,\ \mathrm{Gr} = 1.0 \text{ and } \mathrm{Sc} = 0.22$$

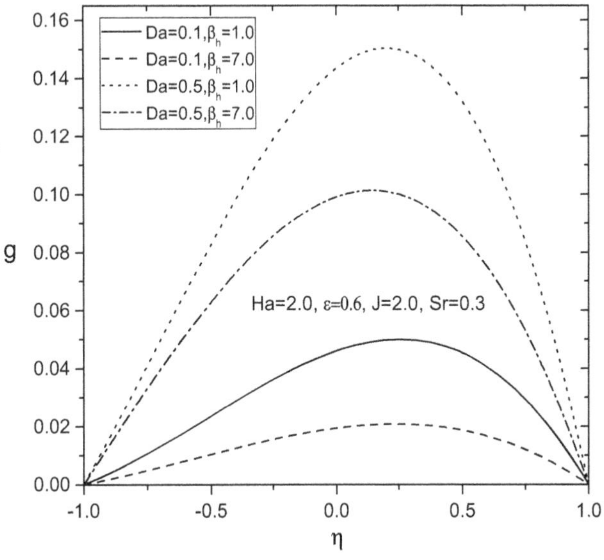


**Fig. 3** Effect of Da and $\beta_h$ on angular velocity profile

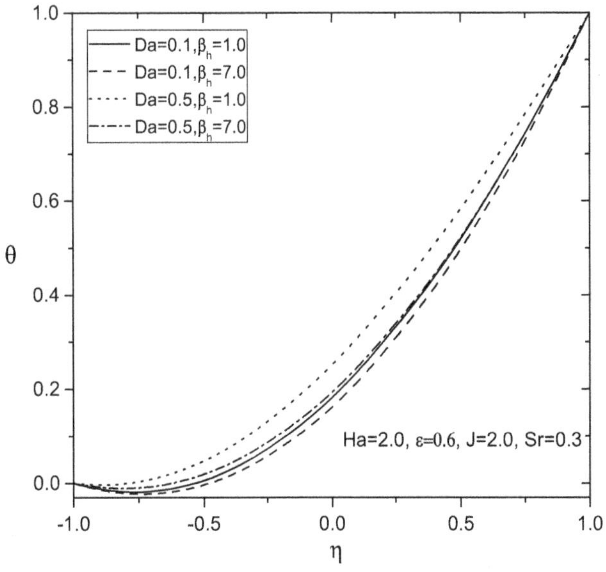


**Fig. 4** Effect of Da and $\beta_h$ on temperature profile

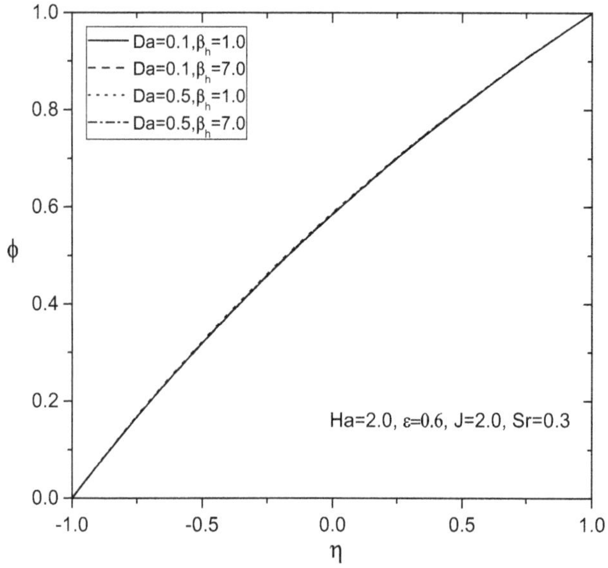


**Fig. 5** Effect of Da and $\beta_h$ on concentration profile

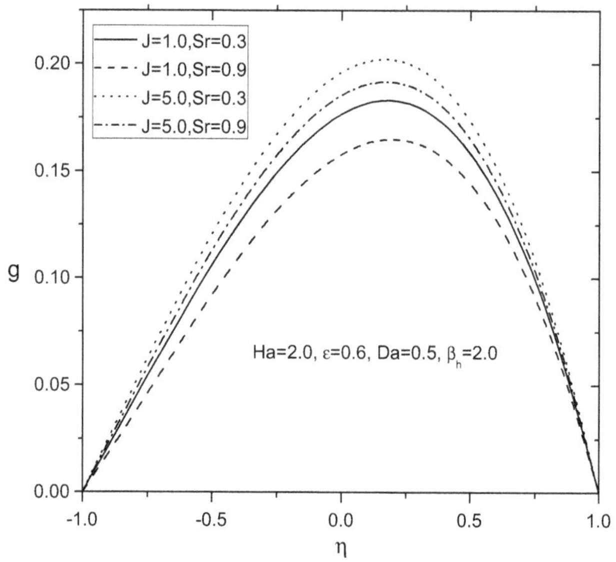


**Fig. 7** Effect of $J$ and Sr on angular velocity profile

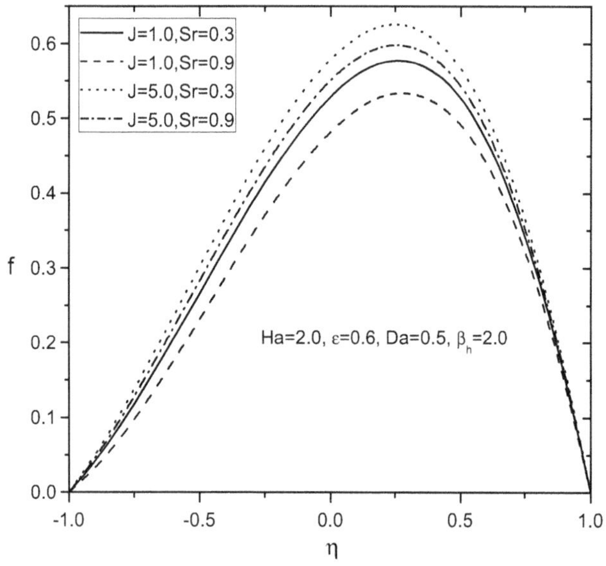


**Fig. 6** Effect of $J$ and Sr on velocity profile

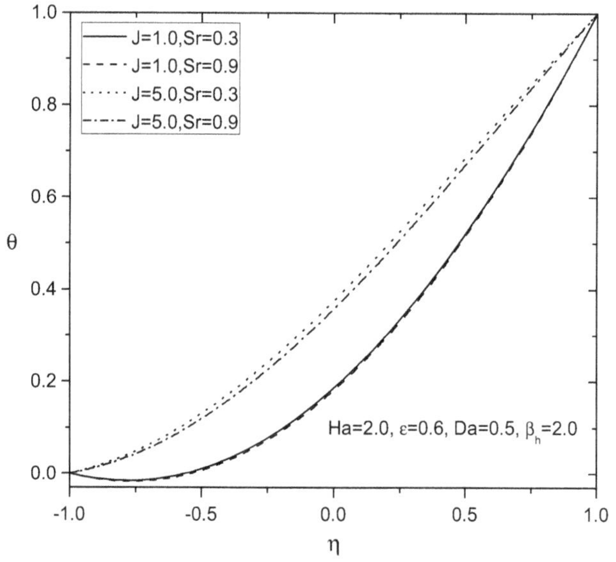


**Fig. 8** Effect of $J$ and Sr on temperature profile

to analyze the effects of the emerging parameters Ha, $\beta_h$, Sr, Da, $J$ and $\epsilon$.

The effect of the hall parameter $\beta_h$ on velocities, temperature and concentration profiles are shown in Figs. 2, 3, 4 and 5. It can be observed from these figures that $f(\eta)$ increases as $\beta_h$ increases. It is noticed from Fig. 3 that the velocity in $z$-direction decreases with an increase in $\beta_h$. This is due to the Hall parameter's inclusion and which reduces the resistive force imposed by the magnetic field due to its effect in reducing the effective conductivity. It can be seen from Fig. 4 that the temperature of the fluid decreases with an increase in the Hall parameter. Finally the influence of Hall number on concentration is shown in Fig. 5. It is observed from the figure that the dimensionless concentration increases with an increase in Hall parameter. This is due to the fact that decrease in temperature accelerates the concentration. In addition to the above, the effect of Darcy parameter on velocities, temperature and concentration profiles have been presented in Figs. 2, 3, 4 and 5. it can be seen from these figures that the flow velocity $f(\eta)$ increases as Darcy number increases. It is observed from Fig. 3 that the induced velocity increases as Da increases. This is due to the fact that the lower permeability enhances the flow, which leads to increase in the velocities

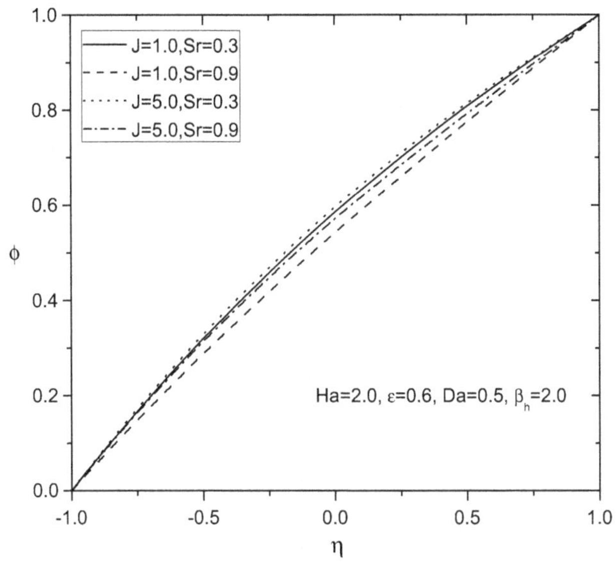


**Fig. 9** Effect of *J* and Sr on concentration profile

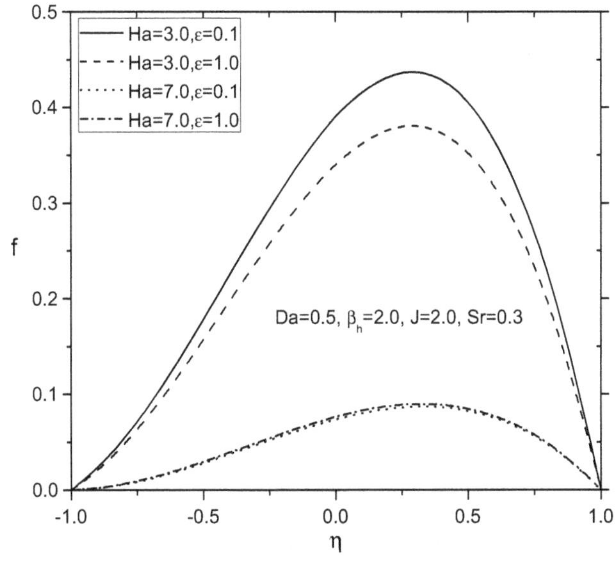


**Fig. 10** Effect of Ha and ε on velocity profile

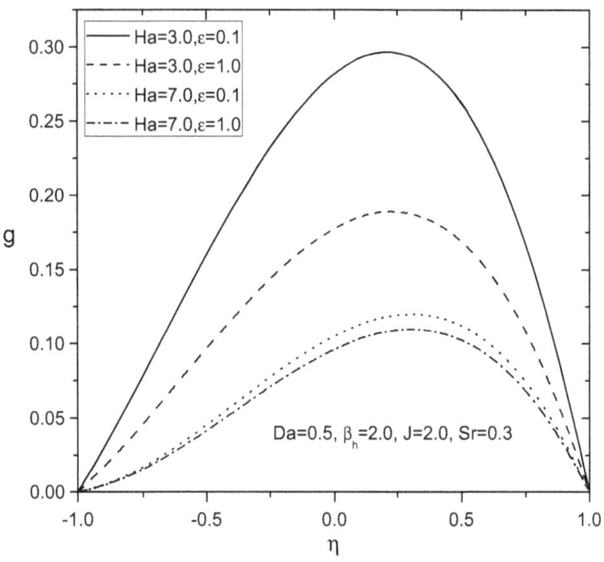


**Fig. 11** Effect of Ha and ε on angular velocity profile

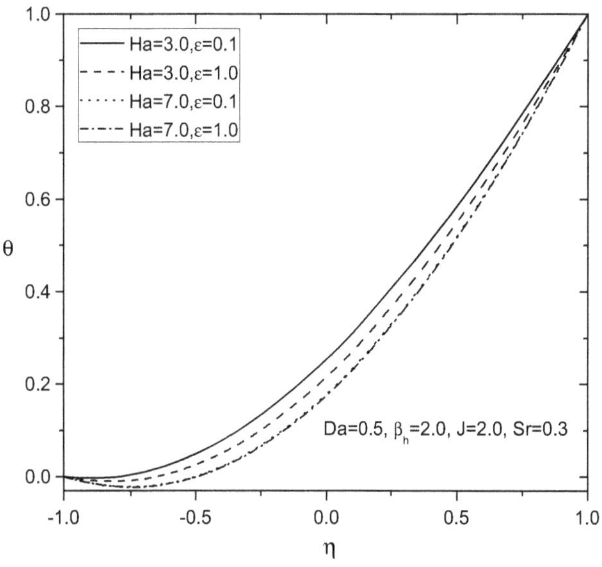


**Fig. 12** Effect of Ha and ε on temperature profile

of the flow. The influence of Da on temperature profile is shown in Fig. 4. It can be noted from this figure that as Da increases the dimensionless temperature increases. Finally the influence of Darcy number on concentration is shown in Fig. 5. It is observed from the figure that the dimensionless concentration increases with an increase in Darcy parameter. Increasing Da increases the porous medium permeability and simultaneously decreases the darcian impedance since progressively less solid fibers are present in the regime.

The influence of thermal diffusion parameter on velocities, temperature and concentration are shown in Figs. 6, 7, 8 and 9. It is clear from Figs. 6 and 7 that the higher values of Soret number Sr decreases the velocities. Figure 8 presents the effect of Sr on dimensionless temperature. It is noticed from this figure that the temperature decreases with the higher values of Sr. Figure 9 explains the nature of the concentration profile for different values of Sr. It can be seen that the concentration of the fluid decreases as Sr increases. Figures 6, 7, 8 and 9 presents the influence of Joule heating parameter on velocities, temperature and concentration profiles. It can be seen from these figures as *J* increases the velocity profiles decreases considerably. It is observed from Figs. 8 and 9 that the dimensionless temperature and concentration profiles decreases with an increase in *J*.

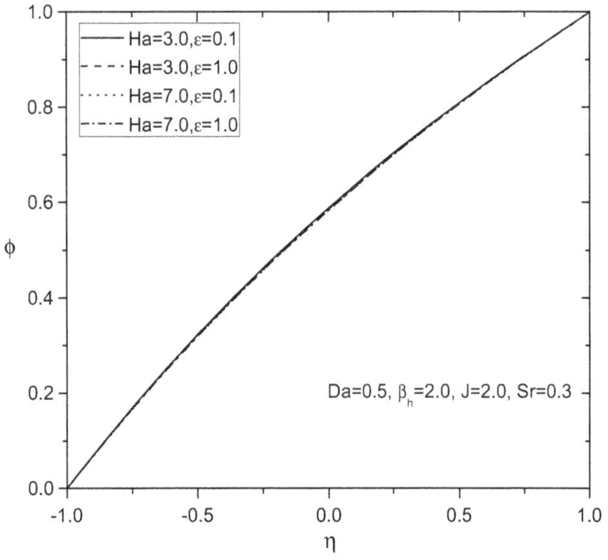


**Fig. 13** Effect of Ha and $\varepsilon$ on concentration profile

Figure 10 prepared to study the influence of the Hartman number Ha and the porosity parameter $\epsilon$ on $f(\eta)$. It is seen from this figure that the flow velocity decreases when Ha increases. Since the flow is resisted by the Lorentz forces generated due the aaplied magnetic field. Figure 11 presents the influence of Ha on the cross flow velocity $g(\eta)$. By nature it has been noticed by many researchers that the cross flow velocity increases with the increase of Ha but it is noticed from the present study that the cross flow velocity decreases as the magnetic parameter increases. Porosity of the flow regime diminishes the concentration of the magnetic effect, which leads to the decreases the velocity in $z$-direction. The nature of dimensionless temperature has been presented in Fig. 12 with the various of Ha. It is observed from this figure that the temperature decreases with an increase in Ha. The effect of magnetic parameter on $\phi(\eta)$ is revealed in Fig. 13. It is seen from this figure that the concentration of the fluid decreases as Ha increases.

As discussed above, the Lorentz force creates resistance in the fluid which leads to the friction between the fluid layers. Hence the temperature and concentration of the fluid decreases with Magnetic parameter. In addition to the effect of magnetic parameter, porosity influence on velocities, temperature and concentrations are also been presented in Figs. 10 and 11. It is clear from the figures that

**Table 1** Nature of skin friction coefficient, heat and mass transfer rates for various values of Sr, Ha, $\beta_h$, $\epsilon$, Da and $J$ when Re $= 2.0, N = 5$, Pr $= 0.71$, Gr $= 1.0$ and Sc $= 0.22$

| Ha | $\epsilon$ | Da | $\beta_h$ | $J$ | Sr | $C_{f_1}$ | $C_{f_2}$ | $Nu_1$ | $Nu_2$ | $Sh_1$ | $Sh_2$ |
|---|---|---|---|---|---|---|---|---|---|---|---|
| 3 | 0.6 | 0.5 | 2 | 2 | 0.3 | 0.3261 | −2.5849 | 0.1050 | −0.9523 | −0.7096 | −0.3626 |
| 5 | 0.6 | 0.5 | 2 | 2 | 0.3 | 0.0349 | −1.1900 | 0.1588 | −1.0228 | −0.7049 | −0.3662 |
| 7 | 0.6 | 0.5 | 2 | 2 | 0.3 | 0.0046 | −0.5981 | 0.1846 | −1.0919 | −0.7018 | −0.3666 |
| 9 | 0.6 | 0.5 | 2 | 2 | 0.3 | 0.0020 | −0.3457 | 0.1954 | −1.1325 | −0.7002 | −0.3663 |
| 2 | 0.1 | 0.5 | 2 | 2 | 0.3 | 1.0086 | −4.2104 | 0.0311 | −0.8778 | −0.7152 | −0.3570 |
| 2 | 0.4 | 0.5 | 2 | 2 | 0.3 | 0.8461 | −3.8124 | 0.0709 | −0.9468 | −0.7113 | −0.3590 |
| 2 | 0.6 | 0.5 | 2 | 2 | 0.3 | 0.7551 | −3.5824 | 0.0901 | −0.9811 | −0.7094 | −0.3600 |
| 2 | 0.8 | 0.5 | 2 | 2 | 0.3 | 0.6767 | −3.3771 | 0.1054 | −1.0089 | −0.7079 | −0.3607 |
| 2 | 0.6 | 0.1 | 0 | 2 | 0.3 | 0.2663 | −2.0681 | 0.1730 | −1.1372 | −0.7011 | −0.3636 |
| 2 | 0.6 | 0.3 | 2 | 2 | 0.3 | 0.6092 | −3.1933 | 0.1177 | −1.0317 | −0.7066 | −0.3613 |
| 2 | 0.6 | 0.5 | 4 | 2 | 0.3 | 0.7551 | −3.5824 | 0.0901 | −0.9811 | −0.7094 | −0.3600 |
| 2 | 0.6 | 1.0 | 6 | 2 | 0.3 | 0.8967 | −3.9377 | 0.0594 | −0.9265 | −0.7124 | −0.3585 |
| 2 | 0.6 | 0.5 | 1 | 2 | 0.3 | 0.5362 | −2.9896 | 0.0779 | −0.8849 | −0.7130 | −0.3617 |
| 2 | 0.6 | 0.5 | 3 | 2 | 0.3 | 0.9281 | −3.9401 | 0.1023 | −1.0385 | −0.7070 | −0.3595 |
| 2 | 0.6 | 0.5 | 5 | 2 | 0.3 | 1.1013 | −4.2506 | 0.1155 | −1.0879 | −0.7048 | −0.3594 |
| 2 | 0.6 | 0.5 | 7 | 2 | 0.3 | 1.1700 | −4.3647 | 0.1208 | −1.1059 | −0.7039 | −0.3594 |
| 2 | 0.6 | 0.5 | 2 | 1 | 0.3 | 0.7146 | −3.5328 | 0.1519 | −1.0987 | −0.7032 | −0.3626 |
| 2 | 0.6 | 0.5 | 2 | 3 | 0.3 | 0.7984 | −3.6353 | 0.0241 | −0.8563 | −0.7160 | −0.3571 |
| 2 | 0.6 | 0.5 | 2 | 5 | 0.3 | 0.8956 | −3.7534 | −0.123 | −0.5806 | −0.7308 | −0.3505 |
| 2 | 0.6 | 0.5 | 2 | 7 | 0.3 | 1.0113 | −3.8928 | −0.300 | −0.2597 | −0.7483 | −0.3422 |
| 2 | 0.6 | 0.5 | 2 | 2 | 0.3 | 0.7551 | −3.5824 | 0.0901 | −0.9811 | −0.7094 | −0.3600 |
| 2 | 0.6 | 0.5 | 2 | 2 | 0.5 | 0.7009 | −3.5222 | 0.0947 | −0.9888 | −0.6853 | −0.3884 |
| 2 | 0.6 | 0.5 | 2 | 2 | 0.7 | 0.6454 | −3.4612 | 0.0992 | −0.9965 | −0.6606 | −0.4171 |
| 2 | 0.6 | 0.5 | 2 | 2 | 0.9 | 0.5888 | −3.3992 | 0.1036 | −1.0040 | −0.6353 | −0.4462 |

the velocity profiles $f(\eta)$, $g(\eta)$, the temperature $(\theta(\eta))$ and the concentration of the fluid flow decreases with an increase in $\epsilon$. This can be attributed to the fact that increasing $\epsilon$ decreases the velocities and, in turn, decreases the viscous dissipation which decreases the temperature and concentration.

Variation of Joule heating parameter ($J$), thermal diffusion parameter (Sr), magnetic parameter Ha, Hall number ($\beta_h$), porosity parameter ($\epsilon$) together with the Darcy parameter (Da) is presented in Table 1 with fixed values of other parameters. It can be seen from this table that the skin friction coefficient decreases at the initial plate and increases at the terminal plate where as the reverse trend is noticed on heat and mass transfer rate with an increase in Ha and $\epsilon$. As magnetic parameter increases, the resistive force slow downs the friction factor at $\eta = -1$. It is observed that as porosity of the medium increases, heat transfer rate and mass transfer rates increases at $\eta = -1$ and decreases at the other plate but the skin friction coefficient presents the reverse trend. It is clear from the table that the friction, heat and mass transfer rates are increases at the initial plates and decreases at the terminal plate with the increase of Hall parameter, Joule heating parameter and Da. Finally, the influence of Sr on friction factor, heat and mass transfer rates are presented in this table. The performance of these parameters are self-evident from the Table 1 and hence are not discussed for brevity. These results are clearly shows that the emerging parameters have remarkable impact on all the profiles.

## Conclusions

This present study investigates the steady manetohydrodynamic flow of newtonian fluid in a vertical channel saturated porous medium in presence of Hall, Joule heating and the Soret effects. Adomian decomposition method is used to solve the final dimensionless governing equations. The significant findings are summarized as:

- Fluid flow velocity and the concentration profiles amplifies where as the flow in $z$-direction and the the temperature profile decreases with an increase in Ha.
- As Da increases, the velocity profiles, temperature profile and concentration profile are increased.
- It is noticed that the presence of Soret and Joule heating parameters in the fluid decreases the velocities, temperature and the concentration of the fluid.
- The velocities, temperature and concentration profiles are decreases with the increase in the magnetic and porosity parameters.

## References


1. Ingham, D.B., Pop, I., Cheng, P.: Combined free and forced convection porous medium between two vertical walls with viscous dissipation. Transp. Porous Media **5**, 381–398 (1990)
2. Paul, T., Jha, B.K., Singh, A.K.: Free-convection between vertical walls partially filled with porous medium. Heat Mass Transf. **33**, 515–519 (1998)
3. Umavathi, J.C.: Free convection of composite porous medium in a vertical channel. Heat Transf. Asian Res. **40**(4), 308–329 (2011)
4. Mishra, A.K., Djam, X.Y., Manjak, N.H.: Effect of radiation on free convection flow due to heat and mass transfer through porous medium bounded by two vertical walls. Int. J. Adv. Technol. Eng. Res. **3**, 120–125 (2013)
5. Afify, A.A.: Similarity solution in MHD: effects of thermal diffusion and diffusion thermo on free convective heat and mass transfer over a stretching surface considering suction or injection. Commun. Nonlinear Sci. Numer. Simul. **14**, 2202–2214 (2009)
6. Srinivasacharya, D., Kaladhar, K.: Soret and Dufour effects on free convection flow of a couple stress fluid in a vertical channel with chemical reaction. Chem. Ind. Chem. Eng. Q. **19**(1), 45–55 (2013)
7. Srinivasacharya, D., Pranitha, J., Ramreddy, C.H., Postelnicu, A.: Soret and Dufour effects on non-Darcy free convection in a power-law fluid in the presence of a magnetic field and stratification. Heat Transf. Asian Res. **43**, 592–606 (2014)
8. Tani, I.: Steady flow of conducting fluids in channels under transverse magnetic fields with consideration of Hall effects. J. Aerosp. Sci. **29**, 297–305 (1962)
9. Srinivasacharya, D., Shiferaw, M.: Hall and Ion-slip effects on the flow of micropolar fluid between parallel plates. Int. J. Appl. Mech. Eng. **13**(1), 251–262 (2008)
10. Srinivasacharya, D., Shiferaw, M.: MHD flow of a micropolar fluid in a rectangular duct with Hall and Ion-slip effects. J. Braz. Soc. Mech. Sci. Eng. **30**(4), 313–318 (2008)
11. Srinivasacharya, D., Shiferaw, M.: MHD flow of a micropolar fluid in a circular pipe with Hall effects. ANZIAM J. **51**, 277–285 (2009)
12. Manglesh, A., Gorla, M.G.: MHD free convection flow through porous medium in the presence of Hall current, radiation and thermal difusion. Indian J. Pure Appl. Math. **44**(6), 743–756 (2013)
13. Garg, B.P., Singh, K.D., Bansal, A.K.: Hall current effect on viscoelastic (Walters liquid model-B) MHD oscillatory convective channel flow through a porous medium with heat radiation. Kragujev. J. Sci. **36**, 19–32 (2014)
14. Chen, C.H.: Combined effects of Joule heating and viscous dissipation on magnetohydrodynamic flow past a permeable, stretching surface with free convection and radiative heat transfer. J. Heat Transf. **132**, 064503-1–064503-5 (2010)
15. Hossain, M.A., Gorla, R.S.R.: Joule heating effect on magnetohydrodynamic mixed convection boundary layer flow with variable electrical conductivity. Int. J. Numer. Methods Heat Fluid Flow **23**(2), 275–288 (2013)
16. Nandkeolyar, R., Motsa, S.S., Sibanda, P.: Viscous and Joule heating in the stagnation point nanofluid flow through a stretching sheet with homogenous- heterogeneous reactions and nonlinear convection. J. Nanotechnol. Eng. Med. **4**, 0410011-1–0410011-9 (2013)

# 6

# Some inequalities associated with the Hermite–Hadamard–Fejér type for convex function

**Mehmet Zeki Sarikaya · Hatice Yaldiz · Samet Erden**

**Abstract** In this paper, we extend some estimates of the right-hand side of a Hermite–Hadamard–Fejér type inequality for functions whose first derivatives' absolute values are convex. The results presented here would provide extensions of those given in earlier works.

**Keywords** Hermite–Hadamard–Fejer inequality · Trapezoid inequality · Convex function · Hölder inequality.

**Mathematics Subject Classification** 26D07 · 26D15

## Introduction

**Definition 1** The function $f : [a, b] \subset \mathbb{R} \to \mathbb{R}$ is said to be convex if the following inequality holds

$$f(\lambda x + (1 - \lambda)y) \leq \lambda f(x) + (1 - \lambda)f(y)$$

for all $x, y \in [a, b]$ and $\lambda \in [0, 1]$. We say that $f$ is concave if $(-f)$ is convex.

The following inequality is well known in the literature as the Hermite–Hadamard integral inequality (see, [2, 4]):

$$f\left(\frac{a+b}{2}\right) \leq \frac{1}{b-a}\int_a^b f(x)dx \leq \frac{f(a)+f(b)}{2} \tag{1.1}$$

where $f : I \subset \mathbb{R} \to \mathbb{R}$ is a convex function on the interval $I$ of real numbers and $a, b \in I$ with $a<b$.

In [1], Dragomir and Agarwal proved the following results connected with the right part of (1.1).

**Lemma 1** Let $f : I^\circ \subseteq \mathbb{R} \to \mathbb{R}$ be a differentiable mapping on $I^\circ$, $a, b \in I^\circ$ with $a<b$. If $f' \in L[a, b]$, then the following equality holds:

$$\frac{f(a)+f(b)}{2} - \frac{1}{b-a}\int_a^b f(x)dx = \frac{b-a}{2}\int_0^1 (1-2t)f'(ta+(1-t)b)dt. \tag{1.2}$$

**Theorem 1** Let $f : I^\circ \subseteq \mathbb{R} \to \mathbb{R}$ be a differentiable mapping on $I^\circ$, $a, b \in I^\circ$ with $a<b$. If $|f'|$ is convex on $[a, b]$, then the following inequality holds:

$$\left|\frac{f(a)+f(b)}{2} - \frac{1}{b-a}\int_a^b f(x)dx\right| \leq \frac{(b-a)}{8}(|f'(a)| + |f'(b)|). \tag{1.3}$$

**Theorem 2** Let $f : I^\circ \subset \mathbb{R} \to \mathbb{R}$ be a differentiable mapping on $I^\circ$, $a, b \in I^\circ$ with $a<b, f' \in L(a, b)$ and $p > 1$. If the mapping $|f'|^{p/(p-1)}$ is convex on $[a, b]$, then the following inequality holds:

$$\left|\frac{f(a)+f(b)}{2} - \frac{1}{b-a}\int_a^b f(x)dx\right| \leq \frac{b-a}{2(p+1)^{1/p}} \times \left(\frac{|f'(a)|^{p/(p-1)} + |f'(b)|^{p/(p-1)}}{2}\right)^{(p-1)/p}. \tag{1.4}$$

The most well-known inequalities related to the integral mean of a convex function are the Hermite–Hadamard

M. Z. Sarikaya · H. Yaldiz (✉)
Department of Mathematics, Faculty of Science and Arts, Düzce University, Konuralp Campus, Düzce, Turkey
e-mail: yaldizhatice@gmail.com

M. Z. Sarikaya
e-mail: sarikayamz@gmail.com

S. Erden
Department of Mathematics, Faculty of Science, Bartın University, Konuralp Campus, Bartin, Turkey
e-mail: erdem1627@gmail.com

inequalities or its weighted versions, the so-called Hermite–Hadamard–Fejér inequalities (see [5–14]). In [3], Fejer gave a weighted generalization of the inequalities (1.1) as the following:

**Theorem 3** $f:[a,b]\to\mathbb{R}$ be a convex function, then the inequality

$$f\left(\frac{a+b}{2}\right)\int_a^b w(x)dx \le \frac{1}{b-a}\int_a^b f(x)w(x)dx \le \frac{f(a)+f(b)}{2}\int_a^b w(x)dx \tag{1.5}$$

holds, where $w:[a,b]\to\mathbb{R}$ is nonnegative, integrable, and symmetric about $x=\frac{a+b}{2}$.

In [5], some inequalities of Hermite–Hadamard–Fejer type for differentiable convex mappings were proved using the following lemma.

**Lemma 2** Let $f:I^\circ\subset\mathbb{R}\to\mathbb{R}$ be a differentiable mapping on $I^\circ$, $a,b\in I^\circ$ with $a<b$, and $w:[a,b]\to[0,\infty)$ be a differentiable mapping. If $f'\in L[a,b]$, then the following equality holds:

$$\frac{f(a)+f(b)}{2}\int_a^b w(x)dx-\int_a^b f(x)w(x)dx = \frac{(b-a)^2}{2}\int_0^1 p(t)f'(ta+(1-t)b)dt \tag{1.6}$$

for each $t\in[0,1]$, where

$$p(t)=\int_t^1 w(as+(1-s)b)ds-\int_0^t w(as+(1-s)b)ds.$$

In this article, using functions whose derivatives' absolute values are convex, we obtained new inequalities of Hermite–Hadamard–Fejer type. The results presented here would provide extensions of those given in earlier works.

## Main results

We will establish some new results connected with the right-hand side of (1.5) and (1.1). Now, we prove our main theorems:

**Theorem 4** Let $f:I^\circ\subseteq\mathbb{R}\to\mathbb{R}$ be a differentiable mapping on $I^\circ$, $a,b\in I^\circ$ with $a<b$ and let $w:[a,b]\to\mathbb{R}$ be continuous on $[a,b]$. If $|f'|$ is convex on $[a,b]$, then for all $x\in[a,b]$, the following inequalities hold:

$$\left|\left(\int_x^b w(s)ds\right)^\alpha f(b)-\left(\int_x^a w(s)ds\right)^\alpha f(a)-\alpha\int_a^b\left(\int_x^t w(s)ds\right)^{\alpha-1}w(t)f(t)dt\right|$$

$$\le \|w\|_{[a,x],\infty}^\alpha\left\{\frac{|f'(a)|}{b-a}\left[\frac{(x-a)^{\alpha+1}(b-x)}{\alpha+1}+\frac{(x-a)^{\alpha+2}}{\alpha+2}\right]+\frac{|f'(b)|}{b-a}\frac{(x-a)^{\alpha+2}}{(\alpha+1)(\alpha+2)}\right\}$$

$$+\|w\|_{[x,b],\infty}^\alpha\left\{\frac{|f'(a)|}{b-a}\frac{(b-x)^{\alpha+2}}{(\alpha+1)(\alpha+2)}+\frac{|f'(b)|}{b-a}\left[\frac{(b-x)^{\alpha+1}(x-a)}{\alpha+1}+\frac{(b-x)^{\alpha+2}}{\alpha+2}\right]\right\}$$

$$\le\frac{\|w\|_{[a,b],\infty}^\alpha}{(b-a)}\left\{|f'(a)|\left[\frac{(x-a)^{\alpha+1}(b-x)}{\alpha+1}+\frac{(b-x)^{\alpha+2}}{(\alpha+1)(\alpha+2)}+\frac{(x-a)^{\alpha+2}}{\alpha+2}\right]\right.$$

$$\left.+|f'(b)|\left[\frac{(b-x)^{\alpha+1}(x-a)}{\alpha+1}+\frac{(x-a)^{\alpha+2}}{(\alpha+1)(\alpha+2)}+\frac{(b-x)^{\alpha+2}}{\alpha+2}\right]\right\}$$

where $\alpha>0$ and $\|w\|_\infty=\sup_{t\in[a,b]}|w(t)|$.

*Proof* By integration by parts, we have the following equalities:

$$\int_a^b\left(\int_x^t w(s)ds\right)^\alpha f'(t)dt$$

$$=\left(\int_x^t w(s)ds\right)^\alpha f(t)\Big|_a^b-\alpha\int_a^b\left(\int_x^t w(s)ds\right)^{\alpha-1}w(t)f(t)dt$$

$$=\left(\int_x^b w(s)ds\right)^\alpha f(b)-\left(\int_x^a w(s)ds\right)^\alpha f(a)-\alpha\int_a^b\left(\int_x^t w(s)ds\right)^{\alpha-1}w(t)f(t)dt. \tag{2.1}$$

We take absolute value of (2.1) and use convexity of $|f'|$, we find that

$$\left|\left(\int_x^b w(s)ds\right)^{\alpha} f(b)-\left(\int_x^a w(s)ds\right)^{\alpha} f(a)\right.$$

$$\left.-\alpha\int_a^b\left(\int_x^t w(s)ds\right)^{\alpha-1} w(t)f(t)dt\right|$$

$$\leq\int_a^x\left(\left|\int_x^t w(s)ds\right|\right)^{\alpha}\left|f^{'}(t)\right|dt+\int_x^b\left(\left|\int_x^t w(s)ds\right|\right)^{\alpha}\left|f^{'}(t)\right|dt$$

$$\leq\|w\|_{[a,x],\infty}^{\alpha}\int_a^x (x-t)^{\alpha}\left|f^{'}(t)\right|dt+\|w\|_{[x,b],\infty}\int_x^b (t-x)^{\alpha}\left|f^{'}(t)\right|dt$$

$$=\|w\|_{[a,x],\infty}^{\alpha}\left[\int_a^x (x-t)^{\alpha}\left|f^{'}(\frac{b-t}{b-a}a+\frac{t-a}{b-a}b)\right|dt\right]$$

$$+\|w\|_{[x,b],\infty}^{\alpha}\left[\int_x^b (t-x)^{\alpha}\left|f^{'}(\frac{b-t}{b-a}a+\frac{t-a}{b-a}b)\right|dt\right]$$

$$\leq\|w\|_{[a,x],\infty}^{\alpha}\left\{\frac{\left|f^{'}(a)\right|}{b-a}\left[\frac{(x-a)^{\alpha+1}(b-x)}{\alpha+1}+\frac{(x-a)^{\alpha+2}}{\alpha+2}\right]\right.$$

$$\left.+\frac{\left|f^{'}(b)\right|}{b-a}\frac{(x-a)^{\alpha+2}}{(\alpha+1)(\alpha+2)}\right\}$$

$$+\|w\|_{[x,b],\infty}^{\alpha}\left\{\frac{\left|f^{'}(a)\right|}{b-a}\frac{(b-x)^{\alpha+2}}{(\alpha+1)(\alpha+2)}\right.$$

$$\left.+\frac{\left|f^{'}(b)\right|}{b-a}\left[\frac{(b-x)^{\alpha+1}(x-a)}{\alpha+1}+\frac{(b-x)^{\alpha+2}}{\alpha+2}\right]\right\}$$

$$\leq\frac{\|w\|_{[a,b],\infty}^{\alpha}}{(b-a)}\left\{\left|f^{'}(a)\right|\left[\frac{(x-a)^{\alpha+1}(b-x)}{\alpha+1}\right.\right.$$

$$\left.+\frac{(b-x)^{\alpha+2}}{(\alpha+1)(\alpha+2)}+\frac{(x-a)^{\alpha+2}}{\alpha+2}\right]$$

$$\left.+\left|f^{'}(b)\right|\left[\frac{(b-x)^{\alpha+1}(x-a)}{\alpha+1}+\frac{(x-a)^{\alpha+2}}{(\alpha+1)(\alpha+2)}+\frac{(b-x)^{\alpha+2}}{\alpha+2}\right]\right\}$$

for all $x\in[a,b]$. Hence, the proof of theorem is completed.

**Corollary 1** Under the same assumptions of Theorem 4 with $w(s)=1$, then the following inequality holds:

$$\left|(b-x)^{\alpha}f(b)-(a-x)^{\alpha}f(a)-\alpha\int_a^b (t-x)^{\alpha-1}f(t)dt\right|$$

$$\leq\frac{1}{(b-a)}\left\{\left|f^{'}(a)\right|\left[\frac{(x-a)^{\alpha+1}(b-x)}{\alpha+1}+\frac{(b-x)^{\alpha+2}}{(\alpha+1)(\alpha+2)}\right.\right.$$

$$\left.+\frac{(x-a)^{\alpha+2}}{\alpha+2}\right]$$

$$\left.+\left|f^{'}(b)\right|\left[\frac{(b-x)^{\alpha+1}(x-a)}{\alpha+1}+\frac{(x-a)^{\alpha+2}}{(\alpha+1)(\alpha+2)}+\frac{(b-x)^{\alpha+2}}{\alpha+2}\right]\right\} \tag{2.2}$$

for all $x\in[a,b]$.

*Remark 1* If we take $\alpha=1$ and $x=\frac{a+b}{2}$ in (2.2), the inequality (2.2) reduces to (1.3).

**Corollary 2** (Fejer Type Inequality) Under the same assumptions of Theorem 4 with $\alpha=1$, then the following inequalities hold:

$$\left|f(b)\int_x^b w(s)ds+f(a)\int_a^x w(s)ds-\int_a^b w(t)f(t)dt\right|$$

$$\leq\left|f^{'}(a)\right|\frac{(x-a)^2(3b-2a-x)\|w\|_{[a,x],\infty}+\|w\|_{[x,b],\infty}(b-x)^3}{6(b-a)}$$

$$+\left|f^{'}(b)\right|\frac{(b-x)^2(x-3a-2b)\|w\|_{[x,b],\infty}+(x-a)^3\|w\|_{[a,x],\infty}}{6(b-a)}$$

$$\leq\left|f^{'}(a)\right|\left[\frac{(x-a)^2(3b-2a-x)+(b-x)^3}{6(b-a)}\right]\|w\|_{[a,b],\infty}$$

$$+\left|f^{'}(b)\right|\left[\frac{(b-x)^2(x-3a-2b)+(x-a)^3}{6(b-a)}\right]\|w\|_{[a,b],\infty}$$

which is proved by Tseng et al. in [8].

**Corollary 3** (Weighted Trapezoid Inequality) Let $w:[a,b]\to\mathbb{R}$ be symmetric to $\frac{a+b}{2}$ and $x=\frac{a+b}{2}$ in Corollary 2. Then the following inequalities hold:

$$\left|\frac{f(a)+f(b)}{2}\int_a^b w(s)ds-\int_a^b w(t)f(t)dt\right|$$

$$\leq\frac{(b-a)^2}{48}\left[5\|w\|^{\alpha}_{[a,\frac{a+b}{2}],\infty}+\|w\|^{\alpha}_{[\frac{a+b}{2},b],\infty}\right]|f'(a)|$$

$$+\left[\|w\|^{\alpha}_{[a,\frac{a+b}{2}],\infty}+5\|w\|^{\alpha}_{[\frac{a+b}{2},b],\infty}\right]|f'(b)|$$

$$\leq(b-a)^2\|w\|^{\alpha}_{[a,b],\infty}\ \frac{|f'(a)|+|f'(b)|}{8}\Bigg)$$

which is proved by Tseng et al. in [8].

**Theorem 5** Let $f: I^\circ\subseteq\mathbb{R}\to\mathbb{R}$ be a differentiable mapping on $I^\circ$, $a,b\in I^\circ$ with $a<b$ and let $w:[a,b]\to\mathbb{R}$ be continuous on $[a,b]$. If $|f'|^q$ is convex on $[a,b]$, $q>1$, then for all $x\in[a,b]$, the following inequalities hold:

$$\left|\left(\int_x^b w(s)ds\right)^{\alpha}f(b)-\left(\int_x^a w(s)ds\right)^{\alpha}f(a)-\alpha\int_a^b\left(\int_x^t w(s)ds\right)^{\alpha-1}w(t)f(t)dt\right|$$

$$\leq\frac{(x-a)^{\alpha+\frac{1}{p}}\|w\|^{\alpha}_{[a,x],\infty}}{(b-a)^{\frac{1}{q}}(\alpha p+1)^{\frac{1}{p}}}\ \left(\frac{(b-a)^2-(b-x)^2}{2}|f'(a)|^q+\frac{(x-a)^2}{2}|f'(b)|^q\right)^{\frac{1}{q}}$$

$$+\frac{(b-x)^{\alpha+\frac{1}{p}}\|w\|^{\alpha}_{[x,b],\infty}}{(b-a)^{\frac{1}{q}}(\alpha p+1)^{\frac{1}{p}}}\ \left(\frac{(b-x)^2}{2}|f'(a)|^q+\frac{(b-a)^2-(x-a)^2}{2}|f'(b)|^q\right)^{\frac{1}{q}}$$

$$\leq\frac{\|w\|^{\alpha}_{[a,b],\infty}}{(b-a)^{\frac{1}{q}}(\alpha p+1)^{\frac{1}{p}}}\Bigg\{(x-a)^{\alpha+\frac{1}{p}}$$

$$\times\ \left(\frac{(b-a)^2-(b-x)^2}{2}|f'(a)|^q+\frac{(x-a)^2}{2}|f'(b)|^q\right)^{\frac{1}{q}}$$

$$+(b-x)^{\alpha+\frac{1}{p}}\ \left[\frac{(b-x)^2}{2}|f'(a)|^q+\frac{(b-a)^2-(x-a)^2}{2}|f'(b)|^q\right]\Bigg)^{\frac{1}{q}}\Bigg\}$$

where $\alpha>0$, $\frac{1}{p}+\frac{1}{q}=1$, and $\|w\|_\infty=\sup_{t\in[a,b]}|w(t)|$.

*Proof* We take absolute value of (2.1). Using Holder's inequality, we find that

$$\left|\left(\int_x^b w(s)ds\right)^{\alpha}f(b)-\left(\int_x^a w(s)ds\right)^{\alpha}f(a)-\alpha\int_a^b\left(\int_x^t w(s)ds\right)^{\alpha-1}w(t)f(t)dt\right|$$

$$\leq\int_a^x\left(\left|\int_x^t w(s)ds\right|\right)^{\alpha}f'(t)dt+\int_x^b\left(\left|\int_x^t w(s)ds\right|\right)^{\alpha}f'(t)dt$$

$$\leq\left(\int_a^x\left|\int_x^t w(s)ds\right|^{\alpha p}dt\right)^{\frac{1}{p}}\left(\int_a^x|f'(t)|^q dt\right)^{\frac{1}{q}}$$

$$+\left(\int_x^b\left|\int_x^t w(s)ds\right|^{\alpha p}dt\right)^{\frac{1}{p}}\left(\int_x^b|f'(t)|^q dt\right)^{\frac{1}{q}}.$$

Since $|f'(t)|^q$ is convex on $[a,b]$

$$\left|f'\left(\frac{b-t}{b-a}a+\frac{t-a}{b-a}b\right)\right|^q\leq\frac{b-t}{b-a}|f'(a)|^q+\frac{t-a}{b-a}|f'(b)|^q. \tag{2.3}$$

From (2.3), it follows that

$$\left|\left(\int_x^b w(s)ds\right)^{\alpha}f(b)-\left(\int_x^a w(s)ds\right)^{\alpha}f(a)-\alpha\int_a^b\left(\int_x^t w(s)ds\right)^{\alpha-1}w(t)f(t)dt\right|$$

$$\leq\|w\|^{\alpha}_{[a,x],\infty}\left(\int_a^x(x-t)^{\alpha p}dt\right)^{\frac{1}{p}}$$

$$\times\left(\int_a^x\left[\frac{b-t}{b-a}|f'(a)|^q+\frac{t-a}{b-a}|f'(b)|^q\right]dt\right)^{\frac{1}{q}}$$

$$+\|w\|^{\alpha}_{[x,b],\infty}$$

$$\times\left(\int_x^b(t-x)^{\alpha p}dt\right)^{\frac{1}{p}}\left(\int_x^b\left[\frac{b-t}{b-a}|f'(a)|^q+\frac{t-a}{b-a}|f'(b)|^q\right]dt\right)^{\frac{1}{q}}$$

$$\leq \frac{(x-a)^{\alpha+\frac{1}{p}}\|w\|^{\alpha}_{[a,x],\infty}}{(b-a)^{\frac{1}{q}}(\alpha p+1)^{\frac{1}{p}}}\left(\frac{(b-a)^2-(b-x)^2}{2}|f'(a)|^q+\frac{(x-a)^2}{2}|f'(b)|^q\right)^{\frac{1}{q}}$$

$$+\frac{(b-x)^{\alpha+\frac{1}{p}}\|w\|^{\alpha}_{[x,b],\infty}}{(b-a)^{\frac{1}{q}}(\alpha p+1)^{\frac{1}{p}}}\left(\left[\frac{(b-x)^2}{2}|f'(a)|^q+\frac{(b-a)^2-(x-a)^2}{2}|f'(b)|^q\right]\right)^{\frac{1}{q}}$$

$$\leq \frac{\|w\|^{\alpha}_{[a,b],\infty}}{(b-a)^{\frac{1}{q}}(\alpha p+1)^{\frac{1}{p}}}\left\{(x-a)^{\alpha+\frac{1}{p}}\left(\frac{(b-a)^2-(b-x)^2}{2}|f'(a)|^q+\frac{(x-a)^2}{2}|f'(b)|^q\right)^{\frac{1}{q}}\right.$$

$$\left.+(b-x)^{\alpha+\frac{1}{p}}\left(\left[\frac{(b-x)^2}{2}|f'(a)|^q+\frac{(b-a)^2-(x-a)^2}{2}|f'(b)|^q\right]\right)^{\frac{1}{q}}\right\}$$

which this completes the proof.

**Corollary 4** Under the same assumptions of Theorem 5 with $w(s)=1$, then the following inequalities hold:

$$\left|(b-x)^{\alpha}f(b)-(a-x)^{\alpha}f(a)-\alpha\int_a^b(t-x)^{\alpha-1}f(t)dt\right|$$

$$\leq \frac{(x-a)^{\alpha+\frac{1}{p}}}{(b-a)^{\frac{1}{q}}(\alpha p+1)^{\frac{1}{p}}}\left(\frac{(b-a)^2-(b-x)^2}{2}|f'(a)|^q+\frac{(x-a)^2}{2}|f'(b)|^q\right)^{\frac{1}{q}}$$

$$+\frac{(b-x)^{\alpha+\frac{1}{p}}}{(b-a)^{\frac{1}{q}}(\alpha p+1)^{\frac{1}{p}}}\left(\frac{(b-x)^2}{2}|f'(a)|^q+\frac{(b-a)^2-(x-a)^2}{2}|f'(b)|^q\right)^{\frac{1}{q}}$$

$$\leq \frac{1}{(b-a)^{\frac{1}{q}}(\alpha p+1)^{\frac{1}{p}}}\left\{(x-a)^{\alpha+\frac{1}{p}}\left(\frac{(b-a)^2-(b-x)^2}{2}|f'(a)|^q+\frac{(x-a)^2}{2}|f'(b)|^q\right)^{\frac{1}{q}}\right.$$

$$\left.+(b-x)^{\alpha+\frac{1}{p}}\left(\left[\frac{(b-x)^2}{2}|f'(a)|^q+\frac{(b-a)^2-(x-a)^2}{2}|f'(b)|^q\right]\right)^{\frac{1}{q}}\right\} \tag{2.4}$$

**Corollary 5** Let the conditions of Corollary 4 hold. If we take $\alpha=1$ and $x=\frac{a+b}{2}$ in (2.4), then the following inequality holds:

$$\left|\frac{f(a)+f(b)}{2}-\frac{1}{b-a}\int_a^b f(t)dt\right|$$

$$\leq \frac{(b-a)}{4(p+1)^{\frac{1}{p}}}\left[\left(\frac{3|f'(a)|^q+|f'(b)|^q}{4}\right)^{\frac{1}{q}}+\left(\frac{|f'(a)|^q+3|f'(b)|^q}{4}\right)^{\frac{1}{q}}\right].$$

**Corollary 6** (Fejer Type Inequality) Under the same assumptions of Theorem 5 with $\alpha=1$, then the following inequalities hold:

$$\left|f(b)\int_x^b w(s)ds+f(a)\int_a^x w(s)ds-\int_a^b w(t)f(t)dt\right|$$

$$\leq \frac{(x-a)^{1+\frac{1}{p}}\|w\|_{[a,x],\infty}}{(b-a)^{\frac{1}{q}}(p+1)^{\frac{1}{p}}}\left(\frac{(b-a)^2-(b-x)^2}{2}|f'(a)|^q+\frac{(x-a)^2}{2}|f'(b)|^q\right)^{\frac{1}{q}}+\frac{(b-x)^{1+\frac{1}{p}}\|w\|_{[x,b],\infty}}{(b-a)^{\frac{1}{q}}(p+1)^{\frac{1}{p}}}\left(\frac{(b-x)^2}{2}|f'(a)|^q\right.$$

$$\left.+\frac{(b-a)^2-(x-a)^2}{2}|f'(b)|^q\right)^{\frac{1}{q}}\leq \frac{\|w\|_{[a,b],\infty}}{(b-a)^{\frac{1}{q}}(p+1)^{\frac{1}{p}}}\left\{(x-a)^{1+\frac{1}{p}}\right.$$

$$\times\left(\frac{(b-a)^2-(b-x)^2}{2}|f'(a)|^q+\frac{(x-a)^2}{2}|f'(b)|^q\right)^{\frac{1}{q}}$$

$$\left.+(b-x)^{1+\frac{1}{p}}\left(\left[\frac{(b-x)^2}{2}|f'(a)|^q+\frac{(b-a)^2-(x-a)^2}{2}|f'(b)|^q\right]\right)^{\frac{1}{q}}\right\}.$$

**Corollary 7** (Weighted Trapezoid Inequality) Let $w:[a,b]\to\mathbb{R}$ be symmetric to $\frac{a+b}{2}$ and $x=\frac{a+b}{2}$ in Corollary 6. Then the following inequalities hold:

$$\left|\frac{f(a)+f(b)}{2}\int_a^b w(s)ds-\int_a^b w(t)f(t)dt\right|$$

$$\leq \frac{(b-a)^2}{4(p+1)^{\frac{1}{p}}}\left[\|w\|_{[a,\frac{a+b}{2}],\infty}\left(\frac{3|f'(a)|^q+|f'(b)|^q}{4}\right)^{\frac{1}{q}}\right.$$

$$\left.+\|w\|_{[\frac{a+b}{2},b],\infty}\left(\frac{|f'(a)|^q+3|f'(b)|^q}{4}\right)^{\frac{1}{q}}\right]$$

$$\leq \frac{(b-a)^2\|w\|_{[a,b],\infty}}{4(p+1)^{\frac{1}{p}}}\left[\left(\frac{3|f'(a)|^q+|f'(b)|^q}{4}\right)^{\frac{1}{q}}\right.$$

$$\left.+\left(\frac{|f'(a)|^q+3|f'(b)|^q}{4}\right)^{\frac{1}{q}}\right].$$

**Theorem 6** Let $f : I^{\circ} \subseteq \mathbb{R} \to \mathbb{R}$ be a differentiable mapping on $I^{\circ}$, $a, b \in I^{\circ}$ with $a<b$ and let $w : [a,b] \to \mathbb{R}$ be continuous on $[a,b]$. If $|f'|^q$ is convex on $[a,b]$, $q>1$, then for all $x \in [a,b]$, the following inequality holds:

$$\left|\left(\int_x^b w(s)ds\right)^{\alpha} f(b) - \left(\int_x^a w(s)ds\right)^{\alpha} f(a) - \alpha\int_a^b \left(\int_x^t w(s)ds\right)^{\alpha-1} w(t)f(t)dt\right|$$

$$\leq \frac{(b-a)^{\frac{1}{q}}\|w\|^{\alpha}_{[a,b],\infty}}{(\alpha p+1)^{\frac{1}{p}}}\left[(x-a)^{\alpha p+1}+(b-x)^{\alpha p+1}\right]^{\frac{1}{p}} \times \left(\frac{|f'(a)|^q+|f'(b)|^q}{2}\right)^{\frac{1}{q}}$$

where $\alpha > 0$, $\frac{1}{p}+\frac{1}{q}=1$, and $\|w\|_{\infty}=\sup_{t\in[a,b]}|w(t)|$.

*Proof* We take absolute value of (2.1). Using Holder's inequality and the convexity of $|f'|^q$, we find that

$$\left|\left(\int_x^b w(s)ds\right)^{\alpha} f(b) - \left(\int_x^a w(s)ds\right)^{\alpha} f(a) - \alpha\int_a^b \left(\int_x^t w(s)ds\right)^{\alpha-1} w(t)f(t)dt\right|$$

$$\leq \left(\int_a^b \left|\int_x^t w(s)ds\right|^{\alpha p} dt\right)^{\frac{1}{p}} \left(\int_a^b |f'(t)|^q dt\right)^{\frac{1}{q}}$$

$$\leq \|w\|^{\alpha}_{[a,b],\infty}\left(\int_a^b |t-x|^{\alpha p}dt\right)^{\frac{1}{p}} \left(\int_a^b \left[\frac{b-t}{b-a}|f'(a)|^q + \frac{t-a}{b-a}|f'(b)|^q\right]dt\right)^{\frac{1}{q}} = \frac{(b-a)^{\frac{1}{q}}\|w\|^{\alpha}_{[a,b],\infty}}{(\alpha p+1)^{\frac{1}{p}}}\left[(x-a)^{\alpha p+1} +(b-x)^{\alpha p+1}\right]^{\frac{1}{p}}\left(\frac{|f'(a)|^q+|f'(b)|^q}{2}\right)^{\frac{1}{q}}$$

which this completes the proof.

**Corollary 8** Under the same assumptions of Theorem 6 with $w(s)=1$, then the following inequality holds:

$$\left|(b-x)^{\alpha}f(b)-(a-x)^{\alpha}f(a)-\alpha\int_a^b (t-x)^{\alpha-1}f(t)dt\right|$$

$$\leq \frac{(b-a)^{\frac{1}{q}}}{(\alpha p+1)^{\frac{1}{p}}}\left[(x-a)^{\alpha p+1}+(b-x)^{\alpha p+1}\right]^{\frac{1}{p}}\left(\frac{|f'(a)|^q+|f'(b)|^q}{2}\right)^{\frac{1}{q}}. \tag{2.5}$$

*Remark 2* Let the conditions of Corollary 8 hold. If we take $\alpha = 1$ and $x=\frac{a+b}{2}$ in (2.5), then the inequality becomes the inequality ( 1.4).

**Corollary 9** (Fejer Type Inequality) Under the same assumptions of Theorem 6 with $\alpha = 1$, then the following inequality holds:

$$\left|f(b)\int_x^b w(s)ds + f(a)\int_a^x w(s)ds - \int_a^b w(t)f(t)dt\right|$$

$$\leq \frac{(b-a)^{\frac{1}{q}}\|w\|_{[a,b],\infty}}{(p+1)^{\frac{1}{p}}}\left[(x-a)^{p+1}+(b-x)^{p+1}\right]^{\frac{1}{p}} \times \left(\frac{|f'(a)|^q+|f'(b)|^q}{2}\right)^{\frac{1}{q}}.$$

**Corollary 10** (Weighted Trapezoid Inequality) Let $w : [a,b] \to \mathbb{R}$ be symmetric to $\frac{a+b}{2}$ and $x=\frac{a+b}{2}$ in Corollary 9. Then the following inequality holds:

$$\left|\frac{f(a)+f(b)}{2}\int_a^b w(s)ds - \int_a^b w(t)f(t)dt\right| \leq \frac{(b-a)^2\|w\|_{[a,b],\infty}}{2(p+1)^{\frac{1}{p}}}\left(\frac{|f'(a)|^q+|f'(b)|^q}{2}\right)^{\frac{1}{q}}.$$

**Theorem 7** Let $f : I^{\circ} \subseteq \mathbb{R} \to \mathbb{R}$ be a differentiable mapping on $I^{\circ}$, $a, b \in I^{\circ}$ with $a<b$ and let $w : [a,b] \to \mathbb{R}$ be continuous on $[a,b]$. If $|f'|^q$ is convex on $[a,b]$, $q>1$, then for all $x \in [a,b]$, the following inequality holds:

$$\left|\left(\int_x^b w(s)ds\right)^\alpha f(b) - \left(\int_x^a w(s)ds\right)^\alpha f(a) - \alpha\int_a^b\left(\int_x^t w(s)ds\right)^{\alpha-1} w(t)f(t)dt\right|$$
$$\leq \frac{\|w\|^\alpha_{[a,b],\infty}}{(\alpha+1)(\alpha+2)^{\frac{1}{q}}(b-a)^{\frac{1}{q}}}\left((x-a)^{\alpha+1}+(b-x)^{\alpha+1}\right)^{\frac{1}{p}}$$
$$\times\Big(\Big((\alpha+1)(b-a)(x-a)^{\alpha+1}+(b-x)\left[(x-a)^{\alpha+1}+(b-x)^{\alpha+1}\right]\Big)|f'(a)|^q$$
$$+\Big((\alpha+1)(b-a)(b-x)^{\alpha+1}+(x-a)\left[(x-a)^{\alpha+1}+(b-x)^{\alpha+1}\right]\Big)|f'(b)|^q\Big)^{\frac{1}{q}}$$

where $\alpha > 0$, $\frac{1}{p}+\frac{1}{q}=1$, and $\|w\|_\infty = \sup_{t\in[a,b]}|w(t)|$.

*Proof* We take absolute value of (2.1). Using Holder's inequality and the convexity of $|f'|^q$, we find that

$$\left|\left(\int_x^b w(s)ds\right)^\alpha f(b) - \left(\int_x^a w(s)ds\right)^\alpha f(a) - \alpha\int_a^b\left(\int_x^t w(s)ds\right)^{\alpha-1} w(t)f(t)dt\right|$$
$$\leq\left(\int_a^b\left|\int_x^t w(s)ds\right|^\alpha dt\right)^{\frac{1}{p}}\left(\int_a^b\left|\int_x^t w(s)ds\right|^\alpha |f'(t)|^q dt\right)^{\frac{1}{q}}$$
$$\leq\|w\|^\alpha_{[a,b],\infty}\left(\int_a^b|t-x|^\alpha dt\right)^{\frac{1}{p}}\left(\int_a^b|t-x|^\alpha\left[\frac{b-t}{b-a}|f'(a)|^q+\frac{t-a}{b-a}|f'(b)|^q\right]dt\right)^{\frac{1}{q}} = \|w\|^\alpha_{[a,b],\infty}\left(\frac{(x-a)^{\alpha+1}+(b-x)^{\alpha+1}}{\alpha+1}\right)^{\frac{1}{p}}$$
$$\times\left(\left(\frac{(b-x)(x-a)^{\alpha+1}}{\alpha+1}+\frac{(x-a)^{\alpha+2}}{\alpha+2}\right)|f'(a)|^q + \left(\frac{(x-a)^{\alpha+2}}{(\alpha+1)(\alpha+2)}\right)|f'(b)|^q + \left(\frac{(b-x)^{\alpha+2}}{(\alpha+1)(\alpha+2)}\right)|f'(a)|^q + \left(\frac{(x-a)(b-x)^{\alpha+1}}{\alpha+1}+\frac{(b-x)^{\alpha+2}}{\alpha+2}\right)|f'(b)|^q\right)^{\frac{1}{q}}$$
$$=\frac{\|w\|^\alpha_{[a,b],\infty}}{(\alpha+1)(\alpha+2)^{\frac{1}{q}}(b-a)^{\frac{1}{q}}}\left((x-a)^{\alpha+1}+(b-x)^{\alpha+1}\right)^{\frac{1}{p}}$$
$$\times\Big(\Big((\alpha+1)(b-a)(x-a)^{\alpha+1}+(b-x)\left[(x-a)^{\alpha+1}+(b-x)^{\alpha+1}\right]\Big)|f'(a)|^q$$
$$+\Big((\alpha+1)(b-a)(b-x)^{\alpha+1}+(x-a)\left[(x-a)^{\alpha+1}+(b-x)^{\alpha+1}\right]\Big)|f'(b)|^q\Big)^{\frac{1}{q}}$$

which this completes the proof.

**Corollary 11** Under the same assumptions of Theorem 7 with $w(s) = 1$, then the following inequality holds:

$$\left|(b-x)^\alpha f(b)-(a-x)^\alpha f(a)-\alpha\int_a^b(t-x)^{\alpha-1}f(t)dt\right|$$
$$\leq\frac{1}{(\alpha+1)(\alpha+2)^{\frac{1}{q}}(b-a)^{\frac{1}{q}}}\left((x-a)^{\alpha+1}+(b-x)^{\alpha+1}\right)^{\frac{1}{p}}$$
$$\times\Big(\Big((\alpha+1)(b-a)(x-a)^{\alpha+1}+(b-x)\left[(x-a)^{\alpha+1}+(b-x)^{\alpha+1}\right]\Big)|f'(a)|^q$$
$$+\Big((\alpha+1)(b-a)(b-x)^{\alpha+1}+(x-a)\left[(x-a)^{\alpha+1}+(b-x)^{\alpha+1}\right]\Big)|f'(b)|^q\Big)^{\frac{1}{q}}. \tag{2.6}$$

**Corollary 12** Let the conditions of Corollary 11 hold. If we take $\alpha = 1$ and $x = \frac{a+b}{2}$ in (2.6), then the following inequality holds:

$$\left|\frac{f(a)+f(b)}{2}-\frac{1}{b-a}\int_a^b f(t)dt\right| \leq \frac{(b-a)}{4}\left(\frac{|f'(a)|^q+|f'(b)|^q}{2}\right)^{\frac{1}{q}}.$$

**Corollary 13** (Fejer Type Inequality) Under the same assumptions of Theorem 7 with $\alpha = 1$, then the following inequality holds:

$$\left|f(b)\int_x^b w(s)ds + f(a)\int_a^x w(s)ds - \int_a^b w(t)f(t)dt\right|$$
$$\leq\frac{\|w\|^\alpha_{[a,b],\infty}}{2\cdot 3^{\frac{1}{q}}(b-a)^{\frac{1}{q}}}\left((x-a)^2+(b-x)^2\right)^{\frac{1}{p}}$$
$$\times\Big(\Big(2(b-a)(x-a)^2+(b-x)\left[(x-a)^2+(b-x)^2\right]\Big)|f'(a)|^q + \Big(2(b-a)(b-x)^2+(x-a)\left[(x-a)^2+(b-x)^2\right]\Big)|f'(b)|^q\Big)^{\frac{1}{q}}.$$

**Corollary 14** (Weighted Trapezoid Inequality) Let $w : [a, b] \to \mathbb{R}$ be symmetric to $\frac{a+b}{2}$ and $x = \frac{a+b}{2}$ in Corollary 13. Then the following inequality holds:

$$\left|\frac{f(a)+f(b)}{2}\int_a^b w(s)ds - \int_a^b w(t)f(t)dt\right| \leq \frac{(b-a)^2\|w\|^\alpha_{[a,b],\infty}}{4}\left(\frac{|f'(a)|^q+|f'(b)|^q}{2}\right)^{\frac{1}{q}}.$$

## References


1. Dragomir, S.S., Agarwal, R.P.: Two inequalities for differentiable mappings and applications to special means of real numbers and to trapezoidal formula. Appl. Math. Lett. **11**(5), 91–95 (1998)
2. Dragomir, S.S., Pearce, C.E.M.: Selected topics on Hermite–Hadamard inequalities and applications. Victoria University, RGMIA Monographs (2000)
3. Fejer, L.: Über die Fourierreihen, II. Math. Naturwiss. Anz Ungar. Akad. Wiss. (Hungarian) 24, 369–390 (1906)
4. Pečarić, J., Proschan, F., Tong, Y.L.: Convex functions, partial ordering and statistical applications. Academic Press, New York (1991)
5. Sarikaya, M.Z.: On new Hermite Hadamard Fejer type integral inequalities. Stud. Universit. Babes Bolyai Math. 57(3), 377–386 (2012)
6. Sarikaya, M.Z., Erden, S.: On the weighted integral inequalities for convex function. RGMIA research report collection. 17(70), 10 (2014)
7. Sarikaya, M.Z., Erden, S.: On the Hermite–Hadamard–Fejér type integral inequality for convex function. RGMIA research report collection. 17(69), 12 (2014)
8. Tseng, K.-L., Yang, G.-S., Hsu, K.-C.: Some inequalities for differentiable mappings and applications to Fejer inequality and weighted trapezoidal formula. Taiwan. J. Math. 15(4), 1737–1747 (2011)
9. Hwang, S.-R., Tseng, K.-L., Hsu, K.-C.: Hermite–Hadamard type and Fejér type inequalities for general weights (I). J. Inequal. Appl. **2013**, 170 (2013)
10. Wang, C.-L., Wang, X.-H.: On an extension of Hadamard inequality for convex functions. Chin. Ann. Math. **3**, 567–570 (1982)
11. Wasowicz, S., Witkonski, A.: On some inequality of Hermite–Hadamard type. Opuscu. Math. 32(2), 591–600 (2012)
12. Wu, S.-H.: On the weighted generalization of the Hermite–Hadamard inequality and its applications. Rocky Mt. J. Math. **39**(5), 1741–1749 (2009)
13. Xi, B.-Y., Qi, F.: Some Hermite–Hadamard type inequalities for differentiable convex functions and applications. Hacet. J. Math. Stat. 42(3), 243–257 (2013)
14. Xi, B.-Y., Qi, F., Hermite–Hadamard type inequalities for functions whose derivatives are of convexities. Nonlinear Funct. Anal. Appl. 18(2), 163–176 (2013)

# 7

# Numerical solutions of fourth-order Volterra integro-differential equations by the Green's function and decomposition method

**Randhir Singh**[1] · **Abdul-Majid Wazwaz**[2]

**Abstract** We propose a reliable technique based on Adomian decomposition method (ADM) for the numerical solution of fourth-order boundary value problems for Volterra integro-differential equations. We use Green's function technique to convert boundary value problem into the integral equation before establishing the recursive scheme for the solution components of a specific solution. The advantage of the proposed technique over the standard ADM or modified ADM is that it provides not only better numerical results but also avoids solving a sequence of transcendental equations for unknown constant. Approximations of the solutions are obtained in the form of series. Convergence and error analysis is also discussed. The accuracy and generality of the proposed scheme are demonstrated by solving some numerical examples.

**Keywords** Integro-differential equations · Boundary value problems · Adomian decomposition method · Green's function · Approximations

**Mathematics Subject Classification** 34B15 · 34B27 · 34B05 · 65L10 · 65L80

✉ Randhir Singh
randhir.math@gmail.com

Abdul-Majid Wazwaz
wazwaz@sxu.edu

1 Department of Applied Mathematics, Birla Institute of Technology, Mesra, Ranchi 835215, India

2 Department of Mathematics, Saint Xavier University, Chicago IL 60655, USA

## Introduction

Consider the following class of fourth-order BVPs for Volterra IDEs [1–5]

$$y^{(iv)}(x) = g(x) + \int_0^x K(x,t)f(y(t))\mathrm{d}t, \quad x \in [0,b], \tag{1}$$

with the boundary conditions

$$y(0) = \alpha_1, \quad y'(0) = \alpha_2, \quad y(b) = \alpha_3, \quad y'(b) = \alpha_4, \tag{2}$$

where $\alpha_i, i = 1,2,3,4$ are any finite real constants, $g(x) \in C[0,b]$, and $K(x,t) \in C([0,b] \times [0,b])$. The IDEs are often involved in the mathematical formulation of physical and engineering phenomena [4–6]. In general, the IDEs with given boundary conditions are difficult to solve analytically. Therefore, these problems must be solved by various approximation and numerical methods. The existence and uniqueness of solutions for such problems can be found in [1].

There is considerable literature on the numerical-approximate treatment of the BVPs for IDEs, for example, the compact finite difference method [7], monotone iterative methods [7, 8], spline collocation method [9], the method of upper and lower solution [10], Haar wavelets [11], and pseudo-spectral method [12]. Though, these numerical techniques have many advantages, a huge amount of computational work is involved that combines some root-finding techniques to obtain an accurate numerical solution especially for nonlinear problems.

Recently, some newly developed semi-numerical methods have also been applied to solve BVPs for IDEs such as, ADM [3], homotopy perturbation method (HPM) [4], and homotopy analysis method (HAM) [13]. In [5], the variational iteration method (VIM) was also used for solving the problem (1)–(2). However, in [14] Wazwaz pointed out that VIM gives good approximations only when the problem is linear or nonlinear with the weak nonlinearity of the form $(y^n, yy', y'^n, \ldots)$, but the VIM suffers when the nonlinearity is of the form $(e^y, \ln y, \sin y, \ldots)$ (for details see [14]).

It is well known that the ADM allows us to solve nonlinear BVPs without restrictive assumptions such as linearization, discretization and perturbation. Many researchers [14–23] have shown interest to study the ADM for different scientific models. According to the ADM, we rewrite the problem (1) in an operator form

$$Ly = g + Ny, \tag{3}$$

where $L = \frac{d^4}{dx^4}$ is a fourth-order linear differential operator, $g$ is a function of $x$ and $Ny = \int_0^x K(x,t) f(y(t))dt$ is a nonlinear term. Inverse integral operator is usually defined as

$$L^{-1}[\cdot] := \int_0^x \int_0^x \int_0^x \int_0^x [\cdot] dx\, dx\, dx\, dx. \tag{4}$$

Operating with $L^{-1}$ on both sides of (3) and using the conditions $y(0) = \alpha_1$ and $y'(0) = \alpha_2$, we obtain

$$y(x) = \alpha_1 + \alpha_2 x + c_1 x^2 + c_2 x^3 + L^{-1}[g + Ny], \tag{5}$$

where $c_1 = \frac{y''(0)}{2!}$ and $c_2 = \frac{y'''(0)}{3!}$ are unknown constants to be determined.

The ADM relies on decomposing $y$ by a series of *components* and nonlinear term $f(y)$ by a series of *Adomian polynomials* as

$$y = \sum_{j=0}^{\infty} y_j(x) \quad \text{and} \quad f(y) = \sum_{j=0}^{\infty} A_j, \tag{6}$$

where $A_j$ are Adomian's polynomials [22], which can be computed as

$$A_n = \frac{1}{n!}\frac{d^n}{d\lambda^n}\left[ f\left(\sum_{k=0}^{\infty} y_k \lambda^k\right)\right]_{\lambda=0}, \quad n = 0, 1, 2, \ldots, \tag{7}$$

Several algorithms have also been given to generate the Adomian polynomial rapidly in [24–26]. Substituting the series (6) in (5), we get

$$\sum_{j=0}^{\infty} y_j(x) = \alpha_1 + \alpha_2 x + c_1 x^2 + c_2 x^3 + L^{-1}[g] + L^{-1}\left\{ \int_0^x K(x,t)\left[\sum_{j=0}^{\infty} A_j\right] dt \right\}. \tag{8}$$

On comparing both sides of equation (8), the ADM is given by

$$\left.\begin{aligned} y_0 &= \alpha_1 + \alpha_2 x + c_1 x^2 + c_2 x^3 + L^{-1}(g), \\ y_j &= L^{-1}\left\{ \int_0^x K(x,t) A_{j-1} dt \right\}, \quad j = 1, 2 \ldots \end{aligned}\right\} \tag{9}$$

Wazwaz [27] suggested a modified ADM (MADM) which is given by

$$\left.\begin{aligned} y_0 &= \alpha_1, \\ y_1 &= \alpha_2 x + c_1 x^2 + c_2 x^3 + L^{-1}[g] + L^{-1}\left\{ \int_0^x K(x,t) A_0 dt \right\}, \\ y_j &= L^{-1}\left\{ \int_0^x K(x,t) A_{j-1} dt \right\}, \quad j = 2, 3, \ldots. \end{aligned}\right\} \tag{10}$$

Hence, the $n$-term approximate series solution is obtained as

$$\phi_n(x, c_1, c_2) = \sum_{j=0}^{n} y_j(x, c_1, c_2). \tag{11}$$

We note that the series solution $\phi_n(x, c_1, c_2)$ depends on the unknown constants $c_1$ and $c_2$. These unknown constants will be determined approximately by imposing the boundary condition at $x = b$ on $\phi_n(x, c_1, c_2)$, which leads a sequence of nonlinear system of equations as

$$\phi_n(b, c_1, c_2) = \alpha_3 \quad \text{and} \quad \phi_n'(b, c_1, c_2) = \alpha_4, \quad n = 1, 2, \ldots. \tag{12}$$

To determine the unknown constants $c_1$ and $c_2$, we require root finding methods such as Newton–Raphson's method which requires additional computational work. But solving the nonlinear equation (12) for $c_1$ and $c_2$ is a difficult task in general. Moreover, in some cases the unknowns $c_1$ and $c_2$ may not be uniquely determined. This may be the main difficulty of the ADM.

In this work, we propose a new recursive scheme which does not involve any unknown constant to be determined. In other words, we introduce a modification of the ADM to overcome the difficulties occurring in ADM or MADM for solving fourth-order BVPs for IDEs.

## The decomposition method with Green's function

Let us first consider homogeneous version of the problem (1) and (2) as

$$\begin{cases} u^{(iv)}(x) = 0, & x \in [0, b], \\ u(0) = \alpha_1, \quad u'(0) = \alpha_2, \quad u(b) = \alpha_3, \quad u'(b) = \alpha_4. \end{cases} \tag{13}$$

Solving (13) analytically, we obtain

$$u(x) = \alpha_1 + \alpha_2 x - \frac{(3\alpha_1 + 2b\alpha_2 - 3\alpha_3 + b\alpha_4)}{b^2}x^2 + \frac{(2\alpha_1 + b\alpha_2 - 2\alpha_3 + b\alpha_4)}{b^3}x^3. \tag{14}$$

We now construct Green's function of the following fourth-order boundary value problem

$$\begin{cases} y^{(iv)}(x) = h(x), & x \in [0, b], \\ y(0) = y'(0) = y(b) = y'(b) = 0. \end{cases} \tag{15}$$

The Green's function of (15) can be easily constructed and it is given by

$$G(x,\xi) = \begin{cases} x^3\left(-\frac{1}{6}+\frac{\xi^2}{2b^2}-\frac{\xi^3}{3b^3}\right)+x^2\left(\frac{\xi}{2}-\frac{\xi^2}{b}+\frac{\xi^3}{2b^2}\right), & 0 \le x \le \xi, \\ \xi^3\left(-\frac{1}{6}+\frac{x^2}{2b^2}-\frac{x^3}{3b^3}\right)+\xi^2\left(\frac{x}{2}-\frac{x^2}{b}+\frac{x^3}{2b^2}\right), & \xi \le x \le b. \end{cases} \tag{16}$$

Using (14) and (16), we transform BVPs for IDEs (1) and (2) into an integral equation as

$$y(x) = u(x) + \int_0^b G(x,\xi)\left\{g(\xi) + \int_0^\xi K(\xi,t)f(y(t))\mathrm{d}t\right\}\mathrm{d}\xi. \tag{17}$$

Substituting the series (6) in (17), we obtain

$$\sum_{j=0}^{\infty} y_j(x) = u(x) + \int_0^b G(x,\xi)\left\{g(\xi) + \int_0^\xi K(\xi,t)\left[\sum_{j=0}^{\infty} A_j\right]\mathrm{d}t\right\}\mathrm{d}\xi. \tag{18}$$

Comparing both sides of (18), *the decomposition with Green's function* (DMGF) is given by the following recursive scheme as

$$\left.\begin{aligned} y_0 &= u(x) + \int_0^b G(x,\xi)g(\xi)\mathrm{d}\xi, \\ y_j &= \int_0^b G(x,\xi)\left\{\int_0^\xi K(\xi,t)A_{j-1}\mathrm{d}t\right\}\mathrm{d}\xi, \quad j = 1,2,3\ldots \end{aligned}\right\} \tag{19}$$

and *the modified decomposition with Green's function* (MDMGF) is given by the following recursive scheme as

$$\left.\begin{aligned} y_0 &= u_0, \\ y_1 &= u_1 + \int_0^b G(x,\xi)\left\{g(\xi) + \int_0^\xi K(\xi,t)A_0\mathrm{d}t\right\}\mathrm{d}\xi, \\ y_j &= \int_0^b G(x,\xi)\left\{\int_0^\xi K(\xi,t)A_{j-1}\mathrm{d}t\right\}\mathrm{d}\xi, \quad j=2,3\ldots \end{aligned}\right\} \tag{20}$$

where $u_0 = \alpha_1$ and $u_1 = \alpha_2 x - \frac{(3\alpha_1+2b\alpha_2-3\alpha_3+b\alpha_4)}{b^2}x^2 + \frac{(2\alpha_1+b\alpha_2-2\alpha_3+b\alpha_4)}{b^3}x^3$. The $n$-terms truncated series solution is obtained as

$$\psi_n = \sum_{j=0}^{n} y_j. \tag{21}$$

## Convergence and error estimate of the scheme (19) or (20)

In this section, we shall show that sequence $\{\psi_n\}$ of the partial sums of series solution defined by (21) converges to the exact solution $y$ of the problem (1), (2).

**Theorem 3.1** (Convergence theorem) *Suppose that* $\mathbb{X} = C[0,b]$ *is a Banach space with the norm* $\|y\| = \max_{x\in I=[0,b]}|y(x)|$, $y \in \mathbb{X}$. *Assume that the function* $f(y)$ *satisfies the Lipschitz condition such that* $|f(y) - f(y^*)| \le l|y - y^*|$ *and denote* $\|K\|_\infty = \max|K(\xi,t)|$ *and* $\|G\|_\infty = \max|G(x,\xi)|$. *Further, we define* $\delta$ *as* $\delta := l\|K\|_\infty\|G\|_\infty b^2$. *Then the sequence* $\{\psi_n\}$ *converges to the exact solution whenever* $\delta < 1$ *and* $\|y_1\| < \infty$.

*Proof* From (19) or (20) and (21), we write

$$\begin{aligned} \psi_n = y_0 + \sum_{j=1}^{n} y_j &= u(x) + \int_0^b G(x,\xi)g(\xi)\mathrm{d}\xi \\ &+ \sum_{j=1}^{n}\left[\int_0^b G(x,\xi)\left\{\int_0^\xi K(\xi,t)A_{j-1}\mathrm{d}t\right\}\mathrm{d}\xi\right] \\ &= u(x) + \int_0^b G(x,\xi)\left\{g(\xi) + \int_0^\xi K(\xi,t)\left[\sum_{j=1}^{n} A_{j-1}\right]\mathrm{d}t\right\}\mathrm{d}\xi. \end{aligned}$$

For all $n, m \in \mathbb{N}$, with $n > m$, consider

$$\|\psi_n - \psi_m\| = \max_{x\in I}\left|\int_0^b G(x,\xi)\left\{\int_0^\xi K(\xi,t)\left[\sum_{j=0}^{n-1} A_j - \sum_{j=0}^{m-1} A_j\right]\mathrm{d}t\right\}\mathrm{d}\xi\right|. \tag{22}$$

Using the relation $\sum_{j=0}^{n} A_j \le f(\psi_n)$ (for details see, [28, pp 944–945]) we have

$$\begin{aligned} \|\psi_n - \psi_m\| &\le \max_{x\in I}\left|\int_0^b G(x,\xi)\left\{\int_0^\xi K(\xi,t)[f(\psi_{n-1}) - f(\psi_{m-1})]\mathrm{d}t\right\}\mathrm{d}\xi\right| \\ &\le \max|G(x,\xi)|\int_0^b\left\{\max|K(\xi,t)|\int_0^\xi l|\psi_{n-1} - \psi_{m-1}|\mathrm{d}t\right\}\mathrm{d}\xi \\ &\le l\|K\|_\infty\|G\|_\infty\|\psi_{n-1} - \psi_{m-1}\|\max_{x\in I}\int_0^b\int_0^\xi \mathrm{d}t\mathrm{d}\xi \\ &\le l\|K\|_\infty\|G\|_\infty b^2\|\psi_{n-1} - \psi_{m-1}\| = \delta\|\psi_{n-1} - \psi_{m-1}\|, \end{aligned}$$

where $\delta = l\|K\|_\infty \|G\|_\infty b^2$.

Setting $n = m+1$ we obtain $\|\psi_{m+1} - \psi_m\| \le \delta\|\psi_m - \psi_{m-1}\|$. Thus, we have $\|\psi_{m+1} - \psi_m\| \le \delta\|\psi_m - \psi_{m-1}\| \le \delta^2\|\psi_{m-1} - \psi_{m-2}\| \le \cdots \le \delta^m\|\psi_1 - \psi_0\|$. Using triangle inequality for any $n, m \in \mathbb{N}$, with $n > m$ we have

$$\begin{aligned}\|\psi_n - \psi_m\| &= \|(\psi_n - \psi_{n-1}) + (\psi_{n-1} - \psi_{n-2}) + \cdots + (\psi_{m+1} - \psi_m)\|\\ &\le \|\psi_n - \psi_{n-1}\| + \|\psi_{n-1} - \psi_{n-2}\| + \cdots + \|\psi_{m+1} - \psi_m\|\\ &\le [\delta^{n-1} + \delta^{n-2} + \cdots + \delta^m]\|\psi_1 - \psi_0\|\\ &= \delta^m[1 + \delta + \delta^2 + \cdots + \delta^{n-m-1}]\|\psi_1 - \psi_0\|\\ &= \delta^m\left(\frac{1-\delta^{n-m}}{1-\delta}\right)\|y_1\|.\end{aligned}$$

Thus, we obtain

$$\|\psi_n - \psi_m\| \le \frac{\delta^m}{1-\delta}\|y_1\|, \tag{23}$$

which converges to zero, i.e., $\|\psi_n - \psi_m\| \to 0$, as $m \to \infty$. This implies that there exits an $\psi$ such that $\lim_{n\to\infty} \psi_n = \psi$. □

In the next theorem, we give the error estimate of the series solution.

**Theorem 3.2** (Error estimate) *The maximum absolute truncation error of the series $\psi_m$ obtained by the scheme (19) or (20) to problem is given as*

$$\|y - \psi_m\| \le \frac{\delta^{m+1}}{l(1-\delta)}\|A_0\|_\infty \tag{24}$$

*where* $\|A_0\|_\infty = \max_{x\in I}|A_0|$, $A_0 = f(y_0)$.

*Proof* Fixing $m$ and letting $n \to \infty$ in the estimate (23) with $n \ge m$, we obtain

$$\|y - \psi_m\| \le \frac{\delta^m}{1-\delta}\|y_1\|. \tag{25}$$

From the scheme (19) we have $y_1 = \int_0^b G(x,\xi)\{\int_0^\xi K(\xi,t)A_0 dt\}d\xi$, and following the steps of theorem (3.1), we have

$$\begin{aligned}\|y_1\| &= \max_{x\in I}\left|\int_0^b G(x,\xi)\left\{\int_0^\xi K(\xi,t)A_0 dt\right\}d\xi\right| \\ &\le \|K\|_\infty\|G\|_\infty\|A_0\|_\infty b^2.\end{aligned} \tag{26}$$

But we know that $\delta = l\|K\|_\infty\|G\|_\infty b^2$. Hence, the inequality (26) becomes

$$\|y_1\| \le \frac{\delta}{l}\|A_0\|_\infty. \tag{27}$$

Combining the estimates (25) and (27), we get the desired result. □

## Numerical results

To demonstrate the efficiency and accuracy of the proposed recursive schemes, we consider four fourth-order BVPs for Volterra IDEs. All symbolic and numerical computations are performed using 'Mathematica' 8.0 software package.

*Example 4.1* Consider the following linear fourth-order BVP for Volterra IDE [2, 3]

$$\left.\begin{aligned}&y^{(iv)}(x) = g(x) + \int_0^x y(t)dt, \quad x \in [0,1]\\ &y(0) = y'(0) = 1, \quad y(1) = 1+e, \quad y'(1) = 2e,\end{aligned}\right\} \tag{28}$$

where $g(x) = -x + 5e^x - 1$, and its exact solution is $y(x) = 1 + xe^x$.

Here, $b = 1$, $\alpha_1 = 1$, $\alpha_2 = 1$, $\alpha_3 = 1+e$, $\alpha_4 = 2e$, $K(x,t) = 1$, $f(y) = y$, and $g(\xi) = -\xi + 5e^\xi - 1$. According to the MDMGF (20), the problem (28) is transformed into the following recursive scheme

$$\left.\begin{aligned}y_0 &= 1,\\ y_1 &= x + (e-2)x^2 + x^3 + \int_0^1 G(x,\xi)\left\{g(\xi) + \int_0^\xi K(\xi,t)y_0 dt\right\}d\xi,\\ y_j &= \int_0^1 G(x,\xi)\left\{\int_0^\xi K(\xi,t)y_{j-1}dt\right\}d\xi, \quad j = 2,3,\ldots\end{aligned}\right\} \tag{29}$$

$G(x,\xi)$ is given by

$$G(x,\xi) = \begin{cases} x^3\left(-\frac{1}{6} + \frac{\xi^2}{2} - \frac{\xi^3}{3}\right) + x^2\left(\frac{\xi}{2} - \frac{\xi^2}{1} + \frac{\xi^3}{2}\right), & 0 \le x \le \xi,\\ \xi^3\left(-\frac{1}{6} + \frac{x^2}{2} - \frac{x^3}{3}\right) + \xi^2\left(\frac{x}{2} - \frac{x^2}{1} + \frac{x^3}{2}\right), & \xi \le x \le 1.\end{cases} \tag{30}$$

Using (29) and (30), we compute the solution components $y_j$ as

$y_0 = 1;$

$y_1 = -5 + 5e^x - 4x - 1.5062x^2 - 0.325258x^3 - 0.0416667x^4;$

$y_2 = -5 + 5e^x - 5x - 2.4938x^2 - 0.84140x^3 - 0.20833x^4 - 0.041666x^5 - 0.00555x^6 + \cdots$

$\vdots$

To check the accuracy and efficiency of the proposed methods, the absolute error function is defined as

$$E_n(x) = |\psi_n(x) - y(x)|, \quad n = 1,2,\ldots$$

where $y$ is the exact solution and $\psi_n$ is the $n$th-stage approximation obtained by the proposed (19) or (20).

In Table 1, we list the numerical results of the absolute errors $|\psi_n - y|$ (obtained by the proposed MDMGF (20) [27])

**Table 1** The absolute error $|\psi_n - y|$ and $|\phi_n - y|$ for $n = 1, 2, 3$ of Example 4.1

| $x$ | MDMGF | | | MADM [27] | | |
|---|---|---|---|---|---|---|
| | $\lvert\psi_1 - y\rvert$ | $\lvert\psi_2 - y\rvert$ | $\lvert\psi_3 - y\rvert$ | $\lvert\phi_1 - y\rvert$ | $\lvert\phi_2 - y\rvert$ | $\lvert\phi_3 - y\rvert$ |
| 0.1 | 5.39E−05 | 3.54E−08 | 2.29E−11 | 4.40E−03 | 4.54E−05 | 2.29E−08 |
| 0.3 | 3.41E−04 | 2.23E−07 | 1.44E−10 | 8.41E−03 | 4.23E−05 | 4.74E−07 |
| 0.5 | 5.66E−04 | 3.63E−07 | 2.34E−10 | 8.66E−03 | 6.63E−05 | 3.37E−07 |
| 0.7 | 4.70E−04 | 2.92E−07 | 1.88E−10 | 7.70E−03 | 6.92E−05 | 4.78E−07 |
| 0.9 | 1.02E−04 | 5.97E−08 | 3.83E−11 | 2.02E−03 | 7.97E−06 | 5.81E−08 |

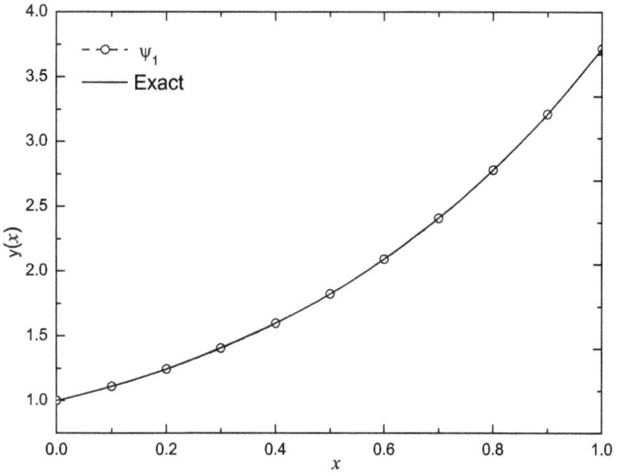


**Fig. 1** Plots of the exact and the approximate solution of Example 4.1

and $|\phi_n - y|$ [obtained by the existing MADM (10)] for $n = 1, 2, 3$. It is observed that the proposed MDMGF provides not only better numerical results but also avoids solving a sequence of transcendental equations for unknown constant.

In Fig. 1, we plot the exact solution $y$ and the approximate solution $\psi_1 = y_0 + y_1$. We observe that only two-term approximations $\psi_1 = y_0 + y_1$ coincide with the exact solution $y$.

*Example 4.2* Consider the following nonlinear fourth-order BVPs for Volterra IDE [3]

$$\left.\begin{aligned} &y^{iv}(x) = g(x) + \int_0^x e^{-t} y^2(t)dt, \quad x \in [0, 1], \\ &y(0) = y'(0) = 1, \quad y(1) = y'(1) = e, \end{aligned}\right\} \tag{31}$$

where $g(x) = 1$. The exact solution is $y(x) = e^x$.

Here, $b = 1$, $\alpha_1 = 1$, $\alpha_2 = 1$, $\alpha_3 = e$, $\alpha_4 = e$, $K(x, t) = e^{-t}$, and $f(y) = y^2(t)$. In view of the MDMGF (20), we transform the problem (31) into the following recursive scheme

$$\left.\begin{aligned} &y_0 = 1, y_1 = x + (2e-5)x^2 - (e-3)x^3 \\ &\quad + \int_0^1 G(x,\xi)\left\{g(\xi) + \int_0^\xi K(\xi,t)A_0 dt\right\}d\xi, \\ &y_j = \int_0^1 G(x,\xi)\left\{\int_0^\xi K(\xi,t)A_{j-1}dt\right\}d\xi, \ j = 2,3,\ldots \end{aligned}\right\} \tag{32}$$

where $g(\xi) = 1$ and $G(x, \xi)$ is given by the Eq. (30). The Adomian's polynomial $f(y) = y^2$ are obtained as

$$A_0 = y_0^2, \quad A_1 = 2y_0y_1, \quad A_2 = y_1^2 + 2y_0y_2, \ldots \tag{33}$$

Using (32) and (33), we obtain the solution components $y_j$ as

$$\begin{aligned} y_0 &= 1, \\ y_1 &= 1 - e^{-x} + 0.991415x^2 + 0.0114132x^3 + 0.0833333x^4, \\ y_2 &= 346.216 + 0.0625e^{-2x} - 184.271x + 43.4637x^2 \\ &\quad - 5.66685x^3 + 0.379275x^4 + \cdots \\ &\vdots \end{aligned}$$

In Table 2, we present the numerical results of the absolute errors $|\psi_n - y|$ (obtained by the proposed MDMGF) and $|\phi_n - y|$ (obtained by MADM) for $n = 1, 2, 3$. It is observed that the proposed DMGF provides not only better numerical results but also avoids solving a sequence of transcendental equations for unknown constant. In Fig. 2, the exact solution $y$ and the approximate solution $\psi_1$ are plotted. From this figure, we observe that only two-term approximations $\psi_1$ coincide with the exact one.

*Example 4.3* Consider the following nonlinear fourth-order BVPs for Volterra IDE

$$\left.\begin{aligned} &y^{(iv)}(x) = g(x) + \int_0^x e^{y(t)}dt, \quad x \in [0, 1], \\ &y(0) = \ln(4), \quad y'(0) = \frac{1}{4}, \quad y(1) = \ln(5), \quad y'(1) = \frac{1}{5} \end{aligned}\right\} \tag{34}$$

**Table 2** The absolute error $|\psi_n - y|$ and $|\phi_n - y|$ for $n = 1, 2, 3$ of Example 4.2

| $x$ | MDMGF | | | MADM [27] | | |
|---|---|---|---|---|---|---|
| | $\lvert\psi_1 - y\rvert$ | $\lvert\psi_2 - y\rvert$ | $\lvert\psi_3 - y\rvert$ | $\lvert\phi_1 - y\rvert$ | $\lvert\phi_2 - y\rvert$ | $\lvert\phi_3 - y\rvert$ |
| 0.1 | 7.44E−05 | 1.43E−05 | 4.32E−08 | 6.44E−04 | 8.48E−05 | 2.32E−06 |
| 0.3 | 4.67E−04 | 9.17E−05 | 2.75E−07 | 4.67E−03 | 9.16E−05 | 7.72E−05 |
| 0.5 | 7.63E−04 | 1.56E−04 | 4.54E−07 | 6.63E−03 | 3.66E−04 | 7.52E−05 |
| 0.7 | 6.22E−04 | 1.34E−04 | 3.72E−07 | 5.22E−03 | 4.54E−04 | 4.72E−05 |
| 0.9 | 1.32E−04 | 3.00E−05 | 7.61E−08 | 3.32E−03 | 3.00E−05 | 7.61E−06 |

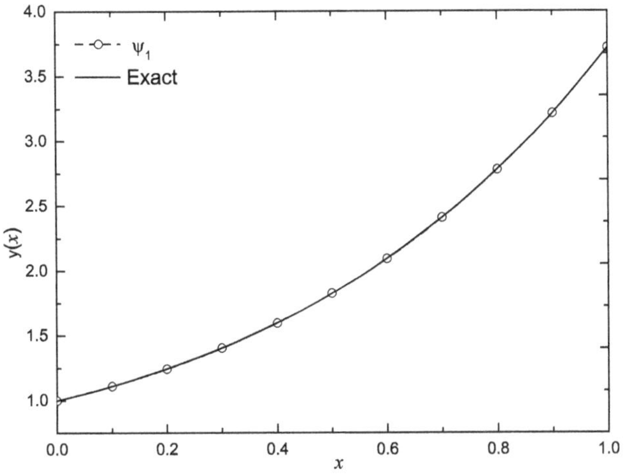


**Fig. 2** Plots of the exact and the approximate solution of Example 4.2

where $g(x) = -\frac{x^2}{2} - 4x - \frac{6}{(x+4)^4}$. The exact solution is $y(x) = \ln(4 + x)$.

Here, we have $b = 1$, $\alpha_1 = \ln(4)$, $\alpha_2 = \frac{1}{4}$, $\alpha_3 = \ln(5)$, $\alpha_4 = \frac{1}{5}$ , $K(x,t) = 1$, and $f(y) = e^{y(t)}$. According to the MDMGF (20), the problem (34) is transformed into the following recursive scheme

$$\left.\begin{aligned} y_0 &= \ln(4), \\ y_1 &= 0.25x - 0.030569x^2 + 0.0037128x^3 \\ &\quad + \int_0^1 G(x,\xi)\left\{ g(\xi) + \int_0^\xi K(\xi,t)A_0 \mathrm{d}t \right\} \mathrm{d}\xi, \\ y_j &= \int_0^1 G(x,\xi)\left\{ \int_0^\xi K(\xi,t)A_{j-1} \mathrm{d}t \right\} \mathrm{d}\xi \quad j = 2,3,\ldots \end{aligned}\right\} \tag{35}$$

where $g(\xi) = -\frac{\xi^2}{2} - 4\xi - \frac{6}{(\xi+4)^4}$ and $G(x,\xi)$ is given by the Eq. (30). The Adomian's polynomials for $f(y) = e^{y(x)}$ are calculated as

$$A_0 = e^{y_0}, \quad A_1 = e^{y_0}y_1, \quad A_2 = \frac{1}{2}e^{y_0}y_1^2 + e^{y_0}y_2(x), \ldots \tag{36}$$

Using (35) and (36), we obtain the components $y_j$ as

$$\begin{aligned} y_0 &= \ln(4), \\ y_1 &= 0.25x - 0.0354167x^2 + 0.010763x^3 - 0.000976x^4 \\ &\quad + 0.0001953x^5 + \cdots, \\ y_2 &= -8.526512 \times 10^{-14} + 7.680826 \times 10^{-13}x \\ &\quad + 0.003971x^2 - 0.005310x^3 + \cdots, \\ &\vdots \end{aligned}$$

In Table 3, we present the numerical results of the absolute errors $|\psi_n - y|$ (obtained by the proposed MDMGF) and $|\phi_n - y|$ (obtained by MADM) for $n = 1, 2, 3$. In this case, we also observe the same trend as was observed in last two examples that the proposed MDMGF gives better numerical results. Moreover, the curves of the exact solution $y$ and the approximate solution $\psi_1$ are plotted in Fig. 3. We observe that only two-term approximations $\psi_1$ and the exact solution overlap each other.

*Example 4.4* Consider the following nonlinear fourth-order BVPs for IDEs

**Table 3** The absolute error $|\psi_n - y|$ and $|\phi_n - y|$ for $n = 1, 2, 3$ of Example 4.3

| $x$ | MDMGF | | | MADM [27] | | |
|---|---|---|---|---|---|---|
| | $\lvert\psi_1 - y\rvert$ | $\lvert\psi_2 - y\rvert$ | $\lvert\psi_3 - y\rvert$ | $\lvert\phi_1 - y\rvert$ | $\lvert\phi_2 - y\rvert$ | $\lvert\phi_3 - y\rvert$ |
| 0.1 | 3.61E−05 | 1.71E−06 | 7.13E−08 | 5.61E−03 | 5.71E−04 | 5.23E−06 |
| 0.3 | 2.26E−04 | 1.09E−05 | 4.64E−07 | 6.56E−03 | 5.19E−04 | 3.44E−06 |
| 0.5 | 3.69E−04 | 1.85E−05 | 8.02E−07 | 7.69E−03 | 3.95E−04 | 9.41E−06 |
| 0.7 | 3.00E−04 | 1.57E−05 | 7.02E−07 | 5.40E−03 | 4.07E−04 | 9.42E−06 |
| 0.9 | 6.31E−05 | 3.49E−06 | 1.61E−07 | 4.31E−04 | 6.49E−04 | 5.71E−06 |

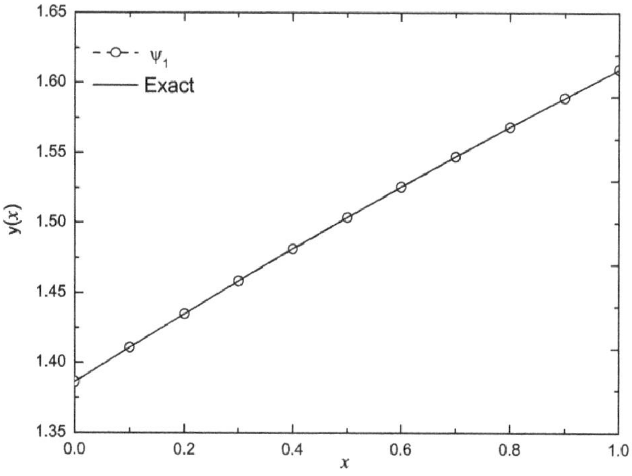


**Fig. 3** Plots of the exact and the approximate solution of Example 4.3

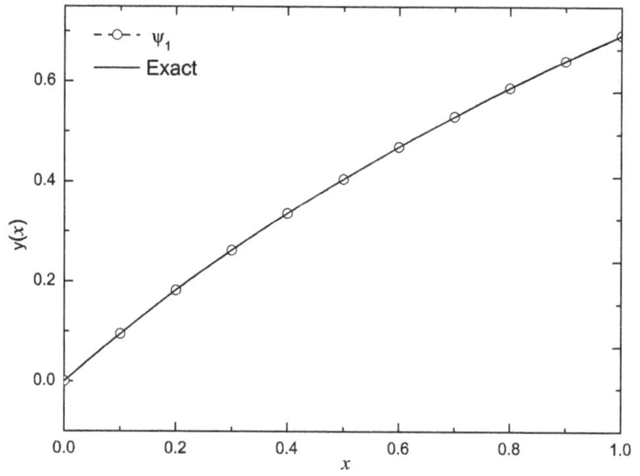


**Fig. 4** Plots of the exact and the approximate solution of Example 4.4

$$\left.\begin{aligned} &y^{(iv)}(x) = g(x) + \int_0^x e^{y(t)}dt, \quad x \in [0,1], \\ &y(0) = 0, \quad y'(0) = 1, \quad y(1) = \ln(2), \quad y'(1) = \frac{1}{2}, \end{aligned}\right\} \tag{37}$$

where $g(x) = -\frac{x^2}{2} - x - \frac{6}{(x+1)^4}$. The exact solution is $y(x) = \ln(1+x)$.

Here, $b = 1$, $\alpha_1 = 0$, $\alpha_2 = 1$, $\alpha_3 = \ln(2)$, $\alpha_4 = \frac{1}{2}$, $K(x,t) = 1$, and $f(y) = e^{y(t)}$. According to MDMGF (20), we transform the problem (37) into the following recursive scheme as

$$\left.\begin{aligned} y_0 &= 0, \\ y_1 &= x - 0.420558x^2 + 0.113706x^3 \\ &\quad + \int_0^1 G(x,\xi)\left\{g(\xi) + \int_0^1 K(\xi,t)A_0dt\right\}d\xi, \\ y_j &= \int_0^1 G(x,\xi)\left\{\int_0^1 K(\xi,t)A_{j-1}dt\right\}d\xi, \quad j = 2,3,\ldots \end{aligned}\right\} \tag{38}$$

where $g(\xi) = -\frac{\xi^2}{2} - \xi - \frac{6}{(\xi+1)^4}$. Using (36) and (38), we obtain the solution components $y_j$ as

$$\begin{aligned} y_0 &= 0, \\ y_1 &= -0.0041666x^2 + 0.005555x^3 - 0.001388x^6 + \ln(1+x), \\ y_2 &= 2.5052108 \times 10^{-8}\big(-x(332640 + 1.355178 \\ &\quad \times 10^{-6}x + 2.7963706 \times 10^{-6}x^2 \\ &\quad + 2.13444 \times 10^{-6}x^3 + 759528x^4 + 66x^6 - 33x^7 + x^{10}) \\ &\quad + 332640(1+x)^5\ln(1+x)\big), \\ &\vdots \end{aligned}$$

In Table 4, we present the numerical results of the absolute errors $|\psi_n - y|$ (obtained by the proposed DMGF) and $|\phi_n - y|$ (obtained by MADM) for $n = 1, 2, 3$. We also plot the curves of the exact $y$ and the approximate solution $\psi_1$ for $n = 1$ in Fig. 4. Like previous examples, it is observed that only two-term approximations $\psi_1$ coincide with the exact solution $y$.

## Conclusions

In this paper, we studied a reliable technique based on the decomposition method and Green's function for the numerical solution of the fourth-order BVPs for Volterra

**Table 4** The absolute error $|\psi_n - y|$ and $|\phi_n - y|$ for $n = 1, 2, 3$ of Example 4.4

| $x$ | MDMGF | | | MADM [27] | | |
|---|---|---|---|---|---|---|
| | $\|\psi_1 - y\|$ | $\|\psi_2 - y\|$ | $\|\psi_3 - y\|$ | $\|\phi_1 - y\|$ | $\|\phi_2 - y\|$ | $\|\phi_3 - y\|$ |
| 0.1 | 3.61E−05 | 5.39E−06 | 6.48E−07 | 5.61E−04 | 8.39E−05 | 8.18E−06 |
| 0.3 | 2.26E−04 | 3.45E−05 | 4.22E−06 | 4.96E−03 | 2.45E−04 | 7.25E−05 |
| 0.5 | 3.69E−04 | 5.83E−05 | 7.29E−06 | 3.69E−03 | 3.83E−04 | 8.20E−05 |
| 0.7 | 3.00E−04 | 4.93E−05 | 6.39E−06 | 5.70E−03 | 6.93E−04 | 7.39E−05 |
| 0.9 | 6.31E−05 | 1.09E−05 | 1.47E−06 | 5.31E−04 | 3.09E−04 | 4.47E−05 |

IDEs. The technique depends on constructing Green's function before establishing the recursive scheme for the solution components. The proposed technique provides a direct recursive scheme for obtaining the approximations to the solutions of BVPs. Unlike the existing ADM or the MADM, the proposed method DMGF or MDMGF avoids unnecessary evaluation of unknown constants and provides better numerical solutions. Convergence and error analysis of the proposed technique have also been discussed. The performance of the proposed recursive scheme have been examined by solving four numerical examples. It has been shown that only two-term series solution is enough to obtain an accurate approximation to the solution.

## References


1. Agarwal, R.P.: Boundary value problems for higher order integro-differential equations. Nonlinear Anal.: Theory Methods Appl. **7**(3), 259–270 (1983)
2. Hashim, I.: Adomian decomposition method for solving BVPs for fourth-order integro-differential equations. J. Comput. Appl. Math. **193**(2), 658–664 (2006)
3. Wazwaz, A.M.: A reliable algorithm for solving boundary value problems for higher-order integro-differential equations. Appl. Math. Comput. **118**(2), 327–342 (2001)
4. Yıldırım, A.: Solution of BVPs for fourth-order integro-differential equations by using homotopy perturbation method. Comput. Math. Appl. **56**(12), 3175–3180 (2008)
5. Sweilam, N.: Fourth order integro-differential equations using variational iteration method. Comput. Math. Appl. **54**(7), 1086–1091 (2007)
6. Wazwaz, A.M.: A comparison study between the modified decomposition method and the traditional methods for solving nonlinear integral equations. Appl. Math. Comput. **181**(2), 1703–1712 (2006)
7. Zhao, J., Corless, R.M.: Compact finite difference method for integro-differential equations. Appl. Math. Comput. **177**(1), 271–288 (2006)
8. Al-Mdallal, Q.M.: Monotone iterative sequences for nonlinear integro-differential equations of second order. Nonlinear Anal.: Real World Appl. **12**(6), 3665–3673 (2011)
9. Brunner, H.: On the numerical solution of nonlinear Volterra integro-differential equations. BIT Numer. Math. **13**(4), 381–390 (1973)
10. Liz, E., Nieto, J.J.: Boundary value problems for second order integro-differential equations of Fredholm type. J. Comput. Appl. Math. **72**(2), 215–225 (1996)
11. Lepik, Ü.: Haar wavelet method for nonlinear integro-differential equations. Appl. Math. Comput. **176**(1), 324–333 (2006)
12. Sweilam, N.H., Khader, M.M., Kota, W.: Numerical and analytical study for fourth-order integro-differential equations using a pseudospectral method. Math. Probl. Eng. **2013**, 1–7 (2013). doi:10.1155/2013/434753
13. Saeidy, M., Matinfar, M., Vahidi, J.: Analytical solution of BVPs for fourth-order integro-differential equations by using homotopy analysis method. Int. J. Nonlinear Sci. **9**(4), 414–421 (2010)
14. Wazwaz, A.M., Rach, R.: Comparison of the Adomian decomposition method and the variational iteration method for solving the Lane-Emden equations of the first and second kinds. Kybernetes **40**(9/10), 1305–1318 (2011)
15. Singh, R., Kumar, J., Nelakanti, G.: Approximate series solution of singular boundary value problems with derivative dependence using Green's function technique. Comput. Appl. Math. **33**(2), 451–467 (2014)
16. Singh, R., Kumar, J.: The Adomian decomposition method with Green's function for solving nonlinear singular boundary value problems. J. Appl. Math. Comput. **44**(1), 397–416 (2014)
17. Al-Hayani, W.: Adomian decomposition method with Green's function for solving tenth-order boundary value problems. Appl. Math. **5**, 1437–1447 (2014)
18. Al-Hayani, W.: Adomian decomposition method with Green's function for solving twelfth-order boundary value problems. Appl. Math. Sci. **9**(8), 353–368 (2015)
19. Singh, R., Kumar, J., Nelakanti, G.: Numerical solution of singular boundary value problems using Green's function and improved decomposition method. J. Appl. Math. Comput. **43**(1), 409–425 (2013)
20. Wazwaz, A.M.: The combined Laplace transform-Adomian decomposition method for handling nonlinear Volterra integro-differential equations. Appl. Math. Comput. **216**(4), 1304–1309 (2010)
21. Wazwaz, A.M.: A reliable study for extensions of the Bratu's problem with boundary conditions. Math. Methods Appl. Sci. **35**(7), 845–856 (2012)
22. Adomian, G., Rach, R.: Inversion of nonlinear stochastic operators. J. Math. Anal. Appl. **91**(1), 39–46 (1983)
23. Adomian, G.: Solving Frontier Problems of Physics: The Decomposition Method. Kluwer Academic Publishers, Boston (1994)
24. Duan, J.: An efficient algorithm for the multivariable Adomian polynomials. Appl. Math. Comput. **217**(6), 2456–2467 (2010)
25. Rach, R.: A convenient computational form for the Adomian polynomials. J. Math. Anal. Appl. **102**(2), 415–419 (1984)
26. Duan, J.: Convenient analytic recurrence algorithms for the Adomian polynomials. Appl. Math. Comput. **217**(13), 6337–6348 (2011)
27. Wazwaz, A.M.: A reliable modification of Adomian decomposition method. Appl. Math. Comput. **102**(1), 77–86 (1999)
28. Rach, R.C.: A new definition of the Adomian polynomials. Kybernetes **37**(7), 910–955 (2008)

# 8

# A directed tabu search method for solving controlled Volterra integral equations

Asyieh Ebrahimzadeh[1] · Raheleh Khanduzi[2]

**Abstract** This paper proposes a pseudo-spectral scheme for obtaining approximate optimal control and state. This computational scheme represents the solution of the optimal control problem (OCP) by an *m*th degree Lagrange interpolating polynomial, using Legendre nodes. Then, OCP of non-linear Volterra integral equation is transformed into an optimization problem. Directed tabu search (DTS) method is utilized to derive the solutions of the optimal control and state as well as the optimal value of the objective function. In the DTS method, two neighborhood-local search strategies based on Nelder–Mead method and adaptive pattern search are applied. In addition, a tabu list with anti-cycling rules, the so-called tabu regions and semi-tabu regions are used. In addition, diversification and intensification search schemes are employed. DTS is able to converge to the global optimum solutions of a set of numerical examples. Therefore, some good balance between the diversification and intensification ensures a faster and efficient convergence to get quality solutions. Numerical examples are provided which confirm the reliability and efficiency of the proposed method. Moreover, a comparison is made with optimal solutions obtained by the other numerical methods in the literature.

✉ Asyieh Ebrahimzadeh
ebrahimzadeh263@gmail.com

Raheleh Khanduzi
raheleh.khanduzi@gmail.com

1 Young Researchers and Elite Club, Najafabad Branch, Islamic Azad University, Najafabad, Iran

2 Department of Mathematics, Gonbad-e-Kavous University, Gonbad-e-Kavous, Iran

**Keywords** Directed tabu search method · Optimal control · Lagrange interpolating polynomial · Non-linear Volterra integral equation · Pseudo-spectral method

## Introduction

The controlled Volterra integral equations (VIE) are very momentous, because these equations arise in modelling many classes of phenomena. Many problems in economics, biology, epidemiology, and memory effects can be modeled as Volterra control problems [34]. Since the analytical methods based on the necessary conditions obtained using the Pontryagin's maximum principle and dynamic programming have less implementation ability, different numerical approaches have been devised to overcome the problems arising from the application of analytical methods [8–12, 34, 35, 40, 46].

In this paper, we have introduced a spectral approach based on collocation method to obtain optimal control of systems governed by VIE. Spectral techniques have been demonstrated to provide effective and flexible methods to solve diverse problems numerically [5–7, 44, 46, 47]. The presented method consists of reducing the OCP to a non-linear programming (NLP) by first expanding the state and control functions in terms of Lagrange interpolating polynomial with unknown coefficients. These polynomials together with numerical integration and collocation approach are utilized to convert OCP to finite-dimensional programming problem. Since the resulted finite-dimensional programming problem is large scale, there are various optimization methods which can be implemented to solve it. It is tried that to find a method which is useable and effective on solving these problems in large-scale setting. Metaheuristic methods are becoming one of the important tools for producing robust and efficient methods that compute approximate solutions of problems with high quality in reasonable computation time. Therefore, metaheuristic methods are applied for solving real-world optimization problems [18, 19, 24, 30, 33, 42, 43]. Among those metaheuristic algorithms, tabu search (TS) [18] is one of the most efficient memory-based methods. However, the original TS cannot completely exert the optimization problems. The modified versions of the TS have been presented to ameliorate its performance, for example, One of the enhanced versions of the TS is referred to as directed tabu search (DTS) [26]. DTS has shown a notable performance when used to solve non-linear continuous optimization problems. Due to its neighborhood mechanisms, DTS had quick convergence, since it detected promising search directions in the neighborhood of a global minimum. Indeed, the slow convergence of TS was overcome by incorporating local search strategies, such as the Nelder–Mead (NM) method [39] and the adaptive pattern search (APS) [25], into the main algorithm. On the other hand, the DTS has been successfully applied to various real-world optimization problems [26, 28, 29, 31, 38, 41]. Therefore, we survey application of DTS to solve OCP and use the values of the parameters that are required in the implementation of the DTS method of Hedar and Fukushima [26].

This paper is arranged as follows: OCP is defined in Sect. 2. In Sect. 3, we illustrate that how the pseudo-spectral method using Lagrange interpolation polynomials converts the continuous OCP to a finite-dimensional optimization problem. Directed tabu search algorithm is exerted to solve this non-linear optimization problem which has global convergence. In Sect. 4, basic convergence results are given. Section 5 contains numerical examples that demonstrate the efficiency and accuracy of the proposed framework. Section 6 ends this paper with a brief conclusion.

## The formulation of optimal control problem

In the present work, we focus on a numerical approach to solve the following OCP of non-linear VIE:

**Problem** $\mathcal{A}$: Specify the real-valued continuous optimal control $u^*(t)$ and the corresponding optimal state $x^*(t)$, $t \in [0, 1]$, that maximize (or minimize) the functional

$$\mathcal{J}(x,u) = \int_0^1 \mathcal{F}(t, x(t), u(t))\mathrm{d}t, \tag{1}$$

subject to state dynamics

$$x(t) = y(t) + \int_0^t \mathcal{K}(s, t, x(s), u(s))\mathrm{d}s. \tag{2}$$

It is assumed that $\mathcal{F}$, $\mathcal{K}$ and $y$ are real valued and continuously differentiable with respect to their arguments, and both $x$ and $u$ belong to Sobolev space $W^{l,\infty}$. The interval [0, 1] can be transformed to an arbitrary interval [$a$, $b$] via a proper change of variable. It is also supposed that the optimal control of this problem is unique. Analytical discussions about the existence and uniqueness for optimal control of systems governed by non-linear VIE 2 can be found in [1, 27] and references there in.

## The method of discretization

Pseudo-spectral approaches are powerful tools that are frequently employed in various fields of numerical analysis. These methods have a higher order of accuracy in the case of smooth solution of any considered problem. In this section, implementation of the pseudo-spectral method on problem $\mathcal{A}$ is given. At first, the control and state functions are approximated in terms of $M$th degree Lagrange interpolating polynomials of the form

$$x(t) \simeq \mathcal{X}^M(t) = \sum_{k=0}^{M-1} \bar{x}_k^M \psi_k(t), \tag{3}$$

$$u(t) \simeq \mathcal{U}^M(t) = \sum_{k=0}^{M-1} \bar{u}_k^M \psi_k(t),$$

where $\psi_k(t)$ for $k = 0, \ldots, M-1$ are the Lagrange interpolating polynomials defined by

$$\psi_k(t) = \prod_{i=0, i\neq k}^{M-1} \left(\frac{t - \tau_i}{\tau_k - \tau_i}\right), \tag{4}$$

in which $\tau_i$, $i = 0, 1, \ldots, M-1$ can be Gauss–Legendre (GL) or Gauss–Chebyshev (GC) nodes. GC nodes are the roots of Chebyshev polynomial of the first kind in $[-1, 1]$ [36]. These nodes in the interval [0, 1] are defined by

$$\tau_i = \frac{1}{2} + \frac{1}{2}\cos\left(\frac{2i+1}{2M}\pi\right), \quad i = 0, 1, \ldots, M-1. \tag{5}$$

GL nodes are the zeros of Legendre polynomial. Explicit formulas for GL nodes are not known. However, they can be computed numerically using existing subroutines [14]. The best choice of interpolation points to ensure uniform convergence is GC nodes (see Section 2.2 [36]). It can be verified that

$$\psi_k(\tau_i) = \delta_{ki} = \begin{cases} 1 & i = k, \\ 0 & i \neq k. \end{cases} \tag{6}$$

The approximation process of problem $\mathcal{A}$ includes the discretization of both the cost function and the controlled integral equation constraint.

### Controlled integral equation discretization

We discretize Eq. 2 using shifted GC nodes, $t_p$, $p = 0, 1, \ldots, M-1$, as follows:

$$x(t_p) = y(t_p) + \int_0^{t_p} \mathcal{K}(s, t_p, x(s), u(s))\mathrm{d}t. \tag{7}$$

We should notice that GC nodes should be transformed into the interval [0, 1]. By substituting 3 in 7, we obtain

$$\mathcal{X}^M(t_p) \approx y(t_p) + \int_0^{t_p} \mathcal{K}(s, t_p, \mathcal{X}^M(s), \mathcal{U}^M(s))\mathrm{d}t. \tag{8}$$

The GL quadrature formula is utilized to approximate the integral term in Eq. 8. For this purpose, this change of variable must be made with the following form:

$$b_p(\tau) = \frac{t_p}{2}(\tau + 1). \tag{9}$$

Then, by applying GL quadrature for approximating the integral, we derive

$$\mathcal{X}^M(t_p) \approx \tag{10}$$

$$y(t_p) + \frac{t_p}{2}\sum_{j=0}^{M-1} w_j \mathcal{K}\left(b_p^j, t_p, \mathcal{X}^M(b_p^j), \mathcal{U}^M(b_p^j)\right),$$

$$p = 0, 2, \ldots, M-1,$$

where $b_p^j = b_p(\tau_j)$, and $\tau_j$s are the GL nodes, in the interval $[-1, 1]$, and $w_j$s are the corresponding weights. The quadrature weight, $w_j$, can be obtained from the following relation:

$$w_j = \frac{2}{(1-\tau_j^2)[\mathcal{L}'_M(\tau_j)]^2}, \quad j = 0, \ldots, M-1. \tag{11}$$

In Eq. 11, $\mathcal{L}_M$ is Legendre polynomial of degree $M$; For more details about these polynomials, see Tohidi and Samadi [48]. Finally, the controlled Volterra integral Eq. 2 is reduced to $M$ non-linear algebraic equations given in Eq. 10.

### Cost functional discretization

For approximating the cost functional stated in Eq. 1, we utilize the GL quadrature after the proper interval transformation

$$\int_0^1 \mathcal{F}(t, x(t), u(t))$$
$$\approx \frac{1}{2}\int_{-1}^{1} \mathcal{F}\left(\frac{\tau+1}{2}, \mathcal{X}^M\left(\frac{\tau+1}{2}\right), \mathcal{U}^M\left(\frac{\tau+1}{2}\right)\right)\mathrm{d}\tau$$
$$\approx \sum_{j=0}^{M-1} w'_j \mathcal{F}\left(\tau'_j, \mathcal{X}^M(\tau'_j), \mathcal{U}^M(\tau'_j)\right), \tag{12}$$

where $w'_j = \frac{1}{2}w_j$ and $\tau'_j = \frac{\tau_j+1}{2}$, and $\tau_j$ and $w_j$ are GL nodes and weights stated in Eq. 11.

### The discretized optimization problem

Finally, problem $\mathcal{A}$ is approximated by the following NLP:

$$\min \mathcal{J}^M(X, U) \tag{13}$$

subject to

$$\mathcal{K}_p(X, U) = -y(t_p), \quad p = 0, 1, \ldots, M-1, \tag{14}$$

where $X = (\overline{x}_0^M, \overline{x}_1^M, \ldots, \overline{x}_{M-1}^M)$ and $U = (\overline{u}_0^M, \overline{u}_1^M, \ldots, \overline{u}_{M-1}^M)$ are the unknown parameters of our discrete problem, and

$$\mathcal{J}^M(X, U) := \sum_{j=0}^{M-1} w'_j \mathcal{F}(\tau'_j, \mathcal{X}^M(\tau'_j), \mathcal{U}^M(\tau'_j)),$$

and

$$\mathcal{K}_p(X, U) := \frac{t_p}{2}\sum_{j=0}^{M-1} w_j \mathcal{K}(b_p^j, t_p, \mathcal{X}^M(b_p^j), \mathcal{U}^M(b_p^j)) - \mathcal{X}^M(t_p), \tag{15}$$

$$p = 0, 1, \ldots, M-1.$$

As shown above, the resulted discrete problem can be reformulated as the following non-linear programming problem:

**Problem** $\mathcal{A}^M$: Find (*X*, *U*) that minimize

$$\mathcal{J}^M(X, U) = \sum_{j=0}^{M-1} \mathcal{F}(\tau'_j, \mathcal{X}^M(\tau'_j), \mathcal{U}^M(\tau'_j)) w'_j, \tag{16}$$

subject to

$$\mathcal{K}_p(X, U) + y(t_p) = 0, \quad p = 0, 1, \ldots, M-1. \tag{17}$$

When the continuous problem $\mathcal{A}$ is discretized, the infinite-dimensional problem $\mathcal{A}$ is reduced to the finite-dimensional non-linear optimization problem $\mathcal{A}^M$. Many well-developed NLP techniques can be used to solve this problem [32]. The method has been used to solve the non-linear constrained optimization problem which is based on DTS method.

## Directed tabu search

DTS is one of the recent development metaheuristics proposed for combinatorial optimization problems. In the DTS method, three search procedures were used: Exploration, Diversification, and Intensification. In the Exploration Search, a local search procedure is applied to produce trial moves, based on the Nelder–Mead method (NM) and the adaptive pattern search method (APS). Besides, memory elements called Tabu Region (TR), Semi-TR, and a multi-ranked Tabu List (TL) were presented to obtain anti-cycling rules. Visited Regions List (VRL) was also presented for the Diversification Search to diversify the search to unvisited areas of the solution region. At last, supposing that one of the best points generated by the Exploration and Diversification Searches was close to a global solution, the Intensification Search was utilized again as the final step to rectify the best solutions visited so far. Indeed, the Exploration and Diversification search processes were congregated to prepare the DTS main loop and were repeated until the termination conditions were satisfied. Therewith, the Exploration Search process was comprised as an inner loop within the diversification loop [26].

Here, the DTS method applies pattern search to get better performance. The main operators used in the DTS are:

*Memory elements* Memory elements are used to provide anti-cycling rules and to diversify the search to unvisited areas of the solution space.

- *TL* Due to the recency and objective function values, solutions are ranked and saved in the TL.
- *VRL* The centers of the visited regions and the frequency of them are saved in the VRL for generating new diverse solutions.
- *TR and Semi-TR* No new trial solution is allowed to be generated in a TR. A Semi-TR is a neighbor region around TR. New trial solutions are generated in a Semi-TRs, so that returning back to a visited TR is avoided.

*APS* In the APS method, the approximate descent direction method is used to find promising directions.

*Exploration search* In exploration search process, trial solutions are generated based on neighborhood region structures and the APS procedure.

*Diversification search* While the exploration search is taking a long time to improve the current solution, then the VRL is called to generate new diverse solutions.

*Intensification search* In this phase, NM method IS applied starting from some of the best solution found till now.

In the DTS method, the exploration, diversification, and intensification search procedures are applied to develop the search process on a global strategy. Indeed, these search procedures are used to get a vast exploration and an extensive diversification. The DTS has been used in this work is stated in Algorithm 1.

**Algorithm 1. Directed Tabu Search**

1. **Initialization.** Select an initial solution, and set the TL and the VRL to be empty.

2. **Exploration-Diversification Search** (Main Loop). Repeat this main loop steps until a predefined number of consecutive main iterations fail to obtain improvement or the main loop iteration counter exceeds a predefined maximum number.

2.1. **Exploration Search** (Inner Loop). Repeat this inner loop steps until a predefined number of consecutive inner iterations fail to obtain improvement or the inner loop iteration counter exceeds a predefine maximum number.

2.1.1. **Search Directions.** Generate search directions based on the APS strategy. If the current solution lies in a Semi-TR, then generate search directions to point outside the TR.

2.1.2. **Neighborhood Search.** Generate neighborhood trial moves based on tabu restriction and aspiration criterion. While a better movement is found during this process, stop generating points and go to Step 2.1.4.

2.1.3. **Local Search.** Apply a predefine number of iterations of the NM method starting from the current solution.

2.1.4. **Parameter Update.** Replace the element with the smallest membership value in TL with the current solution and re-rank the TL elements, and update the VRL.

2.2. **Diversification Search.** Generate a diverse solution, and update the TL and VRL.

3. **Intensification Search.** Apply a complete NM method starting from some elite solutions in the TL.

## Theoretical consideration

In numerical schemes, discussion on convergence of an propounded algorithm is essential. The convergence of discretization methods for OCP is a topic of active research [4, 13, 16, 16, 17, 20–22]. In this section, we investigate the following questions:

1. Does the discretized problem $\mathcal{A}^M$ admit a feasible solution?
2. Does a sequence of discretized optimal solutions of problems $\mathcal{A}^M$ converge to the optimal solution of problem $\mathcal{A}$?

In Theorem 1, it has been proven that the feasibility of discretized problems $\mathcal{A}^M$ is guaranteed if a solution to the continuous-time problem $\mathcal{A}$ exists? It should be noted that the convergence analysis of the proposed scheme can be done using similar idea provided in [48]. First, we need the following lemma and definition.

**Definition 1** [21] A function $\mathcal{Q}:[0,1]\to\Re$ belongs to Sobolev space, $W^{l,p}$, if its $j$th weak derivative, $\mathcal{Q}^{(j)}$, lies in $L^p[0,1]$ for all $0\le j\le l$.

**Lemma 1** [20] *Given any function $\mathcal{Q}(t)\in W^{l,\infty}, t\in[0,1]$, there is a polynomial $\mathcal{P}^M(t)$ of degree M or less, such that*

$$\left|\mathcal{Q}(t)-\mathcal{P}^M(t)\right|\le CC_\circ M^{-l},\quad \forall t\in[0,1], \tag{18}$$

*where C is a constant independent of M and $C_\circ=\|\mathcal{Q}\|_{W^{l,\infty}}$.*

*Remark 1* The computational interval can be transformed from [0, 1] to [*a*, *b*] via an affine transformation.

**Theorem 1** (Existence) [48] *Let (x(t), u(t)) be a feasible solution for problem $\mathcal{A}$ and x(t) and u(t) belong to $W^{l,\infty}$ with $l\ge 2$. Then, there exists a positive integer $M_1$, such that, for any $M>M_1$, problem $\mathcal{A}^M$ has a feasible solution, $(\mathcal{X}^M(t_p),\mathcal{U}^M(t_p))$ and this feasible solution also satisfies in the following inequalities*:

$$\|x(t_p)-\mathcal{X}^M(t_p)\|_\infty\le L_1(M-1)^{1-l}\quad p=0,1,\ldots,M-1 \tag{19}$$

$$\|u(t_p)-\mathcal{U}^M(t_p)\|_\infty\le L_2(M-1)^{1-l}\quad p=0,1,\ldots,M-1 \tag{20}$$

*in which $t_p$ is Gauss–Legendre node or Gauss–Chebyshev node, and $L_1$ and $L_2$ are positive constant independent of M.*

We establish in Theorem 1 the existence of feasible solutions for discretized problem $\mathcal{A}^M$. The next theorem demonstrates that there exists a sequence of optimal solutions of problems $\mathcal{A}^M$ converging to an optimal solution of problem $\mathcal{A}$? Let $X^*=[\overline{x}_0^{M*},\overline{x}_1^{M*},\ldots,\overline{x}_{M-1}^{M*}]^T$ and $U^*=[\overline{u}_0^{M*},\overline{u}_1^{M*},\ldots,\overline{u}_{M-1}^{M*}]^T$ be the optimal solutions of the problem $\mathcal{A}^M$. The approximate optimal control and state are

$$\mathcal{X}^{M^*}(t)=\sum_{i=0}^{M-1}\overline{x}_i^{M*}\psi_i(t),\quad \mathcal{U}^{M^*}(t)=\sum_{i=0}^{M-1}\overline{u}_i^{M*}\psi_i(t). \tag{21}$$

The next theorem will be demonstrated the convergence of the sequence $\{(\mathcal{X}^{M^*}(t_p),\mathcal{U}^{M^*}(t_p))\}_{M=M_1}^{\infty}$.

**Theorem 2** (Convergence of optimal solutions) [48] *Let $\{(\mathcal{X}^{M^*}(t_p),\mathcal{U}^{M^*}(t_p)),0\le p\le M-1\}_{M=M_1}^{\infty}$ be a sequence of optimal solutions to problem $\mathcal{A}$. If $\lim_{M\to\infty}\mathcal{X}^{M^*}(t_0)=\tilde{x}_0$, $\lim_{M\to\infty}\mathcal{U}^{M^*}(t_0)=\tilde{u}_0$ and the function sequence $\{(\mathcal{X}^{M^{*'}}(t),\mathcal{U}^{M^{*'}}(t)),i=0,1,2,\ldots,M-1\}_{M=M_1}^{\infty}$ has a subsequence that uniformly converges to the continuous functions $\{p(t),q(t)\}$ on [0, 1], then $\tilde{x}(t)=\int_{t_0}^{t}p(\tau)d\tau+\tilde{x_0}$ and $\tilde{u}(t)=\int_{t_0}^{t}q(\tau)d\tau+\tilde{u_0}$ are the optimal solution of problem $\mathcal{A}$.*

## Illustrative examples

In this section, two numerical examples are considered to illustrate the effectiveness of the pseudo-spectral method along with DTS scheme. The numerical results obtained from our propounded method is also compared with the results in other works in [35, 48]. The following numerical implementation is performed using Mathematica.

*Example 1* Consider the minimization of functional

$$\mathcal{J}=\int_0^1(x(t)-\sin(t))^2+(u(t)-t)^2\mathrm{d}t, \tag{22}$$

subject to Borzabadi et al. [13]

$$x(t)=y(t)+\int_0^t u(s)(x(s)+t)\mathrm{d}s, \tag{23}$$

where $y(t)=t\cos(t)-\frac{1}{2}t^3$.

The optimal value of cost function is $\mathcal{J}^*=0$. The optimal control $u^*(t)$ and corresponding optimal state $x^*(t)$ are as follows:

$$\begin{cases}x^*(t)=\sin(t),\\ u^*(t)=t.\end{cases}$$

The approximate solutions for both state and control functions together with the exact solutions are depicted in Figs. 1 and 2. The blue lines in the figures indicate the exact optimal solutions and the red lines are the approximate solutions. The absolute errors of cost functional value, $E^{\mathcal{J}}$, for example 1, are compared with those obtained from the methods Maleknejad and Almasieh [35] and Tohidi and Samadi [48] in Table 1.

*Example 2* Consider the minimization of the cost functional

$$\mathcal{J}=\int_0^1(x(t)-e^{-t^2})^2+(u(t)-t)^2\mathrm{d}t,$$

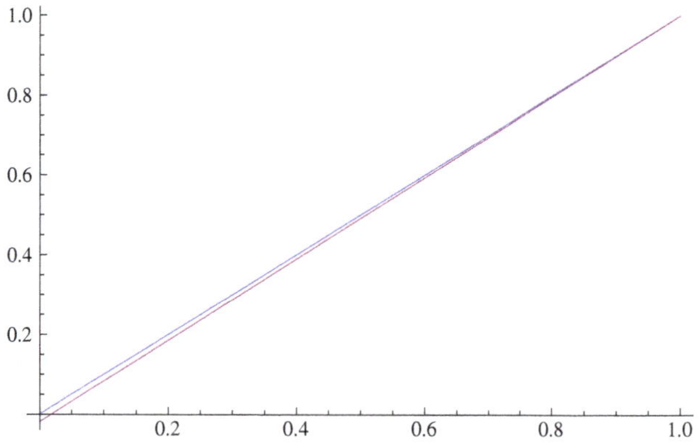


**Fig. 1** Exact and approximate optimal control

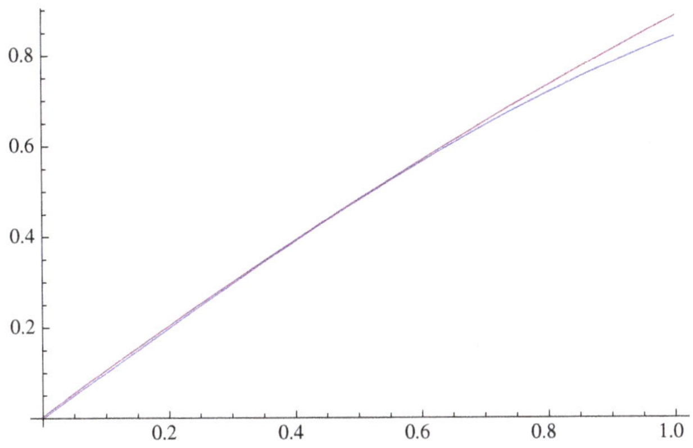


**Fig. 2** Exact and approximate optimal state

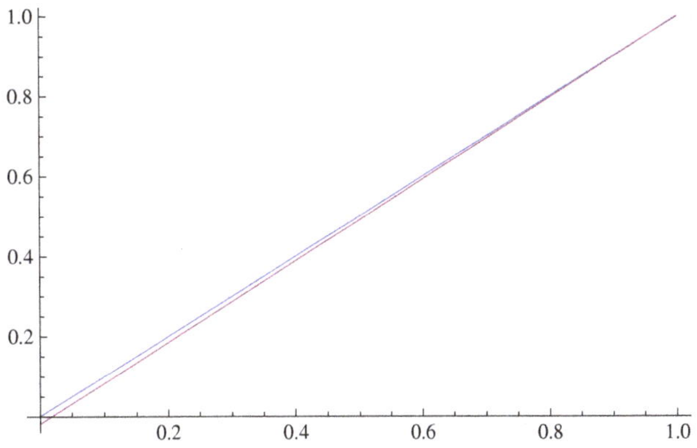


**Fig. 3** Exact and approximate optimal control

**Table 1** Numerical results of Example 1

| $M$ | $E^{\mathcal{J}}$(DTS) | $E^{\mathcal{J}}$ [35] | $E^{\mathcal{J}}$ [48] |
|---|---|---|---|
| 2 | 3.1760E−15 | 2.3296E−07 | 5.9668E−06 |
| 3 | 5.8431E−17 | 5.6973E−09 | 4.9400E−08 |
| 4 | 8.7651E−20 | 3.3217E−12 | 4.2260E−11 |
| 5 | 1.4916E−21 | 6.8699E−15 | 1.9397E−13 |
| 6 | 1.5157E−23 | 1.2832E−17 | 8.2905E−17 |

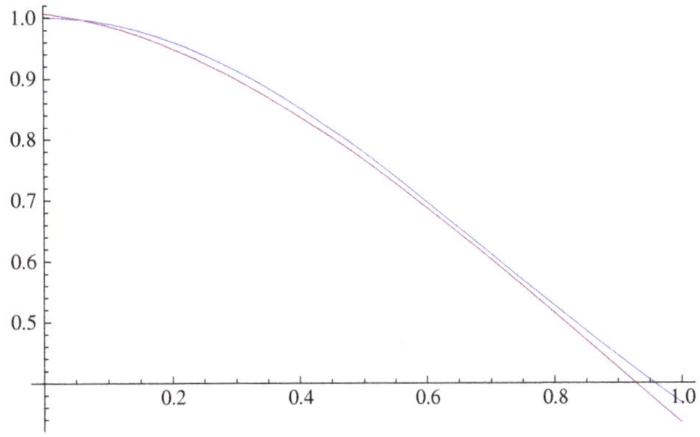


**Fig. 4** Exact and approximate optimal state

subject to the controlled Volterra integral

$$x(t) = y(t) - \int_0^t u(s)tx(s)\mathrm{d}s.$$

where $y(t) = e^{-t^2} + \frac{t(1-e^{-t^2})}{2}$. This problem has the optimal solutions $u^*(t) = t$ and $x^*(t) = e^{-t^2}$. Figures 3 and 4 show the exact and approximate optimal control and state. Table 2 gives the results obtained from our proposed scheme and the methods in Maleknejad and Almasieh [35] and Tohidi and Samadi [48].

At the end, we answer a natural question: are there advantages of the proposed pseudo-spectral method along with DTS method compared with the existing ones? To answer this, we summarize what we have observed from numerical experiments and theoretical results as follows:

- As seen in Examples 1 and 2, the proposed method with a few collocation points has satisfactory results with respect to other methods.
- The proposed orthogonal collocation method leads to rapid convergence, as the number of collocation points increases. The main reason for this fast convergence is that the DTS explore the global search space using local information about promising search direction.

**Table 2** Numerical results of Example 2

| $M$ | $E^{\mathcal{J}}$ (DTS) | $E^{\mathcal{J}}$ [35] | $E^{\mathcal{J}}$ [48] |
|---|---|---|---|
| 2 | 4.7431E−18 | 9.6934E−07 | 5.8680E−06 |
| 3 | 4.8690E−21 | 6.4649E−08 | 5.2080E−07 |
| 4 | 1.1847E−23 | 2.0778E−10 | 1.1391E−09 |
| 5 | 9.1304E−26 | 8.5037E−12 | 1.1597E−10 |
| 6 | 7.4725E−30 | 3.4206E−14 | 2.0071E−13 |

## Conclusion

In this study, an advanced numerical PS method has been proposed for solving optimal control of Volterra integral equation by means of Lagrange polynomials via collocation method. The problem has been reduced to a finite-dimensional parametric optimization, and there exist many effective algorithms which can be applied to solve the NLP. To solve this problem, directed tabu search (DTS) method is applied. This algorithm comes from the incorporation of a modified Tabu Search (TS) with two local optimizer methods. The local optimizers are direct search methods, which aim to find a better solution neighbor to the present best solution by conducting the intensification and diversification along the move direction of the best solution in the area. The results indicate that using DTS could help to get better solutions of problems than available approximate methods. Illustrative examples have shown the validity, applicability, and efficiency of the proposed method. The method is in the case of optimal control of systems governed by VIE which is applicable in the field of practical science and engineering [34].

**Acknowledgments** The first author would like to appreciate the Young Researchers and Elite Club, Islamic Azad University of Najafabad Branch for supporting this research. The second author would like to thank the research council of Gonbad-e-Kavous University for supporting this research work.

## References

1. Angell, T.S.: On the optimal control of systems governed by nonlinear Volterra equations. J. Optim. Theory Appl. **19**, 29–45 (1976)
2. Belbas, S.A.: A new method for optimal control of Volterra integral equations. Appl. Math. Comput. **189**, 1902–1915 (2007)
3. Belbas, S.A.: A reduction method for optimal control of Volterra integral equations. Appl. Math. Comput. **197**, 880–890 (2008)
4. Betts, J.T., Biehn, N., Campbell, S.L.: Convergence of nonconvergent IRK discretizations of optimal control problems with state inequality constraints. SIAM J. Sci. Comput. **23**, 1981–2007 (2002)
5. Bhrawy, A.H.: A Jacobi spectral collocation method for solving multi-dimensional nonlinear fractional sub-diffusion equations. Numer. Algorithms (2016). doi:10.1007/s11075-015- 0087-2
6. Bhrawy, A.H.: An efficient Jacobi pseudospectral approximation for nonlinear complex generalized Zakharov system. Appl. Math. Comput. **247**, 30–46 (2014)
7. Bhrawy, A.H., Zaky, M.A.: A method based on the Jacobi tau approximation for solving multi-term time-space fractional partial differential equations. J. Comput. Phys. **281**, 876–895 (2015)
8. Bhrawy, A.H., Ezz-Eldien, S.S.: A new Legendre operational technique for delay fractional optimal control problems. Calcolo (2016). doi:10.1007/s10092-015-0160-1
9. Bhrawy, A.H., Doha, E.H., Machado, J.A.T., Ezz-Eldien, S.S.: An efficient numerical scheme for solving multi-dimensional fractional optimal control problems with a quadratic performance index. Asian J. Control **17**, 2389–2402 (2015)
10. Bhrawy, A.H., Abdelkawy, M.A., Machado, J.T., Amin, A.Z.M.: Legender–Gauss–Lobatto collocation method for solving multidimensional fredholm integral equations. Computers Math. Appl. (2016). doi:10.1016/j.camwa.2016.04.011
11. Bhrawy, A.H., Tohidi, E., Soleymani, F.: A new Bernoulli matrix method for solving high-order linear and nonlinear Fredholm integro-differential equations with piecewise intervals. Appl. Math. Comput. **219**, 482–497 (2012)
12. Bhrawy, A.H., Doha, E.H., Baleanu, D., Ezz-Eldien, S.S., Abdelkawy, M.A.: An accurate numerical technique for solving fractional optimal control problems. Proc. Rom. Acad. A **16**, 47–54 (2015)
13. Borzabadi, A.H., Abbasi, A., Fard, O.S.: Approximate optimal control for a class of nonlinear Volterra integral equations. J. Am. Sci **6**, 1017–1021 (2010)
14. Canuto, C., Hussaini, M.Y., Quarteroni, A., Zang Jr., T.A.: Spectral Methods in Fluid Dynamics. Springer, New York (1988)
15. Dontchev, A.L.: Discrete Approximations in Optimal Control. Springer, New York (1996)
16. Dontchev, A.L., Hager, W.W.: The Euler approximation in state constrained optimal control. Math. Comput. **70**, 173–203 (2001)
17. Freud, G.: Orthogonal Polynomials. Pergamon Press, Elmsford (1971)
18. Glover, F.: Future paths for integer programming and links to artificial intelligence. Computers Oper. Res. **13**(5), 533–549 (1986)
19. Golden, B.L., Wasil, E.A., Kelly, J.P., Chao, I.M.: Fleet management and logistics. The impact of metaheuristics on solving the vehicle routing problem: algorithms, problem sets, and computational results, pp. 33–57. Kluwer Academic Publishers, Boston (1998)
20. Gong, Q., Ross, I.M., Kang, W., Fahroo, F.: On the pseudospectral covector mapping theorem for nonlinear optimal control. In: 45th IEEE Conference on Decision and Control, pp. 2679–2686 (2006)
21. Gong, Q., Ross, I.M., Kang, W., Fahroo, F.: Connections between the covector mapping theorem and convergence of pseudospectral methods for optimal control. Comput. Optim. Appl. **41**, 307–335 (2008)
22. Hager, W.W.: Runge–Kutta methods in optimal control and the transformed adjoint system. Numer. Math. **87**, 247–282 (2000)
23. Hassan, M.Y., Suharto, M.N., Abdullah, M.P., Majid, M.S., Hussin, F.: Application of particle swarm optimization for solving optimal generation plant location problem. Int. J. Electr. Electr. Syst. Res. **5**, 47–56 (2012)
24. Haupt, R.L., Haupt, S.E.: Practical Genetic Algorithms. Wiley, New York (2004)
25. Hedar, A., Fukushima, M.: Heuristic pattern search and its hybridization with simulated annealing for nonlinear global optimization. Optim. Methods Softw. **19**, 291–308 (2004)
26. Hedar, A., Fukushima, M.: Tabu search directed by direct search methods for nonlinear global optimization. Eur. J. Oper. Res. **170**, 329–349 (2006)
27. Kamien, M.I., Muller, E.: Optimal control with integral state equations. Rev. Econ. Stud. **43**, 469–473 (1976)
28. Karim, F., Seddiki, O.: Synthesis of chirped apodized fiber Bragg grating parameters using direct tabu search algorithm: application to the determination of thermo-optic and thermal expansion coefficients. Optics Commun. **283**, 2109–2116 (2010a)
29. Karim, F., Seddiki, O.: Direct tabu search algorithm for the fiber Bragg grating distributed strain sensing. J. Optics **12**, 095401 (2010b)
30. Khabbazi, A., Atashpaz-Gargari, E., Lucas, C.: Imperialist competitive algorithm for minimum bit error rate beam forming. Int. J. Bio-Inspired Comput. **1**, 125–133 (2009)
31. Khanduzi, R., Peyghami, M.R., Maleki, H.R.: Solving continuous single-objective defensive location problem based on hybrid directed tabu search algorithm. Int J. Adv. Manuf. Technol. **76**, 295–310 (2015)
32. Luenberger, D.G., Ye, Y.: Linear and Nonlinear Programming, 3rd edn. Springer, New York (2008)
33. Mahmoudi, S., lotfi, S.: Modified cuckoo optimization algorithm (MCOA) to solve graph coloring problem. Appl. Soft Comput. **33**, 48–64 (2015)
34. Maleknejad, K., Ebrahimzadeh, A.: The use of rationalized Haar wavelet collocation method for solving optimal control of Volterra integral equations. J. Vib. Control **21**, 1958–1967 (2013)
35. Maleknejad, K., Almasieh, H.: Optimal control of Volterra integral equations via triangular functions. Math. Comput. Model. **53**, 1902–1909 (2011)

36. Mason, J.C., Handscomb, D.C.: Chebyshev Polynomials. CRC Press, Boca Raton (2003)
37. Medhin, N.G.: Optimal processes governed by integral equations. J. Math. Anal. Appl. **120**, 1–12 (1986)
38. Mimis, A.: A geographical information system approach for evaluating the optimum location of point-like facilities in a hierarchical network. Geo-spatial Inf. Sci. **15**(1), 37–42 (2012)
39. Nelder, J.A., Mead, R.: A simplex method for function minimization. Comput. J. **7**, 308–313 (1965)
40. Peyghami, M.R., Hadizadeh, M., Ebrahimzadeh, A.: Some explicit class of hybrid methods for optimal control of Volterra integral equations. J. Inf. Comput. Sci **7**, 253–266 (2012)
41. Ramadas, G.C.V., Fernandes, E.M.G.P.: Self-adaptive combination of global tabu search and local search for nonlinear equations. Int. J. Computer Math. **89**(1314), 1847–1864 (2012)
42. Rao, R.V., Savsani, V.J., Vakharia, D.P.: Teachinglearning-based optimization: an optimization method for continuous non-linear large scale problem. Inf. Sci. **183**, 1–15 (2012)
43. Ribeiro, C.C., Hansen, P.: Essays and Surveys in Metaheuristics. Kluwer Academic Publishers, Boston (2002)
44. Ross, I.M., Fahroo, F.: Convergence of pseudo-spectral discretizations of optimal control problems. In: Proceedings of 40th IEEE Conference on Decision and Control, vol. 4, pp. 3175–3177 (2001)
45. Schmidt, W.H.: Volterra integral processes with state constraints. SAMS **9**, 213–224 (1992)
46. Tohidi, E., Lotfi Noghabi, S.: An efficient Legendre pseudo-spectral method for solving nonlinear quasi bang-bang optimal control problems. J. Appl. Math. Stat. Inf. **8**, 73–85 (2012)
47. Tohidi, E., Pasban, A., Kilicman, A., Lotfi Noghabi, S.: An efficient pseudo-spectral method for solving a class of nonlinear optimal control problems. Abstr. Appl. Anal. **7** (2013) (Article ID 357931)
48. Tohidi, E., Samadi, O.R.N.: Optimal control of nonlinear Volterra integral equations via Legendre polynomials. IMA J. Math. Control Inf. **30**, 67–83 (2013)

9

# Weighted approximation by double singular integral operators with radially defined kernels

Gumrah Uysal[1] · Ertan Ibikli[2]

**Abstract** In this study, we present some results on the weighted pointwise convergence of a family of singular integral operators with radial kernels given in the following form:

$$L_\lambda(f;x,y) = \iint_{\mathbb{R}^2} f(t,s)H_\lambda(t-x,s-y)\mathrm{d}s\,\mathrm{d}t, \quad (x,y)\in\mathbb{R}^2, \quad \lambda\in\Lambda,$$

where $\Lambda$ is a set of non-negative numbers with accumulation point $\lambda_0$, and the function $f$ is measurable on $\mathbb{R}^2$ in the sense of Lebesgue.

**Keywords** Generalized Lebesgue point · Radial kernel · Pointwise convergence · Rate of convergence

**Mathematics Subject Classification** Primary: 41A35 · Secondary: 41A25 · 45P05 · 47A58 · 47B65

## Introduction

The approximation of functions by integral operators with positive definite kernels is widely used in many branches of mathematics, such as approximation theory, representation theory, theory of differential equations, Fourier analysis, and singular integral theory. Besides, it is well known that Fourier analysis is used and has many applications in medicine and engineering; more specifically, magnetic resonance imaging (MRI) and fingerprint identification are the familiar examples in those areas, respectively. Particulary, the great importance of singular integral theory, which originated in Fourier analysis, must be emphasized here. In the construction stage of Fourier series of the functions, the following integral is obtained at the end of consecutive operations, that is

$$L_\lambda(f;x) = \int_{-\pi}^{\pi} f(t)K_\lambda(t,x)\mathrm{d}t, \quad x\in[-\pi,\pi], \quad \lambda\in\mathbb{N}, \qquad (1)$$

where $K_\lambda(t)$ denotes a kernel satisfying some conditions similar to usual approximate identities. Singular integrals consist of the different settings of the operators of type (1) with an appropriate singularity assumption on the kernel. For mentioned applications and related other applications concerning the usage of approximation theory in natural and applied sciences, the authors refer to [1–11].

The pointwise approximation problem may be seen as a problem of representing functions at some characteristic points, such as point of continuity, Lebesgue point, generalized Lebesgue point, and $\mu$-generalized Lebesgue point. $\mu$-generalized Lebesgue point, among others, comes to the fore. Actually, depending on the choice of the function $\mu(t)$, definitions of the remaining points can be easily obtained. In practice, there are two major investigation methods related to the pointwise convergence of integral-type operators, such as operators of type (1).

The first method can be described as fixing the variable $x$ within the operator of type (1). In other words, we pick a point in the domain of integration and it represents all other

✉ Gumrah Uysal
guysal@karabuk.edu.tr

Ertan Ibikli
ertan.ibikli@ankara.edu.tr

1 Department of Computer Technologies, Division of Technology of Information Security, Karabuk University, Karabuk 78050, Turkey

2 Faculty of Science, Department of Mathematics, Ankara University, Anadolu, Ankara 06100, Turkey

points of the same kind. Therefore, the convergence of the operator is investigated almost everywhere in the domain of integration. This method is used in many works, such as Rydzewska [12], Mamedov [13], Butzer and Nessel [14], and Uysal et al. [15].

The second method also known as Fatou-type convergence can be described as restricting the pointwise convergence to some subsets of the plane [16]. Therefore, a sensitive convergence analysis is obtained. For some studies related to Fatou-type convergence, the authors refer to [16–18].

Now, we summarize some of the works in which this method is harnessed.

In [19], Taberski, who indicated the importance of singular integrals in Fourier series in his works, investigated the pointwise approximation of periodic and integrable functions on $\langle -\pi, \pi \rangle$, where $\langle -\pi, \pi \rangle$ is an arbitrary closed, semi-closed, or open interval. The work used a two parameter family of singular integral operators of the form:

$$L_\lambda(f;x) = \int_{-\pi}^{\pi} f(t) K_\lambda(t-x)\,\mathrm{d}t, \quad x \in \langle -\pi, \pi \rangle, \quad \lambda \in \Lambda, \tag{2}$$

where $K_\lambda : \mathbb{R} \to \mathbb{R}_0^+$ denotes a family of periodic kernels satisfying suitable conditions, and $\Lambda$ is a given set of non-negative numbers with accumulation point $\lambda_0$.

Taberski [20], which gave enthusiasm to researchers, advanced his analysis to double singular integral operators of the form:

$$L_\lambda(f;x,y) = \iint_Q f(t,s) H_\lambda(t-x, s-y)\,\mathrm{d}s\,\mathrm{d}t, \quad (x,y) \in Q, \tag{3}$$

where $Q = \langle -\pi, \pi \rangle \times \langle -\pi, \pi \rangle$ is an arbitrary closed, semi-closed, or open region, $H_\lambda : \mathbb{R}^2 \to \mathbb{R}_0^+$ stands for a family of kernels, and $\lambda \in \Lambda$ is a set of non-negative numbers with accumulation point $\lambda_0$. Indicated paper also contains two-dimensional generalization of well-known Natanson's lemma. Then, Siudut [21, 22] presented considerable theorems by the aid of these results. Note that Rydzewska [23] also improved her previous work [12] using the results of [20], and she obtained the rate of convergence of the operators of type (3). Later on, Taberski [24] obtained the weighted pointwise approximation of some integral operators using a weight function satisfying some conditions. Moreover, this study was seen as a continuation and two-dimensional analogue of [25]. In recent papers [26–28], the kernel functions within the operators of type (3) were defined as radial functions and the domains of integration were replaced by an arbitrary region $\langle a,b \rangle \times \langle c,d \rangle$. As concerns the study of integral operators in several settings, the reader may also see [29–35].

This study is a continuation and further generalization of [26]. Besides, the current manuscript deals with Fatou-type pointwise convergence of a family of singular integral operators with radial kernels given in the following form:

$$L_\lambda(f;x,y) = \iint_{\mathbb{R}^2} f(t,s) H_\lambda(t-x, s-y)\,\mathrm{d}s\mathrm{d}t, \quad (x,y) \in \mathbb{R}^2, \quad \lambda \in \Lambda, \tag{4}$$

where $H_\lambda(t-x, s-y) = K_\lambda\left(\sqrt{(t-x)^2 + (s-y)^2}\right)$, and $\Lambda$ is a set of non-negative numbers with accumulation point $\lambda_0$. Here, $f \in L_{1,\varphi}(\mathbb{R}^2)$ and $L_{1,\varphi}(\mathbb{R}^2)$ are the space of all measurable functions for which $\left|\frac{f}{\varphi}\right|$ is integrable provided $\varphi : \mathbb{R}^2 \to \mathbb{R}^+$ is a weight function which is bounded on any bounded subset of $\mathbb{R}^2$.

The paper is organized as follows: In Sect. 2, we introduce the fundamental definitions. In Sect. 3, we prove the pointwise convergence of $L_\lambda(f;x,y)$ to $f(x_0, y_0)$. In Sect. 4, we establish the rate of convergence of the operators of type (4).

## Preliminaries

In this section, we introduce the main definitions used in this paper.

**Definition 1** A function $H \in L_1(\mathbb{R}^2)$ is said to be radial if there exists a function $K : \mathbb{R}_0^+ \to \mathbb{R}$, such that $H(t,s) = K\left(\sqrt{t^2+s^2}\right)$ almost everywhere [36].

Now, we give another characterization of $\mu$-generalized Lebesgue point using the $\mu$-generalized Lebesgue point definition given in [23].

**Definition 2** Let $\delta_0 > 0$ be an arbitrary fixed real number. A $\mu$-generalized Lebesgue point of a locally integrable function $g : \mathbb{R}^2 \to \mathbb{R}$ is a point $(x_0, y_0) \in \mathbb{R}^2$ satisfying

$$\lim_{(h,k)\to(0,0)} \frac{1}{\mu_1(h)\mu_2(k)} \int_0^h \int_0^k |g(t+x_0, s+y_0) - g(x_0,y_0)|\,\mathrm{d}s\mathrm{d}t = 0,$$

where $\mu_1(h) = \int_0^h \rho_1(t)\mathrm{d}t > 0$, $0<h<\delta_0$ and $\rho_1(t)$ is an integrable and non-negative function on $[0, \delta_0]$, and similarly, $\mu_2(k) = \int_0^k \rho_2(s)\mathrm{d}s > 0$, $0<k<\delta_0$ and $\rho_2(s)$ is an integrable and non-negative function on $[0, \delta_0]$.

*Example 1* Let $f : \mathbb{R}^2 \to \mathbb{R}$ is given as follows:

$$f(t,s) = \begin{cases} \dfrac{1}{\sqrt{|t||s|}}, & \text{if } ts \neq 0, \\ 1, & \text{if } ts = 0. \end{cases}$$

and $\varphi : \mathbb{R}^2 \to \mathbb{R}^+$ is given by $\varphi(t,s) = (1+|t|)(1+|s|)$. Therefore, we have

$$\frac{f(t,s)}{\varphi(t,s)} = \begin{cases} \dfrac{1}{(1+|t|)(1+|s|)}, & \text{if } ts = 0, \\ \dfrac{1}{\sqrt{|t|}(1+|t|)\sqrt{|s|}(1+|s|)}, & \text{if } ts \neq 0. \end{cases}$$

Using the definition of $\mu$-generalized Lebesgue point and taking $\rho_1(t) = \left\{t^{\frac{1}{4}}e^t\right\}'_t$ and $\rho_2(s) = \left\{s^{\frac{1}{4}}e^s\right\}'_s$, we see that origin is a $\mu$-generalized Lebesgue point of $f \in L_{1,\varphi}(\mathbb{R}^2)$. On the other hand, one can check that origin is not a generalized Lebesgue point for any choice of $\alpha \in [0,1)$. Therefore, this example shows that the nature of $\mu$-generalized Lebesgue point depends on $\rho_1(t)$ and $\rho_2(s)$. For the analysis of one-dimensional counterpart of the function $\frac{f}{\varphi}$, we refer the reader to see [37].

**Definition 3** $(Class\,A_\varphi)$ Let $H_\lambda : \mathbb{R}^2 \to \mathbb{R}_0^+$ be a radial function, i.e., there exists a function $K_\lambda : \mathbb{R}_0^+ \to \mathbb{R}_0^+$, such that $H_\lambda(t,s) := K_\lambda(\sqrt{t^2+s^2})$ holds for almost everywhere on $\mathbb{R}^2$ for each fixed $\lambda \in \Lambda$. Furthermore, let $H_\lambda : \mathbb{R}^2 \to \mathbb{R}_0^+$ be a family of radial kernels, which are integrable on $\mathbb{R}^2$ and the weight function $\varphi : \mathbb{R}^2 \to \mathbb{R}^+$, which is bounded on arbitrary bounded subsets of $\mathbb{R}^2$, satisfies the following inequality:

$$\varphi(u+t, v+s) \leq \varphi(u,v)\varphi(t,s), (u,v) \in \mathbb{R}^2, (t,s) \in \mathbb{R}^2, \tag{5}$$

and there hold:

a. For any given $(x_0, y_0) \in \mathbb{R}^2$

$$\lim_{(x,y,\lambda)\to(x_0,y_0,\lambda_0)} \frac{1}{\varphi(x_0,y_0)} \iint_{\mathbb{R}^2} \varphi(t,s)K_\lambda\left(\sqrt{(t-x)^2+(s-y)^2}\right)\mathrm{d}s\mathrm{d}t = 1.$$

b. $\forall \xi > 0$,

$$\lim_{\lambda\to\lambda_0} \sup_{\xi \leq \sqrt{(t^2+s^2}} \left[\varphi(t,s)K_\lambda\left(\sqrt{t^2+s^2}\right)\right] = 0.$$

c. $\forall \xi > 0$,

$$\lim_{\lambda\to\lambda_0} \left[\iint_{\xi \leq \sqrt{t^2+s^2}} \varphi(t,s)K_\lambda\left(\sqrt{t^2+s^2}\right)\mathrm{d}s\mathrm{d}t\right] = 0.$$

d. $H_\lambda(t,s)$ is monotonically increasing with respect to $s$ on $(-\infty,0]$, and similarly, $H_\lambda(t,s)$ is monotonically increasing with respect to $t$ on $(-\infty,0]$ for any $\lambda \in \Lambda$. Analogously, $H_\lambda(t,s)$ is bimonotonically increasing with respect to *(t,s)* on $[0,\infty)\times[0,\infty)$ and $(-\infty,0]\times(-\infty,0]$ and bimonotonically decreasing with respect to *(t,s)* on $[0,\infty)\times(-\infty,0]$ and $(-\infty,0]\times[0,\infty)$ for any $\lambda \in \Lambda$.

e. $\|\varphi K_\lambda\|_{L_1(\mathbb{R}^2)} \leq M < \infty$, for all $\lambda \in \Lambda$.

f. For fixed $(t_0,s_0) \in \mathbb{R}^2$, $H_\lambda(t_0,s_0)$ tends to infinity, as $\lambda$ tends to $\lambda_0$.

Note that throughout this paper, we suppose that the function $H_\lambda(t,s)$ belongs to class $A_\varphi$.

*Remark 1* If the function $g : \mathbb{R}^2 \to \mathbb{R}$ is bimonotonically increasing on $[\alpha_1,\alpha_2]\times[\beta_1,\beta_2] \subset \mathbb{R}^2$, then the following equality

$$V(g;[\alpha_1,\alpha_2]\times[\beta_1,\beta_2]) = \bigvee_{\alpha_1}^{\alpha_2}\bigvee_{\beta_1}^{\beta_2}(g(t,s)) = g(\alpha_1,\beta_1) - g(\alpha_1,\beta_2) - g(\alpha_2,\beta_1) + g(\alpha_2,\beta_2)$$

holds. On the other hand, if the function $g : \mathbb{R}^2 \to \mathbb{R}$ is bimonotonically decreasing on $[\alpha_1,\alpha_2]\times[\beta_1,\beta_2] \subset \mathbb{R}^2$, then the following equality

$$V(g;[\alpha_1,\alpha_2]\times[\beta_1,\beta_2]) = \bigvee_{\alpha_1}^{\alpha_2}\bigvee_{\beta_1}^{\beta_2}(g(t,s)) = g(\alpha_1,\beta_2) - g(\alpha_1,\beta_1) - g(\alpha_2,\beta_2) + g(\alpha_2,\beta_1)$$

holds [20, 38].

## Convergence at characteristic points

The following lemma gives the existence of the operators defined by (4).

**Lemma 1** *If $f \in L_{1,\varphi}(\mathbb{R}^2)$, then the operator $L_\lambda(f;x,y)$ defines a continuous transformation acting on $L_{1,\varphi}(\mathbb{R}^2)$.*

*Proof* Since $L_\lambda(f;x,y)$ is linear, it is sufficient to show that the expression given by

$$\|L_\lambda\|_\varphi = \sup_{f\neq 0}\frac{\|L_\lambda(f;x,y)\|_{L_{1,\varphi}(\mathbb{R}^2)}}{\|f\|_{L_{1,\varphi}(\mathbb{R}^2)}}$$

is bounded.

The following expression

$$\|f\|_{L_{1,\varphi}(\mathbb{R}^2)} = \iint_{\mathbb{R}^2}\left|\frac{f(t,s)}{\varphi(t,s)}\right|\mathrm{d}s\,\mathrm{d}t, \tag{6}$$

defines a norm in the space $L_{1,\varphi}(\mathbb{R}^2)$ [24]. Using inequality (5) and Fubini's theorem [14], we have

$$\|L_\lambda(f;x,y)\|_{L_{1,\varphi}(\mathbb{R}^2)}=\iint_{\mathbb{R}^2}\frac{1}{\varphi(x,y)}\left|\iint_{\mathbb{R}^2}f(t,s)H_\lambda(t-x,s-y)\mathrm{d}s\,\mathrm{d}t\right|\mathrm{d}y\,\mathrm{d}x$$

$$=\iint_{\mathbb{R}^2}\frac{1}{\varphi(x,y)}\left|\iint_{\mathbb{R}^2}f(t,s)K_\lambda\left(\sqrt{(t-x)^2+(s-y)^2}\right)\mathrm{d}s\mathrm{d}t\right|\mathrm{d}y\,\mathrm{d}x$$

$$\le\iint_{\mathbb{R}^2}\frac{1}{\varphi(x,y)}\left(\iint_{\mathbb{R}^2}|f(t+x,s+y)|K_\lambda\left(\sqrt{t^2+s^2}\right)\mathrm{d}s\,\mathrm{d}t\right)\mathrm{d}y\,\mathrm{d}x$$

$$=\iint_{\mathbb{R}^2}K_\lambda\left(\sqrt{t^2+s^2}\right)\left(\iint_{\mathbb{R}^2}\left|\frac{f(t+x,s+y)}{\varphi(x,y)}\right|\mathrm{d}y\,\mathrm{d}x\right)\mathrm{d}s\,\mathrm{d}t$$

$$=\iint_{\mathbb{R}^2}K_\lambda\left(\sqrt{t^2+s^2}\right)\left(\iint_{\mathbb{R}^2}\left|\frac{f(t+x,s+y)}{\varphi(x,y)}\right|\frac{\varphi(t+x,s+y)}{\varphi(t+x,s+y)}\mathrm{d}y\,\mathrm{d}x\right)\mathrm{d}s\,\mathrm{d}t$$

$$\le\iint_{\mathbb{R}^2}K_\lambda\left(\sqrt{t^2+s^2}\right)\left(\iint_{\mathbb{R}^2}\left|\frac{f(t+x,s+y)}{\varphi(t+x,s+y)}\right|\frac{\varphi(x,y)\varphi(t,s)}{\varphi(x,y)}\mathrm{d}y\,\mathrm{d}x\right)\mathrm{d}s\,\mathrm{d}t$$

$$=\iint_{\mathbb{R}^2}K_\lambda\left(\sqrt{t^2+s^2}\right)\varphi(t,s)\mathrm{d}s\,\mathrm{d}t\iint_{\mathbb{R}^2}\left|\frac{f(t+x,s+y)}{\varphi(t+x,s+y)}\right|\mathrm{d}y\,\mathrm{d}x$$

$$\le M\|f\|_{L_{1,\varphi}(\mathbb{R}^2)}.$$

Thus, the proof is completed. □

The following theorem gives a Fatou-type pointwise convergence of the integral operators of type (4) at $\mu$-generalized Lebesgue point of $f\in L_{1,\varphi}(\mathbb{R}^2)$.

**Theorem 1** *If $(x_0,y_0)$ is a $\mu$-generalized Lebesgue point of $f\in L_{1,\varphi}(\mathbb{R}^2)$, then*

$$\lim_{(x,y,\lambda)\to(x_0,y_0,\lambda_0)}L_\lambda(f;x,y)=f(x_0,y_0)$$

*on any set Z on which the function*

$$\int_{x_0-\delta}^{x_0+\delta}\int_{y_0-\delta}^{y_0+\delta}K_\lambda\left(\sqrt{(t-x)^2+(s-y)^2}\right)\rho_1(|x_0-t|)\rho_2(|y_0-s|)\mathrm{d}s\,\mathrm{d}t$$

$$+\,2\mu_2(|y_0-y|)\int_{x_0-\delta}^{x_0+\delta}K_\lambda(|t-x|)\rho_1(|x_0-t|)\mathrm{d}t$$

$$+\,2\mu_1(|x_0-x|)\int_{y_0-\delta}^{y_0+\delta}K_\lambda(|s-y|)\rho_2(|y_0-s|)\mathrm{d}s$$

$$+\,4K_\lambda(0)\mu_1(|x_0-x|)\mu_2(|y_0-y|)$$

*is bounded as $(x,y,\lambda)$ tends to $(x_0,y_0,\lambda_0)$.*

*Proof* Let $|x_0-x|<\frac{\delta}{2}$ and $|y_0-y|<\frac{\delta}{2}$, for any $0<\delta<\delta_0$. Furthermore, let $0<x_0-x<\frac{\delta}{2}$ and $0<y_0-y<\frac{\delta}{2}$ for any $0<\delta<\delta_0$. Since $(x_0,y_0)\in\mathbb{R}^2$ is a $\mu$-generalized Lebesgue point of function $f\in L_{1,\varphi}(\mathbb{R}^2)$, for all given $\varepsilon>0$, there exists $\delta>0$, such that for all $h$ and $k$ satisfying $0<h,k\le\delta$, we have the following inequality:

$$\int_{x_0}^{x_0+h}\int_{y_0-k}^{y_0}\left|\frac{f(t,s)}{\varphi(t,s)}-\frac{f(x_0,y_0)}{\varphi(x_0,y_0)}\right|\mathrm{d}s\,\mathrm{d}t<\varepsilon\mu_1(h)\mu_2(k). \tag{7}$$

Write

$$|L_\lambda(f;x,y)-f(x_0,y_0)|=\left|\iint_{\mathbb{R}^2}f(t,s)H_\lambda(t-x,s-y)\mathrm{d}s\,\mathrm{d}t-f(x_0,y_0)\right|.$$

Adding and subtracting the expression given by $\frac{f(x_0,y_0)}{\varphi(x_0,y_0)}\iint_{\mathbb{R}^2}\varphi(t,s)H_\lambda(t-x,s-y)\mathrm{d}s\,\mathrm{d}t$ to the right-hand side of the above equality, we have

$$|L_\lambda(f;x,y)-f(x_0,y_0)|$$

$$=\left|\iint_{\mathbb{R}^2}f(t,s)\frac{\varphi(t,s)}{\varphi(t,s)}H_\lambda(t-x,s-y)\mathrm{d}s\,\mathrm{d}t-f(x_0,y_0)\frac{\varphi(x_0,y_0)}{\varphi(x_0,y_0)}\right.$$

$$+\frac{f(x_0,y_0)}{\varphi(x_0,y_0)}\iint_{\mathbb{R}^2}\varphi(t,s)H_\lambda(t-x,s-y)\mathrm{d}s\,\mathrm{d}t$$

$$\left.-\frac{f(x_0,y_0)}{\varphi(x_0,y_0)}\iint_{\mathbb{R}^2}\varphi(t,s)H_\lambda(t-x,s-y)\mathrm{d}s\,\mathrm{d}t\right|$$

$$\le\iint_{\mathbb{R}^2}\left|\frac{f(t,s)}{\varphi(t,s)}-\frac{f(x_0,y_0)}{\varphi(x_0,y_0)}\right|\varphi(t,s)H_\lambda(t-x,s-y)\mathrm{d}s\,\mathrm{d}t$$

$$+\left|\frac{f(x_0,y_0)}{\varphi(x_0,y_0)}\right|\left|\iint_{\mathbb{R}^2}\varphi(t,s)H_\lambda(t-x,s-y)\mathrm{d}s\,\mathrm{d}t-\varphi(x_0,y_0)\right|$$

$$=I_1+I_2.$$

Since $H_\lambda$ is a radial function, we may write

$$I_2=\left|\frac{f(x_0,y_0)}{\varphi(x_0,y_0)}\right|\left|\iint_{\mathbb{R}^2}\varphi(t,s)H_\lambda(t-x,s-y)\mathrm{d}s\,\mathrm{d}t-\varphi(x_0,y_0)\right|$$

$$=\left|\frac{f(x_0,y_0)}{\varphi(x_0,y_0)}\right|\left|\iint_{\mathbb{R}^2}\varphi(t,s)K_\lambda\left(\sqrt{(t-x)^2+(s-y)^2}\right)\mathrm{d}s\,\mathrm{d}t-\varphi(x_0,y_0)\right|.$$

In view of condition $(a)$ of class $A_\varphi$, $I_2\to 0$ as $(x,y,\lambda)$ tends to $(x_0,y_0,\lambda_0)$. The integral $I_1$ can be written in the form:

$$I_1=\left\{\iint_{\mathbb{R}^2\setminus B_\delta}+\iint_{B_\delta}\right\}\left|\frac{f(t,s)}{\varphi(t,s)}-\frac{f(x_0,y_0)}{\varphi(x_0,y_0)}\right|\varphi(t,s)H_\lambda(t-x,s-y)\mathrm{d}s\,\mathrm{d}t$$

$$=I_{11}+I_{12},$$

where

$$B_\delta = \left\{(t,s) : (t-x_0)^2+(s-y_0)^2<\delta^2, (x_0,y_0) \in \mathbb{R}^2\right\}.$$

In view of definition of $H_\lambda$, and using inequality (5), we have

$$\begin{aligned} I_{11} &= \iint\limits_{\mathbb{R}^2\setminus B_\delta} \left|\frac{f(t,s)}{\varphi(t,s)} - \frac{f(x_0,y_0)}{\varphi(x_0,y_0)}\right| \varphi(t,s) H_\lambda(t-x,s-y)\mathrm{d}s\mathrm{d}t \\ &\leq \varphi(x,y) \iint\limits_{\mathbb{R}^2\setminus B_\delta} \left|\frac{f(t,s)}{\varphi(t,s)} - \frac{f(x_0,y_0)}{\varphi(x_0,y_0)}\right| \varphi(t-x,s-y) \\ &\quad \times K_\lambda\left(\sqrt{(t-x)^2+(s-y)^2}\right)\mathrm{d}s\mathrm{d}t. \end{aligned}$$

Now, using the initial assumptions given as $0<|x_0-x|<\frac{\delta}{2}$ and $0<|y_0-y|<\frac{\delta}{2}$, we may define the following set:

$$A_\delta = \left\{(x,y) : (x-x_0)^2+(y-y_0)^2<\frac{\delta^2}{2}, (x_0,y_0) \in \mathbb{R}^2\right\}.$$

Taking into account the geometric representations of the sets $B_\delta$ and $A_\delta$ gives the inclusion relation $\mathbb{R}^2\setminus B_\delta \subseteq \mathbb{R}^2\setminus C_\delta$, where

$$C_\delta = \left\{(t,s) : (t-x)^2+(s-y)^2<\frac{\delta^2}{2}, (x,y) \in A_\delta\right\}.$$

Therefore, we have the following inequality:

$$\begin{aligned} I_{11} &\leq \varphi(x,y) \iint\limits_{\mathbb{R}^2\setminus C_\delta} \left|\frac{f(t,s)}{\varphi(t,s)} - \frac{f(x_0,y_0)}{\varphi(x_0,y_0)}\right| \varphi(t-x,s-y) \\ &\quad \times K_\lambda\left(\sqrt{(t-x)^2+(s-y)^2}\right)\mathrm{d}s\,\mathrm{d}t \\ &= \varphi(x,y) \iint\limits_{\frac{\delta}{\sqrt{2}}\leq\sqrt{u^2+v^2}} \left|\frac{f(u+x,v+y)}{\varphi(u+x,v+y)} - \frac{f(x_0,y_0)}{\varphi(x_0,y_0)}\right| \varphi(u,v) \\ &\quad \times K_\lambda\left(\sqrt{u^2+v^2}\right)\mathrm{d}v\,\mathrm{d}u \\ &\leq \varphi(x,y) \sup_{\frac{\delta}{\sqrt{2}}\leq\sqrt{u^2+v^2}} \left[\varphi(u,v)K_\lambda\left(\sqrt{u^2+v^2}\right)\right] \|f\|_{L_{1,\varphi}(\mathbb{R}^2)} \\ &\quad + \varphi(x,y)\left|\frac{f(x_0,y_0)}{\varphi(x_0,y_0)}\right| \iint\limits_{\frac{\delta}{\sqrt{2}}\leq\sqrt{u^2+v^2}} \varphi(u,v)K_\lambda\left(\sqrt{u^2+v^2}\right)\mathrm{d}v\,\mathrm{d}u. \end{aligned}$$

Consequently, by conditions $(b)$ and $(c)$ of class $A_\varphi$, and using boundedness of $\varphi$, $I_{11} \to 0$ as $(x,y,\lambda) \to (x_0,y_0,\lambda_0)$.

Now, we prove that $I_{12}$ tends to zero, as $(x,y,\lambda)$ tends to $(x_0,y_0,\lambda_0)$. Since $\varphi(t,s)$ is bounded on $B_\delta$, it is easy to see that the following inequality

$$I_{12} \leq \sup_{(t,s)\in B_\delta} \varphi(t,s) \iint\limits_{B_\delta} \left|\frac{f(t,s)}{\varphi(t,s)} - \frac{f(x_0,y_0)}{\varphi(x_0,y_0)}\right| H_\lambda(t-x,s-y)\mathrm{d}s\,\mathrm{d}t$$

holds for $I_{12}$. Thus, we have

$$\begin{aligned} I_{12} &\leq \sup_{(t,s)\in B_\delta} \varphi(t,s) \left\{ \int_{x_0}^{x_0+\delta}\int_{y_0-\delta}^{y_0} + \int_{x_0-\delta}^{x_0}\int_{y_0-\delta}^{y_0} \right\} \\ &\quad \times \left|\frac{f(t,s)}{\varphi(t,s)} - \frac{f(x_0,y_0)}{\varphi(x_0,y_0)}\right| H_\lambda(t-x,s-y)\mathrm{d}s\,\mathrm{d}t \\ &\quad + \sup_{(t,s)\in B_\delta} \varphi(t,s) \left\{ \int_{x_0-\delta}^{x_0}\int_{y_0}^{y_0+\delta} + \int_{x_0}^{x_0+\delta}\int_{y_0}^{y_0+\delta} \right\} \\ &\quad \times \left|\frac{f(t,s)}{\varphi(t,s)} - \frac{f(x_0,y_0)}{\varphi(x_0,y_0)}\right| H_\lambda(t-x,s-y)\mathrm{d}s\,\mathrm{d}t \\ &= \sup_{(t,s)\in B_\delta} \varphi(t,s)(I_{121}+I_{122}+I_{123}+I_{124}). \end{aligned}$$

Let us consider the integral $I_{121}$.

Let us define the function $V(t,s)$ by

$$V(t,s) := \int_{x_0}^{t}\int_{s}^{y_0} \left|\frac{f(u,v)}{\varphi(u,v)} - \frac{f(x_0,y_0)}{\varphi(x_0,y_0)}\right| \mathrm{d}v\,\mathrm{d}u.$$

In view of inequality (7), the following expression

$$|V(t,s)| \leq \varepsilon\mu_1(t-x_0)\mu_2(y_0-s), \tag{8}$$

where $0<t-x_0\leq\delta$ and $0<y_0-s\leq\delta$, holds. From Theorem 2.6 in [20], we can write

$$\begin{aligned} I_{121} &= (L)\int_{x_0}^{x_0+\delta}\int_{y_0-\delta}^{y_0} \left|\frac{f(t,s)}{\varphi(t,s)} - \frac{f(x_0,y_0)}{\varphi(x_0,y_0)}\right| H_\lambda(t-x,s-y)\mathrm{d}s\,\mathrm{d}t \\ &= (LS)\int_{x_0}^{x_0+\delta}\int_{y_0-\delta}^{y_0} H_\lambda(t-x,s-y)d[-V(t,s)], \end{aligned}$$

where LS denotes Lebesgue–Stieltjes integral. Applying integration by parts (see Theorem 2.2, p. 100 in [20]) to the Lebesgue–Stieltjes integral, we have

$$\begin{aligned} |I_{121}| &= \left|\int_{x_0}^{x_0+\delta}\int_{y_0-\delta}^{y_0} H_\lambda(t-x,s-y)d[-V(t,s)]\right| \\ &\leq \int_{x_0}^{x_0+\delta}\int_{y_0-\delta}^{y_0} |V(t,s)||dH_\lambda(t-x,s-y)| \\ &\quad + \int_{x_0}^{x_0+\delta} |V(t,y_0-\delta)||dH_\lambda(t-x,y_0-\delta-y)| \\ &\quad + \int_{y_0-\delta}^{y_0} |V(x_0+\delta,s)||dH_\lambda(x_0-x+\delta,s-y)| \\ &\quad + |V(x_0+\delta,y_0-\delta)|H_\lambda(x_0+\delta-x,y_0-\delta-y). \end{aligned}$$

If we apply inequality (8) to the last inequality and make change of variables, then we have

$$|I_{121}| \leq \varepsilon \int_{x_0-x}^{x_0+\delta-x} \int_{y_0-\delta-y}^{y_0-y} \mu_1(t+x-x_0)\mu_2(y_0-s-y)|dH_\lambda(t,s)|$$
$$+ \varepsilon\mu_2(\delta) \int_{x_0-x}^{x_0+\delta-x} \mu_1(t+x-x_0)|dH_\lambda(t,y_0-\delta-y)|$$
$$+ \varepsilon\mu_1(\delta) \int_{y_0-\delta-y}^{y_0-y} \mu_2(y_0-s-y)|dH_\lambda(x_0+\delta-x,s)|$$
$$+ \varepsilon\mu_1(\delta)\mu_2(\delta)H_\lambda(x_0+\delta-x,y_0-\delta-y).$$

Let us define the following variations:

$$P_1(t,s) := \begin{cases} \bigvee_t^{x_0+\delta-x} \bigvee_{y_0-\delta-y}^{s} (H_\lambda(u,v)), & x_0-x \leq t < x_0+\delta-x \\ & y_0-\delta-y < s \leq y_0-y \\ 0, & \text{otherwise.} \end{cases}$$

$$P_2(t) := \begin{cases} \bigvee_t^{x_0+\delta-x} (H_\lambda(u,y_0-\delta-y)), & x_0-x \leq t < x_0+\delta-x \\ 0, & \text{otherwise.} \end{cases}$$

$$P_3(s) := \begin{cases} \bigvee_{y_0-\delta-y}^{s} (H_\lambda(x_0-x+\delta,v)), & y_0-\delta-y < s \leq y_0-y \\ 0, & \text{otherwise.} \end{cases}$$

Taking above variations into account and applying bivariate integration by parts method to the last inequality, we have

$$|I_{121}| \leq -\varepsilon \int_{x_0-x}^{x_0+\delta-x} \int_{y_0-\delta-y}^{y_0-y} [P_1(t,s)+P_2(t)+P_3(s)$$
$$+ H_\lambda(x_0-x+\delta,y_0-\delta-y)]$$
$$\times \{\mu_1(t-x_0+x)\}_t'\{\mu_2(y_0-s-y)\}_s' \mathrm{d}s\mathrm{d}t$$
$$= \varepsilon(i_1+i_2+i_3+i_4).$$

Using Remark 1 and condition $(d)$ of class $A_\varphi$, we get

$$i_1+i_2+i_3+i_4 = -\int_{x_0-x}^{x_0+\delta-x} \int_{y_0-\delta-y}^{0} K_\lambda\left(\sqrt{t^2+s^2}\right)$$
$$\{\mu_1(t+x-x_0)\}_t'\{\mu_2(y_0-s-y)\}_s' \mathrm{d}s\,\mathrm{d}t$$
$$+ \int_{x_0-x}^{x_0+\delta-x} \int_{0}^{y_0-y} \left(K_\lambda\left(\sqrt{t^2+s^2}\right) - 2K_\lambda(|t|)\right)$$
$$\{\mu_1(t+x-x_0)\}_t' \times \{\mu_2(y_0-s-y)\}_s' \mathrm{d}s\,\mathrm{d}t.$$

Hence, the following inequality holds for $I_{121}$ (for the similar situation, see [20, 23]):

$$|I_{121}| \leq \varepsilon \int_{x_0}^{x_0+\delta} \int_{y_0-\delta}^{y_0} K_\lambda\left(\sqrt{(t-x)^2+(s-y)^2}\right)$$
$$\times \rho_1(t-x_0)\rho_2(|y_0-s|)\mathrm{d}s\,\mathrm{d}t$$
$$+ 2\varepsilon\mu_2(|y_0-y|) \int_{x_0}^{x_0+\delta} K_\lambda(|t-x|)\rho_1(t-x_0)\mathrm{d}t.$$

Analogous computations for $I_{122}$, $I_{123}$, and $I_{124}$ give us

$$|I_{122}| \leq \varepsilon \int_{x_0-\delta}^{x_0} \int_{y_0-\delta}^{y_0} K_\lambda\left(\sqrt{(t-x)^2+(s-y)^2}\right)\rho_1(x_0-t)\rho_2(y_0-s)\mathrm{d}s\,\mathrm{d}t$$
$$+ 2\varepsilon\mu_2(|y_0-y|) \int_{x_0-\delta}^{x_0} K_\lambda(|t-x|)\rho_1(x_0-t)\mathrm{d}t$$
$$+ 2\varepsilon\mu_1(|x_0-x|) \int_{y_0-\delta}^{y_0} K_\lambda(|s-y|)\rho_2(y_0-s)\mathrm{d}s$$
$$+ 4\varepsilon K_\lambda(0)\mu_1(|x_0-x|)\mu_2(|y_0-y|),$$

$$|I_{123}| \leq \varepsilon \int_{x_0-\delta}^{x_0} \int_{y_0}^{y_0+\delta} K_\lambda\left(\sqrt{(t-x)^2+(s-y)^2}\right)$$
$$\times \rho_1(x_0-t)\rho_2(s-y_0)\mathrm{d}s\,\mathrm{d}t$$
$$+ 2\varepsilon\mu_1(|x_0-x|) \int_{y_0}^{y_0+\delta} K_\lambda(|s-y|)\rho_2(s-y_0)\mathrm{d}s,$$

$$|I_{124}| \leq \varepsilon \int_{x_0}^{x_0+\delta} \int_{y_0}^{y_0+\delta} K_\lambda\left(\sqrt{(t-x)^2+(s-y)^2}\right)$$
$$\times \rho_1(t-x_0)\rho_2(s-y_0)\mathrm{d}s\,\mathrm{d}t.$$

Hence, the following inequality is obtained for $I_{12}$:

$$|I_{12}| \leq \varepsilon \sup_{(t,s)\in B_\delta} \varphi(t,s) \Bigg\{ \int_{x_0-\delta}^{x_0+\delta} \int_{y_0-\delta}^{y_0+\delta} K_\lambda\left(\sqrt{(t-x)^2+(s-y)^2}\right)$$
$$\times \rho_1(|x_0-t|)\rho_2(|y_0-s|)\mathrm{d}s\,\mathrm{d}t.$$
$$+ 2\mu_2(|y_0-y|) \int_{x_0-\delta}^{x_0+\delta} K_\lambda(|t-x|)\rho_1(|x_0-t|)\mathrm{d}t$$
$$+ 2\mu_1(|x_0-x|) \int_{y_0-\delta}^{y_0+\delta} K_\lambda(|s-y|)\rho_2(|y_0-s|)\mathrm{d}s$$
$$+ 4K_\lambda(0)\mu_1(|x_0-x|)\mu_2(|y_0-y|)\Bigg\}.$$

In addition, the last inequality is obtained for other cases of the assumptions $|x_0-x| < \frac{\delta}{2}$ and $|y_0-y| < \frac{\delta}{2}$. The remaining part of the proof is obvious by the hypotheses. Thus, the proof is completed. □

## Rate of convergence

In this section, we give a theorem concerning the rate of pointwise convergence.

**Theorem 2** *Assume that the hypotheses of Theorem 1 are satisfied. Let*

$$\Delta(\lambda,\delta,x,y)=\int_{x_0-\delta}^{x_0+\delta}\int_{y_0-\delta}^{y_0+\delta}K_\lambda\left(\sqrt{(t-x)^2+(s-y)^2}\right)\times\rho_1(|x_0-t|)\rho_2(|y_0-s|)\mathrm{d}s\,\mathrm{d}t$$
$$+2\mu_2(|y_0-y|)\int_{x_0-\delta}^{x_0+\delta}K_\lambda(|t-x|)\rho_1(|x_0-t|)\mathrm{d}t$$
$$+2\mu_1(|x_0-x|)\int_{y_0-\delta}^{y_0+\delta}K_\lambda(|s-y|)\rho_2(|y_0-s|)\mathrm{d}s$$
$$+4K_\lambda(0)\mu_1(|x_0-x|)\mu_2(|y_0-y|)$$

*for* $0<\delta<\delta_0$, *and the following conditions are satisfied:*

1. $\Delta(\lambda,\delta,x,y)\to 0$ *as* $(x,y,\lambda)\to(x_0,y_0,\lambda_0)$ *for some* $\delta>0$.
2. *Letting* $(x,y,\lambda)\to(x_0,y_0,\lambda_0)$, *we have*
$$\left|\frac{1}{\varphi(x_0,y_0)}\iint_{\mathbb{R}^2}\varphi(t,s)K_\lambda\left(\sqrt{(t-x)^2+(s-y)^2}\right)\mathrm{d}s\,\mathrm{d}t-1\right|$$
$$=o(\Delta(\lambda,\delta,x,y)).$$
3. *For every* $\xi>0$
$$\sup_{\xi\le\sqrt{t^2+s^2}}\left[\varphi(t,s)K_\lambda\left(\sqrt{t^2+s^2}\right)\right]=o(\Delta(\lambda,\delta,x,y))$$
*as* $(x,y,\lambda)\to(x_0,y_0,\lambda_0)$.
4. *For every* $\xi>0$
$$\iint_{\xi\le\sqrt{t^2+s^2}}\varphi(t,s)K_\lambda\left(\sqrt{t^2+s^2}\right)\mathrm{d}s\,\mathrm{d}t=o(\Delta(\lambda,\delta,x,y))$$

*as* $(x,y,\lambda)\to(x_0,y_0,\lambda_0)$.

Then, at each $\mu$-generalized Lebesgue point of $f\in L_{1,\varphi}(\mathbb{R}^2)$, we have

$$|L_\lambda(f;x,y)-f(x_0,y_0)|=o(\Delta(\lambda,\delta,x,y)).$$

as $(x,y,\lambda)\to(x_0,y_0,\lambda_0)$.

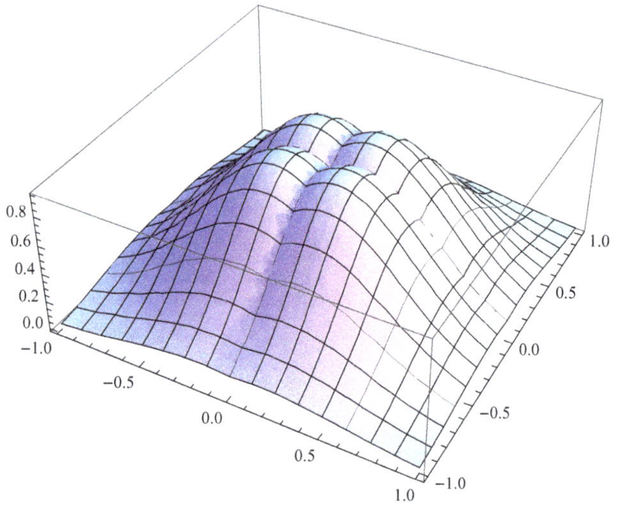


**Fig. 1** Illustration of condition (3)

*Proof* The result is obvious by the hypotheses of Theorem 1. □

*Example 2* Let $\Lambda=(0,\infty)$, $\lambda_0=0$, the weight function $\varphi:\mathbb{R}^2\to\mathbb{R}^+$ is given by $\varphi(t,s)=(1+|t|)(1+(|s|)$ and the kernel function $H_\lambda:\mathbb{R}^2\to\mathbb{R}_0^+$ is given by $H_\lambda(t,s)=\frac{1}{4\pi\lambda}e^{\frac{-(t^2+s^2)}{4\lambda}}$. To verify that $H_\lambda(t,s)$ satisfies the hypotheses of Theorem 1, see [21]. Let $(x_0,y_0)=(0,0)$, $\rho_1(t)=1$ and $\rho_2(s)=1$. Therefore, $\mu_1(t)=t$ and $\mu_2(s)=s$. First, we give the graphical illustrations of the conditions (3) and (4). The assumptions in Fig. 1 are as follows: $\lambda=0.1$ and $-1\le t,s\le 1$. Therefore, $\varphi(t,s)H_{0.1}(t,s)=\frac{1}{0.4\pi}e^{\frac{-(t^2+s^2)}{0.4}}(1+|t|)(1+(|s|)$.

The assumptions in Fig. 2 are as follows: $\lambda=0.1$ and $-10\le t,s\le 10$.

The Fig. 1 and Fig. 2 are generated by using the software Wolfram Mathematica 7. Note that

$$\lim_{\lambda\to 0}\iint_{\mathbb{R}^2}\frac{1}{4\lambda\pi}e^{\frac{-(t^2+s^2)}{4\lambda}}(1+|t|)(1+(|s|)\mathrm{d}s\,\mathrm{d}t=1.$$

Now, we focus on condition (1). Computing $\Delta(\lambda,\delta,x,y)$ gives

$$\Delta(\lambda,\delta,x,y)=\frac{1}{\pi\lambda}|x||y|+\frac{1}{2\sqrt{\lambda\pi}}\left(\mathrm{Erf}\left(\frac{\delta-y}{2\sqrt{\lambda}}\right)+\mathrm{Erf}\left(\frac{\delta+y}{2\sqrt{\lambda}}\right)\right)$$
$$=\frac{1}{4}\left(\mathrm{Erf}\left(\frac{\delta-x}{2\sqrt{\lambda}}\right)+\mathrm{Erf}\left(\frac{\delta+x}{2\sqrt{\lambda}}\right)\right)$$
$$\times\left(\mathrm{Erf}\left(\frac{\delta-y}{2\sqrt{\lambda}}\right)+\mathrm{Erf}\left(\frac{\delta+y}{2\sqrt{\lambda}}\right)\right)$$
$$+\frac{1}{2\sqrt{\lambda\pi}}\left(\mathrm{Erf}\left(\frac{\delta-x}{2\sqrt{\lambda}}\right)+\mathrm{Erf}\left(\frac{\delta+x}{2\sqrt{\lambda}}\right)\right).$$

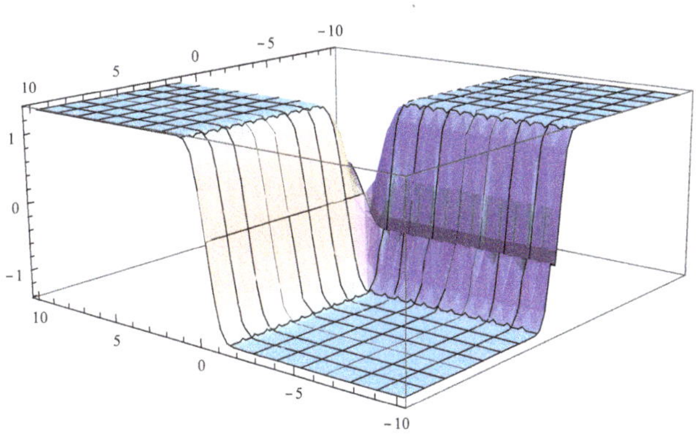


**Fig. 2** Illustration of condition (4)

Here, the function $\mathrm{Erf}(x)$ is given by

$$\mathrm{Erf}(x) = \frac{2}{\sqrt{\pi}} \int_0^x e^{-t^2} dt.$$

To find the numbers $\delta > 0$, we let $\Delta(\lambda, \delta, x, y) \to 0$ as $(x, y, \lambda) \to (0, 0, 0)$. Therefore, if $\delta = o(\sqrt{\lambda})$ and $|x| = |y| = o(\sqrt{\lambda})$, then we obtain

$$\lim_{(x,y,\lambda)\to(0,0,0)} \Delta(\lambda, \delta, x, y) = 0.$$

Hence, $\Delta(\lambda, \delta, x, y) = O(\lambda^c)$, $c > \frac{1}{2}$. Using these results and evaluating the supremum value of the function $\varphi(t,s)K_\lambda(\sqrt{t^2+s^2})$ in terms of $\lambda$, we have

$$\lim_{\lambda\to 0}\left[\sup_{\frac{\delta}{\sqrt{2}} \le \sqrt{t^2+s^2}} \left[\varphi(t,s)K_\lambda\left(\sqrt{t^2+s^2}\right)\right]\right]$$
$$= \left(1 + \frac{1}{2}(-1 + \sqrt{1+8\lambda})\right) e^{-\frac{(-1+\sqrt{1+8\lambda})^2}{16\lambda}}$$
$$= 0.$$

Therefore, we obtain the desired result for the condition (3), that is

3. $\sup_{\xi \le \sqrt{t^2+s^2}} \varphi(t,s)K_\lambda(\sqrt{t^2+s^2}) = o(\lambda^c), c > \frac{1}{2}$ as $(x, y, \lambda) \to (x_0, y_0, \lambda_0)$.

The conditions (2) and (4) are verified with the same method. Finally, we obtain

$$|L_\lambda(f; x, y) - f(x_0, y_0)| = o(\lambda^c), c > \frac{1}{2}.$$

## Concluding remark

In this paper, we investigated the weighted pointwise convergence and the rate of convergence for the family of double singular integral operators of the form (4). This study may be seen as a continuation and generalization of the previous studies, such as [26]. For this aim, we used two-dimensional counterparts of some concepts given in one-dimensional case, such as monotonicity and integration by parts method. Next, since the approximation results and the character of the kernel function are related, a special class of kernel functions, called class $A_\varphi$, has been defined. Therefore, the main result is presented as Theorem 1. Using this theorem, we obtained the rate of pointwise convergence and gave an example including graphical illustration.

**Author contribution statement** All authors contributed equally to the writing of this paper. All authors read and approved the final manuscript.

**Acknowledgments** The authors thank to the referees for their valuable comments and suggestions for the improvement of the manuscript. This research has been supported by Ankara University Scientific Research Projects Coordination Unit (Project Number: 15L0430010, 2016). The authors would like to thank Ankara University.

**Compliance with ethical standards**

**Conflict of interest** The authors declare that they have no competing interests.

## References

1. Carasso, A.S.: Singular integrals, image smoothness, and the recovery of texture in image deblurring. SIAM J. Appl. Math. **64**(5), 1749–1774 (2004)
2. Bracewell, R.: The Fourier transform and its applications, 3rd edn. McGraw-Hill Science, New York (1999)
3. Gallagher, T.A., Nemeth, A.J., Bey, L.H.: An introduction to the Fourier transform: relationship to MRI. AJR. **190**(5), 1396–1405 (2008)
4. Ruiz, P., Orozco, H.M., Mateos, J., Vergara, O.O.V., Molina, R., Katsaggelos, A.K.: Combining Poisson singular integral and total variation prior models in image restoration. Signal Process. **103**(10), 296–308 (2014)
5. Jianjiang, F., Anil, K. J.: FM model based fingerprint reconstruction from minutiae template. In: Advances in biometrics. Lecture notes in computer science, vol 5558, Springer, Berlin, pp. 544–553 (2009)
6. Bardaro, C., Musielak, J., Vinti, G.: Nonlinear integral operators and applications. Gruyter series in nonlinear analysis and applications. Walter de Gruyter, Berlin (2003)
7. Yan, H. (ed.): Signal processing in magnetic resonance imaging and spectroscopy. Marcel Dekker Inc., New York (2002)
8. Bardaro, C, Mantellini, I, Stens, R, Vautz, J, Vinti, G: Generalized sampling approximation for multivariate discontinuous signals and applications to image processing. In: New perspectives on approximation and sampling theory. Appl. Numer. Harmon. Anal., Birkhä user/Springer, Cham. 87–114 (2014)
9. Kobylin, O., Lyashenko, V.: Comparison of standard image edge detection tecniques and of method based on wavelet transform. Int. J. Adv. Res. **8**(2), 572–580 (2014)
10. Unser, M., Sage, D., Van De Ville, D.: Multiresolution monogenic signal analysis using the Riesz–Laplace wavelet transform. IEEE Trans. Image Process. **18**(11), 2402–2418 (2009)

11. Unser, M., Tafti, P.D.: An introduction to sparse stochastic processes. Cambridge University Press, Cambridge (2014)
12. Rydzewska, B.: Approximation des fonctions par des int égrales singulières ordinaires. Fasc. Math. **7**, 71–81 (1973)
13. Mamedov, R.G.: On the order of convergence of m-singular integrals at generalized Lebesgue points and in the space Lp(-∞, ∞ ). Izv. Akad. Nauk SSSR Ser. Mat **27**(2), 287–304 (1963)
14. Butzer, P.L., Nessel, R.J.: Fourier analysis and approximation, vol. I. Academic Press, New York (1971)
15. Uysal, G., Yilmaz, M.M., Ibikli, E.: Approximation by radial type multidimensional singular integral operators. Palest. J. Math. **5**(2), 61–70 (2016)
16. Karsli, H.: Fatou type convergence of nonlinear m-singular integral operators. Appl. Math. Comput. **246**, 221–228 (2014)
17. Carlsson, M.: Fatou-type theorems for general approximate identities. Math. Scand. **102**(2), 231–252 (2008)
18. Siudut, S.: On the Fatou type convergence of abstract singular integrals. Comment. Math. Prace Mat. **30**(1), 171–176 (1990)
19. Taberski, R.: Singular integrals depending on two parameters. Prace Mat. **7**, 173–179 (1962)
20. Taberski, R.: On double integrals and Fourier series. Ann. Polon. Math. **15**, 97–115 (1964)
21. Siudut, S.: On the convergence of double singular integrals. Comment. Math. Prace Mat. **28**(1), 143–146 (1988)
22. Siudut, S.: A theorem of Romanovski type for double singular integrals. Comment. Math. Prace Mat. **29**, 277–289 (1989)
23. Rydzewska, B.: Approximation des fonctions de deux variables par des intégrales singulières doubles. Fasc. Math. **8**, 35–45 (1974)
24. Taberski, R.: On double singular integrals. Prace Mat. **19**, 155–160 (1976)
25. Alexits, G.: Konvergenzprobleme der Orthogonalreihen. Verlag der Ungarischen Akademie der Wissenschaften, Budapest (1960)
26. Uysal, G., Yilmaz, M.M., Ibikli, E.: A study on pointwise approximation by double singular integral operators. J. Inequal. Appl. **2015**, 94 (2015)
27. Uysal, G., Yilmaz, M.M.: Some theorems on the approximation of non-integrable functions via singular integral operators. Proc. Jangjeon Math. Soc. **18**(2), 241–251 (2015)
28. Yilmaz, M.M., Uysal, G., Ibikli, E.: A note on rate of convergence of double singular integral operators. Adv. Differ. Equ. **2014**, 287 (2014)
29. Gadjiev, AD: The order of convergence of singular integrals which depend on two parameters. In: Special problems of functional analysis and their applications to the theory of differential equations and the theory of functions. Izdat. Akad. Nauk Azerbaidžan. SSR., Baku 40–44 (1968)
30. Bardaro, C.: On approximation properties for some classes of linear operators of convolution type. Atti Sem. Mat. Fis. Univ. Modena **33**(2), 329–356 (1984)
31. Bardaro, C., Karsli, H., Vinti, G.: On pointwise convergence of linear Integral operators with homogeneous kernel. Integral Transform. Spec. Funct. **19**(6), 429–439 (2008)
32. Vinti, G., Zampogni, L.: A unifying approach to convergence of linear sampling type operators in Orlicz spaces. Adv. Differ. Equ. **16**, 573–600 (2011)
33. Yilmaz, M.M.: On convergence of singular integral operators depending on three parameters with radial kernels. Int. J. Math. Anal. **4**(39), 1923–1928 (2010)
34. Karsli, H., Ibikli, E.: On convergence of convolution type singular integral operators depending on two parameters. Fasc. Math. **38**, 25–39 (2007)
35. Karsli, H.: On Fatou type convergence of convolution type double singular integral operators. Anal. Theory Appl. **31**(3), 307–320 (2015)
36. Bochner, S., Chandrasekharan, K.: Fourier transforms. Ann. Math. Stud. No. 19, Princeton University Press, Princeton. Oxford University Press, London, ix + 219 (1949)
37. Esen, S: Convergence and the order of convergence of family of nonconvolution type integral operators at characteristic points, Ph.D. Thesis, Ankara University, Graduate School of Applied Science, Ankara (2002)
38. Ghorpade, S.R., Limaye, B.V.: A course in multivariable calculus and analysis, p. xii + 475. Springer, New York (2010)

# 10

# Ranking $p$-norm generalised fuzzy numbers with different left height and right height using integral values

**Rituparna Chutia · Rekhamoni Gogoi · D. Datta**

**Abstract** This paper considers ranking of generalised fuzzy numbers with different left height and right height using integral values. With the advances in new type of fuzzy number (generalised fuzzy number with different left height and right height) methods should be developed to compare them. Keeping this in view a new modified method has been proposed.

**Keywords** Generalised fuzzy number · Ranking · $p$-Norm · Left height · Right height

## Introduction

Decision making in engineering, medical and any other real-life problems may be interpreted in terms of fuzzy. This demands ranking or ordering of fuzzy quantities to make a transparent decision. With the advances in fuzzy set theory, different ranking methods are developed. This concept was first proposed by Jain [7]. Some of the literatures that describe different approach of ranking fuzzy quantities are [1, 3, 4, 10, 13–15]. Recently, ranking of trapezoidal fuzzy numbers based on the shadow length has been discussed by Pour et al. [12]. Also, ranking triangular fuzzy numbers by Pareto approach based on two dominance stages is discussed by Bahri et al. [2].

The literatures that are available on ranking fuzzy quantities based on the integral values are [6–9]. These type of methods of ranking fuzzy numbers are based on the convex combination of right and left integral values through an index of optimism found in Liou and Wang [11] and Kim and Park [8]. This concept was further generalised to rank non-normal $p$-norm trapezoidal fuzzy numbers [6]. However, this method was found insufficient to rank non-normal $p$-norm fuzzy numbers with different height; keeping this in mind, Kumar et al. [9] developed an approach to overcome those shortcoming's. With the advances of generalised fuzzy numbers (GFNs) with different left height and right height [6], Kumar's approach fails to rank them. Hence, Kumar's approach is only sufficient for ranking fuzzy numbers or non-normal $p$-norm fuzzy numbers with different height, but the method is insufficient for ranking GFNs with different left height and right height.

Keeping this in view, Kumar's approach has been modified in this paper to rank $p$-norm GFNs with different left height and right height. This modified method thus handle both normal and non-normal trapezoidal fuzzy number with different height. The modified method can also rank non-normal $p$-norm trapezoidal fuzzy numbers with different height.

The structure of the paper is as follows. In Sect. 2, some general concept of the GFN is put forwarded. Membership function of GFN is defined. Also the membership function of $p$-norm GFN with different left height and right height is defined. Section 3 starts with definitions of different integral values of $p$-norm GFN with different left height and right height, And finally, some properties related to them are discussed in this section. Section 4 describes the

R. Chutia (✉)
Department of Chemistry, Indian Institute of Technology, Guwahati 781039, Assam, India
e-mail: rituparnachutia7@gmail.com

R. Gogoi
North Lakhimpur Girls' Higher Secondary School, Lakhimpur 787001, Assam, India

D. Datta
Health Physics Division, Bhabha Atomic Research Division, Mumbai 400085, India

proposed modified method along with some numerical examples. Finally, in Sect. 4, conclusions are made.

## Definitions and notations

In this section, brief review of some concepts of generalised fuzzy number with different left height and right height are put forward.

### Generalised fuzzy number

Let $\tilde{A}$ be represented by $(a, b, c, d; h_L, h_R)$ on the real line $\mathbb{R}$ such that $-\infty < a \leq b \leq c \leq d < \infty$ is called a GFN with different left height and right height which is bounded and convex. The values $a, b, c$ and $d$ are real, $h_L$ is called the left height of the GFN $\tilde{A}$, $h_R$ is called the right height of the GFN, $h_L \in [0, 1]$ and $h_R \in [0, 1]$ [5]. For now, let $\mathbb{F}(\mathbb{R})$ be the set of all GFNs with different left height and right height. If $h_L = h_R = 1$ then the GFN reduces to a standard trapezoidal fuzzy number.

The membership function of GFN $\tilde{A}$ with different left height and right height is as given below

$$\mu_{\tilde{A}}(x) = \begin{cases} \mu_1(x), & \text{if } a \leq x \leq b; \\ \mu_2(x), & \text{if } b \leq x \leq c; \\ \mu_3(x), & \text{if } c \leq x \leq d; \\ 0, & \text{otherwise} \end{cases} \tag{1}$$

where $\mu_1 : [a, b] \longrightarrow [0, h_L]$, $\mu_2 : [b, c] \longrightarrow [h_L, h_R]$ (or $[h_R, h_L]$) and $\mu_3 : [c, d] \longrightarrow [0, h_R]$ are continuous. The functions $\mu_1(x)$ and $\mu_3(x)$ are strictly increasing and strictly decreasing, respectively. The function $\mu_2(x)$ is strictly increasing when $h_L < h_R$ and strictly decreasing when $h_L > h_R$. Then the inverse of $\mu_{\tilde{A}}(x)$ is

$$\mu^{-1}_{\tilde{A}}(y) = \begin{cases} \mu_1^{-1}(y), & \text{if } 0 \leq y \leq h_L; \\ \mu_2^{-1}(y), & \text{if } h_L \leq y \leq h_R, \text{or } (h_R \leq y \leq h_L); \\ \mu_3^{-1}(y), & \text{if } 0 \leq y \leq h_R; \\ 0, & \text{otherwise} \end{cases} \tag{2}$$

where $\mu_1^{-1} : [0, h_L] \longrightarrow [a, b]$, $\mu_2^{-1} : [h_L, h_R]$ (or $[h_R, h_L]$) $\longrightarrow [b, c]$ and $\mu_3^{-1} : [0, h_R] \longrightarrow [c, d]$ are continuous. The function $\mu_1^{-1}(x)$ and $\mu_3^{-1}(x)$ are strictly increasing and strictly decreasing, respectively. The function $\mu_2^{-1}(x)$ is strictly increasing when $h_L < h_R$ and strictly decreasing when $h_L > h_R$.

Let $\tilde{A} = (a, b, c, d; h_L, h_R)$ be a trapezoidal GFN with different left height and right height then the membership function is defined as

$$\mu_{\tilde{A}}(x) = \begin{cases} \dfrac{h_L(x-a)}{b-a}, & \text{if } a \leq x \leq b; \\ \dfrac{h_L(c-b) + (h_R - h_L)(x-b)}{c-b}, & \text{if } b \leq x \leq c; \\ \dfrac{h_R(x-d)}{c-d}, & \text{if } c \leq x \leq d; \\ 0, & \text{otherwise.} \end{cases} \tag{3}$$

**Definition 2.1.1** A GFN $\tilde{A}_p = (a, b, c, d; h_L, h_R)_p$ is said to be a $p$-norm GFN with different left height and right height if its membership function is given by

$$\mu_{\tilde{A}_p}(x) = \begin{cases} f^L_{\tilde{A}_p}(x) = h_L\left[1 - \left(\dfrac{x-b}{a-b}\right)^p\right]^{\frac{1}{p}}, & \text{if } a \leq x \leq b; \\ f^M_{\tilde{A}_p}(x) = h_L + (h_R - h_L)\left[1 - \left(\dfrac{x-c}{b-c}\right)^p\right]^{\frac{1}{p}}, & \text{if } b \leq x \leq c; \\ f^R_{\tilde{A}_p}(x) = h_R\left[1 - \left(\dfrac{x-c}{d-c}\right)^p\right]^{\frac{1}{p}}, & \text{if } c \leq x \leq d; \\ 0, & \text{otherwise} \end{cases} \tag{4}$$

where $p$ is a positive integer. The functions $f^L_{\tilde{A}_p} : [a, b] \longrightarrow [0, h_L]$ and $f^R_{\tilde{A}_p} : [c, d] \longrightarrow [0, h_R]$ are both continuous as well as strictly increasing and strictly decreasing functions, respectively. The function $f^M_{\tilde{A}_p} : [b, c] \longrightarrow [h_L, h_R]$ (or $[h_R, h_L]$) is strictly increasing (decreasing) when $h_L < h_R (h_R < h_L)$. When $p$ is one, the $p$-norm GFN with different left height and right height reduces to trapezoidal GFN with different left height and right height as defined by Eq. (3).

**Definition 2.1.2** The inverse function of $\mu_{\tilde{A}_p}(x)$ as given by membership function in (4) is given by

$$\mu^{-1}_{\tilde{A}_p}(y) = \begin{cases} g^L_{\tilde{A}_p}(y) = b + (a-b)\left[1 - \left(\dfrac{y}{h_L}\right)^p\right]^{\frac{1}{p}}, & \text{if } 0 \leq y \leq h_L; \\ g^M_{\tilde{A}_p}(y) = c + (b-c)\left[1 - \left(\dfrac{y-h_L}{h_R-h_L}\right)^p\right]^{\frac{1}{p}}, & \text{if } h_L \leq y \leq h_R, \text{or } (h_R \leq y \leq h_L); \\ g^R_{\tilde{A}_p}(y) = c + (d-c)\left[1 - \left(\dfrac{y}{h_R}\right)^p\right]^{\frac{1}{p}}, & \text{if } 0 \leq y \leq h_R; \end{cases} \tag{5}$$

The functions $g^{\mathrm{L}}_{\tilde{A}_p}:[0,h_{\mathrm{L}}]\longrightarrow[a,b]$ and $g^{\mathrm{R}}_{\tilde{A}_p}:[0,h_{\mathrm{R}}]\longrightarrow[c,d]$ are both continuous as well as strictly increasing and strictly decreasing functions, respectively. The function $g^{\mathrm{M}}_{\tilde{A}_p}:[h_{\mathrm{L}},h_{\mathrm{R}}](\text{or }[h_{\mathrm{R}},h_{\mathrm{L}}])\longrightarrow[b,c]$ is strictly increasing (decreasing) when $h_{\mathrm{L}}<h_{\mathrm{R}}(h_{\mathrm{R}}<h_{\mathrm{L}})$.

## Total integral value

Convex combination of right and left integral values through an index of optimism is called the total integral value [8, 11]. The middle integral value is zero for normal and non-normal $p$-norm trapezoidal fuzzy numbers with different height. However, for a $p$-norm GFNs with different left height and right height this integral value has to be counted for a transparent decision. Keeping this in view, following definitions are put forwarded.

**Definition 3.1** If $\tilde{A}$ is a fuzzy number with different left height and right height as defined by the membership function (1) and the inverse membership function given by (2) then the left integral value of $\tilde{A}$ is defined as

$$I_{\mathrm{L}}(\tilde{A})=\int_0^{h_{\mathrm{L}}}\mu_1^{-1}(y)\mathrm{d}y \tag{6}$$

**Definition 3.2** If $\tilde{A}$ is a fuzzy number with different left height and right height as defined by the membership function (1) and the inverse membership function given by (2) then the right integral value of $\tilde{A}$ is defined as

$$I_{\mathrm{R}}(\tilde{A})=\int_0^{h_{\mathrm{R}}}\mu_3^{-1}(y)\mathrm{d}y \tag{7}$$

**Definition 3.3** If $\tilde{A}$ is a fuzzy number with different left height and right height as defined by the membership function (1) and the inverse membership function given by (2) then the middle integral value of $\tilde{A}$ is defined as

$$I_{\mathrm{M}}(\tilde{A})=\int_{h_{\mathrm{L}}}^{h_{\mathrm{R}}}\mu_2^{-1}(y)\mathrm{d}y \quad \text{or} \quad I_{\mathrm{M}}(\tilde{A})=\int_{h_{\mathrm{R}}}^{h_{\mathrm{L}}}\mu_2^{-1}(y)\mathrm{d}y \tag{8}$$

**Definition 3.4** If $\tilde{A}$ is a fuzzy number with different left height and right height as defined by the membership function (1), then the total integral value with index of optimism $\alpha$ is defined as

$$I^{\alpha}_{\mathrm{T}}(\tilde{A})=\alpha(I_{\mathrm{R}}(\tilde{A})+I_{\mathrm{M}}(\tilde{A}))+(1-\alpha)(I_{\mathrm{L}}(\tilde{A})+I_{\mathrm{M}}(\tilde{A}))$$

**Proposition 3.1** *Let $\tilde{A}_p=(a,b,c,d;h_{\mathrm{L}},h_{\mathrm{R}})_p$ be a p-norm GFN with different left height and right height with membership function* (4), *where p is a positive integer. Then:*

1. *The left membership function $f^{\mathrm{L}}_{\tilde{A}_p}(x)$ is continuous and strictly increasing function and its left integral value is*

$$I_{\mathrm{L}}(\tilde{A}_p)=bh_{\mathrm{L}}+\frac{a-b}{p}h_{\mathrm{L}}\frac{\Gamma\left(\frac{1}{p}+1\right)\Gamma\left(\frac{1}{p}\right)}{\Gamma\left(\frac{2}{p}+1\right)} \tag{9}$$

*where $\Gamma(x)$ is Euler's gamma function, defined by $\int_0^{\infty}y^{x-1}\mathrm{e}^{-y}\mathrm{d}y$.*

2. *The right membership function $f^{\mathrm{R}}_{\tilde{A}_p}(x)$ is continuous and strictly decreasing function and its right integral value is*

$$I_{\mathrm{R}}(\tilde{A}_p)=ch_{\mathrm{R}}+\frac{d-c}{p}h_{\mathrm{R}}\frac{\Gamma\left(\frac{1}{p}+1\right)\Gamma\left(\frac{1}{p}\right)}{\Gamma\left(\frac{2}{p}+1\right)} \tag{10}$$

3. *The middle membership function $f^{\mathrm{M}}_{\tilde{A}_p}(x)$ is continuous and strictly increasing and strictly decreasing when $h_{\mathrm{L}}<h_{\mathrm{R}}$ and $h_{\mathrm{R}}<h_{\mathrm{L}}$, respectively. The middle integral value is given by*

$$I_{\mathrm{M}}(\tilde{A}_p)=(h_{\mathrm{R}}-h_{\mathrm{L}})c+\frac{b-c}{p}(h_{\mathrm{R}}-h_{\mathrm{L}})\times\frac{\Gamma\left(\frac{1}{p}+1\right)\Gamma\left(\frac{1}{p}\right)}{\Gamma\left(\frac{2}{p}+1\right)} \tag{11}$$

4. The total integral value with optimism $\alpha$ *is*

$$\begin{aligned}I^{\alpha}_{\mathrm{T}}(\tilde{A}_p)&=\alpha ch_{\mathrm{R}}+(1-\alpha)bh_{\mathrm{L}}+(h_{\mathrm{R}}-h_{\mathrm{L}})c\\&+\{\alpha h_{\mathrm{R}}(d-c)+(1-\alpha)h_{\mathrm{L}}(a-b)\\&+(h_{\mathrm{R}}-h_{\mathrm{L}})(b-c)\}\frac{\Gamma\left(\frac{1}{p}+1\right)\Gamma\left(\frac{1}{p}\right)}{p\times\Gamma\left(\frac{2}{p}+1\right)}\end{aligned} \tag{12}$$

*Proof* Continuity of the left membership function $f^{\mathrm{L}}_{\tilde{A}_p}(x)$ is trivial. Also, this function is strictly increasing and its integral values are inherited from [6]. Similarly, for the right membership function $f^{\mathrm{L}}_{\tilde{A}_p}(x)$.

Trivially, the function $f^{\mathrm{M}}_{\tilde{A}_p}(x)$ is continuous. Now,

$$\begin{aligned}\frac{\mathrm{d}}{\mathrm{d}x}f^{\mathrm{M}}_{\tilde{A}_p}(x)&=\frac{\mathrm{d}}{\mathrm{d}x}\left\{h_{\mathrm{L}}+(h_{\mathrm{R}}-h_{\mathrm{L}})\left[1-\left(\frac{x-c}{b-c}\right)^p\right]^{\frac{1}{p}}\right\}\\&=(h_{\mathrm{R}}-h_{\mathrm{L}})\frac{1}{p}\left[1-\left(\frac{x-c}{b-c}\right)^p\right]^{\frac{1}{p}-1}\\&\times\left[-p\left(\frac{x-c}{b-c}\right)^{p-1}\left(\frac{1}{b-c}\right)\right]\end{aligned}$$

Since $0\leq\left(\frac{x-c}{b-c}\right)^p\leq1$, it trivially follows that $\frac{\mathrm{d}}{\mathrm{d}x}f^{\mathrm{M}}_{\tilde{A}_p}(x)\geq0$ if $h_{\mathrm{R}}-h_{\mathrm{L}}\geq0$ and $\frac{\mathrm{d}}{\mathrm{d}x}f^{\mathrm{M}}_{\tilde{A}_p}(x)\leq0$ if $h_{\mathrm{R}}-h_{\mathrm{L}}\leq0$. Hence the function $f^{\mathrm{M}}_{\tilde{A}_p}(x)$ is strictly increasing when $h_{\mathrm{R}}\geq h_{\mathrm{L}}$ and strictly decreasing when $h_{\mathrm{R}}\leq h_{\mathrm{L}}$. Also,

$$I_{\mathrm{M}}(\tilde{A}_p) = \int_{h_{\mathrm{L}}(\text{or } h_{\mathrm{R}})}^{h_{\mathrm{R}}(\text{or } h_{\mathrm{L}})} \left[ c + (b-c)\left[1 - \left(\frac{y-h_{\mathrm{L}}}{h_{\mathrm{R}}-h_{\mathrm{L}}}\right)^p\right]^{\frac{1}{p}} \right] dx$$

$$= (h_{\mathrm{R}} - h_{\mathrm{L}})c + \frac{b-c}{p}(h_{\mathrm{R}} - h_{\mathrm{L}}) \frac{\Gamma\left(\frac{1}{p}+1\right)\Gamma\left(\frac{1}{p}\right)}{\Gamma\left(\frac{2}{p}+1\right)}.$$

Now, the total integral value with optimism $\alpha$ is

$$I_{\mathrm{T}}^{\alpha}(\tilde{A}_p) = \alpha\{I_{\mathrm{R}}(\tilde{A}_p) + I_{\mathrm{M}}(\tilde{A}_p)\} + (1-\alpha)\{I_{\mathrm{L}}(\tilde{A}_p) + I_{\mathrm{M}}(\tilde{A}_p)\}$$
$$= \alpha c h_{\mathrm{R}} + (1-\alpha) b h_{\mathrm{L}} + (h_{\mathrm{R}} - h_{\mathrm{L}})c$$
$$+ \{\alpha h_{\mathrm{R}}(d-c) + (1-\alpha)h_{\mathrm{L}}(a-b)$$
$$+ (h_{\mathrm{R}} - h_{\mathrm{L}})(b-c)\} \frac{\Gamma\left(\frac{1}{p}+1\right)\Gamma\left(\frac{1}{p}\right)}{p \times \Gamma\left(\frac{2}{p}+1\right)}$$

□

*Remark* 3.1 For a pessimistic decision maker $\alpha = 0$, for an optimistic decision maker $\alpha = 1$ and for a moderate decision maker $\alpha = 0.5$.

**Corollary 3.1** *For an pessimistic decision maker* $(\alpha = 0)$ *the total integral* $I_{\mathrm{T}}^{\alpha}(\tilde{A}) = I_{\mathrm{L}}(\tilde{A}) + I_{\mathrm{M}}(\tilde{A})$, *for optimistic decision maker* $(\alpha = 1)$ *the total integral value* $I_{\mathrm{T}}^{\alpha}(\tilde{A}) = I_{\mathrm{R}}(\tilde{A}) + I_{\mathrm{M}}(\tilde{A})$ *and for moderate decision maker* $(\alpha = 0.5)$ *the total integral value* $I_{\mathrm{T}}^{\alpha}(\tilde{A}) = I_{\mathrm{M}}(\tilde{A}) + \frac{1}{2}\{I_{\mathrm{R}}(\tilde{A}) + I_{\mathrm{L}}(\tilde{A})\}$.

**Note 1** The above total integral values for pessimistic decision maker, optimistic decision maker and moderate decision maker reduce to $I_{\mathrm{T}}^{\alpha}(\tilde{A}) = I_{\mathrm{L}}(\tilde{A})$, $I_{\mathrm{T}}^{\alpha}(\tilde{A}) = I_{\mathrm{R}}(\tilde{A})$ and $I_{\mathrm{T}}^{\alpha}(\tilde{A}) = \frac{1}{2}\{I_{\mathrm{R}}(\tilde{A}) + I_{\mathrm{L}}(\tilde{A})\}$, respectively, for a non-normal $p$-norm trapezoidal fuzzy number. This is because $I_{\mathrm{M}}(\tilde{A}) = 0$ when $h_{\mathrm{L}} = h_{\mathrm{R}}$.

## Arithmetic operations

The arithmetic of $p$-norm GFNs with different left height and right height are reviewed from Chen [5]. Let $\tilde{A}_p = (a, b, c, d; h_{\mathrm{L}}, h_{\mathrm{R}})_p$ and $\tilde{B}_p = (q, r, s, t; h'_{\mathrm{L}}, h'_{\mathrm{R}})_p$ be $p$-norm GFNs with different left height and right height. Then

1. $\tilde{A}_p \oplus \tilde{B}_p = (a+q, b+r, c+s, d+t; \min(h_{\mathrm{L}}, h'_L), \min(h_{\mathrm{R}}, h'_R))_p$
2. $\tilde{A}_p \ominus \tilde{B}_p = (a-t, b-s, c-r, d-q; \min(h_{\mathrm{L}}, h'_L), \min(h_{\mathrm{R}}, h'_R))_p$
3. 

$$\lambda\tilde{A} = \begin{cases} (\lambda a, \lambda b, \lambda c, \lambda d; h_{\mathrm{L}}, h_{\mathrm{R}})_p & \lambda > 0 \\ (\lambda d, \lambda c, \lambda b, \lambda a; h_{\mathrm{L}}, h_{\mathrm{R}})_p & \lambda < 0 \end{cases}$$

## The proposed method

In this section, a method of ranking $p$-norm GFNs with different left height and right height is presented. The method calculates total integral value on the basis of left integral value, right integral value and middle integral value. Ranking is done on the basis of these evaluated total integral values. Let $\tilde{A}_p = (a, b, c, d; h_{\mathrm{L}}, h_{\mathrm{R}})_p$ and $\tilde{B}_p = (q, r, s, t; h'_{\mathrm{L}}, h'_{\mathrm{R}})_p$ be $p$-norm GFNs with different left height and right height. Then

1. $\tilde{A}_p \succ \tilde{B}_p$ if $I_{\mathrm{T}}^{\alpha}(\tilde{A}_p) > I_{\mathrm{T}}^{\alpha}(\tilde{B}_p)$,
2. $\tilde{A}_p \prec \tilde{B}_p$ if $I_{\mathrm{T}}^{\alpha}(\tilde{A}_p) < I_{\mathrm{T}}^{\alpha}(\tilde{B}_p)$,
3. $\tilde{A}_p \sim \tilde{B}_p$ if $I_{\mathrm{T}}^{\alpha}(\tilde{A}_p) = I_{\mathrm{T}}^{\alpha}(\tilde{B}_p)$,

The following are the steps involved in this ranking method:

Step 1. Find $h_1 = \min(h_{\mathrm{L}}, h'_L)$ and $h_2 = \min(h_{\mathrm{R}}, h'_R)$.
Step 2. Find $I_{\mathrm{L}}(\tilde{A}_p)$, $I_{\mathrm{R}}(\tilde{A}_p)$, $I_{\mathrm{M}}(\tilde{A}_p)$ and $I_{\mathrm{L}}(\tilde{B}_p)$, $I_{\mathrm{R}}(\tilde{B}_p)$, $I_{\mathrm{M}}(\tilde{B}_p)$, such that

$$I_{\mathrm{L}}(\tilde{A}_p) = \int_0^{h_1} g_{\tilde{A}_p}^{\mathrm{L}}(x)\mathrm{d}x, \quad \text{where}$$
$$g_{\tilde{A}_p}^{\mathrm{L}}(x) = b + (a-b)\left[1 - \left(\frac{x}{h_1}\right)^p\right]^{\frac{1}{p}}, \tag{13}$$
$$= bh_1 + \frac{a-b}{p}h_1 \frac{\Gamma\left(\frac{1}{p}+1\right)\Gamma\left(\frac{1}{p}\right)}{\Gamma\left(\frac{2}{p}+1\right)}$$

$$I_{\mathrm{R}}(\tilde{A}_p) = \int_0^{h_2} g_{\tilde{A}_p}^{\mathrm{R}}(x)\mathrm{d}x, \quad \text{where}$$
$$g_{\tilde{A}_p}^{\mathrm{R}}(x) = c + (d-c)\left[1 - \left(\frac{x}{h_2}\right)^p\right]^{\frac{1}{p}}, \tag{14}$$
$$= ch_2 + \frac{d-c}{p}h_2 \frac{\Gamma\left(\frac{1}{p}+1\right)\Gamma\left(\frac{1}{p}\right)}{\Gamma\left(\frac{2}{p}+1\right)}$$

$$I_{\mathrm{M}}(\tilde{A}_p) = \int_{h_1(\text{or } h_2)}^{h_2(\text{or } h_1)} g_{\tilde{A}_p}^{\mathrm{M}}(x)\mathrm{d}x,$$
$$\text{where} \quad g_{\tilde{A}_p}^{\mathrm{M}}(x) = c + (b-c)\left[1 - \left(\frac{x-h_1}{h_2-h_1}\right)^p\right]^{\frac{1}{p}}, \tag{15}$$
$$= c(h_2 - h_1) + \frac{b-c}{p}(h_2 - h_1) \frac{\Gamma\left(\frac{1}{p}+1\right)\Gamma\left(\frac{1}{p}\right)}{\Gamma\left(\frac{2}{p}+1\right)}$$

$$I_{\mathrm{L}}(\tilde{B}_p) = \int_0^{h_1} g_{\tilde{B}_p}^{\mathrm{L}}(x)\mathrm{d}x,$$
$$\text{where} \quad g_{\tilde{B}_p}^{\mathrm{L}}(x) = r + (q-r)\left[1 - \left(\frac{x}{h_1}\right)^p\right]^{\frac{1}{p}}, \tag{16}$$
$$= rh_1 + \frac{q-r}{p}h_1 \frac{\Gamma\left(\frac{1}{p}+1\right)\Gamma\left(\frac{1}{p}\right)}{\Gamma\left(\frac{2}{p}+1\right)}$$

$$I_{\mathrm{R}}(\tilde{B}_p) = \int_0^{h_2} g^{\mathrm{R}}_{\tilde{B}_p}(x)\mathrm{d}x, \quad \text{where}$$

$$g^{\mathrm{R}}_{\tilde{B}_p}(x) = s + (t-s)\left[1-\left(\frac{x}{h_2}\right)^p\right]^{\frac{1}{p}}, \tag{17}$$

$$= sh_2 + \frac{t-s}{p}h_2\frac{\Gamma\left(\frac{1}{p}+1\right)\Gamma\left(\frac{1}{p}\right)}{\Gamma\left(\frac{2}{p}+1\right)}$$

$$I_{\mathrm{M}}(\tilde{B}_p) = \int_{h_1(\text{or } h_2}^{h_2(\text{or } h_1)} g^{\mathrm{M}}_{\tilde{B}_p}(x)\mathrm{d}x, \quad \text{where}$$

$$g^{\mathrm{M}}_{\tilde{B}_p}(x) = s + (r-s)\left[1-\left(\frac{x-h_1}{h_2-h_1}\right)^p\right]^{\frac{1}{p}},$$

$$= s(h_2-h_1) + \frac{r-s}{p}(h_2-h_1)\frac{\Gamma\left(\frac{1}{p}+1\right)\Gamma\left(\frac{1}{p}\right)}{\Gamma\left(\frac{2}{p}+1\right)} \tag{18}$$

Step 3. Find $I^{\alpha}_{\mathrm{T}}(\tilde{A}_p)$ and $I^{\alpha}_{\mathrm{T}}(\tilde{B}_p)$, which are given by

$$\begin{aligned} I^{\alpha}_{\mathrm{T}}(\tilde{A}_p) &= \alpha ch_2 + (1-\alpha)bh_1 + (h_2-h_1)c \\ &\quad + \{\alpha h_2(d-c) + (1-\alpha)h_1(a-b) \\ &\quad + (h_2-h_1)(b-c)\}\frac{\Gamma\left(\frac{1}{p}+1\right)\Gamma\left(\frac{1}{p}\right)}{p\times\Gamma\left(\frac{2}{p}+1\right)}, \end{aligned} \tag{19}$$

$$\begin{aligned} I^{\alpha}_{\mathrm{T}}(\tilde{B}_p) &= \alpha sh_2 + (1-\alpha)rh_1 + (h_2-h_1)s \\ &\quad + \{\alpha h_2(t-s) + (1-\alpha)h_1(q-r) \\ &\quad + (h_2-h_1)(r-s)\}\frac{\Gamma\left(\frac{1}{p}+1\right)\Gamma\left(\frac{1}{p}\right)}{p\times\Gamma\left(\frac{2}{p}+1\right)}. \end{aligned} \tag{20}$$

Step 4. Check $I^{\alpha}_{\mathrm{T}}(\tilde{A}_p) > I^{\alpha}_{\mathrm{T}}(\tilde{B}_p)$ or $I^{\alpha}_{\mathrm{T}}(\tilde{A}_p) < I^{\alpha}_{\mathrm{T}}(\tilde{B}_p)$ or $I^{\alpha}_{\mathrm{T}}(\tilde{A}_p) = I^{\alpha}_{\mathrm{T}}(\tilde{B}_p)$.

Case (i) If $I^{\alpha}_{\mathrm{T}}(\tilde{A}_p) > I^{\alpha}_{\mathrm{T}}(\tilde{B}_p)$ then $\tilde{A}_p \succ \tilde{B}_p$.
Case (ii) If $I^{\alpha}_{\mathrm{T}}(\tilde{A}_p) < I^{\alpha}_{\mathrm{T}}(\tilde{B}_p)$ then $\tilde{A}_p \prec \tilde{B}_p$.
Case (iii) If $I^{\alpha}_{\mathrm{T}}(\tilde{A}_p) = I^{\alpha}_{\mathrm{T}}(\tilde{B}_p)$ then $\tilde{A}_p \sim \tilde{B}_p$.

*Remark* 4.1 For any two arbitrary generalised fuzzy numbers with different left height and right height, $\tilde{A}_p$ and $\tilde{B}_p$, we have

$$I^{\alpha}_{\mathrm{T}}(\tilde{A}_p + \tilde{B}_p) = I^{\alpha}_{\mathrm{T}}(\tilde{A}_p) + I^{\alpha}_{\mathrm{T}}(\tilde{B}_p)$$

**Proposition 4.1** *Let* $\tilde{A} = (a,b,c,d;h_{AL},h_{AR})$ *and* $\tilde{B} = (a,e,d;h_B)$ *be GFN with different left height and right height and non-normal triangular fuzzy number, respectively, such that* $-\infty < a \leq b \leq e \leq c \leq d < \infty$. *Then*

(i) $I_{\mathrm{L}}(\tilde{A}) \leq I_{\mathrm{L}}(\tilde{B})$,
(ii) $I_{\mathrm{R}}(\tilde{A}) \geq I_{\mathrm{R}}(\tilde{B})$,
(iii) $I_{\mathrm{M}}(\tilde{A}) = I_{\mathrm{M}}(\tilde{B})$, *if* $b+c=2e$,
(iv) $I_{\mathrm{M}}(\tilde{A}) > I_{\mathrm{M}}(\tilde{B})$, *if* $h_1 < h_2$ *and* $b+c > 2e$ *or* $h_1 > h_2$ *and* $b+c < 2e$,
(v) $I_{\mathrm{M}}(\tilde{A}) < I_{\mathrm{M}}(\tilde{B})$, *if* $h_1 < h_2$ *and* $b+c < 2e$ *or* $h_1 > h_2$ *and* $b+c > 2e$,
(vi) $I^{\alpha}_{\mathrm{T}}(\tilde{A}) > I^{\alpha}_{\mathrm{T}}(\tilde{B})$ *if* $\alpha h_2(c-e) + (1-\alpha)h_1(b-e) + (h_2-h_1)(b+c-2e) > 0$,
(vii) $I^{\alpha}_{\mathrm{T}}(\tilde{A}) < I^{\alpha}_{\mathrm{T}}(\tilde{B})$ *if* $\alpha h_2(c-e) + (1-\alpha)h_1(b-e) + (h_2-h_1)\ (b+c-2e) < 0$ *and*
(viii) $I^{\alpha}_{\mathrm{T}}(\tilde{A}) = I^{\alpha}_{\mathrm{T}}(\tilde{B})$ *if* $\alpha h_2(c-e) + (1-\alpha)h_1(b-e) + (h_2-h_1)\ (b+c-2e) = 0$.

*Proof* From Eqs. (13), (14), (15) and (19), on appropriate substitutions of the variables the following could be easily obtained:

$$I_{\mathrm{L}}(\tilde{A}) = \frac{h_1(a+b)}{2}, \quad I_{\mathrm{R}}(\tilde{A}) = \frac{h_2(d+c)}{2},$$

$$I_{\mathrm{M}}(\tilde{A}) = \frac{(h_2-h_1)(b+c)}{2}, I_{\mathrm{L}}(\tilde{B}) = \frac{h_1(a+e)}{2},$$

$$I_{\mathrm{R}}(\tilde{B}) = \frac{h_2(e+d)}{2}, \quad I_{\mathrm{M}}(\tilde{B}) = e(h_2-h_1),$$

$$\begin{aligned} I^{\alpha}_{\mathrm{T}}(\tilde{A}) &= \frac{1}{2}\{\alpha h_2(c+d) + (1-\alpha)h_1(a+b) \\ &\quad + (h_2-h_1)(c+b)\} \quad \text{and} \end{aligned}$$

$$I^{\alpha}_{\mathrm{T}}(\tilde{B}) = \frac{1}{2}\{\alpha h_2(e+d) + (1-\alpha)h_1(a+e) + 2e(h_2-h_1)\}$$

where $h_1 = \min(h_{AL}, h_B)$ and $h_2 = \min(h_{AR}, h_B)$.

Now $I_{\mathrm{L}}(\tilde{A}) - I_{\mathrm{L}}(\tilde{B}) = \frac{h_1}{2}(b-e) \leq 0$ as $b \leq e \leq c$, hence inequality (i) is deduced. Similarly, inequality (ii) could be deduced. Again, we have $I_{\mathrm{M}}(\tilde{A}) - I_{\mathrm{M}}(\tilde{B}) = \frac{h_2-h_1}{2}(b+c-2e)$ hence the inequalities (iii), (iv) and (v) follow immediately. Also we have

$$\begin{aligned} I^{\alpha}_{\mathrm{T}}(\tilde{A}) - I^{\alpha}_{\mathrm{T}}(\tilde{B}) &= \frac{1}{2}\{\alpha h_2(c-e) + (1-\alpha)h_1(b-e) \\ &\quad + (h_2-h_1)(c+b-2e)\}. \end{aligned}$$

Hence the inequalities (vi), (vii) and (viii) can be deduced easily. □

**Corollary 4.1** [9] *Let* $\tilde{A} = (a,b,c,d;h_A)$ *and* $\tilde{B} = (a,e,d;h_B)$ *be non-normal trapezoidal and triangular fuzzy numbers, respectively, where* $-\infty < a \leq b \leq e \leq c \leq d < \infty$. *Then*

(i) $I_{\mathrm{L}}(\tilde{A}) \leq I_{\mathrm{L}}(\tilde{B})$,
(ii) $I_{\mathrm{R}}(\tilde{A}) \geq I_{\mathrm{R}}(\tilde{B})$,
(iii) $I^{\alpha}_{\mathrm{T}}(\tilde{A}) > I^{\alpha}_{\mathrm{T}}(\tilde{B})$ *if* $e < c\alpha + (1-\alpha)b$,
(iv) $I^{\alpha}_{\mathrm{T}}(\tilde{A}) = I^{\alpha}_{\mathrm{T}}(\tilde{B})$ *if* $e = c\alpha + (1-\alpha)b$ *and*
(v) $I^{\alpha}_{\mathrm{T}}(\tilde{A}) < I^{\alpha}_{\mathrm{T}}(\tilde{B})$ *if* $e > c\alpha + (1-\alpha)b$.

These inequalities are particular case of the inequalities in the Proposition 4.1. These can be obtained by appropriate substitutions on the inequalities in the Proposition 4.1.

**Proposition 4.2** *Let $\tilde{A}=(a,b,c,d;h_{AL},h_{AR})$ and $\tilde{B}_2=(a,b,c,d;h_{BL},h_{BR})_2$ be GFN and 2-norm GFN with different left height and right height, respectively. Then*

(i) $I_L(\tilde{A})\geq I_L(\tilde{B}_2)$,
(ii) $I_L(\tilde{A})\leq I_L(\tilde{B}_2)$,
(iii) $I_M(\tilde{A})\geq(\leq)I_M(\tilde{B}_2)$ *if* $h_1<(>)h_2$,
(iv) $I^{\alpha}_{M}(\tilde{A})>I^{\alpha}_{M}(\tilde{B}_2)$ *if* $\alpha h_2(d-c)+h_1(1-\alpha)(a-b)+(h_2-h_1)(b-c)<0$,
(v) $I^{\alpha}_{M}(\tilde{A})<I^{\alpha}_{M}(\tilde{B}_2)$ *if* $\alpha h_2(d-c)+h_1(1-\alpha)(a-b)+(h_2-h_1)(b-c)>0$ *and*
(vi) $I^{\alpha}_{M}(\tilde{A})=I^{\alpha}_{M}(\tilde{B}_2)$ *if* $\alpha h_2(d-c)+h_1(1-\alpha)(a-b)+(h_2-h_1)(b-c)=0$.

*Proof* $\tilde{A}$ and $\tilde{B}_2$ are GFN and 2-norm GFN with different left height and right height. Hence by the proposed method $I_L(\tilde{A})$, $I_R(\tilde{A})$, $I_M(\tilde{A})$, $I^{\alpha}_{T}(\tilde{A})$ $I_L(\tilde{B}_2)$, $I_R(\tilde{B}_2)$, $I_M(\tilde{B}_2)$ and $I^{\alpha}_{T}(\tilde{B}_2)$ are obtained by using Eqs. (13), (14), (15) and (19) as:

$$I_L(\tilde{A})=\frac{h_1(a+b)}{2},\quad I_R(\tilde{A})=\frac{h_2(d+c)}{2},$$
$$I_M(\tilde{A})=\frac{(h_2-h_1)(b+c)}{2},\ I_L(\tilde{B}_2)=bh_1+\frac{a-b}{4}h_1\pi,$$
$$I_R(\tilde{B}_2)=ch_2+\frac{d-c}{4}h_2\pi,\quad I_M(\tilde{B}_2)=\frac{b+3c}{4}(h_2-h_1)\pi,$$
$$I^{\alpha}_{T}(\tilde{A})=\frac{1}{2}\{\alpha h_2(c+d)+(1-\alpha)h_1(a+b)+(h_2-h_1)(c+b)\}\quad\text{and}$$
$$I^{\alpha}_{T}(\tilde{B}_2)=\frac{1}{4}[(h_2-h_1)\{4c+\pi(b-c)\}+(1-\alpha)\times h_1\{4b+\pi(a-b)\}+\alpha h_2\{4c+\pi(d-c)\}]$$

where $h_1=\min(h_{AL},h_{BL})$ and $h_2=\min(h_{AR},h_{BR})$.

Now, $I_L(\tilde{A})-I_L(\tilde{B}_2)=\frac{2-\pi}{4}h_1(a-b)\geq 0$ and $I_R(\tilde{A})-I_R(\tilde{B}_2)=\frac{2-\pi}{4}h_2(d-c)\leq 0$. Thus the desired inequalities (i) and (ii) are obtained. $(b-c)(2-\pi)$ is always greater than or equal to zero, thus $I_M(\tilde{A})-I_M(\tilde{B}_2)=(h_2-h_1)\frac{(b-c)(2-\pi)}{4}\geq(\leq)0$ if $h_1<(>)h_2$, which prove the inequality (iii). For the inequalities (iv), (v) and (iv), we have

$$I^{\alpha}_{M}(\tilde{A})-I^{\alpha}_{M}(\tilde{B}_2)=\frac{2-\pi}{4}\{\alpha h_2(d-c)+(1-\alpha)h_1(a-b)+(h_2-h_1)(b-c)\}.$$

Hence the inequalities (iv), (v) and (vi) follow immediately. □

**Corollary 4.2** [9] *Let $\tilde{A}=(a,b,c,d;h_A)$ and $\tilde{B}_2=(a,b,c,d;h_B)$ be non-normal trapezoidal fuzzy number and non-normal 2-norm trapezoidal fuzzy number, respectively, then*

(i) $I_L(\tilde{A})\geq I_L(\tilde{B}_2)$,
(ii) $I_L(\tilde{A})\leq I_L(\tilde{B}_2)$,
(iii) $I^{\alpha}_{T}<I^{\alpha}_{T}(\tilde{B})$ *if* $\alpha(d-c)+(1-\alpha)(a-b)>0$,
(iv) $I^{\alpha}_{T}=I^{\alpha}_{T}(\tilde{B})$ *if* $\alpha(d-c)+(1-\alpha)(a-b)=0$ *and*
(v) $I^{\alpha}_{T}>I^{\alpha}_{T}(\tilde{B})$ *if* $\alpha(d-c)+(1-\alpha)(a-b)<0$.

These inequalities are particular case of the inequalities in the Proposition 4.2. These can be obtained by appropriate substitutions on the inequalities in the Proposition 4.2.

**Proposition 4.3** *Let $\tilde{A}=(a,a,a,a;1,1)$ and $\tilde{B}=(b,b,b,b;1,1)$ be GFNs with height 1. Then*

(i) $I_L(\tilde{A})\geq(\leq)I_L(\tilde{B})$, *if* $a\geq(\leq)b$,
(ii) $I_M(\tilde{A})\geq(\leq)I_M(\tilde{B}_2)$ *if* $a\geq(\leq)b$ *and*
(iii) $I^1_T(\tilde{A})>(<)I^1_T(\tilde{B})$ *if* $a>(<)b$.

The Proposition 4.3 validates that the proposed method can also be applied for real numbers.

*Example* 4.1 Let $\tilde{A}=(5,7,8,9;0.5,0.6)$ and $\tilde{B}_2=(5,7,8,9;0.7,0.6)_2$ be GFN and 2-norm GFN with different left height and right height, which are depicted in Fig. 1. But, according to the proposed modified method $h_1=\min(0.5,0.6)$ and $h_2=\min(0.7,0.6)$. Also, $I^{0.5}_{T}(\tilde{A})=4.8000$ and $I^{0.5}_{T}(\tilde{B}_2)=4.7144$. Thus, $I^{0.5}_{T}(\tilde{A})>I^{0.5}_{T}(\tilde{B}_2)$.

*Example* 4.2 Let $\tilde{A}=(0.1659,0.2803,0.7463,1.154;0.5,0.6)$, $\tilde{B}=(0.1611,0.2475,0.5696,0.8187;0.4,0.5)$ and $\tilde{C}=(0.1645,0.2445,0.5869,0.8894;0.5,0.6)$, are GFNs with different left height and right height. Figure 2 depicts the membership function of the above fuzzy numbers.

Here $I^{0.5}_{T}(\tilde{A})=0.3335$, $I^{0.5}_{T}(\tilde{B})=0.2553$ and $I^{0.5}_{T}(\tilde{C})=0.2670$, and $I^1_T(\tilde{A})=0.1406$, $I^1_T(\tilde{B})=0.1226$ and $I^1_T(\tilde{C})=0.1234$. Thus a moderate decision and a pessimistic decision maker rank them as $\tilde{A}>\tilde{C}>\tilde{B}$.

*Example* 4.3 Let $\tilde{A}_2=(-2,-1,0,1;0.5,0.5)_2$, $\tilde{B}_2=(-1.5,-0.5,0.5,1.5;0.5,0.6)_2$ and $\tilde{C}_2=(1,1.5,2,2.5;0.6,0.5)_2$, are GFNs with different left height and right height. The membership functions of the fuzzy numbers are depicted in Fig. 3. Here, $I^{0.5}_{T}(\tilde{A}_2)=-0.2500$, $I^{0.5}_{T}(\tilde{B}_2)=0.000$ and $I^{0.5}_{T}(\tilde{C}_2)=0.6787$, and $I^{0}_{T}(\tilde{A}_2)=-0.8927$, $I^{0}_{T}(\tilde{B}_2)=-0.6427$ and $I^{0}_{T}(\tilde{C}_2)=0.5537$. A moderate decision maker ($\alpha=0.5$) ranks $\tilde{A}$, $\tilde{B}$ and $\tilde{C}$ as $\tilde{C}>\tilde{B}>\tilde{A}$ and also a pessimistic decision maker ranks them in the same order.

**Fig. 1** Fuzzy numbers $\tilde{A} = (5, 7, 8, 9; 0.5, 0.6)$ and $\tilde{B}_2 = (5, 7, 8, 9; 0.7, 0.6)_2$

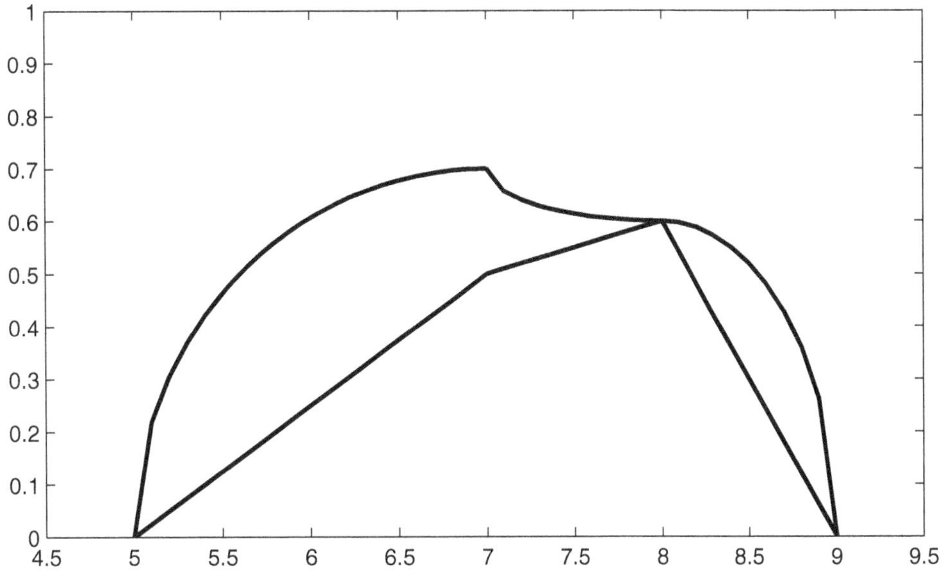

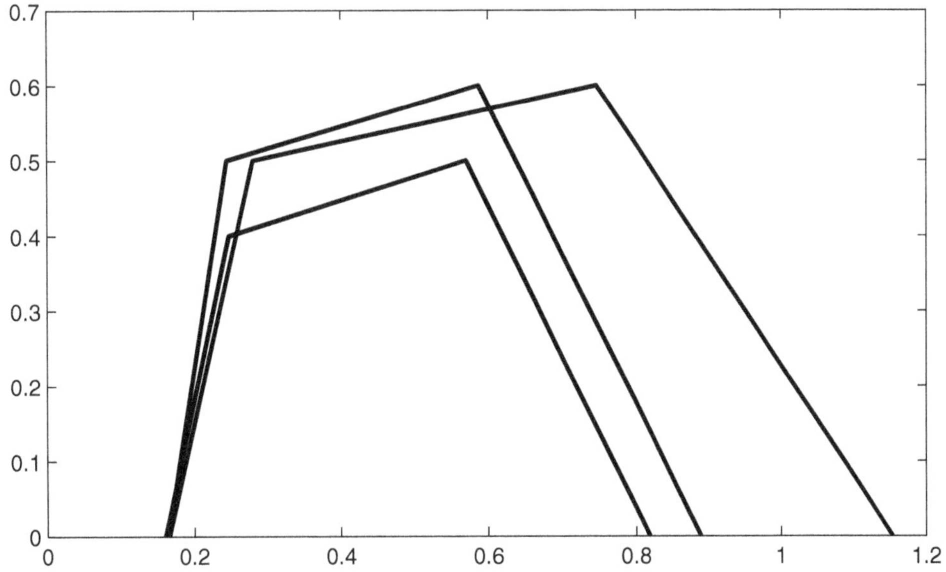

**Fig. 2** Fuzzy numbers $\tilde{A} = (0.1659, 0.2803, 0.7463, 1.154; 0.5, 0.6)$, $\tilde{B} = (0.1611, 0.2475, 0.5696, 0.8187; 0.4, 0.5)$ and $\tilde{C} = (0.1645, 0.2445, 0.5869, 0.8894; 0.5, 0.6)$

*Example* 4.4 Let $\tilde{A} = (1, 1, 1, 1; 1, 1)$ and $\tilde{B} = (2, 2, 2, 2; 1, 1)$ be GFNs which are actually real numbers. Now by Proposition 4.3 $\tilde{B} > \tilde{A}$ trivially.

*Example* 4.5 Consider the following set of fuzzy numbers,

Set A $\tilde{A} = (0.4, 0.5, 0.5, 0.6; 1, 1)$,
$\tilde{B} = (0.2, 0.4, 0.6, 0.8; 1, 1)$.

Set B $\tilde{A} = (0.2, 0.3, 0.3, 0.6; 1, 1)$,
$\tilde{B} = (0.2, 0.4, 0.4, 0.6; 1, 1)$,
$\tilde{C} = (0.2, 0.5, 0.5, 0.6; 1, 1)$.

Set C $\tilde{A} = (0.2, 0.4, 0.4, 0.6; 0.9, 0.9)$,
$\tilde{B} = (0.2, 0.4, 0.4, 0.6; 1, 1)$,
$\tilde{C} = (0.2, 0.4, 0.4, 0.6; 0.5, 0.5)$.

Set D $\tilde{A} = (0.4, 0.5, 0.5, 0.6; 0.5, 0.5)$,
$\tilde{B} = (0.2, 0.4, 0.6, 0.8; 0.6, 0.6)$.

Set E $\tilde{A} = (0.4, 0.5, 0.5, 0.6; 0.6, 0.7)$,
$\tilde{B} = (0.2, 0.4, 0.6, 0.8; 0.5, 0.6)$.

Set F $\tilde{A} = (0.4, 0.5, 0.5, 0.6; 0.6, 0.7)$,
$\tilde{B} = (0.2, 0.4, 0.6, 0.8; 0.5, 0.6)_2$.

**Fig. 3** Fuzzy numbers $\tilde{A}_2 = (-2, -1, 0, 1; 0.5, 0.5)_2$, $\tilde{B}_2 = (-1.5, -0.5, 0.5, 1.5; 0.5, 0.6)_2$ and $\tilde{C}_2 = (1, 1.5, 2, 2.5; 0.6, 0.5)_2$

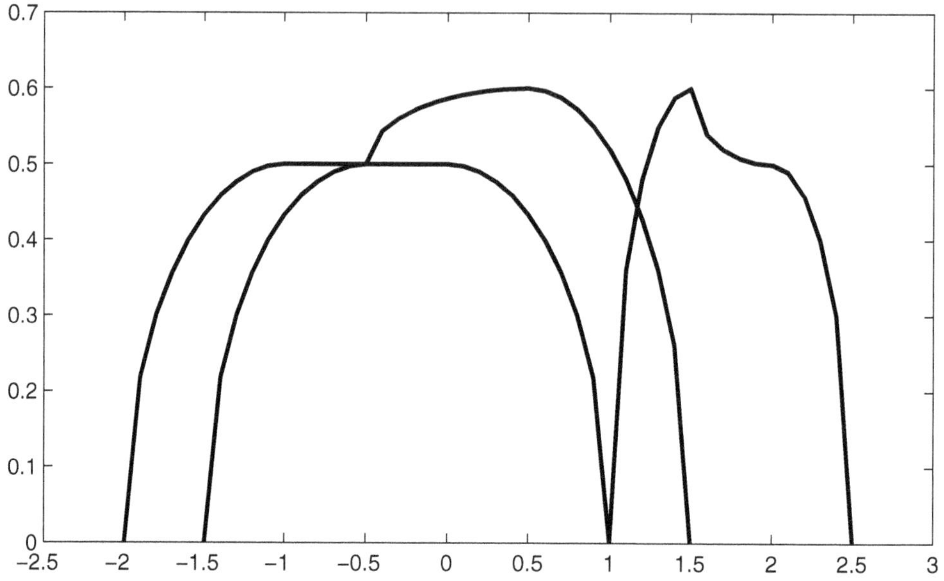


**Table 1** Comparison of the proposed method with Kim and Park [8] and Kumar et al. [9] methods

| Test sets | Index | Kim and Park | | | Kumar's method | | | Proposed method | | |
|---|---|---|---|---|---|---|---|---|---|---|
| | | $\tilde{A}$ | $\tilde{B}$ | $\tilde{C}$ | $\tilde{A}$ | $\tilde{B}$ | $\tilde{C}$ | $\tilde{A}$ | $\tilde{B}$ | $\tilde{C}$ |
| Set A | $\alpha = 1.0$ | 0.5714 | 0.75 | – | 0.55 | 0.70 | – | 0.55 | 0.70 | – |
| | $\alpha = 0.5$ | 0.5000 | 0.50 | – | 0.50 | 0.50 | – | 0.50 | 0.50 | – |
| | $\alpha = 0.0$ | 0.4286 | 0.25 | – | 0.45 | 0.30 | – | 0.45 | 0.30 | – |
| Set B | $\alpha = 1.0$ | 0.5714 | 0.6666 | 0.8000 | 0.45 | 0.50 | 0.55 | 0.45 | 0.50 | 0.55 |
| | $\alpha = 0.5$ | 0.3857 | 0.4999 | 0.6143 | 0.35 | 0.40 | 0.45 | 0.35 | 0.40 | 0.45 |
| | $\alpha = 0.0$ | 0.2000 | 0.3333 | 0.4286 | 0.25 | 0.30 | 0.35 | 0.25 | 0.30 | 0.35 |
| Set C | $\alpha = 1.0$ | – | – | – | 0.45 | 0.50 | 0.25 | 0.45 | 0.50 | 0.25 |
| | $\alpha = 0.5$ | – | – | – | 0.36 | 0.40 | 0.20 | 0.36 | 0.40 | 0.20 |
| | $\alpha = 0.0$ | – | – | – | 0.27 | 0.30 | 0.15 | 0.27 | 0.30 | 0.15 |
| Set D | $\alpha = 1.0$ | – | – | – | 0.275 | 0.35 | – | 0.275 | 0.35 | – |
| | $\alpha = 0.5$ | – | – | – | 0.250 | 0.25 | – | 0.250 | 0.25 | – |
| | $\alpha = 0.0$ | – | – | – | 0.225 | 0.15 | – | 0.225 | 0.15 | – |
| Set E | $\alpha = 1.0$ | – | – | – | – | – | – | 0.3800 | 0.470 | – |
| | $\alpha = 0.5$ | – | – | – | – | – | – | 0.3275 | 0.335 | – |
| | $\alpha = 0.0$ | – | – | – | – | – | – | 0.2750 | 0.200 | – |
| Set F | $\alpha = 1.0$ | – | – | – | – | – | – | 0.3800 | 0.4985 | – |
| | $\alpha = 0.5$ | – | – | – | – | – | – | 0.3275 | 0.3321 | – |
| | $\alpha = 0.0$ | – | – | – | – | – | – | 0.2750 | 0.1658 | – |

The results of the sets A, B, C, D and E are depicted in the Table 1. Sets A and B consist of normal fuzzy numbers, hence the ranking order by the three methods is same. Sets C and D consist of non-normal fuzzy numbers, Kim and Park [8] give no option for ranking such type of fuzzy number. The method of Kumar et al. [9] and the proposed method's ranking order are same for the sets C and D. However, sets E and F which consist of $p$-norm generalised fuzzy numbers with different left height and right height can be ranked only by the proposed method.

Validation of the proposed modified ranking method

For the validation of the proposed ranking method, the following reasonable axioms that Wang and Kerre [13] have proposed for fuzzy numbers' ranking are considered. Let *RM* be an ordering method, *S* the set of fuzzy numbers for which the method *RM* can be applied, and $\mathcal{A}$ and $\mathcal{A}'$ finite subsets of *S*. The statements of two elements $\tilde{A}_p$ and $\tilde{B}_p$ in $\mathcal{A}$ satisfy that $\tilde{A}_p$ has a higher ranking than $\tilde{B}_p$ when *RM* is applied to the fuzzy numbers in $\mathcal{A}$ will be written as $\tilde{A}_p \succ \tilde{B}_p$ by *RM* on $\mathcal{A}$. $\tilde{A}_p \sim \tilde{B}_p$ by *RM* on $\mathcal{A}$, and $\tilde{A}_p \succeq \tilde{B}_p$ by *RM* on $\mathcal{A}$ are similarly interpreted. The following axioms show the reasonable properties of the ordering approach *RM*.

$A_1$ For $\tilde{A}_p \in \mathcal{A}$, $\tilde{A}_p \preceq \tilde{A}_p$ by *RM* on $\mathcal{A}$.

$A_2$ For $(\tilde{A}_p, \tilde{B}_p) \in \mathcal{A}^2, \tilde{A}_p \preceq \tilde{B}_p$ and $\tilde{B}_p \preceq \tilde{A}_p$ by *RM* on $\mathcal{A}$, we should have $\tilde{A}_p \sim \tilde{B}_p$ by *RM* on $\mathcal{A}$.

$A_3$ For $(\tilde{A}_p, \tilde{B}_p, \tilde{C}_p) \in \mathcal{A}^3, \tilde{A}_p \preceq \tilde{B}_p$ and $\tilde{B}_p \preceq \tilde{C}_p$ by *RM* on $\mathcal{A}$, we should have $\tilde{A}_p \preceq \tilde{C}_p$ by *RM* on $\mathcal{A}$.

$A_4$ For $(\tilde{A}_p, \tilde{B}_p) \in \mathcal{A}^2$, $\inf \operatorname{supp}(\tilde{B}_p) > \sup \operatorname{supp}(\tilde{A}_p)$, we should have $\tilde{A}_p \preceq \tilde{B}_p$ by *RM* on $\mathcal{A}$.

$A'_4$ For $(\tilde{A}_p, \tilde{B}_p) \in \mathcal{A}^2$, $\inf \operatorname{supp}(\tilde{B}_p) > \sup \operatorname{supp}(\tilde{A}_p)$, we should have $\tilde{A}_p \prec \tilde{B}_p$ by *RM* on $\mathcal{A}$.

$A_5$ Let $(\tilde{A}_p, \tilde{B}_p) \in (\mathcal{A} \cap \mathcal{A}')^2$. We obtain the ranking order $\tilde{A}_p \preceq \tilde{B}_p$ by *RM* on $\mathcal{A}'$ if and only if $\tilde{A}_p \preceq \tilde{B}_p$ by *RM* on $\mathcal{A}$.

$A_6$ Let $\tilde{A}_p$, $\tilde{B}_p$, $\tilde{A}_p + \tilde{C}_p$ and $\tilde{B}_p + \tilde{C}_p$ be elements of $S$. If $\tilde{A}_p \succeq \tilde{B}_p$ by *RM* on $\{\tilde{A}_p, \tilde{B}_p\}$, then $\tilde{A}_p + \tilde{C}_p \succeq \tilde{B}_p + \tilde{C}_p$ by *RM* on $\{\tilde{A}_p + \tilde{C}_p, \tilde{B}_p + \tilde{C}_p\}$.

$A'_6$ Let $\tilde{A}_p$, $\tilde{B}_p$, $\tilde{A}_p + \tilde{C}_p$ and $\tilde{B}_p + \tilde{C}_p$ be elements of $S$. If $\tilde{A}_p \succ \tilde{B}_p$ by *RM* on $\{\tilde{A}_p, \tilde{B}_p\}$, then $\tilde{A}_p + \tilde{C}_p \succ \tilde{B}_p + \tilde{C}_p$ by *RM* on $\{\tilde{A}_p + \tilde{C}_p, \tilde{B}_p + \tilde{C}_p\}$ when $\tilde{C}_p \neq \phi$.

**Proposition 4.4** *The proposed ranking method RM has the properties* $A_1$, $A_2$, $A_3$, $A_4$, $A'_4$, $A_5$, $A_6$ *and* $A'_6$.

*Proof* It is easy to verify that properties $A_1$–$A_5$ are hold. For the proof of $A_6$, consider the generalised fuzzy numbers with different left height and right height as $\tilde{A}_p = (a, b, c, d; h_L, h_R)_p$, $\tilde{B}_p = (q, r, s, t; h'_L, h'_R)_p$ and $\tilde{C}_p = (l, m, n, o, ; h''_L, h''_R)_p$. Let $\tilde{A}_p \succeq \tilde{B}_p$ by *RM*, hence

$$I_T^\alpha(\tilde{A}_p) \geq I_T^\alpha(\tilde{B}_p),$$

by adding $I_T^\alpha(\tilde{C}_p)$

$$I_T^\alpha(\tilde{A}_p) + I_T^\alpha(\tilde{C}_p) \geq I_T^\alpha(\tilde{B}_p) + I_T^\alpha(\tilde{C}_p),$$

and by Remark 4.1

$$I_T^\alpha(\tilde{A}_p + \tilde{C}_p) \geq I_T^\alpha(\tilde{B}_p + \tilde{C}_p).$$

Therefore, $\tilde{A}_p + \tilde{C}_p \succeq \tilde{B}_p + \tilde{C}_p$. Similarly $A'_6$ also holds. □

## Conclusions

In this paper, ranking of $p$-norm GFNs with different left height and right height is proposed. The proposed method is generalization of Kumar's approach. Kumar's approach can only deal with non-normal $p$-norm trapezoidal fuzzy numbers. The proposed method can handle non-normal $p$-norm trapezoidal fuzzy numbers as well as $p$-norm GFNs with different left height and right height.

**Acknowledgments** The authors would like to thank the anonymous referees for their valuable comments and suggestions which improved the paper form technical as well as clarity point of view. The author RC would like to thank Indian Institute of Technology, Guwahati for funding the research work.

## References


1. Abbasbandy, S., Hajjari, T.: A new approach for ranking of trapezoidal fuzzy numbers. Comput. Math. Appl. **57**, 413–419 (2009)
2. Bahri, O., Amor, N.B., El-Ghazali, T.: New Pareto approach for ranking triangular fuzzy numbers. Inf. Process. Manag. Uncertain. Knowl. Based Syst. Commun. Comput. Inf. Sci. **443**, 264–273 (2014)
3. Chen, S.M., Chen, J.H.: Fuzzy risk analysis based on ranking generalised fuzzy numbers with different heights and different spreads. Expert Syst. Appl. **36**, 6833–6842 (2009)
4. Cheng, C.H.: A new approach for ranking fuzzy numbers by distance method. Fuzzy Sets Syst. **95**, 307–317 (1998)
5. Chen, S.M., Munif, A., Chen, G.S., Liu, H.C., Kuo, B.C.: Fuzzy risk analysis based on ranking generalised fuzzy numbers with different left heights and right heights. Expert Syst. Appl. **39**, 6320–6334 (2012)
6. Chen, C.C., Tang, H.C.: Ranking non-normal $p$-norm trapezoidal fuzzy number with integral value. Comput. Math. Appl. **56**, 2340–2346 (2008)
7. Jain, R.: Decision-making in the presence of fuzzy variables. IEEE Trans. Syst. Man Cybern. **6**, 698–703 (1976)
8. Kim, K., Park, K.S.: Ranking fuzzy numbers with index of optimism. Fuzzy Sets Syst. **35**, 143–150 (1990)
9. Kumar, A., Singh, P., Kaur, A., Kaur, P.: A new approach for ranking nonnormal $p$-norm trapezoidal fuzzy numbers. Comput. Math. Appl. **61**, 881–887 (2011)
10. Lee, E.S., Li, R.J.: Comparison of fuzzy numbers based on the probability measure of fuzzy events. Comput. Math. Appl. **15**, 887–896 (1988)
11. Liou, T.S., Wang, M.J.J.: Ranking fuzzy numbers with integral value. Fuzzy Sets Syst. **50**, 247–255 (1992)
12. Pour, N.S., Moghaddam, R.T., Basiri, M.: A new method for trapezoidal fuzzy numbers ranking based on the shadow length and its application to manager's risk taking. J. Intell. Fuzzy Syst. **26**, 77–89 (2014)
13. Wang, X., Kerre, E.E.: Reasonable properties for the ordering of fuzzy quantities (I). Fuzzy Sets Syst. **118**, 375–385 (2001)
14. Yager, R.R.: A procedure for ordering fuzzy subsets of the unit interval. Inf. Sci. **24**, 143–161 (1981)
15. Yoon, K.P.: A probabilistic approach to rank complex fuzzy numbers. Fuzzy Sets Syst. **80**, 167–176 (1996)

11

# Utilizing artificial neural network approach for solving two-dimensional integral equations

**B. Asady · F. Hakimzadegan · R. Nazarlue**

**Abstract** This paper surveys the artificial neural networks approach. Researchers believe that these networks have the wide range of applicability, they can treat complicated problems as well. The work described here discusses an efficient computational method that can treat complicated problems. The paper intends to introduce an efficient computational method which can be applied to approximate solution of the linear two-dimensional Fredholm integral equation of the second kind. For this aim, a perceptron model based on artificial neural networks is introduced. At first, the unknown bivariate function is replaced by a multilayer perceptron neural net and also a cost function to be minimized is defined. Then a famous learning technique, namely, the steepest descent method, is employed to adjust the parameters (the weights and biases) to optimize their behavior. The article also examines application of the method which turns to be so accurate and efficient. It concludes with a survey of an example in order to investigate the accuracy of the proposed method.

**Keywords** Two-dimensional integral equations · Neural networks · Learning algorithm · Cost function · Approximate solution

## Introduction

Recently, integral equations have been extensively investigated theoretically and numerically. Note that they occur in a wide variety of physical applications, various fields of neural sciences and numerous applications such as electrical engineering, economics, elastically, plasticity, etc. Since these equations usually cannot be solved explicitly, it is going to be obtained in approximate solutions. There are several numerical methods for approximating solution of Fredholm and Volterra integral equations in one- and two-dimensions. For example, Tricomi in his book [25] introduced the classical method of successive approximations for integral equations. Variational iteration method [15] was effective and convenient for solving integral equations. The Homotopy analysis method (HAM) was proposed by Liao [16] and then has been applied in [1]. The Taylor expansion approach was presented for solving integral equations by Kanwal and Liu [14] and then has been extended in [17]. In addition, Jafari et al. [12] applied Legendre wavelets method to find numerical solution of linear integral equations. In [13] an architecture of artificial neural networks (NNs) was suggested to approximate solution of linear Fredholm integral equations systems. For this aim, first the truncation of the Taylor expansions for unknown functions was substituted in the origin system. Then the purposed neural network has been applied for adjusting the real coefficients of given expansions in resulting system. In [9], a numerical method based on feed-forward neural networks has been presented for solving Fredholm integral equations of the second kind. The Bernstein polynomials have frequently been applied in the solution of integral equations and approximation theory [5–7, 19, 20]. Also, there are many articles which deal with the solution and analysis of two-dimensional Fredholm and Volterra integral equations. Mirzaei and Dehghan [22] described a numerical scheme based on the moving least squares (MLS) method for solving integral equations in one- and two-dimensional spaces. The method was a meshless method, since it did not require any background

B. Asady (✉) · F. Hakimzadegan · R. Nazarlue
Department of Mathematics, Islamic Azad University, Science and Branch, Arāk, Iran
e-mail: babakmz2002@yahoo.com; b-asadi@iau-arak.ac.ir

interpolation or approximation cells and it did not depend on the geometry of domain. Hadizadeh and Asgary [11] using the bivariate Chebyshev collocation method solved the linear Volterra–Fredholm integral equations of the second kind. Alipanah and Esmaeili [2] approximated the solution of the two-dimensional Fredholm integral equation using Gaussian radial basis function based on Legendre–Gauss–Lobatto nodes and weights. Two-dimensional orthogonal triangular functions are used in [3, 18] as a new set of basis functions to approximate solutions of nonlinear two-dimensional integral equations. Babolian et al. [4] applied two-dimensional rationalized Haar functions for finding the numerical solution of nonlinear second kind two-dimensional integral equations. They reduced the present problem to solve a nonlinear system of algebraic equations using bivariate collocation method and Newton–Cotes nodes. Moreover, some different valid methods for solving these kind of equations have been developed.

This paper focuses on constructing a new algorithm with the use of feed-forward neural networks to reach an approximate solution of the linear two-dimensional Fredholm integral equation. For this purpose, first unknown two-variable function in the problem is replaced by a three-layer perceptron neural network. Supposedly, the limits of integrations are partitioned into set points, this architecture of neural networks can calculate the output corresponding to input vector. Now a cost function to be minimized is defined on the set points. Consequently, the suggested neural net using a learning algorithm that is based on the gradient descent method adjusts parameters (the weights and biases) to any desired degree of accuracy. Here is an outline of the paper. In "Preliminaries", the basic notations and definitions of the integral equations and the artificial neural networks are briefly presented. "The general method" describes how to find approximate solution of the given two-dimensional integral equations using proposed approach. Finally in "An example", an numerical example is provided and results are compared with the analytical solutions to demonstrate the validity and applicability of the method.

## Preliminaries

In this section we will focus on the basic definitions and introductory concepts in integral equations. In addition the basic principles of artificial neural network (ANN) approach are presented and reviewed for solving linear second kind two-dimensional integral equations (2D-IEs).

### Integral equations

Integral equations appear in many scientific and engineering applications, especially when initial value problems for boundary value problems are converted to integral equations. As stated before, we will review some integral equations and linear two-dimensional integral equations of the second kind as well.

**Definition 2.1** Let $f : [a,b] \to \mathbb{R}$. For each partition $P = \{t_0, t_1, \ldots, t_n\}$ of $[a,b]$ and for arbitrary $\xi_i \in [t_{i-1}, t_i]$ $(1 \le i \le n)$, suppose

$$R_P = \sum_{i=1}^{n} f(\xi_i)(t_i - t_{i-1}),$$

$$\Delta := max\{|t_i - t_{i-1}|, i = 1, \ldots, n\}.$$

The definite integral of $f(t)$ over $[a,b]$ is

$$\int_a^b f(t)\mathrm{d}t = \lim_{\Delta \to 0} R_P$$

provided that this limit exists in the metric $D$ [25].

**Definition 2.2** The linear two-dimensional Fredholm integral equation (2D-FIE) of the second kind is presented by the form [2]

$$F(x,y) = f(x,y) + \lambda \int_c^d \int_a^b k(x,y,s,t)F(s,t)\mathrm{d}s\mathrm{d}t, \quad (x,y) \in [a,b] \times [c,d] \tag{1}$$

where $\lambda$ is a constant parameter, the kernel $k$ and $f$ are given analytic functions on $L^2([a,b] \times [c,d])$. The two-variable unknown function $F$ that must be determined appears inside and outside the integral signs. This is a characteristic feature of a second kind integral equation. It is important to point out that if the unknown function appears only inside the integral signs, the resulting equation is of first kind.

If the kernel function satisfies $k(x,y,s,t) = 0,\ s > x,\ t > y$ in Eq. (1), we obtain the linear two-dimensional Volterra integral equation (2D-VIE) [24]

$$F(x,y) = f(x,y) + \lambda \int_c^y \int_a^x k(x,y,s,t)F(s,t)\mathrm{d}s\mathrm{d}t, \quad (x,y) \in [a,b] \times [c,d]. \tag{2}$$

It should be noted that, if one of the limits of integration varies, the integral equation is called a Volterra–Fredholm integral equation. It is clear that, two-dimensional integral equations appear in many forms. Three distinct ways that depend on the limits of integration are used to characterize these equations which have been are briefly introduced. Notice that, if the function $f(x,y)$ in the present integral equations is identically zero, the equation is called homogeneous. Otherwise it is called inhomogeneous. These three concepts play a major role in the structure of the solution.

Artificial neural networks

Artificial neural networks (ANNs) can be considered as simplified computational structures that are inspired by observed process in natural networks of biological neurons in the brain. They are nonlinear mapping architectures based on the function of the human brain, therefore can be considered as powerful tools for modeling, especially when the underlying data relationship is unknown. A very important feature of these networks is their adaptive nature, where "learning by example" replaces "programming" in solving problems. In other words, in contrast to conventional methods, which are used to perform specific task, most neural networks are more versatile. This feature raises a very appealing computational model which can be applied to solve variety of problems.

The multilayer feed-forward neural network or multilayer perceptron (MLP) that had been proposed by Rosenblatt [23] is very popular and is used more than other neural network type for a wide variety of tasks. The present network learned by back-propagation algorithm is based on supervised procedure. In other words, the network constructs a model based on examples of data with known output.

In this subsection, an architecture of MLP model is discussed here briefly. We intend to give a short review on learning of the given neural network. First consider a three-layer ANN with two input units, $N$ neurons in hidden layer and one output unit. Mathematical representation of the present neural network is given in Fig. 1. Using the figure, input–output relation of each unit and calculated output $u_N(x,y)$ can be written as follows:

*Input units:*

The input neurons make no change in their inputs, so:

$$o_1 = x, \tag{3}$$

$$o_2 = y.$$

*Hidden units:*

Input into a node in hidden layer is a weighted sum of outputs from nodes connected to it. Each unit takes its net input and applies an activation function to it. The input/output relation is normally given as follows:

$$O_p = g(net(p)), \tag{4}$$

$$net(p) = \sum_{i=1}^{2}(w_{pi}.o_i) + b_p, \quad p = 1,\ldots,N.$$

where $net(p)$ describes the result of the net outputs $o_i$ impacting on unit $p$. Also, $w_{pi}$ are weights connecting neuron $i$ to neuron $p$ and $b_p$ is a bias for neuron $p$. Bias term is baseline input to a node in absence of any other inputs.

*Output unit:*

$$u_N(x,y) = \sum_{p=1}^{N} net(p). \tag{5}$$

## The general method

In this section, we intend to use the MLP method to get a new numerical approach for solving the linear two-dimensional Fredholm integral equation of the second kind. In other words, how to apply this method to make a series approximation for the solution $F(x,y)$ in (1) will be

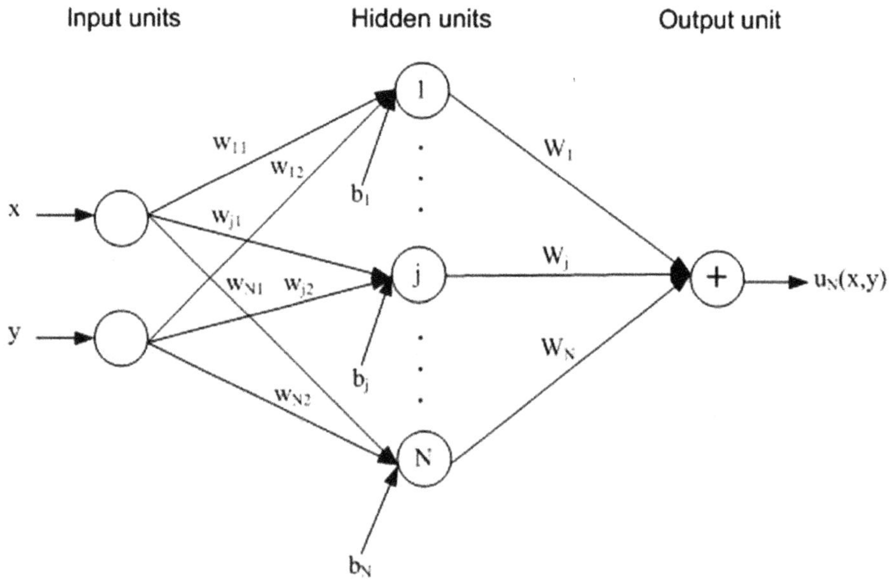


**Fig. 1** Schematic diagram of the proposed MLP

described. The output of two-layer MLP network that is defined in Eq. (1) can be rewritten as follows:

$$u_N(x,y) = \sum_{p=1}^{N}\sum_{i=1}^{2} g(w_{pi}.o_i + b_p). \tag{6}$$

In order to approximate function $u$, first the intervals $[a\,b]$ and $[c\,d]$ are partitioned into set points $x_i$ and $y_j$, respectively. Thus, the following set of equations will be obtained:

$$u_N(x_i,y_j) = \sum_{p=1}^{N} g\left(w_{p1}.x_i + w_{p2}.y_j + b_p\right). \tag{7}$$

Cost function

First suppose that $u_N(x,y)$ is the approximate solution with the adjustable parameters (wights and biases) for the unknown $F(x,y)$. After substituting this solution instead of the unknown function in the given 2D-FIE, the Eq. (1) can be transformed to a sum squared error minimization problem corresponding to the proposed neural network. So, the error function is regarded as a function on the weights and biases space of the net for $x = x_i$ and $y = y_j$ as follows:

$$E^{i,j}(w,W,b) := \frac{1}{2}\left(E_N^{i,j}(w,W,b)\right)^2, \tag{8}$$

where

$$E_N^{i,j}(w,W,b) = u_N(x_i,y_j) - f(x_i,y_j) - \lambda\int_c^d\int_a^b k(x_i,y_j,s,t)u_N(s,t)\mathrm{d}s\mathrm{d}t.$$

Now the total error of the network is defined as:

$$E(w,W,b) = \sum_{i,j} E^{i,j}(w,W,b). \tag{9}$$

The goal then is to minimize this function; therefore, we must deduce a back-propagation learning algorithm using the present cost function.

Proposed learning algorithm

Multilayer feed-forward neural network is learned by back-propagation algorithm that is based on supervised procedure. In other words, the MLP network is trained using a supervised learning algorithm which uses the training data to adjust the network weights and biases. Now let $w_{p,q}$, $W_p$ and $b_p$ (for $p = 1,\ldots,N;\ q = 1,2$) are initialized at small random values for input signals. For parameter $w_{p,q}$ adjustment rule can be written as follows:

$$w_{p,q}(r+1) = w_{p,q}(r) + \Delta w_{p,q}(r),\ p = 1,\ldots,N;\ q = 1,2, \tag{10}$$

$$\Delta w_{p,q}(r) = -\eta.\frac{\partial E^{i,j}}{\partial w_{p,q}} + \alpha.\Delta w_{p,q}(r-1), \tag{11}$$

where $r$ is the number of adjustments, $\eta$ is the learning rate and $\alpha$ is the momentum term constant. Similarly this adjustment rule can be written for other weight parameters. Thus, our problem is to calculate the derivative $\frac{\partial E^{i,j}}{\partial w_{p,q}}$ in (11). The derivative can be calculated as follows:

$$\frac{\partial E^{i,j}}{\partial w_{p,q}} = \frac{\partial E^{i,j}}{\partial E_N^{i,j}}\cdot\frac{\partial E_N^{i,j}}{\partial w_{p,q}}, \tag{12}$$

where

$$\frac{\partial u_N(x_i,y_j)}{\partial w_{p,q}} = \left(\frac{\partial u_N(x_i,y_j)}{\partial O(p)}\cdot\frac{\partial O(p)}{\partial net(p)}\cdot\frac{\partial net(p)}{\partial w_{p,q}}\right).$$

Consequently,

$$\frac{\partial E^{i,j}}{\partial w_{p,q}} = \tag{13}$$

$$\begin{cases} E_N^{i,j}.W_p.(x_i.g'(net(p)) - \lambda\int_c^d\int_a^b s.k(x_i,y_j,s,t).g'(w_{p,1}s + w_{p,2}t + b_p)\mathrm{d}s\mathrm{d}t), & \text{q=1} \\ E_N^{i,j}.W_p.(y_j.g'(net(p)) - \lambda\int_c^d\int_a^b t.k(x_i,y_j,s,t).g'(w_{p,1}s + w_{p,2}t + b_p)\mathrm{d}s\mathrm{d}t), & \text{q=2} \end{cases}.$$

Using a similar procedure as mentioned above, we have the correspondingly corollary for parameters $W_p$ and $b_p$, in which we are refrained from going through proof details. So, we have:

$$\frac{\partial E^{i,j}}{\partial W_p} = E_N^{i,j}.\left(net(p) - \lambda\int_c^d\int_a^b k(x_i,y_j,s,t).g(w_{p,1}s + w_{p,2}t + b_p)\mathrm{d}s\mathrm{d}t\right), \tag{14}$$

and

$$\frac{\partial E^{i,j}}{\partial b_p} = E_N^{i,j}.W_p.\left(g'(net(p)) - \lambda\int_c^d\int_a^b k(x_i,y_j,s,t).\times g'(w_{p,1}s + w_{p,2}t + b_p)\mathrm{d}s\mathrm{d}t\right). \tag{15}$$

The MLP neural nets are the sample of regular networks, therefore they can approximate any continuous function on a compact set to arbitrary accuracy [10]. Now the learning algorithm can be summarized as follows:

*Learning process*

*Step 1:* $\eta > 0$, $\alpha > 0$ and $Emax > 0$ are chosen. Then quantities $w_{p,q}$, $W_p$ and $b_p$ $(p = 1,\ldots,N;\ q = 1,2)$ are initialized at small random values.

*Step 2:* Let $r := 0$ where $r$ is the number of iterations of the learning algorithm. Then the running error $E$ is set to 0.

*Step 3:* Let $r := r + 1$. Repeat below procedure for different values of $i$ and $j$:

i Forward calculation: Calculate the output vector $u_N(x_i, y_j)$ by presenting the input vectors $x_i$ and $y_j$.
ii Back propagation: Adjust the parameters $w_{p,q}$, $W_p$ *and* $b_p$ using the cost function (8).

*Step 4:* Cumulative cycle error is computed by adding the present error to $E$.
*Step 5:* The training cycle is completed. For $E < Emax$ terminate the training session. If $E > Emax$ then $E$ is set to 0 and we initiate a new training cycle by going back to *Step 3*.

## An example

In this section, in order to investigate the accuracy of the proposed method, we have chosen an example of linear two-dimensional integral equations of the second kind. For the example, the computed values of the approximate solution are calculated over a number of iterations and the cost function is plotted. Also, to show the efficiency of the present method for our problem, results will be compared with the exact solution.

*Example 4.1* Consider the linear 2D-FIE

$$F(x,y) = f(x,y) + \int_0^1 \int_0^1 (s.sin(t) + 1)F(s,t)\mathrm{d}s\mathrm{d}t, \qquad (16)$$

where

$$f(x,y) = x.cos(y) - \frac{1}{6} sin(1)(3 + sin(1)),$$

with the exact solution $F(x,y) = x.cos(y)$. In this example, we illustrate the use of the FNN technique to approximate the solution of this integral equation. In the following simulations, we use the specifications as follows:

1. The number of hidden units: $N = 3$,
2. Learning rate $\eta = 0.5$,
3. Momentum constant $\alpha = 0.05$.

Numerical result can be found in Table 1, and Fig. 2 shows the cost function in the 20 iterations. Figures 3, 4, 5, 6 show the convergence behaviors for computed values of the weight parameters $w_{p,q}$ and $W_p$, bias $b_p$ for different number of iterations.

There is no magic formula for selecting the optimum number of hidden neurons. However, some thumb rules are available for calculating number of hidden neurons. A rough approximation can be obtained by the geometric pyramid rule proposed by Masters [21]. For a three-layer network with $n$ input and $m$ output neurons, the hidden layer would have at least $[\sqrt{nm}] + 1$ neurons.

To show convergence of the proposed method we solve Example 4.1 using shifted Legendre collocation method.

**Table 1** Numerical results for example 4.1 by FNN technique

| (x, y) = (0.1r, 0.1r) | Exact solution | Approximate solution | | Error | |
|---|---|---|---|---|---|
| | | $N = 3$ | $N = 8$ | $N = 3$ | $N = 8$ |
| $r = 1$ | 0.09950 | 0.09823 | 0.09905 | 0.00127 | 0.000444 |
| $r = 2$ | 0.19601 | 0.19445 | 0.19585 | 0.00155 | 0.000143 |
| $r = 3$ | 0.28660 | 0.28657 | 0.28654 | 0.00073 | 0.000056 |
| $r = 4$ | 0.36842 | 0.36810 | 0.36836 | 0.00030 | 0.000037 |
| $r = 5$ | 0.43879 | 0.43857 | 0.43867 | 0.00013 | 0.000023 |
| $r = 6$ | 0.49520 | 0.49515 | 0.49518 | 0.00005 | 0.000013 |
| $r = 7$ | 0.53539 | 0.53525 | 0.53528 | 0.00005 | 0.000013 |
| $r = 8$ | 0.55737 | 0.55725 | 0.55728 | 0.00005 | 0.000013 |
| $r = 9$ | 0.55945 | 0.55935 | 0.55938 | 0.00005 | 0.000013 |

The reason for choosing shifted Legandre collocation method is its simplicity. The details of shifted Legandre collocation method are as follows.

*Shifted Legandre collocation method*

The Legendre polynomials, $P_n(x), n = 0, 1, \ldots,$ are the eigenfunctions of the singular Sturm–Liouville problem

$$((1 - x^2)P_n'(x))' + n(n + 1)P_n(x) = 0.$$

Also, they are orthogonal with respect to $L^2$ inner product on the interval $[-1, 1]$ with the weight function $w(x) = 1$, that is

$$\int_{-1}^{1} P_n(x)P_m(x)\mathrm{d}x = \frac{2}{2n+1}\delta_{nm},$$

where $\delta_{nm}$ is the Kronecker delta. The Legendre polynomials satisfy the recursion relation

$$P_{n+1}(x) = \frac{2n+1}{n+1}xP_n(x) - \frac{n}{n+1}P_{n-1}(x),$$

where $P_0(x) = 1$ and $P_1(x) = x$. If $P_n(x)$ is normalized so that $P_n(1) = 1$, then for any $n$, the Legendre polynomials in terms of power of $x$ are

$$P_n(x) = \frac{1}{2^n}\sum_{m=0}^{[\frac{n}{2}]}(-1)^m\binom{n}{m}\binom{2n-2m}{n}x^{n-2m},$$

where $[\frac{n}{2}]$ denotes the integer part of $\frac{n}{2}$.

The Legendre–Gauss–Lobatto (LGL) collocation points $-1 = x_0 < x_1 < \cdots < x_N = 1$ are the roots of $P_N'(x)$ together with the points $-1$ and 1. Explicit formulas for the LGL points are not known. The LGL points have the property that

$$\int_{-1}^{1} p(x)\mathrm{d}x = \sum_{i=0}^{N} w_i p(x_i),$$

which is exact for polynomials of degree at most $2N - 1$, where $w_i$, $0 \leqslant i \leqslant N$, are LGL quadrature weights. For more details about Legendre polynomials, see [8].

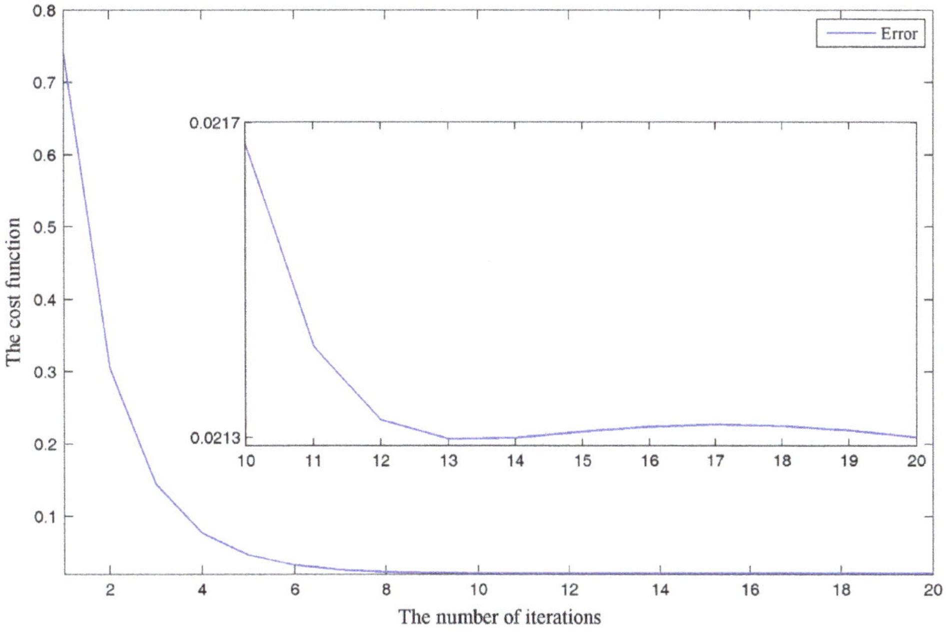


**Fig. 2** The cost function for Example 4.1 on the number of iterations

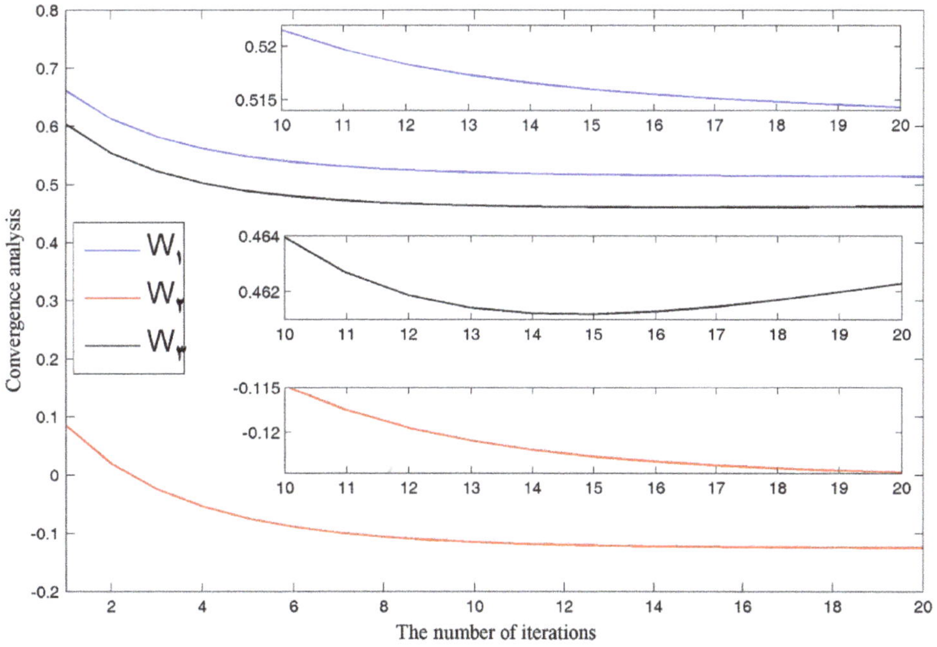


**Fig. 3** Convergence of the weights $W_r$ for Example 4.1

The shifted Legendre polynomials (ShLP) on the interval $t \in [0, 1]$ are defined by

$$\widehat{P}_n(t) = P_n(2t - 1), \quad n = 0, 1, \ldots,$$

which are obtained by an affine transformation from the Legendre polynomials. The set of ShLP is a complete $L^2[0, 1]$-orthogonal system with the weight function $w(t) = 1$. Thus, any function $f \in L^2[0, 1]$ can be expanded in terms of ShLP.

The ShLGL (Shifted Legendre–Gauss–Lobatto) collocation points $0 = t_0 < t_1 < \cdots < t_N = 1$ on the interval $[0, 1]$ are obtained by shifting the LGL points, $x_i$, using the transformation

$$t_i = \frac{1}{2}(x_i + 1), \quad i = 0, 1, \ldots, N. \tag{17}$$

Thanks to the property of the standard LGL quadrature, it follows that for any polynomial $p$ of degree at most $2N - 1$ on $(0, 1)$,

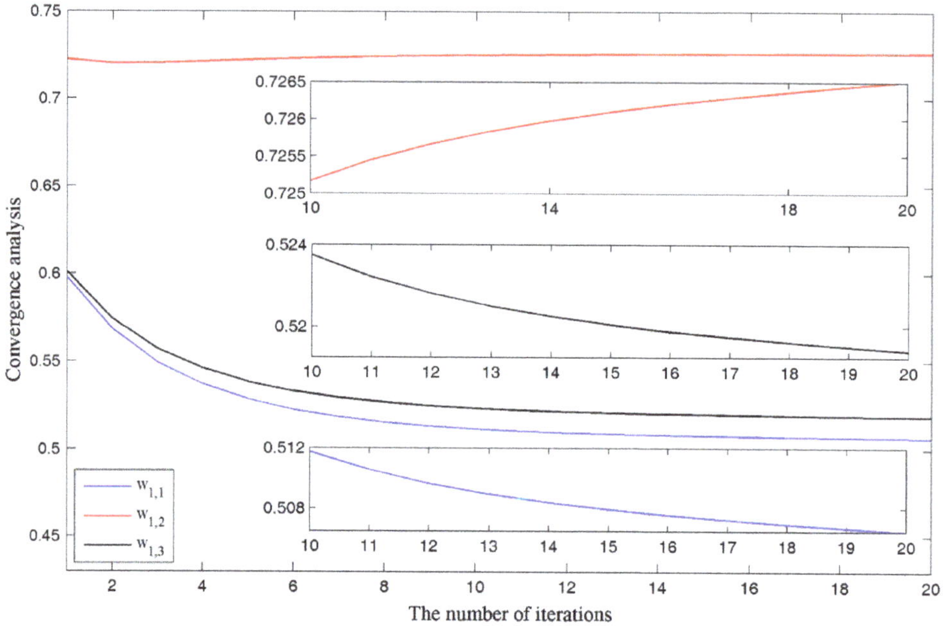


**Fig. 4** Convergence of the weights $w_{1,r}$ for Example 4.1

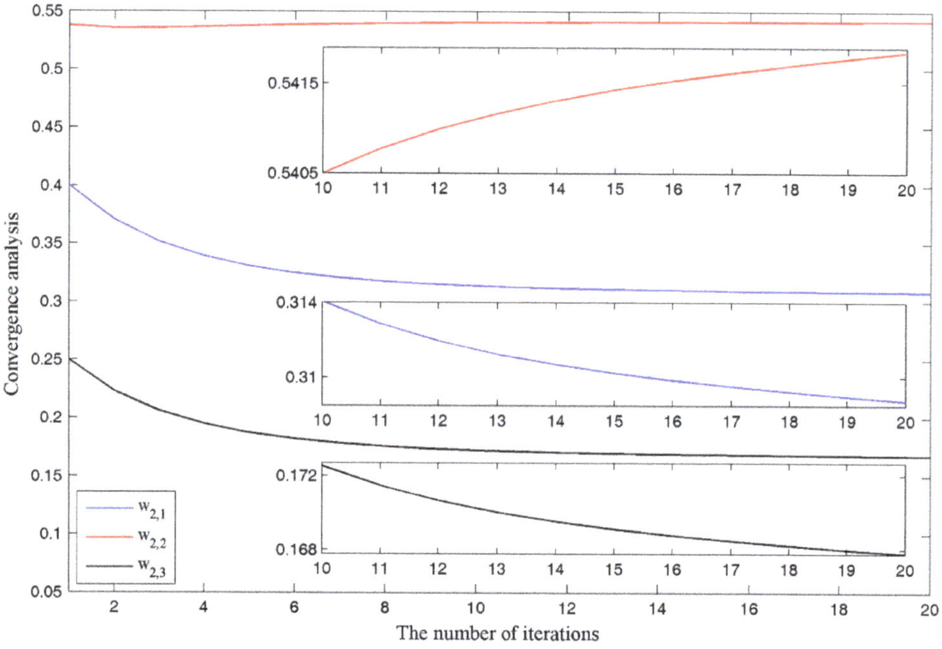


**Fig. 5** Convergence of the weights $w_{2,r}$ for Example 4.1

$$\int_0^1 p(t)\mathrm{d}t = \frac{1}{2}\int_{-1}^1 p\left(\frac{1}{2}(x+1)\right)\mathrm{d}x$$
$$= \frac{1}{2}\sum_{i=0}^N w_i p\left(\frac{1}{2}(x_i+1)\right) = \sum_{i=0}^N \hat{w}_i p(t_i),$$

where $\hat{w}_i = \frac{1}{2}w_i, 0 \leqslant i \leqslant N$, are ShLGL quadrature weights. The results stated above are also satisfied for Legendre–Gauss and Legendre–Gauss–Radau quadrature rules.

The function $F(x, y)$ is approximated by a ShLP of degree at most $N$ as

$$F(x,y) = \sum_{i=0}^N \sum_{j=0}^N \alpha_{ij}\widehat{P}_i(x)\widehat{P}_j(y) \tag{18}$$

Now, by substituting (18) and collocation points (17) in (16), we have

$$\sum_{i=0}^N \sum_{j=0}^N \alpha_{ij}\widehat{P}_i(t_k)\widehat{P}_j(t_l)\bar{f}(t_k,t_l) + \int_0^1\int_0^1 (s.sin(t)+1)$$
$$\sum_{i=0}^N \sum_{j=0}^N \alpha_{ij}\widehat{P}_i(s)\widehat{P}_j(t)\mathrm{d}s\mathrm{d}t, \quad k,l = 0,1,\ldots,N$$

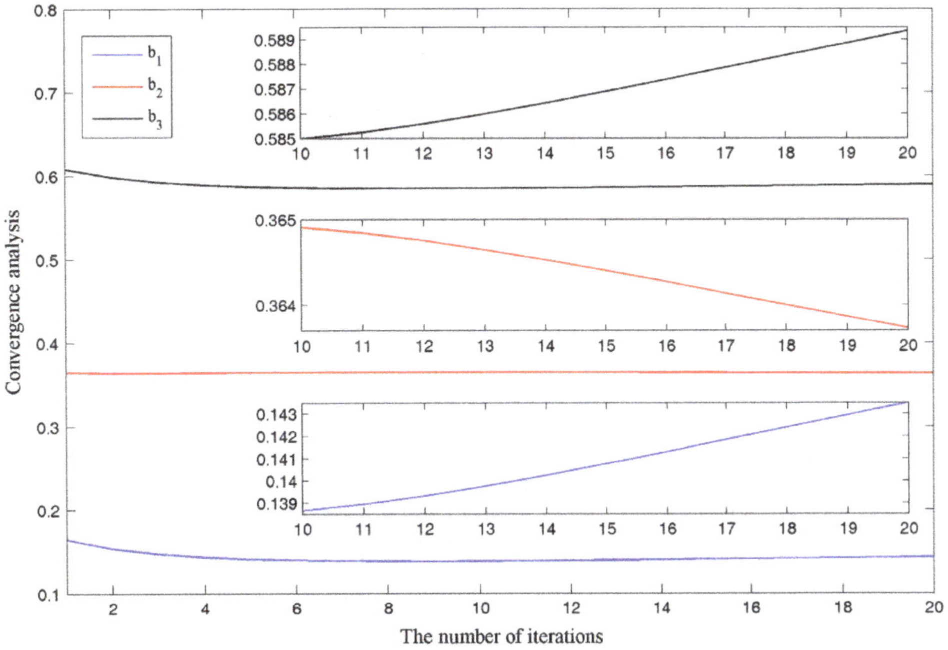


**Fig. 6** Convergence of the weights $b_r$ for Example 4.1

**Table 2** Numerical results for Example 4.1 by SHLGL

| (x, y) = (0.1*r*, 0.1*r*) | Exact solution | Approximate solution | | Error | |
|---|---|---|---|---|---|
| | | $N = 2$ | $N = 3$ | $N = 2$ | $N = 3$ |
| $r = 1$ | 0.09950 | 0.10497 | 0.09951 | 0.00547 | 0.00001 |
| $r = 2$ | 0.19601 | 0.19855 | 0.19575 | 0.00254 | 0.00026 |
| $r = 3$ | 0.28660 | 0.28268 | 0.28610 | 0.00392 | 0.00050 |
| $r = 4$ | 0.36842 | 0.35734 | 0.36798 | 0.01108 | 0.00044 |
| $r = 5$ | 0.43879 | 0.42256 | 0.43878 | 0.01623 | 0.00001 |
| $r = 6$ | 0.49520 | 0.47832 | 0.49589 | 0.01688 | 0.00069 |
| $r = 7$ | 0.53539 | 0.52462 | 0.53671 | 0.01077 | 0.00132 |
| $r = 8$ | 0.55737 | 0.56147 | 0.55865 | 0.00410 | 0.00128 |
| $r = 9$ | 0.55945 | 0.58885 | 0.55910 | 0.02940 | 0.00035 |

By solving this linear system we can find $\alpha_{ij},\ i,j = 0,1,\ldots,N$, and then approximate the solution $F(x,y)$.

We solved the Example 4.1 using the method described for $N = 2$ and $N = 3$. Results are shown in Table 2. By comparing Tables 1 and 2 we find that obtained results in Table 1 are in concordance.

## Conclusions

This paper suggested a new computational method to solve a two-dimensional Fredholm integral equations. So, a feed-forward artificial neural network has been proposed. This network is able of estimating approximate solution of assumed equation using the learning algorithm which is based on steepest descent rule. Clearly, in order to obtain accurate solution, many learning procedure should be considered. The analyzed examples illustrated the ability and reliability of the present approach. The obtained solutions, in comparison with exact solutions admit a remarkable accuracy. Extensions to the case of more general of integral equations are left for future studies.

## References


1. Abbasbandy, S.: Numerical solution of integral equation: Homotopy perturbation method and Adomians decomposition method. Appl. Math. Comput. **173**, 493–500 (2006)
2. Alipanah, A., Esmaeili, Sh: Numerical solution of the two-dimensional Fredholm integral equations using Gaussian radial basis function. J. Comput. Appl. Math. **235**, 5342–5347 (2011)
3. Babolian, E., Maleknejad, K., Roodaki, M., Almasieh, H.: Two-dimensional triangular functions and their applications to nonlinear 2D Volterra-Fredholm integral equations. Comput. Math. Appl. **60**, 1711–1722 (2010)
4. Babolian, E., Bazm, S., Lima, P.: Numerical solution of nonlinear two-dimensional integral equations using rationalized Haar functions. Commun. Nonlinear Sci. Numer. Simulat. **16**, 1164-1175 (2011)
5. Bhatta, D.D., Bhatti, M.I.: Numerical solution of KdV equation using modied Bernsein polynomials. Appl. Math. Comput. **174**, 1255–1268 (2006)
6. Bhatti, M.I., Bracken, P.: Solutions of dierential equations in a Bernstein polynomial basis. J. Comput. Appl. Math. (2007). doi:10.1016/j.cam2006.05.002
7. Bhattacharya, S., Mandal, B.N.: Use of Bernstein polynomials in numerical solution of Volterra integral equations. Appl. Math. Sci. **36**(2), 1773–1787 (2008)
8. Canuto, C., Hussaini, M.Y., Quarteroni, A., Zang, T.A.: Spectral methods: fundamentals in single domains. Springer, Berlin (2006)
9. Effati, S., Buzhabadi, R.: A neural network approach for solving

Fredholm integral equations of the second kind. Neural Comput. Appl. doi:10.1007/s00521-010-0489-y
10. Fuller, R.: Neural Fuzzy Systems. Abo Akademi University, Department of Information Thechnologies (1995)
11. Hadizadeh, M., Asgary, M.: An ecient numerical approximation for the linear class of mixed integral equations. Appl. Math. Comput. **167**, 1090–1100 (2005)
12. Jafari, H., Hosseinzadeh, H., Mohamadzadeh, S.: Numerical solution of system of linear integral equations by using Legendre wavelets. Int. J.Open Prob. Comput. Math. **5**, 63–71 (2010)
13. Jafarian, A., Measoomy Nia, S.: Utilizing feed-back neural network approach for solving linear Fredholm integral equations system. Appl. Math. Mode. (2012). doi:10.1016/j.apm
14. Kanwal, R.P., Liu, K.C.: A Taylor expansion approach for solving integral equations. Int. J. Math. Educ. Sci. Technol. **20**, 411–414 (1989)
15. Lan, X.: Variational iteration method for solving integral equations. Comput. Math. Appl. **54**, 1071–1078 (2007)
16. Liao, S.J.: Beyond perturbation: introduction to the homotopy analysis method. Chapman Hall/CRC Press, Boca Raton (2003)
17. Maleknejad, K., Aghazadeh, N.: Numerical solution of Volterra integral equations of the second kind with convolution kernel by using Taylor-series expansion method. Appl. Math. Comput. **161**, 915–922 (2005)
18. Maleknejad, K., Jafari Behbahani, Z.: Applications of two-dimensional triangular functions for solving nonlinear class of mixed Volterra-Fredholm integral equations. Math. Comput. Mode (2011). doi:10.1016/j.mcm.2011.11.041
19. Maleknejad, K., Basirat, B., Hashemizadeh, E.: A Bernstein operational matrix approach for solving a system of high order linear Volterra–Fredholm integro-differential equations. Math. Comput. Mode **55**, 1363–1372 (2012)
20. Mandal, B.N., Bhattacharya, S.: Numerical solution of some classes of integral equations using Bernstein polynomials. Appl. Math. Comput. **190**, 1707–1716 (2007)
21. Masters, T.: Practical neural network recipes in C++. Academic press, NewYork (1993)
22. Mirzaei, D., Dehghan, M.: A meshless based method for solution of integral equations. Appl. Numer. Math. **60**, 245–262 (2010)
23. Rosenblatt, F.: The perceptron: a probabilistic model for information storage ang organization in the brain. Psychol. Rev. **65**, 386–408 (1958)
24. Tari, A., Rahimib, M.Y., Shahmorad, S., Talati, F.: Solving a class of two-dimensional linear and nonlinear Volterra integral equations by the differential transform method. J. Comput. Appl. Math. **228**, 70–76 (2009)
25. Tricomi, F.G.: Integral equations. Dover Publications, New York (1982)

# 12

# Quadrature iterative method for numerical solution of two-dimensional linear fuzzy Fredholm integral equations

H. Nouriani [1] · R. Ezzati[1]

**Abstract** In this paper, first, we propose an iterative method based on quadrature formula for solving two-dimensional linear fuzzy Fredholm integral equations (2DLFFIE). Then, we prove the error estimation of the method. In addition, we show the numerical stability analysis of the method with respect to the choice of the first iteration. Finally, supporting examples are also provided.

**Keywords** Fuzzy-number-valued functions · Numerical method · Two-dimensional linear fuzzy Fredholm integral equations · Quadrature iterative method

## Introduction

The study of fuzzy differential and integral equations begins with the research of Kaleva [13] and Seikkala [21]. As we know, showing the existence and uniqueness solution of these equations is very important. So, to this end, many authors applied the Banach's fixed point theorem and the method of successive approximations. Recently, many researchers proposed numerical methods for solving fuzzy Fredholm integral equations (FFIEs). To see more details about solving FFIE, one can refer to [2, 4–6, 8–12, 15, 16, 18, 19, 22, 23].

Solving linear and nonlinear FFIEs based on iterative method is done by many authors but for the first time, solving 2DLFFIEs was studied by authors of [18] that the authors proved the existence and uniqueness solution of these equations by using Banach's fixed point theorem. In addition, in [19], authors presented a numerical method for solving two-dimensional nonlinear FFIEs by using an iterative method. Recently, authors of [7, 14, 17, 20] proposed some numerical approaches to solve 2DLFFIEs.

Here, we propose a numerical method for solving (3.1). In addition, we present the estimation error and the numerical stability analysis. The rest of the paper is organized as follows: In "Preliminaries", we review some properties for fuzzy-number-valued functions, such as continuity, boundedness, fuzzy Henstock and fuzzy Riemann integrability and fuzzy quadrature rules . In "2D linear fuzzy Fredholm integral equations", we propose a new approach based on quadrature rule and iterative method to solve 2DLFFIEs. In "Convergence analysis", we investigate convergence analysis of the proposed method. Considering the numerical stability analysis of the proposed method is done in "Numerical stability analysis". Finally, some numerical examples are presented in "Numerical examples".

## Preliminaries

At first, we present some basic definitions and necessary results about fuzzy set theory.

**Definition 2.1** [1] A fuzzy number is a function $u : \mathbb{R} \to [0, 1]$ with the following properties:

1. $u$ is normal, i.e. $\exists x_0 \in \mathbb{R}$; $u(x_0) = 1$.
2. $u(\eta x + (1 - \eta)y) \geq \min\{u(x), u(y)\} \forall x, y \in \mathbb{R}$, $\forall \eta \in [0, 1]$ ($u$ is called a convex fuzzy subset).
3. $u$ is upper semicontinuous on $\mathbb{R}$, i.e., $\forall x_0 \in \mathbb{R}$ and $\forall \epsilon > 0$, $\exists$ neighborhood $V(x_0) : u(x) \leq u(x_0) + \epsilon$, $\forall x \in V(x_0)$.

✉ R. Ezzati
ezati@kiau.ac.ir

[1] Department of Mathematics, Karaj Branch, Islamic Azad University, Karaj, Iran

4. The set $\overline{supp(u)}$ is compact in $\mathbb{R}$ (where $supp(u) := \{x \in \mathbb{R}; u(x) > 0\}$).

The set of all fuzzy numbers is denoted by $\mathbb{R}_F$.

**Definition 2.2** [1] For $0 < r \leq 1$ and $u \in \mathbb{R}_F$ define $[u]^r := \{x \in \mathbb{R} : u(x) \geq r\}$ and

$$[u]^0 := \overline{\{x \in \mathbb{R} : u(x) > 0\}}.$$

It is well known that for each $r \in [0, 1]$, $[u]^r$ is a closed and bounded interval of $\mathbb{R}$. For $\tilde{u}, \tilde{v} \in \mathbb{R}_F$ and $\lambda \in \mathbb{R}$, we define uniquely the sum $\tilde{u} \oplus \tilde{v}$ and the product $\lambda \odot \tilde{u}$ by

$$[\tilde{u} \oplus \tilde{v}]^r = [\tilde{u}]^r + [\tilde{v}]^r, [\lambda \odot \tilde{u}]^r = \lambda[\tilde{u}]^r, \forall r \in [0, 1],$$

where $[\tilde{u}]^r + [\tilde{v}]^r$ means the usual addition of two intervals (as subsets of $\mathbb{R}$). Also, $\lambda[\tilde{u}]^r$ means the usual product between a scalar and a subset of $\mathbb{R}$. Notice $1 \odot \tilde{u} = \tilde{u}$ and it holds $\tilde{u} \oplus \tilde{v} = \tilde{v} \oplus \tilde{u}$, $\lambda \odot \tilde{u} = \tilde{u} \odot \lambda$. If $0 \leq r_1 \leq r_2 \leq 1$ then $[\tilde{u}]^{r_2} \subseteq [\tilde{u}]^{r_1}$. Actually $[\tilde{u}]^r = [\tilde{u}_-^{(r)}, \tilde{u}_+^{(r)}]$, where $\tilde{u}_-^{(r)} \leq \tilde{u}_+^{(r)}$, $\tilde{u}_-^{(r)}$, $\tilde{u}_+^{(r)} \in \mathbb{R}$, $\forall r \in [0, 1]$. For $\lambda > 0$ one has $\lambda \tilde{u}_\pm^{(r)} = (\lambda \odot \tilde{u})_\pm^{(r)}$, respectively.

**Definition 2.3** [1] Define $D : \mathbb{R}_F \times \mathbb{R}_F \to \mathbb{R}_+$ by

$$\begin{aligned} D(\tilde{u}, \tilde{v}) &:= \sup_{r \in [0,1]} \max \left\{ \left|\tilde{u}_-^{(r)} - \tilde{v}_-^{(r)}\right|, \left|\tilde{u}_+^{(r)} - \tilde{v}_+^{(r)}\right| \right\} \\ &= \sup_{r \in [0,1]} \text{ Hausdorff distance } ([\tilde{u}]^r, [\tilde{v}]^r), \end{aligned}$$

where $[\tilde{v}]^r = [\tilde{v}_-^{(r)}, \tilde{v}_+^{(r)}]$; $\tilde{u}, \tilde{v} \in \mathbb{R}_F$. Clearly, $D$ is a metric on $\mathbb{R}_F$. Also $(\mathbb{R}_F, D)$ is a complete metric space, with the following properties [1]:

$$\begin{aligned} D(\tilde{u} \oplus \tilde{w}, \tilde{v} \oplus \tilde{w}) &= D(\tilde{u}, \tilde{v}), \forall \tilde{u}, \tilde{v}, \tilde{w} \in \mathbb{R}_F, \\ D(k' \odot \tilde{u}, k' \odot \tilde{v}) &= |k'| D(\tilde{u}, \tilde{v}), \forall \tilde{u}, \tilde{v} \in \mathbb{R}_F, \forall k' \in \mathbb{R}, \\ D(\tilde{u} \oplus \tilde{v}, \tilde{w} \oplus \tilde{e}) &\leq D(\tilde{u}, \tilde{w}) + D(\tilde{v}, \tilde{e}), \forall \tilde{u}, \tilde{v}, \tilde{w}, \tilde{e} \in \mathbb{R}_F. \end{aligned}$$

**Definition 2.4** [1] Let $f, g : \mathbb{R} \to \mathbb{R}_F$ be fuzzy number valued functions. The distance between $f$, $g$ is defined by

$$D^*(f, g) := \sup_{x \in \mathbb{R}} D(f(x), g(x)).$$

**Lemma 2.5** [1]

1. *If we denote* $\tilde{0} := \chi_{\{0\}}$, *then* $\tilde{0} \in \mathbb{R}_F$ *is the neutral element with respect to* $\oplus$, *i.e.,* $\tilde{u} \oplus \tilde{0} = \tilde{0} \oplus \tilde{u} = \tilde{u}$, $\forall \tilde{u} \in \mathbb{R}_F$.
2. With respect to $\tilde{0}$, *none of* $\tilde{u} \in \mathbb{R}_F$, $\tilde{u} \neq \tilde{0}$ *has opposite in* $\mathbb{R}_F$.
3. Let $\alpha$, $\beta \in \mathbb{R} : \alpha.\beta \geq 0$, *and any* $\tilde{u} \in \mathbb{R}_F$, *we have* $(\alpha + \beta) \odot \tilde{u} = \alpha \odot \tilde{u} \oplus \beta \odot \tilde{u}$. *For general* $\alpha$, $\beta \in \mathbb{R}$, *the above property is false.*
4. For any $\gamma \in \mathbb{R}$ *and any* $\tilde{u}$, $\tilde{v} \in \mathbb{R}_F$, *we have* $\gamma \odot (\tilde{u} \oplus \tilde{v}) = \gamma \odot \tilde{u} \oplus \gamma \odot \tilde{v}$.
5. For any $\gamma$, $\eta \in \mathbb{R}$ *and any* $\tilde{u} \in \mathbb{R}_F$, *we have* $\gamma \odot (\eta \odot \tilde{u}) = (\gamma \odot \eta) \odot \tilde{u}$.

*If we denote* $\|\tilde{u}\|_F := D(\tilde{u}, \tilde{0})$, $\forall \tilde{u} \in \mathbb{R}_F$, *then* $\|.\|_F$ *has the properties of a usual norm on* $\mathbb{R}_F$, *i.e.,*

$$\begin{aligned} \|\tilde{u}\|_F &= 0 \text{ iff } \tilde{u} = \tilde{0}, \|\lambda \odot \tilde{u}\|_F = |\lambda|.\|\tilde{u}\|_F, \\ \|\tilde{u} \oplus \tilde{v}\|_F &\leq \|\tilde{u}\|_F + \|\tilde{v}\|_F, \|\tilde{u}\|_F - \|\tilde{v}\|_F \leq D(\tilde{u}, \tilde{v}). \end{aligned}$$

*Notice that* $(\mathbb{R}_F, \oplus, \odot)$ *is not a linear space over* $\mathbb{R}$, *and consequently* $(\mathbb{R}_F, \|.\|_F)$ *is not a normed space. Here* $\sum^*$ *denotes the fuzzy summation.*

**Definition 2.6** [13] A fuzzy valued function $f : [a, b] \to \mathbb{R}_F$ is said to be continuous at $x_0 \in [a, b]$, if for each $\epsilon > 0$ there exists $\delta > 0$ such that $D(f(x), f(x_0)) < \epsilon$, whenever $x \in [a, b]$ and $|x - x_0| < \delta$. We say that $f$ is fuzzy continuous on $[a, b]$ if $f$ is continuous at each $x_0 \in [a, b]$, and denotes the space of all such functions by $C_F[a, b]$.

**Definition 2.7** [3] Let $f : [a, b] \to \mathbb{R}_F$ be a bounded mapping. Then the function $\omega_{[a,b]}(f, .) : \mathbb{R}_+ \cup \{0\} \to \mathbb{R}_+$

$$\omega_{[a,b]}(f, \delta) = \sup\{D(f(x), f(y)); x, y \in [a, b], |x - y| \leq \delta\},$$

is called the modulus of oscillation of $f$ on $[a, b]$.

If $f \in C_F[a, b]$ (i.e. $f : [a, b] \to \mathbb{R}_F$ is continuous on $[a, b]$), then $\omega_{[a,b]}(f, \delta)$ is called uniform modulus of continuity of $f$.

The following properties will be very useful in what follows.

**Theorem 2.8** [3] *The following statements, concerning the modulus of oscillation, are true:*

1. $D(f(x), f(y)) \leq \omega_{[a,b]}(f, |x - y|)$, $\forall x, y \in [a, b]$,
2. $\omega_{[a,b]}(f, \delta)$ *is a nondecreasing mapping in* $\delta$,
3. $\omega_{[a,b]}(f, 0) = 0$,
4. $\omega_{[a,b]}(f, \delta_1 + \delta_2) \leq \omega_{[a,b]}(f, \delta_1) + \omega_{[a,b]}(f, \delta_2)$, $\forall \delta_1, \delta_2 \geq 0$,
5. $\omega_{[a,b]}(f, n\delta) \leq n\omega_{[a,b]}(f, \delta)$, $\forall \delta \geq 0$, $n \in \mathbb{N}$,
6. $\omega_{[a,b]}(f, \eta\delta) \leq (\eta + 1)\omega_{[a,b]}(f, \delta)$, $\forall \delta$, $\eta \geq 0$.

**Definition 2.9** [1] Let $f : [a, b] \to \mathbb{R}_F$. We say that $f$ is fuzzy-Riemann integrable to $I \in \mathbb{R}_F$ if for any $\epsilon > 0$, there exists $\delta > 0$ such that for any division $P = \{[u, v]; \xi\}$ of $[a, b]$ with the norms $\Delta(p) < \delta$, we have

$$D\left(\sum_p^* (v - u) \odot f(\xi), I\right) < \epsilon,$$

where $\sum^*$ denotes the fuzzy summation. We choose to write

$$I := (FR) \int_a^b f(x) \mathrm{d}x.$$

We also call an $f$ as above $(FR)$-integrable.

**Lemma 2.10** [1] *If $f, g : [a,b] \subseteq \mathbb{R} \to \mathbb{R}_F$ are fuzzy continuous functions, then the function $F : [a,b] \to \mathbb{R}_+$ defined by $F(x) := D(f(x), g(x))$ is continuous on* $[a, b]$, and

$$D\left((FR)\int_a^b f(x)\mathrm{d}x, (FR)\int_a^b g(x)\mathrm{d}x\right) \le \int_a^b D(f(x), g(x))\mathrm{d}x.$$

**Theorem 2.11** [3] *Let $f : [a,b] \to \mathbb{R}_F$ be a Henstock integrable, bounded mapping. Then, for any division $a = x_0 < x_1 < \cdots < x_n = b$ and any points $\xi_i \in [x_{i-1}, x_i]$ we have*

$$D\left((FH)\int_a^b f(t)dt, \sum_{i=1}^{n}(x_i - x_{i-1}) \odot f(\xi_i)\right) \le \sum_{i=1}^{n}(x_i - x_{i-1})\omega_{[x_{i-1},x_i]}(f, x_i - x_{i-1}).$$

By the above theorem, the following result holds:

**Corollary 2.12** [3] *Let $f : [a,b] \to \mathbb{R}_F$ be a Henstock integrable, bounded mapping. Then*

$$D\left((FH)\int_a^b f(t)dt, \frac{b-a}{6} \odot \left(f(a) \oplus 4 \odot f\left(\frac{a+b}{2}\right) \oplus f(b)\right)\right) \le 3(b-a)\omega_{[a,b]}\left(f, \frac{b-a}{6}\right).$$

**Definition 2.13** [18] Suppose that $f : [a,b] \times [c,d] \to \mathbb{R}_f$ is a bounded mapping. The function $\omega_{[a,b]\times[c,d]}(f, .) : \mathbb{R}_+ \cup \{0\} \to \mathbb{R}_+$ defined by

$$\omega_{[a,b]\times[c,d]}(f, \delta) = \sup\{D(f(x,y), f(s,t)); x, s \in [a,b]; y, t \in [c,d]; \sqrt{(x-s)^2 + (y-t)^2} \le \delta\},$$

is called modules of oscillation of $f$ on $[a,b] \times [c,d]$. In addition, if $f \in C_F([a,b] \times [c,d])$, then $\omega_{[a,b]\times[c,d]}(f, \delta)$ is called uniform modules of continuity of $f$.

**Theorem 2.14** [18] *The following properties hold:*

1. $D(f(x,y), f(s,t)) \le \omega_{[a,b]\times[c,d]}(f, \sqrt{(x-s)^2 + (y-t)^2})$, $\forall x, s \in [a,b], y, t \in [c,d]$;
2. $\omega_{[a,b]\times[c,d]}(f, \delta)$ *is a nondecreasing mapping in* $\delta$;
3. $\omega_{[a,b]\times[c,d]}(f, 0) = 0$;
4. $\omega_{[a,b]\times[c,d]}(f, \delta_1 + \delta_2) \le \omega_{[a,b]\times[c,d]}(f, \delta_1) + \omega_{[a,b]\times[c,d]}(f, \delta_2)$, $\forall \delta_1, \delta_2 \ge 0$;
5. $\omega_{[a,b]\times[c,d]}(f, n\delta) \le n\omega_{[a,b]\times[c,d]}(f, \delta)$, $\forall \delta \ge 0$, $n \in \mathbb{N}$;
6. $\omega_{[a,b]\times[c,d]}(f, \lambda\delta) \le (\lambda + 1)\omega_{[a,b]\times[c,d]}(f, \delta)$, $\forall \lambda, \delta \ge 0$.

**Corollary 2.15** [18] *We have*

$$\underline{(FR)\int_c^d \int_a^b f(s,t;r)dsdt} = \int_c^d \int_a^b \underline{f}(s,t,r)dsdt,$$

$$\overline{(FR)\int_c^d \int_a^b f(s,t;r)dsdt} = \int_c^d \int_a^b \overline{f}(s,t,r)dsdt.$$

**Theorem 2.16** [18] *If $f$ and $g$ are Henstock integrable mapping on $[a,b] \times [c,d]$ and if $D(f(s, t), g(s, t))$ is Lebesgue integrable, then*

$$D\left((FH)\int_c^d \int_a^b f(s,t)dsdt, (FH)\int_c^d \int_a^b g(s,t)dsdt\right) \le (L)\int_c^d \int_a^b D(f(s,t), g(s,t))dsdt. \tag{2.1}$$

**Definition 2.17** [18] A function $f : [a,b] \times [c,d] \to \mathbb{R}_F$ is said to be L-Lipschitz, if

$$D(f(x,y), f(s,t)) \le L\sqrt{(x-s)^2 + (y-t)^2}, \tag{2.2}$$

$\forall x, s \in [a,b]$, $y, t \in [c,d]$.

**Theorem 2.18** [18] *Let $f : [a,b] \times [c,d] \to \mathbb{R}_F$ be Henstock integrable, bounded mappings. Then, for any divisions $a = x_0 < x_1 < \cdots < x_n = b$ and $c = y_0 < y_1 < \cdots < y_n = d$ and any points $\xi_i \in [x_{i-1}, x_i]$ and $\eta_j \in [y_{j-1}, y_j]$, one has*

$$D\left((FH)\int_c^d (FH)\int_a^b f(s,t)dsdt, \sum_{j=1}^{n}\sum_{i=1}^{n}(x_i - x_{i-1})(y_j - y_{j-1}) \odot f(\xi_i, \eta_j)\right) \le \sum_{i=1}^{n}(x_i - x_{i-1})(y_j - y_{j-1})\omega_{[x_{i-1},x_i]\times[y_{j-1},y_j]}\left(f, \sqrt{(x_i - x_{i-1})^2 + (y_j - y_{j-1})^2}\right).$$

**Corollary 2.19** [18] *Let $f : [a,b] \times [c,d] \to \mathbb{R}_F$ be a two-dimensional Henstock integrable, bounded mapping. Then*

$$\begin{aligned} &D\Bigg((FH)\int_c^d (FH)\int_a^b f(s,t)dsdt, \frac{(b-a)(d-c)}{36} \odot \Bigg(f(a,c) \oplus f(a,d) \\ &\oplus 4 \odot f\left(a, \frac{c+d}{2}\right) \oplus 4 \odot f\left(\frac{a+b}{2}, c\right) \oplus 16 \odot f\left(\frac{a+b}{2}, \frac{a+c}{2}\right) \\ &\oplus 4 \odot f\left(\frac{a+b}{2}, d\right) \oplus 4 \odot f\left(b, \frac{c+d}{2}\right) \oplus f(b,c) \oplus f(b,d)\Bigg)\Bigg) \\ &\le (b-a)(d-c)\omega_{[a,b]\times[c,d]}\left(f, \frac{(b-a)(d-c)}{36}\right). \end{aligned} \tag{2.3}$$

## 2D linear fuzzy Fredholm integral equations

Consider 2DLFFIE as follows

$$F(s,t) = f(s,t) \oplus \lambda \odot (FR)\int_c^d (FR) \times \int_a^b k(s,t,x,y) \odot F(x,y)\mathrm{d}x\mathrm{d}y, \tag{3.1}$$

where $\lambda > 0$, $k(s, t, x, y)$ is an arbitrary positive function on $[a,b] \times [c,d] \times [a,b] \times [c,d]$ and $f : [a,b] \times [c,d] \to \mathbb{R}_F$. We assume that $k$ is continuous and, therefore, it is uniformly continuous with respect to $(s, t)$. So, there exists $M > 0$ such that $M = \max_{s,x\in[a,b],t,y\in[c,d]} |k(s,t,x,y)|$. Here, we present a quadrature iterative method for solving this equation.

**Theorem 3.1** [18] *Let the function k(s, t, x, y) be continuous and positive for* $s, x \in [a, b]$, *and* $t, y \in [c, d]$, *and let* $f : [a, b] \times [c, d] \to \mathbb{R}_F$ *be continuous on* $[a, b] \times [c, d]$. *Also, suppose f(x,t) that is not zero. If* $B = \lambda M(b-a)(d-c) < 1$ *then the fuzzy integral equation* (3.1) *has a unique solution* $F^* \in X$ *where*

$$X = \{f : [a, b] \times [c, d] \to \mathbb{R}_F; f \text{ is continous}\},$$

*is the space of two-dimensional fuzzy continuous functions with the metric*

$$D^*(f, g) = \sup_{s\in[a,b], t\in[c,d]} D(f(s,t), g(s,t)),$$

*and it can be obtained by the following successive approximations method*

$$F_0(s,t) = f(s,t),$$

$$F_m(s,t) = f(s,t) \oplus \lambda \odot (FR)\int_c^d (FR)\int_a^b k(s,t,x,y) \odot F_{m-1}(x,y)\mathrm{d}x\mathrm{d}y, \quad \forall m \geq 1. \tag{3.2}$$

*Moreover, the sequence of successive approximations,* $(F_m)_{m\geq 1}$ *converges to the solution* $F^*$. *Furthermore, the following error bound holds*

$$D^*(F^*, F_m) \leq \frac{B^{m+1}}{1-B} M_1, \forall m \geq 1, \tag{3.3}$$

*where* $M_1 = \sup_{s\in[a,b], t\in[c,d]} \|F(s,t)\|_F$.

Now, we propose a numerical method to solve (3.1). In this way, we assume 2DLFFIE (3.1) and uniform partitions of the interval $[a, b] \times [c, d]$

$$\Delta_x : a = s_0 < s_1 < \cdots < s_{2n-1} < s_{2n} = b, \tag{3.4}$$

with $s_i = a + ih$, where $h = \frac{b-a}{2n}$, $i = 0, \cdots, 2n$ and

$$\Delta_y : c = t_0 < t_1 < \cdots < t_{2n-1} < t_{2n} = d, \tag{3.5}$$

with $t_j = c + jh'$, where $h' = \frac{d-c}{2n}$, $j = 0, \cdots, 2n$. So, the following iterative procedure gives the approximate solution of (3.1) in point (s, t) as follows

$$\begin{aligned}
u_0(s,t) &= f(s,t)\\
u_m(s,t) &= f(s,t) \oplus \frac{\lambda hh'}{9} \odot \sum_{j=1}^{n}\sum_{i=1}^{n}\Big(k(s,t,s_{2i-2},t_{2j-2}) \odot u_{m-1}(s_{2i-2},t_{2j-2})\\
&\oplus k(s,t,s_{2i-2},t_{2j}) \odot u_{m-1}(s_{2i-2},t_{2j}) \oplus k(s,t,s_{2i},t_{2j-2}) \odot u_{m-1}(s_{2i},t_{2j-2})\\
&\oplus k(s,t,s_{2i},t_{2j}) \odot u_{m-1}(s_{2i},t_{2j}) \oplus 4\Big(k(s,t,s_{2i-2},t_{2j-1}) \odot u_{m-1}(s_{2i-2},t_{2j-1})\\
&\oplus k(s,t,s_{2i-1},t_{2j-2}) \odot u_{m-1}(s_{2i-1},t_{2j-2}) \oplus k(s,t,s_{2i},t_{2j-1}) \odot u_{m-1}(s_{2i},t_{2j-1})\\
&\oplus k(s,t,s_{2i-1},t_{2j}) \odot u_{m-1}(s_{2i-1},t_{2j})\Big) \oplus 16k(s,t,s_{2i-1},t_{2j-1}) \odot u_{m-1}(s_{2i-1},t_{2j-1})\Big).
\end{aligned} \tag{3.6}$$

## Convergence analysis

In this section, we obtain an error estimate between the exact solution and the approximate solution of 2DLFFIE (3.1).

**Theorem 4.1** *Under the hypotheses of Theorem* 3.1 *and* $\lambda > 0$, *the iterative procedure* (3.6) *converges to the unique solution of* (3.1), $F^*$, *and its error estimate is as follows*

$$\begin{aligned}
D^*(F^*, u_m) \leq\; & \frac{B^{m+1}}{1-B} M_1\\
&+ \frac{10B}{9(1-B)}(\omega_{[a,b]\times[c,d]}(f, hh')\\
&+ \frac{B}{M(1-B)}\|f\|\omega_{st}(k, hh')\\
&+ \frac{1}{M(1-B)}\|f\|\omega_{xy}(k, hh')).
\end{aligned} \tag{4.1}$$

*where*

$$\begin{aligned}
\omega_{st}(k,\delta) = \sup\{&|k(s_1,t_1,x,y) - k(s_2,t_2,x,y)| :\\
&\sqrt{(s_2-s_1)^2 + (t_2-t_1)^2} \leq \delta\},\\
&\forall \delta \geq 0, \forall s_1, s_2 \in [a,b], t_1, t_2 \in [c,d]
\end{aligned} \tag{4.2}$$

$$\begin{aligned}
\omega_{xy}(k,\delta) = \sup\{&|k(s,t,x_1,y_1) - k(s,t,x_2,y_2)| :\\
&\sqrt{(x_2-x_1)^2 + (y_2-y_1)^2} \leq \delta\},\\
&\forall \delta \geq 0, x_1, x_2 \in [a,b], y_1, y_2 \in [c,d]
\end{aligned} \tag{4.3}$$

*Proof* Considering iterative procedure (3.6), for all $(s,t) \in [a,b] \times [c,d]$, we have

$$\begin{aligned}
&D(F_1(s,t), u_1(s,t))\\
&= D\Big(f(s,t) \oplus \lambda \odot (FR)\int_c^d (FR)\int_a^b k(s,t,x,y)\\
&\odot F_0(x,y)\mathrm{d}x\mathrm{d}y, f(s,t) \oplus \frac{\lambda hh'}{9}\\
&\odot \sum_{j=1}^{n}\sum_{i=1}^{n}\Big((k(s,t,s_{2i-2},t_{2j-2}) \odot u_0(s_{2i-2},t_{2j-2})\\
&\oplus k(s,t,s_{2i-2},t_{2j}) \odot u_0(s_{2i-2},t_{2j}) \oplus k(s,t,s_{2i},t_{2j-2})
\end{aligned}$$

$$\odot u_0(s_{2i},t_{2j-2})$$
$$\oplus k(s,t,s_{2i},t_{2j})\odot u_0(s_{2i},t_{2j}))$$
$$\oplus 4(k(s,t,s_{2i-2},t_{2j-1})\odot u_0(s_{2i-2},t_{2j-1})$$
$$\oplus k(s,t,s_{2i-1},t_{2j-2})\odot u_0(s_{2i-1},t_{2j-2})$$
$$\oplus k(s,t,s_{2i},t_{2j-1})\odot u_0(s_{2i},t_{2j-1})$$
$$\oplus k(s,t,s_{2i-1},t_{2j})\odot u_0(s_{2i-1},t_{2j}))$$
$$\oplus 16k(s,t,s_{2i-1},t_{2j-1})\odot u_0(s_{2i-1},t_{2j-1})\Big)\Big)$$

$$= D\Big(\lambda\odot\sum_{j=1}^{n}\sum_{i=1}^{n}(FR)\int_{t_{2j-2}}^{t_{2j}}(FR)\int_{s_{2i-2}}^{s_{2i}}k(s,t,x,y)\odot f(x,y)\mathrm{d}x\mathrm{d}y,$$

$$\frac{\lambda hh'}{9}\odot\sum_{j=1}^{n}\sum_{i=1}^{n}\Big(k(s,t,s_{2i-2},t_{2j-2})\odot f(s_{2i-2},t_{2j-2})$$
$$\oplus k(s,t,s_{2i-2},t_{2j})\odot f(s_{2i-2},t_{2j})\oplus k(s,t,s_{2i},t_{2j-2})\odot f(s_{2i},t_{2j-2})$$
$$\oplus k(s,t,s_{2i},t_{2j})\odot f(s_{2i},t_{2j})$$
$$\oplus 4(k(s,t,s_{2i-2},t_{2j-1})\odot f(s_{2i-2},t_{2j-1})$$
$$\oplus k(s,t,s_{2i-1},t_{2j-2})\odot f(s_{2i-1},t_{2j-2})$$
$$\oplus k(s,t,s_{2i},t_{2j-1})\odot f(s_{2i},t_{2j-1})$$
$$\oplus k(s,t,s_{2i-1},t_{2j})\odot f(s_{2i-1},t_{2j}))$$
$$\oplus 16k(s,t,s_{2i-1},t_{2j-1})\odot f(s_{2i-1},t_{2j-1})\Big)\Big)$$

$$\leq \lambda\sum_{j=1}^{n}\sum_{i=1}^{n}D\Big(\int_{t_{2j-2}}^{t_{2j}}\int_{s_{2i-2}}^{s_{2i}}k(s,t,x,y)\odot f(x,y)\mathrm{d}x\mathrm{d}y,$$
$$\frac{hh'}{9}\odot\Big(k(s,t,s_{2i-2},t_{2j-2})\odot f(s_{2i-2},t_{2j-2})$$
$$\oplus k(s,t,s_{2i-1},t_{2j})\odot f(s_{2i-2},t_{2j})\oplus k(s,t,s_{2i},t_{2j-2})\odot f(s_{2i},t_{2j-2})$$
$$\oplus k(s,t,s_{2i},t_{2j})\odot f(s_{2i},t_{2j})$$
$$\oplus 4(k(s,t,s_{2i-2},t_{2j-1})\odot f(s_{2i-2},t_{2j-1})$$
$$\oplus k(s,t,s_{2i-1},t_{2j-2})\odot f(s_{2i-1},t_{2j-2})$$
$$\oplus k(s,t,s_{2i},t_{2j-1})\odot f(s_{2i},t_{2j-1})$$
$$\oplus k(s,t,s_{2i-1},t_{2j})\odot f(s_{2i-1},t_{2j}))$$
$$\oplus 16k(s,t,s_{2i-1},t_{2j-1})\odot f(s_{2i-1},t_{2j-1})\Big)\Big).$$

Using Corollary 2.19, we have

$$D(F_1(s,t),u_1(s,t))\leq\lambda\sum_{j=1}^{n}\sum_{i=1}^{n}(s_{2i}-s_{2i-2})(t_{2j}-t_{2j-2})$$
$$\times\omega_{[s_{2i-2},s_{2i}]\times[t_{2j-2},t_{2j}]}\Big(kf,\frac{hh'}{9}\Big),$$
$$\leq\frac{10}{9}\lambda\sum_{j=1}^{n}\sum_{i=1}^{n}(s_{2i}-s_{2i-2})(t_{2j}-t_{2j-2})$$
$$\times\omega_{[s_{2i-2},s_{2i}]\times[t_{2j-2},t_{2j}]}(kf,hh').$$

Since $(\alpha,\gamma),(\beta,\eta)\in[s_{2i-2},s_{2i}]\times[t_{2j-2},t_{2j}]$ with $\sqrt{(\alpha-\beta)^2+(\gamma-\eta)^2}\leq hh'$, we have

$$D\Big(f(\alpha,\gamma)\odot k(s,t,\alpha,\gamma),f(\beta,\eta)\odot k(s,t,\beta,\eta)\Big)$$
$$\leq D\Big(f(\alpha,\gamma)\odot k(s,t,\alpha,\gamma),f(\beta,\eta)\odot k(s,t,\alpha,\gamma)\Big)$$
$$+D\Big(f(\beta,\eta)\odot k(s,t,\alpha,\gamma),f(\beta,\eta)\odot k(s,t,\beta,\eta)\Big)$$
$$\leq|k(s,t,\alpha,\gamma)|D\Big(f(\alpha,\gamma),f(\beta,\eta)\Big)$$
$$+|k(s,t,\alpha,\gamma)-k(s,t,\beta,\eta)|D\Big(f(\beta,\eta),\tilde{0}\Big)$$
$$\leq M\omega_{[s_{2i-2},s_{2i}]\times[t_{2j-2},t_{2j}]}(f,hh')+\omega_{xy}(k,hh')\|f\|.$$

Taking supremum from above inequality, we conclude that,

$$\omega_{[s_{2i-2},s_{2i}]\times[t_{2j-2},t_{2j}]}(kf,hh')\leq M\omega_{[s_{2i-2},s_{2i}]\times[t_{2j-2},t_{2j}]}(f,hh')+\omega_{xy}(k,hh')\|f\|. \tag{4.4}$$

Therefore,

$$D^*(F_1,u_1)\leq\frac{10B}{9}\omega_{[a,b]\times[c,d]}(f,hh')+\frac{10B}{9M}\omega_{xy}(k,hh')\|f\|.$$

Now, for $m=2$, it follows that

$$D(F_2(s,t),u_2(s,t))=D(f(s,t),f(s,t))$$
$$+D\Big(\lambda\odot(FR)\int_c^d(FR)$$
$$\times\int_a^b k(s,t,x,y)\odot F_1(x,y)\mathrm{d}x\mathrm{d}y,$$
$$\frac{\lambda hh'}{9}\sum_{j=1}^{n}\sum_{i=1}^{n}\Big((k(s,t,s_{2i-2},t_{2j-2})$$
$$\odot u_1(s_{2i-2},t_{2j-2})\oplus k(s,t,s_{2i-2},t_{2j})$$
$$\odot u_1(s_{2i-2},t_{2j})\oplus k(s,t,s_{2i},t_{2j-2})$$
$$\odot u_1(s_{2i},t_{2j-2})$$
$$\oplus k(s,t,s_{2i},t_{2j})\odot u_1(s_{2i},t_{2j}))$$
$$\oplus 4(k(s,t,s_{2i-2},t_{2j-1})\odot u_1(s_{2i-2},t_{2j-1})$$
$$\oplus k(s,t,s_{2i-1},t_{2j-2})\odot u_1(s_{2i-1},t_{2j-2})$$
$$\oplus k(s,t,s_{2i},t_{2j-1})\odot u_1(s_{2i},t_{2j-1})$$
$$\oplus k(s,t,s_{2i-1},t_{2j})\odot u_1(s_{2i-1},t_{2j}))$$
$$\oplus 16k(s,t,s_{2i-1},t_{2j-1})\odot u_1(s_{2i-1},t_{2j-1})\Big)\Big)$$

$$\leq\lambda\sum_{j=1}^{n}\sum_{i=1}^{n}\Big(D\Big(\int_{t_{2j-2}}^{t_{2j}}\int_{s_{2i-2}}^{s_{2i}}k(s,t,x,y)$$
$$\odot F_1(x,y)\mathrm{d}x\mathrm{d}y,$$
$$\frac{hh'}{9}\odot\Big((k(s,t,s_{2i-2},t_{2j-2})\odot F_1(s_{2i-2},t_{2j-2})$$

$$
\begin{aligned}
&\oplus k(s,t,s_{2i-2},t_{2j})\odot F_1(s_{2i-2},t_{2j})\\
&\oplus k(s,t,s_{2i},t_{2j-2})\odot F_1(s_{2i},t_{2j-2})\\
&\oplus k(s,t,s_{2i},t_{2j})\odot F_1(s_{2i},t_{2j}))\\
&\oplus 4(k(s,t,s_{2i-2},t_{2j-1})\odot F_1(s_{2i-2},t_{2j-1})\\
&\oplus k(s,t,s_{2i-1},t_{2j-2})\odot F_1(s_{2i-1},t_{2j-2})\\
&\oplus k(s,t,s_{2i},t_{2j-1})\odot F_1(s_{2i},t_{2j-1})\\
&\oplus k(s,t,s_{2i-1},t_{2j})\odot F_1(s_{2i-1},t_{2j}))\\
&\oplus 16k(s,t,s_{2i-1},t_{2j-1})\odot F_1(s_{2i-1},t_{2j-1})\Big)\Big)\\
&+D\bigg(\frac{hh'}{9}((k(s,t,s_{2i-2},t_{2j-2})\odot F_1(s_{2i-2},t_{2j-2})\\
&\oplus k(s,t,s_{2i-2},t_{2j})\odot F_1(s_{2i-2},t_{2j})\\
&\oplus k(s,t,s_{2i},t_{2j-2})\odot F_1(s_{2i},t_{2j-2})\\
&\oplus k(s,t,s_{2i},t_{2j})\odot F_1(s_{2i},t_{2j}))\\
&\oplus 4(k(s,t,s_{2i-2},t_{2j-1})\odot F_1(s_{2i-2},t_{2j-1})\\
&\oplus k(s,t,s_{2i-1},t_{2j-2})\odot F_1(s_{2i-1},t_{2j-2})\\
&\oplus k(s,t,s_{2i},t_{2j-1})\odot F_1(s_{2i},t_{2j-1})\\
&\oplus k(s,t,s_{2i-1},t_{2j})\odot F_1(s_{2i-1},t_{2j}))\\
&\oplus 16k(s,t,s_{2i-1},t_{2j-1})\odot F_1(s_{2i-1},t_{2j-1})),\\
&\frac{hh'}{9}((k(s,t,s_{2i-2},t_{2j-2})\odot u_1(s_{2i-2},t_{2j-2})\\
&\oplus k(s,t,s_{2i-2},t_{2j})\odot u_1(s_{2i-2},t_{2j})\\
&\oplus k(s,t,s_{2i},t_{2j-2})\odot u_1(s_{2i},t_{2j-2})\\
&\oplus k(s,t,s_{2i},t_{2j})\odot u_1(s_{2i},t_{2j}))\\
&\oplus 4(k(s,t,s_{2i-2},t_{2j-1})\odot u_1(s_{2i-2},t_{2j-1})\\
&\oplus k(s,t,s_{2i-1},t_{2j-2})\odot u_1(s_{2i-1},t_{2j-2})\\
&\oplus k(s,t,s_{2i},t_{2j-1})\odot u_1(s_{2i},t_{2j-1})\\
&\oplus k(s,t,s_{2i-1},t_{2j})\odot u_1(s_{2i-1},t_{2j}))\\
&\oplus 16k(s,t,s_{2i-1},t_{2j-1})\odot u_1(s_{2i-1},t_{2j-1})\Big)\\
&\le \frac{10B}{9}\omega_{[a,b]\times[c,d]}(F_1,hh')+\frac{10B}{9M}\omega_{xy}(k,hh')\|F_1\|\\
&\quad+\frac{\lambda hh'}{9}\sum_{j=1}^{n}\sum_{i=1}^{n}\Big(|k(s,t,s_{2i-2},t_{2j-2})|\\
&\quad\times D(F_1(s_{2i-2},t_{2j-2}),u_1(s_{2i-2},t_{2j-2}))\\
&\quad+|k(s,t,s_{2i-2},t_{2j})|D(F_1(s_{2i-2},t_{2j}),u_1(s_{2i-2},t_{2j}))\\
&\quad+|k(s,t,s_{2i},t_{2j-2})|D(F_1(s_{2i},t_{2j-2}),u_1(s_{2i},t_{2j-2}))\\
&\quad+|k(s,t,s_{2i},t_{2j})|D(F_1(s_{2i},t_{2j}),u_1(s_{2i},t_{2j}))\\
&\quad+4|k(s,t,s_{2i-2},t_{2j-1})|D(F_1(s_{2i-2},t_{2j-1}),u_1(s_{2i-2},t_{2j-1}))\\
&\quad+4|k(s,t,s_{2i-1},t_{2j-2})|D(F_1(s_{2i-1},t_{2j-2}),u_1(s_{2i-1},t_{2j-2}))\\
&\quad+4|k(s,t,s_{2i},t_{2j-1})|D(F_1(s_{2i},t_{2j-1}),u_1(s_{2i},t_{2j-1}))\\
&\quad+4|k(s,t,s_{2i-1},t_{2j})|D(F_1(s_{2i-1},t_{2j}),u_1(s_{2i-1},t_{2j}))\\
&\quad+16|k(s,t,s_{2i-1},t_{2j-1})|D(F_1(s_{2i-1},t_{2j-1}),u_1(s_{2i-1},t_{2j-1}))\Big)
\end{aligned}
$$

So, we have the following result:

$$
\begin{aligned}
D^*(F_2,u_2)\le{}&\frac{10B}{9}\omega_{[a,b]\times[c,d]}(F_1,hh')+\frac{10B}{9M}\omega_{xy}(k,hh')\|F_1\|\\
&+\lambda M(b-a)(d-c)D^*(F_1,u_1)
\end{aligned}
$$

By induction, for $m\ge 3$, we obtain

$$
\begin{aligned}
D^*(F_m,u_m)&\le\frac{10B}{9}\omega_{[a,b]\times[c,d]}(F_{m-1},hh')\\
&\quad+\frac{10B}{9M}\omega_{xy}(k,hh')\|F_{m-1}\|\\
&\quad+BD^*(F_{m-1},u_{m-1}),\\
D^*(F_{m-1},u_{m-1})&\le\frac{10B}{9}\omega_{[a,b]\times[c,d]}(F_{m-2},hh')\\
&\quad+\frac{10B}{9M}\omega_{xy}(k,hh')\|F_{m-2}\|\\
&\quad+BD^*(F_{m-2},u_{m-2}),\\
D^*(F_{m-2},u_{m-2})&\le\frac{10B}{9}\omega_{[a,b]\times[c,d]}(F_{m-3},hh')\\
&\quad+\frac{10B}{9M}\omega_{xy}(k,hh')\|F_{m-3}\|\\
&\quad+BD^*(F_{m-3},u_{m-3}),\\
&\vdots\quad\vdots\quad\vdots\\
D^*(F_1,u_1)&\le\frac{10B}{9}\omega_{[a,b]\times[c,d]}(F_0,hh')\\
&\quad+\frac{10B}{9M}\omega_{xy}(k,hh')\|F_0\|\\
&\quad+BD^*(F_0,u_0).
\end{aligned}
$$

Then

$$
\begin{aligned}
D^*(F_m,u_m)\quad&\le\frac{10B}{9}(\omega_{[a,b]\times[c,d]}(F_{m-1},hh')+B\omega_{[a,b]\times[c,d]}(F_{m-2},hh')\\
&\quad+B^2\omega_{[a,b]\times[c,d]}(F_{m-3},hh')+\cdots+B^{m-1}\omega_{[a,b]\times[c,d]}(f,hh'))\\
&\quad+\frac{10B}{9M}\omega_{xy}(k,hh')(\|F_{m-1}\|+B\|F_{m-2}\|+\cdots+B^{m-1}\|F_0\|).
\end{aligned}\tag{4.5}
$$

On the other hand, we have:

$$
\begin{aligned}
D(F_m(s_1,t_1),F_m(s_2,t_2))&=D(f(s_1,t_1)\oplus\lambda\odot(FR)\\
&\quad\times\int_c^d(FR)\int_a^b k(s_1,t_1,x,y)\\
&\quad\odot F_{m-1}(x,y)\mathrm{d}x\mathrm{d}y,\\
&\quad\times f(s_2,t_2)\oplus\lambda\odot(FR)\int_c^d(FR)\\
&\quad\int_a^b k(s_2,t_2,x,y)\odot F_{m-1}(x,y)\mathrm{d}x\mathrm{d}y)\\
&\le D(f(s_1,t_1),f(s_2,t_2))\\
&\quad+\lambda\int_c^d\int_a^b|k(s_1,t_1,x,y)-k(s_2,t_2,x,y)|\\
&\quad\times D(F_{m-1}(x,y),\tilde{0})\mathrm{d}x\mathrm{d}y,
\end{aligned}
$$

therefore, we see that

$$\omega_{[a,b]\times[c,d]}(F_m, hh') \leq \omega_{[a,b]\times[c,d]}(f, hh') + \frac{B}{M}\omega_{st}(k, hh')\|F_{m-1}\|, \tag{4.6}$$

for any $(s_1, t_1), (s_2, t_2) \in [a, b] \times [c, d]$ with $\sqrt{(s_1 - s_2)^2 + (t_1 - t_2)^2} \leq hh'$.

By using above inequality and (4.5), we obtain

$$\begin{aligned} D^*(F_m, u_m) \leq & \frac{10B}{9}\omega_{[a,b]\times[c,d]}(f, hh')(1 + B + B^2 + \cdots + B^{m-1}) \\ & + \frac{10B^2}{9M}\omega_{st}(k, hh')(\|F_{m-2}\| + B\|F_{m-3}\| + B^2\|F_{m-4}\| \\ & + \cdots + B^{m-2}\|F_0\|) \\ & + \frac{10B}{9M}\omega_{xy}(k, hh')(\|F_{m-1}\| + B\|F_{m-2}\| + B^2\|F_{m-3}\| \\ & + \cdots + B^{m-1}\|F_0\|). \end{aligned}$$

Now, by using (3.2), we conclude that

$$\begin{aligned} D(F_m(s,t), F_{m-1}(s,t)) = & D(f(s,t) \oplus \lambda \odot \\ & (FR)\int_c^d (FR)\int_a^b k(s,t,x,y) \odot F_{m-1}(x,y)\mathrm{d}x\mathrm{d}y, \\ & f(s,t) \oplus \lambda \odot (FR)\int_c^d (FR)\int_a^b k(s,t,x,y) \odot \\ & F_{m-2}(x,y)\mathrm{d}x\mathrm{d}y) \\ & \leq \lambda \int_c^d \int_a^b |k(s,t,x,y)| D(F_{m-1}(x,y), F_{m-2}(x,y))\mathrm{d}x\mathrm{d}y \\ & \leq \lambda M(d-c)(b-a)D^*(F_{m-1}, F_{m-2}) \\ & \leq B^{m-1}D^*(F_1, F_0). \end{aligned}$$

Consequently,

$$\begin{aligned} D(F_m(s,t), F_0(s,t)) \leq & D(F_m(s,t), F_{m-1}(s,t)) \\ & + \cdots + D(F_1(s,t), F_0(s,t)). \end{aligned}$$

Taking supremum for $s, t \in [a, b] \times [c, d]$ from the above inequality, we have

$$\begin{aligned} D^*(F_m, F_0) & \leq (B^{m-1} + B^{m-2} + \cdots + B + 1)D^*(F_1, F_0) \\ & \leq \frac{1}{1-B}D^*(F_1, F_0). \end{aligned}$$

It is obvious that

$$\begin{aligned} D(F_1(s,t), F_0(s,t)) = & D(f(s,t) \oplus \lambda(FR)\int_c^t (FR)\int_a^b k(s,t,x,y) \odot \\ & F_0(x,y)\mathrm{d}x\mathrm{d}y, f(s,t)) \\ & \leq \lambda \int_c^d \int_a^b |k(s,t,x,y)| D(f(x,y), \tilde{0})\mathrm{d}x\mathrm{d}y \\ & \leq \lambda(d-c)(b-a)M\|f\| = B\|f\|, \end{aligned}$$

and

$$\begin{aligned} D(F_m(s,t), \tilde{0}) & \leq D(F_m(s,t), F_0(s,t)) + D(F_0(s,t), \tilde{0}) \\ & \leq \frac{1}{1-B}D^*(F_1, F_0) + \|f\| \\ & \leq \frac{1}{1-B}\|f\|. \end{aligned}$$

Therefore, we see that

$$\begin{aligned} D^*(F_m, u_m) \leq & \frac{10B}{9(1-B)}(\omega_{[a,b]\times[c,d]}(f, hh') \\ & + \frac{B}{M(1-B)}\|f\|\omega_{st}(k, hh') \\ & + \frac{1}{M(1-B)}\|f\|\omega_{xy}(k, hh')). \end{aligned}$$

□

## Numerical stability analysis

To show the numerical stability analysis of the proposed method in previous section, we consider another starting approximation $g(s,t) = Y_0(s,t)$ such that $\exists \epsilon > 0$ for which $D(F_0(s,t), Y_0(s,t)) < \epsilon, \forall s, t \in [a, b] \times [c, d]$. The obtained sequence of successive approximations is

$$\begin{aligned} Y_m(s,t) = & g(s,t) \oplus \lambda \odot (FR)\int_c^d (FR) \\ & \times \int_a^b k(s,t,x,y) \odot Y_{m-1}(x,y)\mathrm{d}x\mathrm{d}y, \end{aligned}$$

and using the same iterative method, the terms of produced sequence are

$$v_0(s,t) = Y_0(s,t) = g(s,t),$$

$$\begin{aligned} v_m(s,t) = g(s,t) \oplus \frac{\lambda hh'}{9} \odot \sum_{j=1}^{n}\sum_{i=1}^{n} \Bigg( & k(s,t,s_{2i-2},t_{2j-2}) \\ & \odot v_{m-1}(s_{2i-2},t_{2j-2}) \\ & \oplus k(s,t,s_{2i-2},t_{2j}) \odot v_{m-1}(s_{2i-2},t_{2j}) \\ & \oplus k(s,t,s_{2i},t_{2j-2}) \odot v_{m-1}(s_{2i},t_{2j-2}) \\ & \oplus k(s,t,s_{2i},t_{2j}) \odot v_{m-1}(s_{2i},t_{2j}) \\ & \oplus 4\Big(k(s,t,s_{2i-2},t_{2j-1}) \odot v_{m-1}(s_{2i-2},t_{2j-1}) \\ & \oplus k(s,t,s_{2i-1},t_{2j-2}) \odot v_{m-1}(s_{2i-1},t_{2j-2}) \\ & \oplus k(s,t,s_{2i},t_{2j-1}) \odot v_{m-1}(s_{2i},t_{2j-1}) \\ & \oplus k(s,t,s_{2i-1},t_{2j}) \odot v_{m-1}(s_{2i-1},t_{2j})\Big) \\ & \oplus 16k(s,t,s_{2i-1},t_{2j-1}) \odot v_{m-1}(s_{2i-1},t_{2j-1})\Bigg). \end{aligned}$$

**Definition 5.1** The proposed algorithm based on iterative method applied to solve 2DLFFIE (3.1) is said to be numerically stable with respect to the choice of the first iteration if there exist four independent constants $k_1, k_2, k_3, k_4 > 0$ such that

$$D^*(u_m, v_m) \le k_1\epsilon + k_2\Big(\omega_{[a,b]\times[c,d]}(f, hh') + \omega_{[a,b]\times[c,d]}(g, hh')\Big) + k_3\omega_{st}(k, hh') + k_4\omega_{xy}(k, hh'), \tag{5.1}$$

where $h = \frac{b-a}{2n}$, $h' = \frac{d-c}{2n}$,

$$k_1 = \frac{1}{1-B}, k_2 = \frac{10B}{9(1-B)}, k_3 = \frac{10B^2}{9M(1-B)^2}(\|f\| + \|g\|), k_4 = \frac{10B}{9M(1-B)^2}(\|f\| + \|g\|).$$

**Theorem 5.2** *Under the assumptions of Theorem 4.1, the presented method (3.6) is numerically stable with respect to the choice of the first iteration.*

*Proof* At first, we obtain that

$$\begin{aligned} D(u_m(s,t), v_m(s,t)) &\le D(u_m(s,t), F_m(s,t)) + D(F_m(s,t), Y_m(s,t)) \\ &\quad + D(Y_m(s,t), v_m(s,t)) \\ &\le \frac{10B}{9(1-B)}\Big(\omega_{[a,b]\times[c,d]}(f, hh') \\ &\quad + \frac{B}{M(1-B)}\|f\|\omega_{st}(k, hh') \\ &\quad + \frac{1}{M(1-B)}\|f\|\omega_{xy}(k, hh')\Big) \\ &\quad + D(F_m(s,t), Y_m(s,t)) \\ &\quad + \frac{10B}{9(1-B)}\Big(\omega_{[a,b]\times[c,d]}(g, hh') \\ &\quad + \frac{B}{M(1-B)}\|g\|\omega_{st}(k, hh') \\ &\quad + \frac{1}{M(1-B)}\|g\|\omega_{xy}(k, hh')\Big). \end{aligned}$$

However,

$$\begin{aligned} D(F_m(s,t), Y_m(s,t)) &= D\Big(f(s,t) \oplus \lambda \odot (FR)\int_c^d (FR)\int_a^b k(s,t,x,y) \odot \\ &\quad F_{m-1}(x,y)\mathrm{d}x\mathrm{d}y, \\ &\quad g(s,t) \oplus \lambda \odot (FR)\int_c^d (FR)\int_a^b k(s,t,x,y) \odot \\ &\quad Y_{m-1}(x,y)\mathrm{d}x\mathrm{d}y\Big) \\ &\le D(f(s,t), g(s,t)) \\ &\quad + \lambda D\Big((FR)\int_c^d (FR)\int_a^b k(s,t,x,y) \odot F_{m-1}(x,y)\mathrm{d}x\mathrm{d}y, \\ &\quad (FR)\int_c^d (FR)\int_a^b k(s,t,x,y) \odot Y_{m-1}(x,y)\mathrm{d}x\mathrm{d}y\Big) \\ &\le \epsilon + \lambda (FR)\int_c^d (FR)\int_a^b |k(s,t,x,y)| \\ &\quad D(F_{m-1}(x,y), Y_{m-1}(x,y))\mathrm{d}x\mathrm{d}y. \end{aligned}$$

We conclude that

$$\begin{aligned} D^*(F_m, Y_m) &\le \epsilon + \lambda \int_c^d \int_a^b MD^*(F_{m-1}, Y_{m-1})\mathrm{d}x\mathrm{d}y \\ &= \epsilon + BD^*(F_{m-1}, Y_{m-1}), \end{aligned}$$

and thus

$$\begin{aligned} D^*(F_m, Y_m) &\le \epsilon + BD^*(F_{m-1}, Y_{m-1}) \\ D^*(F_{m-1}, Y_{m-1}) &\le \epsilon + BD^*(F_{m-2}, Y_{m-2}) \\ D^*(F_{m-2}, Y_{m-2}) &\le \epsilon + BD^*(F_{m-3}, Y_{m-3}) \\ &\vdots \quad \vdots \\ D^*(F_1, Y_1) &\le \epsilon + BD^*(F_0, Y_0). \end{aligned}$$

So,

$$\begin{aligned} D^*(F_m, Y_m) &\le \epsilon + B\Big(\epsilon + BD^*(F_{m-2}, Y_{m-2})\Big) \\ &\le \epsilon + B\epsilon + B^2\Big(\epsilon + BD^*(F_{m-3}, Y_{m-3})\Big) \\ &\le \epsilon + B\epsilon + B^2\epsilon + B^3\Big(\epsilon + BD^*(F_{m-4}, Y_{m-4})\Big) \\ &\vdots \\ &\le \epsilon + B\epsilon + B^2\epsilon + B^3\epsilon + \cdots + B^m D^*(F_0, Y_0) \\ &\le \epsilon\Big(1 + B + B^2 + B^3 + \cdots + B^m\Big) \le \frac{\epsilon}{1-B}. \end{aligned}$$

Therefore,

$$\begin{aligned} D^*(u_m, v_m) &\le \frac{10B}{9(1-B)}\Big(\omega_{[a,b]\times[c,d]}(f, hh') \\ &\quad + \frac{B}{M(1-B)}\|f\|\omega_{st}(k, hh') \\ &\quad + \frac{1}{M(1-B)}\|f\|\omega_{xy}(k, hh')\Big) + \frac{\epsilon}{1-B} \\ &\quad + \frac{10B}{9(1-B)}\Big(\omega_{[a,b]\times[c,d]}(g, hh') \\ &\quad + \frac{B}{M(1-B)}\|g\|\omega_{st}(k, hh') \\ &\quad + \frac{1}{M(1-B)}\|g\|\omega_{xy}(k, hh')\Big), \end{aligned}$$

where

$$k_1 = \frac{1}{1-B}, k_2 = \frac{10B}{9(1-B)},$$

$$k_3 = \frac{10B^2}{9M(1-B)^2}(\|f\| + \|g\|),$$

$$k_4 = \frac{10B}{9M(1-B)^2}(\|f\| + \|g\|).$$

□

## Numerical examples

In this section, we apply the proposed method in "2D linear fuzzy Fredholm integral equations" for solving two examples. In addition, we compare the absolute errors of the obtained results with the results of the method [18].

*Example 6.1* [14] Consider the linear integral equation

$$F(s,t)=f(s,t)\oplus(FR)\int_0^1(FR)\int_0^1 k(s,t,x,y)\odot F(x,y)\mathrm{d}x\mathrm{d}y, \quad (6.1)$$

with

$$k(s,t,x,y)=s^2tx,$$
$$\underline{f}(s,t,r)=(r^2+r)s\sin\left(\frac{t}{2}\right),$$
$$\overline{f}(s,t,r)=(4-r^3-r)s\sin\left(\frac{t}{2}\right).$$

The exact solution of this example is

$$\underline{F}^*(s,t,r)=(r^2+r)\left(s\sin\left(\frac{t}{2}\right)-\frac{16}{21}\left(\cos\left(\frac{1}{2}\right)-1\right)s^2t\right),$$
$$\overline{F}^*(s,t,r)=(4-r^3-r)\left(s\sin\left(\frac{t}{2}\right)-\frac{16}{21}\left(\cos\left(\frac{1}{2}\right)-1\right)s^2t\right).$$

By using the proposed method and the method [18] for $h=h'=\frac{1}{10},\frac{1}{20}$, $m=5,7$ and $r\in\{0.00,0.25,0.50,0.75,1.00\}$ in $(s_0,t_0)=(0.5,0.5)$, we present the absolute errors in Tables 1, 2, 3, and 4

**Table 1** The absolute errors on the level sets with $h=h'=\frac{1}{10}$, $m=5$ for Example 6.1 by using the method [18] in $(s_0,t_0)=(0.5,0.5)$

| $r-level$ | $e_-^r=\vert\underline{F}^*(s_0,t_0,r)-\underline{u}_m(s_0,t_0,r)\vert$ | $e_+^r=\vert\overline{F}^*(s_0,t_0,r)-\overline{u}_m(s_0,t_0,r)\vert$ |
|---|---|---|
| 0.00 | 0 | $2.88943\times10^{-4}$ |
| 0.25 | $2.25737\times10^{-5}$ | $2.69755\times10^{-4}$ |
| 0.50 | $5.41768\times10^{-5}$ | $2.43796\times10^{-4}$ |
| 0.75 | $9.48094\times10^{-5}$ | $2.04292\times10^{-4}$ |
| 1.00 | $1.44471\times10^{-4}$ | $1.44471\times10^{-4}$ |

**Table 2** The absolute errors on the level sets with $h=h'=\frac{1}{20}$, $m=7$ for Example 6.1 by using the method [18] in $(s_0,t_0)=(0.5,0.5)$

| $r-level$ | $e_-^r=\vert\underline{F}^*(s_0,t_0,r)-\underline{u}_m(s_0,t_0,r)\vert$ | $e_+^r=\vert\overline{F}^*(s_0,t_0,r)-\overline{u}_m(s_0,t_0,r)\vert$ |
|---|---|---|
| 0.00 | 0 | $7.25207\times10^{-5}$ |
| 0.25 | $5.66568\times10^{-6}$ | $6.77049\times10^{-5}$ |
| 0.50 | $1.35976\times10^{-5}$ | $6.11894\times10^{-5}$ |
| 0.75 | $2.37959\times10^{-5}$ | $5.12744\times10^{-5}$ |
| 1.00 | $3.62604\times10^{-5}$ | $3.62604\times10^{-5}$ |

**Table 3** The absolute errors on the level sets with $h=h'=\frac{1}{10}$, $m=5$ for Example 6.1 by using the proposed method in $(s_0,t_0)=(0.5,0.5)$

| $r-level$ | $e_-^r=\vert\underline{F}^*(s_0,t_0,r)-\underline{u}_m(s_0,t_0,r)\vert$ | $e_+^r=\vert\overline{F}^*(s_0,t_0,r)-\overline{u}_m(s_0,t_0,r)\vert$ |
|---|---|---|
| 0.00 | 0 | $1.42157\times10^{-6}$ |
| 0.25 | $1.11060\times10^{-7}$ | $1.32717\times10^{-6}$ |
| 0.50 | $2.66545\times10^{-7}$ | $1.19945\times10^{-6}$ |
| 0.75 | $4.66454\times10^{-7}$ | $1.00510\times10^{-6}$ |
| 1.00 | $7.10787\times10^{-7}$ | $7.10787\times10^{-7}$ |

**Table 4** The absolute errors on the level sets with $h=h'=\frac{1}{20}$, $m=7$ for Example 6.1 by using the proposed method in $(s_0,t_0)=(0.5,0.5)$

| $r-level$ | $e_-^r=\vert\underline{F}^*(s_0,t_0,r)-\underline{u}_m(s_0,t_0,r)\vert$ | $e_+^r=\vert\overline{F}^*(s_0,t_0,r)-\overline{u}_m(s_0,t_0,r)\vert$ |
|---|---|---|
| 0.00 | 0 | $2.21362\times10^{-8}$ |
| 0.25 | $1.72939\times10^{-9}$ | $2.06662\times10^{-8}$ |
| 0.50 | $4.15054\times10^{-9}$ | $1.86774\times10^{-8}$ |
| 0.75 | $7.26344\times10^{-9}$ | $1.56510\times10^{-8}$ |
| 1.00 | $1.10681\times10^{-8}$ | $1.10681\times10^{-8}$ |

**Table 5** The absolute errors on the level sets with $h=h'=\frac{1}{10}$, $m=5$ for Example 6.2 by using the method [18] in $(s_0,t_0)=(0.5,0.5)$, respectively

| $r-level$ | $e_-^r=\vert\underline{F}^*(s_0,t_0,r)-\underline{u}_m(s_0,t_0,r)\vert$ | $e_+^r=\vert\overline{F}^*(s_0,t_0,r)-\overline{u}_m(s_0,t_0,r)\vert$ |
|---|---|---|
| 0.00 | 0 | $4.48970\times10^{-5}$ |
| 0.25 | $5.61213\times10^{-6}$ | $3.92849\times10^{-5}$ |
| 0.50 | $1.12243\times10^{-5}$ | $3.36728\times10^{-5}$ |
| 0.75 | $1.68364\times10^{-5}$ | $2.80606\times10^{-5}$ |
| 1.00 | $2.24485\times10^{-5}$ | $2.24485\times10^{-5}$ |

**Table 6** The absolute errors on the level sets with $h=h'=\frac{1}{10}$, $m=5$ for Example 6.2 by using the proposed method in $(s_0,t_0)=(0.5,0.5)$, respectively

| $r-level$ | $e_-^r=\vert\underline{F}^*(s_0,t_0,r)-\underline{u}_m(s_0,t_0)^r\vert$ | $e_+^r=\vert\overline{F}^*(s_0,t_0,r)-\overline{u}_m(s_0,t_0)_+^r\vert$ |
|---|---|---|
| 0.00 | 0 | $9.20708\times10^{-13}$ |
| 0.25 | $1.15088\times10^{-13}$ | $8.05633\times10^{-13}$ |
| 0.50 | $2.30177\times10^{-13}$ | $6.90559\times10^{-13}$ |
| 0.75 | $3.45279\times10^{-13}$ | $5.75429\times10^{-13}$ |
| 1.00 | $4.60354\times10^{-13}$ | $4.60354\times10^{-13}$ |

*Example 6.2* [14] Consider the following fuzzy Fredholm integral equation (6.1) with

$$\underline{f}(s,t,r) = r\left(st + \frac{1}{676}(s^2 + t^2 - 2)\right),$$
$$\overline{f}(s,t,r) = (2-r)\left(st + \frac{1}{676}(s^2 + t^2 - 2)\right),$$

and kernel

$$k(s,t,x,y) = \frac{1}{169}(s^2 + t^2 - 2)(x^2 + y^2 - 2), 0 \leq s,t,x,y \leq 1.$$

The exact solution is

$$\underline{F}^*(s,t,r) = rst,$$
$$\overline{F}^*(s,t,r) = (2-r)st.$$

To compare the absolute errors for the proposed method and the method [18], see Tables 5, and 6

## Conclusions

To approximate the solution of 2DLFFIEs, we developed a quadrature iterative method. We prove the convergence analysis (Theorem 4.1) and the numerical stability analysis (Theorem 5.2) of the proposed method with respect to the choice of the first iteration. Obtained results show that the proposed method can be a suitable method for solving 2DLFFIEs.

## References


1. Anastassiou, G.A.: Fuzzy mathematics: approximation theory. Springer, Berlin (2010)
2. Baghmisheh, M., Ezzati, R.: Numerical solution of nonlinear fuzzy Fredholm integral equations of the second kind using hybrid of block-pulse functions and Taylor series. Adv. Diff. Equ. **2015**, 15 (2015)
3. Bede, B., Gal, S.G.: Quadrature rules for integrals of fuzzy-number-valued functions. Fuzzy Sets and Syst. **145**, 359–380 (2004)
4. Behzadi, ShS, Allahviranloo, T., Abbasbandy, S.: The use of fuzzy expansion method for solving fuzzy linear Volterra–Fredholm integral equations. J. Intell. Fuzzy Syst. **26**, 1817–1822 (2014)
5. Bica, A.M.: Error estimation in the approximation of the solution of nonlinear fuzzy Fredholm integral equations. Inf. Sci. **178**, 1279–1292 (2008)
6. Bica, A.M., Popescu, C.: Approximating the solution of nonlinear Hammerstein fuzzy integral equations. Fuzzy Sets and Syst. **245**, 1–17 (2014)
7. Ezzati, R., Sadatrasoul, S.M.: Application of bivariate fuzzy Bernstein polynomials to solve two-dimensional fuzzy integral equations. Soft Comput. **2016**, 11 (2016)
8. Ezzati, R., Ziari, S.: Numerical solution and error estimation of fuzzy Fredholm integral equation using fuzzy Bernstein polynomials. Aust. J. Basic Appl. Sci. **5**, 2072–2082 (2011)
9. Ezzati, R., Ziari, S.: Numerical solution of nonlinear fuzzy Fredholm integral equations using iterative method. Appl. Math. Comput. **225**, 33–42 (2013)
10. Jelodar, S.S.F., Allahviranloo, T., Abbasbandy, S.: Application of fuzzy expansion methods for solving fuzzy Fredholm-Volterra integral equations of the first kind. Int. J. Math. Model. Comput. **3**, 59–70 (2013)
11. Friedman, M., Ma, M., Kandel, A.: Solution to fuzzy integral equations with arbitrary kernel. Int. J. Approx. Reason. **20**, 249–262 (1999)
12. Friedman, M., Ma, M., Kandel, A.: Numerical solutions of fuzzy differential and integral equations. Fuzzy Sets and Syst. **106**, 35–48 (1999)
13. Kaleva, O.: Fuzzy differential equations. Fuzzy Sets and Syst. **24**, 301–317 (1987)
14. Mirzaee, F., Komak yari, M., Hadadiyan, E.: Numerical solution of two-dimensional fuzzy Fredholm integral equations of the second kind using triangular functions. Beni-Suef Univ. J. Basic Appl. Sci. **4**, 109–118 (2015)
15. Mokhtarnejad, F., Ezzati, R.: The numerical solution of nonlinear Hammerstein fuzzy integral equations by using fuzzy wavelet like operator. J. Intell. Fuzzy Syst. **28**, 1617–1626 (2015)
16. Mordeson, J., Newman, W.: Fuzzy integral equations. Inform. Sci. **87**, 215–229 (1995)
17. Rivaz, A., Yousefi, F.: Modified homotopy perturbation method for solving two-dimensional fuzzy Fredholm integral equation. Int. J. Appl. Math. **25**, 591–602 (2012)
18. Sadatrasoul, S.M., Ezzati, R.: Quadrature rules and iterative method for numerical solution of two-dimensional fuzzy integral equations. In: Abstract and applied analysis. Hindawi Publishing Corporation, vol. 2014, p. 18 (2014)
19. Sadatrasoul, S.M., Ezzati, R.: Iterative method for numerical solution of two-dimensional nonlinear fuzzy integral equations. Fuzzy Sets and Syst. **280**, 91–106 (2015)
20. Sadatrasoul, S.M., Ezzati, R.: Numerical solution of two-dimensional nonlinear Hammerstein fuzzy integral equations based on optimal fuzzy quadrature formula. J. Comput. Appl. Math. **292**, 430–446 (2016)
21. Seikkala, S.: On the fuzzy initial value problem. Fuzzy Sets and Syst. **24**, 319–330 (1987)
22. Ziari, S., Bica, A.M.: New error estimate in the iterative numerical method for nonlinear fuzzy HammersteinFredholm integral equations. Fuzzy Sets and Syst. (In Press, Corrected Proof Note to users)
23. Ziari, S., Ezzati, R., Abbasbandy, S.: Numerical solution of linear fuzzy Fredholm integral equations of the second kind using fuzzy Haar wavelet. In: Advances in Computational Intelligence. Communications in Computer and Information Science, vol. 299 (2012)

# 13

# Singularly perturbed convection-diffusion boundary value problems with two small parameters using nonpolynomial spline technique

**Pooja Khandelwal[1] · Arshad Khan[1]**

**Abstract** In this paper, a new nonpolynomial cubic spline method is developed for solving two-parameter singularly perturbed boundary value problems. Convergence analysis is briefly discussed. Numerical examples and computational results illustrate and guarantee a higher accuracy by this technique. Comparisons are made to confirm the reliability and accuracy of the proposed technique.

**Keywords** Singular perturbation · Nonpolynomial cubic spline · Convergence analysis · Boundary value problem · Convection-diffusion

## Introduction

We consider the two-parameter singularly perturbed convection-diffusion boundary value problems of the form:

$$Ly(x) \equiv -\epsilon y''(x) + \mu p(x)y'(x) + f(x)y(x) = g(x), \quad x \in (a, b) \tag{1.1}$$

subjected to the boundary conditions:

$$y(0) = \alpha_0, \quad y(1) = \alpha_1.$$

with two small positive parameters $0 < \epsilon \ll 1$, $0 < \mu \ll 1$, where $p(x)$, $f(x)$, and $g(x)$ are sufficiently smooth real valued functions with $p(x) \geq p^* > 0, f(x) \geq f^* > 0$, and $g(x) \geq g^* > 0$ for $x \in (a, b)$. Under these assumptions, problem (1.1) is characterized into two cases:

1. For $\mu = 0$, problem (1.1) becomes reaction-diffusion problem.
2. For $\mu = 1$, problem (1.1) becomes convection-diffusion problem.

This type of problem arises in the fields like engineering, mathematical physics, and in many areas of applied mathematics. We often come across boundary value problems in which one or small positive parameter multiplies with the derivatives. A large number of research papers have been found in the literature for single parameter convection-diffusion and reaction-diffusion problems [2, 8, 9, 12, 16]. However, only a very few authors have discussed two-parameter singularly perturbed boundary value problems [4, 6, 7, 10, 11, 14, 16, 18–20]. The nature of two parameters is asymptotically examined by O' Malley [14]. Different numerical methods have been proposed by various authors for two-parameter singularly perturbed problems such as exponentially fitted cubic spline method [7], finite difference, finite element, and B-spline collocation method [6, 11], Haar wavelet method [16], and exponential spline technique [18]. For more information about SPPs, readers are referred to books [13, 15] and references therein.

In this paper, we introduce a new nonpolynomial cubic spline method as an alternative to existing methods. The paper is organised into five sections. In Sect. 2, we give a brief derivation of nonpolynomial parameters cubic spline. In Sect. 3, we presented the formulation of the method. Convergence analysis is briefly discussed in Sect. 4. Finally, in Sect. 5, numerical examples and comparison

✉ Pooja Khandelwal
pooja2n@gmail.com

Arshad Khan
akhan1234in@rediffmail.com

[1] Department of Mathematics, Jamia Millia Islamia, New Delhi 110025, India

with the existing methods are given that demonstrate the practical applicability and superiority of the proposed method.

## Nonpolynomial spline function

We consider a uniform mesh $\Delta$ with nodal points $x_i$ on $[a, b]$, such that $\Delta : a = x_0 < x_1 < x_2 <, \cdots, < x_n - 1 < x_n = b$, where $x_i = a + ih, i = 0, 1, \ldots, n$, and $h = \frac{(b-a)}{n}$. A nonpolynomial spline function $S_\Delta(x)$ of class $C^2[a, b]$ which interpolates $y(x)$ at mesh points $x_i, i = 0(1)n$ depends on a parameter $k$, if we take $k \to 0$, then it reduces to ordinary cubic spline in $[a, b]$.

For each segment $[x_i, x_{i+1}], i = 0, 1, 2 \ldots n - 1$, we consider the nonpolynomial cubic spline $S_\Delta(x)$ of the form:

$$\begin{aligned} S_\Delta(x) =& a_i \sin k(x - x_i) + b_i \cos k(x - x_i) \\ &+ c_i e^{k(x-x_i)} + d_i e^{-k(x-x_i)}, \quad i = 0, 1, \ldots, n, \end{aligned} \tag{2.1}$$

where $a_i, b_i, c_i$, and $d_i$ are unknown coefficients and $k$ is a free parameter which will be used to raise the accuracy of the method.

Let $y(x)$ be the exact solution and $y_i$ be an approximation to $y(x_i)$, obtained by the segment $S_i(x)$ of the mixed splines function passing through the points $(x_i, y_i)$ and $(x_{i+1}, y_{i+1})$. To determine the coefficients of Eq. (2.1) in terms of $y_i, y_{i+1}, M_i, M_{i+1}$, we first define:

$$\left.\begin{aligned} S_\Delta(x_i) = y_i, \quad S_\Delta(x_{i+1}) = y_{i+1}, \\ S''_\Delta(x_i) = M_i, \quad S''_\Delta(x_{i+1}) = M_{i+1}. \end{aligned}\right\} \tag{2.2}$$

We obtain via a long but straightforward calculation

$$\begin{aligned} a_i =& \frac{(k^2 y_{i+1} - M_{i+1}) - \cos\theta(k^2 y_i - M_i)}{2k^2 \sin\theta}, \\ b_i =& \frac{(k^2 y_i - M_i)}{2k^2}, \\ c_i =& \frac{e^\theta(k^2 y_{i+1} + M_{i+1}) - (k^2 y_i + M_i)}{2k^2(e^{2\theta} - 1)}, \\ d_i =& \frac{e^{2\theta}(k^2 y_i + M_i) - e^\theta(k^2 y_{i+1} + M_{i+1})}{2k^2(e^{2\theta} - 1)}, \\ & \theta = kh \quad \text{and} \quad i = 0(1)n - 1. \end{aligned}$$

Using the continuity of the first derivative at the point $x = x_i$, we obtain the following tridiagonal system for $i = 1, 2, \ldots, n - 1$:

$$y_{i-1} + \gamma y_i + y_{i+1} = h^2(\alpha M_{i-1} + \beta M_i + \alpha M_{i+1}), \tag{2.3}$$

where

$$\begin{aligned} \alpha =& \frac{(e^{2\theta} - 2e^\theta \sin\theta - 1)}{\theta^2(e^{2\theta} + 2e^\theta \sin\theta - 1)}, \\ \beta =& 2\frac{[e^{2\theta}(\sin\theta - \cos\theta) - (\sin\theta + \cos\theta)]}{\theta^2(e^{2\theta} + 2e^\theta \sin\theta - 1)}, \\ \gamma =& -2\frac{[e^{2\theta}(\sin\theta + \cos\theta) + (\sin\theta - \cos\theta)]}{\theta^2(e^{2\theta} + 2e^\theta \sin\theta - 1)}. \end{aligned}$$

If $\theta \to 0$, then $(\alpha, \beta, \gamma) \to (\frac{1}{6}, \frac{4}{6}, -2)$, and then spline defined by (2.3) reduces to a ordinary cubic spline relation [13]:

$$(y_{i-1} - 2y_i + y_{i+1}) = \frac{h^2}{6}(M_{i-1} + 4M_i + M_{i+1}). \tag{2.4}$$

The relation (2.3) gives $(n - 1)$ linear algebraic equations in $(n - 1)$ unknowns $y_i, i = 1, 2, \ldots, n - 1$.

## The method

At the grid point $x_i$, the proposed two-parameter singularly perturbed boundary value problem (1.1) can be discretized as follows:

$$-\epsilon y''(x_i) + \mu p(x_i) y'(x_i) + f(x_i) y(x_i) = g(x_i). \tag{3.1}$$

Using spline's second derivative, we have

$$\begin{aligned} M_i =& \frac{\mu p_i y'_i + f_i y_i - g_i}{\epsilon}, \\ M_{i-1} =& \frac{\mu p_{i-1} y'_{i-1} + f_{i-1} y_{i-1} - g_{i-1}}{\epsilon}, \\ M_{i+1} =& \frac{\mu p_{i+1} y'_{i+1} + f_{i+1} y_{i+1} - g_{i+1}}{\epsilon}, \end{aligned}$$

where

$$\begin{aligned} y'_i =& \frac{y_{i+1} - y_{i-1}}{2h}, \\ y'_{i-1} =& \frac{-y_{i+1} + 4y_i - 3y_{i-1}}{2h}, \quad y'_{i+1} = \frac{3y_{i+1} - 4y_i + y_{i-1}}{2h}, \end{aligned}$$

$p_i = p(x_i), f_i = f(x_i)$ and $g_i = g(x_i)$.

Substituting the values of $M_j (j = i, i \pm 1)$ in Eq. (2.3), we have

$$\begin{aligned} &\left[-\epsilon + \frac{\mu h}{2}(-3\alpha p_{i-1} - \beta p_i + \alpha p_{i+1}) + h^2 \alpha f_{i-1}\right] y_{i-1} \\ &+ \left[-\gamma\epsilon + \frac{\mu h}{2}(4\alpha p_{i-1} - 4\alpha p_{i+1}) + h^2 \beta f_i\right] y_i \\ &+ \left[-\epsilon + \frac{\mu h}{2}(-\alpha p_{i-1} + \beta p_i + 3\alpha p_{i+1}) + h^2 \alpha f_{i+1}\right] y_{i+1} \\ &= h^2(\alpha g_{i-1} + \beta g_i + \alpha g_{i+1}), \quad i = 1(1)n - 1. \end{aligned} \tag{3.2}$$

Finally, we arrive at the following system:

$$\begin{cases} \left[-\gamma\epsilon + \frac{\mu h}{2}V_1 + h^2\beta f_1\right]y_1 + \left[-\epsilon + \frac{\mu h}{2}W_1 + h^2\alpha f_2\right]y_2 \\ = h^2[\alpha(g_0 - f_0\alpha_0) + \beta g_1 + \alpha g_2] + \epsilon\alpha_0 - \frac{\mu h}{2}\alpha_0 U_1, & i = 1, \\ \left[-\epsilon + \frac{\mu h}{2}U_i + h^2\alpha f_{i-1}\right]y_{i-1} + \left[-\gamma\epsilon + \frac{\mu h}{2}V_i + h^2\beta f_i\right]y_i + \left[-\epsilon + \frac{\mu h}{2}W_i + h^2\alpha f_{i+1}\right]y_{i+1} \\ = h^2(\alpha g_{i-1} + \beta g_i + \alpha g_{i+1}), & 2 \le i \le n-2, \\ \left[-\epsilon + \frac{\mu h}{2}U_{n-1} + h^2\alpha f_{n-2}\right]y_{n-2} + \left[-\gamma\epsilon + \frac{\mu h}{2}V_{n-1} + h^2\beta f_{n-1}\right]y_{n-1} \\ = h^2[\alpha(g_n - f_n\alpha_n) + \alpha g_{n-2} + \beta g_{n-1}] + \epsilon\alpha_1 - \frac{\mu h}{2}\alpha_n W_{n-1}, & i = n-1, \end{cases} \tag{3.3}$$

where

$$U_i = (-3\alpha p_{i-1} - \beta p_i + \alpha p_{i+1}), \quad V_i = (4\alpha p_{i-1} - 4\alpha p_{i+1}),$$
$$W_i = (-\alpha p_{i-1} + \beta p_i + 3\alpha p_{i+1}), \quad i = 1(1)n-1.$$

## Convergence analysis

In this section, we investigate the convergence analysis of the proposed method. For this, let $Y = y(x_i), \bar{Y} = (y_i), C = (c_i), T = (t_i), E = (e_i) = Y - \bar{Y}, i = 1, 2, \ldots, n-1$ be an exact column vectors, where $Y, \bar{Y}, T,$ and $E$ are exact, approximate, local truncation error, and discretization error, respectively.

We can write the standard matrix equation for the method developed in the following form:

$$M\bar{Y} = C, \tag{4.1}$$

where $M$ is a matrix of order $(n-1)$ with

$$M = (A_0 + A_1 + h^2 A_2 F). \tag{4.2}$$

The tridiagonal matrices $A_0, A_1,$ and $A_2$ have the form:

$$A_0 = \begin{bmatrix} -\gamma\epsilon & -\epsilon & & & \\ -\epsilon & -\gamma\epsilon & -\epsilon & & \\ & & \ddots & & \\ & & & \ddots & \\ & & -\epsilon & -\gamma\epsilon & -\epsilon \\ & & & -\epsilon & -\gamma\epsilon \end{bmatrix}, \tag{4.3}$$

$$A_1 = \begin{bmatrix} \frac{\mu h}{2}V_1 & \frac{\mu h}{2}W_1 & & & & \\ \frac{\mu h}{2}U_2 & \frac{\mu h}{2}V_2 & \frac{\mu h}{2}W_2 & & & \\ & & \ddots & & & \\ & & & \ddots & & \\ & & & \frac{\mu h}{2}U_{n-2} & \frac{\mu h}{2}V_{n-2} & \frac{\mu h}{2}W_{n-2} \\ & & & & \frac{\mu h}{2}U_{n-1} & \frac{\mu h}{2}V_{n-1} \end{bmatrix}, \tag{4.4}$$

$$A_2 = \begin{bmatrix} \beta & \alpha & & & & \\ \alpha & \beta & \alpha & & & \\ & & \ddots & & & \\ & & & \ddots & & \\ & & & \alpha & \beta & \alpha \\ & & & & \alpha & \beta \end{bmatrix}, \tag{4.5}$$

and

$$F = \begin{bmatrix} f_1 \\ f_2 \\ \vdots \\ \vdots \\ f_{n-2} \\ f_{n-1} \end{bmatrix}. \tag{4.6}$$

For the $(n-1)$ column vector $C$, we have

$$c_i = \begin{cases} h^2[\alpha(g_0 - f_0\alpha_0) + \beta g_1 + \alpha g_2] + \epsilon\alpha_0 - \dfrac{\mu h}{2}\alpha_0 U_1, & i = 1, \\ h^2(\alpha g_{i-1} + \beta g_i + \alpha g_{i+1}), & 2 \le i \le n-2, \\ h^2[\alpha(g_n - f_n\alpha_n) + \alpha g_{n-2} + \beta g_{n-1}] + \epsilon\alpha_1 - \dfrac{\mu h}{2}\alpha_n W_{n-1}, & i = n-1. \end{cases} \tag{4.7}$$

Now, considering the above system with exact solution $Y = [y(x_1), y(x_2), \ldots, y(x_{n-1})]$, we have

$$MY = T(h) + C, \tag{4.8}$$

where $T(h) = [t_1(h), t_2(h), \ldots, t_{n-1}(h)]^T$ is the local truncation error vector, where

$$t_i(h) = -(2+\gamma)\epsilon y_i + (2\alpha + \beta - 1)\epsilon h^2 y^{(2)}(\xi_i) + \left(\alpha - \frac{1}{12}\right) \epsilon h^4 y^{(4)}(\xi_i) + \left(\frac{\alpha}{12} - \frac{1}{360}\right)\epsilon h^6 y^{(6)}(\xi_i), \quad x_{i-1} < \xi_i < x_{i+1}. \tag{4.9}$$

for any arbitrary choice of $\alpha, \beta$, and $\gamma$ except $\alpha = 1/6, \beta = 4/6$ and $\gamma = -2$.

If we choose $\alpha = 1/6, \beta = 4/6$, and $\gamma = -2$,

$$t_i(h) = \frac{\epsilon h^4}{12} y^{(4)}(\xi_i), \quad x_{i-1} < \xi_i < x_{i+1}. \tag{4.10}$$

If we choose $\alpha = 1/12, \beta = 10/12$, and $\gamma = -2$,

$$t_i(h) = \frac{\epsilon h^6}{240} y^{(6)}(\xi_i), \quad x_{i-1} < \xi_i < x_{i+1}. \tag{4.11}$$

From Eqs. (4.1) and (4.8), we get

$$M(Y - \bar{Y}) = T(h)$$

or

$$ME = T(h), \tag{4.12}$$

where $E = (Y - \bar{Y}) = [e_1, e_2, \ldots, e_{n-1}]^T$.

Clearly, the row sums $M_1, M_2, \ldots, M_{n-1}$ of $M$ are

$$M_1 = -\gamma\epsilon - \epsilon + \frac{\mu h}{2}(3\alpha p_0 + \beta p_1 - \alpha p_2) + h^2(\beta f_1 + \alpha f_2), \quad i = 1,$$

$$M_i = -\gamma\epsilon - 2\epsilon + h^2(\alpha f_{i-1} + \beta f_i + \alpha f_{i+1}), \quad i = 2(1)n-2,$$

$$M_{n-1} = -\gamma\epsilon - \epsilon + \frac{\mu h}{2}(\alpha p_{n-2} - \beta p_{n-1} - 3\alpha p_n) + h^2(\alpha f_{n-2} + \beta f_{n-1}), \quad i = n-1,$$

If we choose $h$ sufficiently small, matrix $M$ becomes irreducible and monotone [5]. It follows that $M^{-1}$ exists and its elements are nonnegative. Hence, from Eq. (4.12), we have

$$E = M^{-1}T(h). \tag{4.13}$$

Let $m_{k,i}^{-1}$ is the $(k,i)^{\text{th}}$ element of the matrix $M^{-1}$. We define

$$\|m_{k,i}^{-1}\| = \max_{1 \le k \le n} \sum_{i=1}^{n-1} |m_{k,i}^{-1}| \tag{4.14}$$

and

$$\|T\| = \max_{1 \le k \le n} |t_k|. \tag{4.15}$$

In addition, from the theory of matrices, we have

$$\sum_{i=1}^{n-1} m_{k,i}^{-1} M_i = 1, \quad k = 1, 2, \ldots, n-1. \tag{4.16}$$

Therefore

$$m_{k,i}^{-1} \le \frac{1}{\min_{1 \le i \le n-1} M_i} = \frac{1}{h^2 Q_{i_o}}, \tag{4.17}$$

where $Q_{i_o} = \frac{1}{h^2}\min_i M_i > 0$, for some $i_o$ between 1 to $n-1$.

From Eqs. (4.9), (4.13), and (4.14), we have

$$e_i = \sum_{i=1}^{n-1} m_{k,i}^{-1} T_i(h), \quad k = 1, 2, \ldots, n-1 \tag{4.18}$$

and therefore

$$|e_i| \leq \frac{Kh^2}{|Q_{i_o}|}, \quad i = 1, 2, \ldots, n-1, \tag{4.19}$$

where $K$ is a constant independent of $h$. It follows that $\|E\| = O(h^2)$.

However, for the choice of parameters, $\alpha = 1/12, \beta = 10/12$, and $\gamma = -2$,

$$|e_i| \leq \frac{Kh^4}{|Q_{i_o}|}, \quad i = 1, 2, \ldots, n-1, \tag{4.20}$$

where $K$ is a constant independent of $h$. It follows that $\|E\| = O(h^4)$.

We summarize the above result in the following theorem:

**Theorem 4.1** *Let $y(x)$ be the exact solution of two-parameter singularly perturbed boundary value problem* (1.1) *and let $y_i$ be the numerical solution obtained from the difference scheme* (4.1). *Then, for sufficiently small h, scheme gives a second-order convergent solution for any arbitrary choice of $\alpha$ and $\beta$ with $\gamma = -2$ and a fourth-order convergent solution for $\alpha = 1/12, \beta = 10/12$, and $\gamma = -2$.*

## Numerical examples

To test the viability of the proposed method based on nonpolynomial cubic spline, two numerical examples are considered. All the computations were performed using MATLAB. We also compare our method with the existing methods which shown improvement.

*Example 1* Consider the following two-parameter singularly perturbed boundary value problem, which is discussed in [12, 19]:

$$-\epsilon y'' + \mu y' + y = 1, \quad x \in (0, 1), \tag{5.1}$$

subjected to the boundary conditions:

$$y(0) = 0, \quad y(1) = 0. \tag{5.2}$$

The exact solution of the above problem is

$$y(x) = \frac{(e^{\lambda_2} - 1)e^{\lambda_1 x}}{e^{\lambda_1} - e^{\lambda_2}} + \frac{(1 - e^{\lambda_1})e^{\lambda_2 x}}{e^{\lambda_1} - e^{\lambda_2}} + 1, \tag{5.3}$$

where

$$\lambda_1 = \frac{1 + \sqrt{1 + 4\epsilon}}{2\epsilon}, \quad \lambda_2 = \frac{1 - \sqrt{1 + 4\epsilon}}{2\epsilon}.$$

*Example 2* Consider the following two-parameter singularly perturbed boundary value problem, which is discussed in [6, 16, 19]:

$$-\epsilon y'' + \mu y' + y = \cos(\pi x), \quad x \in (0, 1), \tag{5.4}$$

subjected to the boundary conditions:

$$y(0) = 0, \quad y(1) = 0. \tag{5.5}$$

The exact solution of the above problem is

$$u(x) = \rho_1 \cos(\pi x) + \rho_2 \sin(\pi x) + \psi_1 e^{\lambda_1 x} + \psi_2 e^{-\lambda_2 (1-x)}, \tag{5.6}$$

where

$$\rho_1 = \frac{\epsilon\pi^2 + 1}{\epsilon^2\pi^2 + (\epsilon\pi^2 + 1)^2}, \quad \rho_2 = \frac{\epsilon\pi}{\epsilon^2\pi^2 + (\epsilon\pi^2 + 1)^2},$$

$$\psi_1 = -\rho_1 \frac{1 + e^{-\lambda_2}}{1 - e^{\lambda_1 - \lambda_2}}, \quad \psi_2 = \rho_1 \frac{1 + e^{\lambda_1}}{1 - e^{\lambda_1 - \lambda_2}},$$

$$\lambda_1 = \frac{\mu - \sqrt{\mu^2 + 4\epsilon}}{2\epsilon}, \quad \lambda_2 = \frac{\mu + \sqrt{\mu^2 + 4\epsilon}}{2\epsilon}.$$

The numerical results corresponding to the Examples 1 and 2 are briefly summarized in Tables 1, 2, 3, and 4, and Figs. 1, 2, 3, and 4. Comparison with other existing methods are also listed in Tables 1, 2, 3 and 4. These tables show that method is more accurate than the existing methods.

**Table 1** Comparison of pointwise errors, Example 1

| **x** ↓ | $\epsilon$=0.1, $\mu$=1, n=32<br>Lin et al. [12] | Our method | $\epsilon$=0.1, $\mu$=1, n=128<br>Lin et al. [12] | Our method |
|---|---|---|---|---|
| 1/16 | 2.74(−2) | 3.27(−6) | 6.8(−3) | 1.15(−6) |
| 2/16 | 2.59(−2) | 5.44(−6) | 6.4(−3) | 2.18(−6) |
| 4/16 | 2.30(−2) | 3.98(−6) | 5.7(−3) | 3.72(−6) |
| 6/16 | 2.04(−2) | 1.61(−5) | 5.0(−3) | 3.92(−6) |
| 12/16 | 2.50(−2) | 8.52(−4) | 4.0(−4) | 4.45(−5) |
| 14/16 | 3.30(−2) | 1.70(−3) | 9.4(−3) | 9.61(−5) |

**Table 2** Comparison of pointwise errors, Example 1

| x ↓ | $\epsilon$=0.01, $\mu$=1, n=32 Lin et al. [12] | Our method | $\epsilon$=0.01, $\mu$=1, n=128 Lin et al. [12] | Our method |
|---|---|---|---|---|
| 1/16 | 2.95(−2) | 4.55(−6) | 7.3(−3) | 2.84(−7) |
| 2/16 | 2.78(−2) | 8.55(−6) | 6.9(−3) | 5.35(−7) |
| 4/16 | 2.45(−2) | 1.51(−5) | 6.1(−3) | 9.45(−7) |
| 6/16 | 2.17(−2) | 2.00(−5) | 5.4(−3) | 1.25(−6) |
| 12/16 | 1.50(−2) | 3.07(−5) | 3.7(−3) | 1.73(−6) |
| 14/16 | 1.29(−2) | 1.41(−3) | 3.3(−3) | 7.31(−7) |

**Table 3** Comparison of maximum absolute errors, Example 2

| $\mu$ ↓ | $\epsilon$=$10^{-2}$, n=128 Kadalbajoo et al. [6] | Zahra et al. [19] | Pandit et al. [16] | Our method |
|---|---|---|---|---|
| $10^{-3}$ | 8.3832(−5) | 4.1924(−5) | 4.2303(−5) | 6.0243(−6) |
| $10^{-4}$ | 8.2686(−5) | 4.1296(−5) | 4.1318(−5) | 6.1827(−7) |
| $10^{-5}$ | 8.2572(−5) | 4.1232(−5) | 4.1220(−5) | 1.1455(−7) |
| $10^{-6}$ | 8.2561(−5) | 4.1226(−5) | 4.1210(−5) | 7.2269(−8) |
| $10^{-7}$ | 8.25596(−5) | 4.1225(−5) | 4.1209(−5) | 6.8266(−8) |

**Table 4** Comparison of maximum absolute errors, Example 2

| $\mu$ ↓ | $\epsilon$=$10^{-4}$, n=128 Kadalbajoo et al. [6] | Zahra et al. [19] | Pandit et al. [16] | Our method |
|---|---|---|---|---|
| $10^{-3}$ | 9.4446(−3) | 4.7598(−3) | 5.1964(−3) | 6.2154(−3) |
| $10^{-4}$ | 9.0436(−3) | 4.2856(−3) | 4.1710(−3) | 1.8330(−3) |
| $10^{-5}$ | 9.0036(−3) | 4.2295(−3) | 4.0754(−3) | 1.1412(−3) |
| $10^{-6}$ | 8.9996(−3) | 4.2238(−3) | 4.0659(−3) | 1.3699(−3) |
| $10^{-7}$ | 8.9992(−3) | 4.2232(−3) | 4.0650(−3) | 1.3656(−3) |

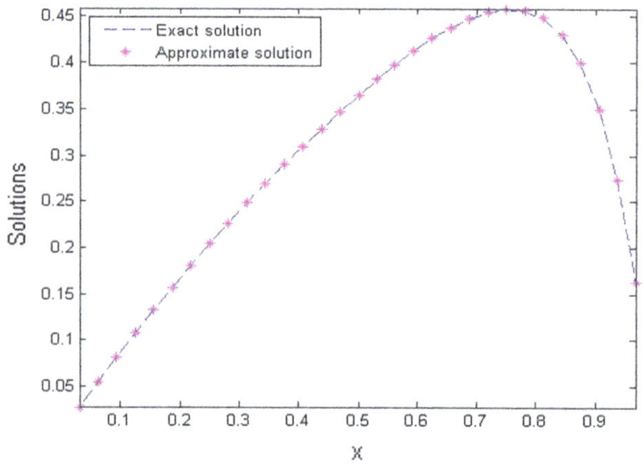


**Fig. 1** Physical behaviour of numerical solution of Example 1 for $\epsilon = 0.1, \mu = 1, and\ n = 32$

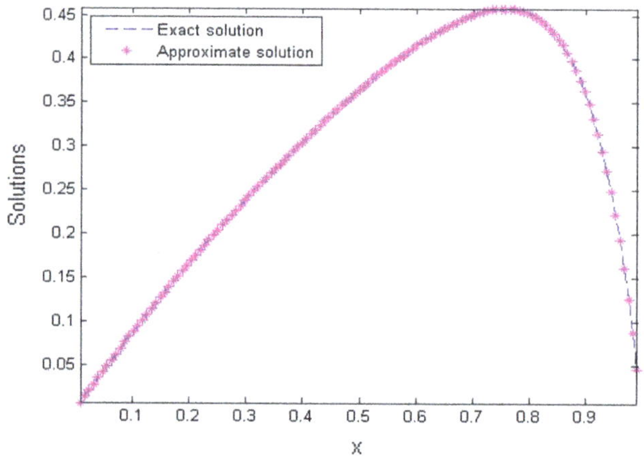


**Fig. 2** Physical behaviour of numerical solution of Example 1 for $\epsilon = 0.1, \mu = 1, and\ n = 128$

Tables 1, 2 show the pointwise errors at different values of $n$ and for small values of $\epsilon$. Tables 3, 4 show the maximum absolute errors of the Example 2 for different values of $\epsilon$ and $\mu$. Figures 1, 2 compare the exact and approximate solutions of Example 1 for $\epsilon = 0.1$ and $\mu = 1$, while Figs. 3 and 4 report the exact and approximate solutions of Example 2 for different values of $\epsilon$ and $\mu$.

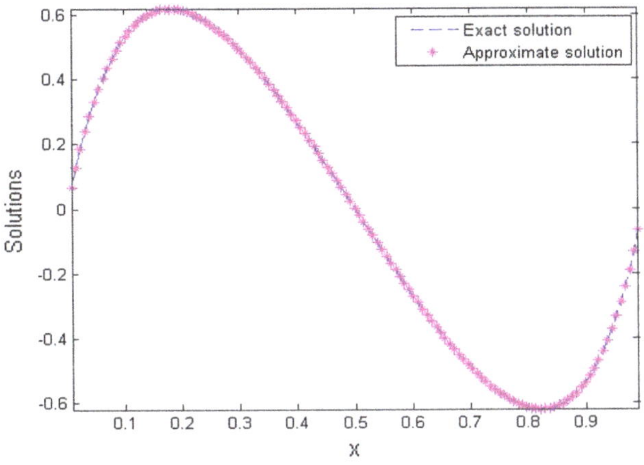


**Fig. 3** Physical behaviour of numerical solution of Example 2 for $\epsilon = 10^{-2}, \mu = 10^{-3}, and\ n = 128$

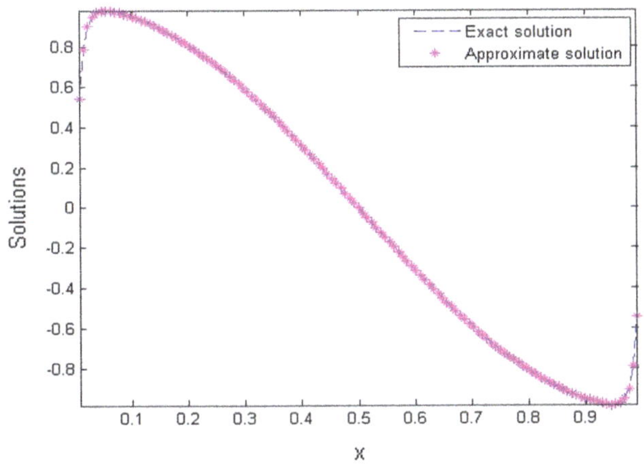


**Fig. 4** Physical behaviour of numerical solution of Example 2 for $\epsilon = 10^{-4}, \mu = 10^{-5}, and\ n = 128$

## Concluding remarks

In this paper, nonpolynomial cubic spline function is used for finding the numerical solution of two-parameter convection-diffusion singularly perturbed boundary value problems. The computations associated with the examples discussed above were performed using MATLAB. The proposed method is computationally efficient and the algorithm can be easily implemented on a computer. Comparison of the method is also depicted through Tables 1, 2, 3, and 4 which shown that our methods perform better in the sense of accuracy and applicability. The solution profiles for the considered examples for different values of $\epsilon$ and $\mu$ are given in Figs. 1, 2, 3, and 4.

**Acknowledgements** The authors are thankful to referee for their valuable suggestions which improved the quality of the paper. The first author is also thankful to the NBHM, Department of Atomic Energy, Government of India, for its financial assistance vide letter no. $2/40(32)/2012/R\&D - II/11626$ to carry out this research work.

## References


1. Ahlberg, J.H., Nilson, E.N., Walsh, J.L.: The Theory of Splines and Their Applications. Academic Press, New York (1967)
2. Aziz, T., Khan, A.: A spline method for second-order singularly perturbed boundary-value problems. J. Comput. Appl. Math. **147**, 445–452 (2002)
3. De Boor, C.: A practical guide to splines. In: Mathematics of computation, vol. 27(149). Springer-verlag (1978). doi:10.2307/2006241
4. Gracia, J.L., O'Riordan, E., Pickett, M.L.: A parameter robust second order numerical method for a singularly perturbed two-parameter problem. Appl. Numer. Math. **56**, 962–980 (2006)
5. Henrici, P.: Discrete Variable Methods in Ordinary Differential Equations. Wiley, New York (1962)
6. Kadalbajoo, M.K., Yadaw, A.S.: B-spline collocation method for a two-parameter singularly perturbed convection-diffusion boundary value problems. Appl. Math. Comput. **201**, 504–513 (2008)
7. Kadalbajoo, M.K., Jha, A.: Exponentially fitted cubic spline for two-parameter singularly perturbed boundary value problems. Int. J. Comput. Math. **89**(6), 836–850 (2012)
8. Khan, A., Khan, I., Aziz, T.: Sextic spline solution of singularly perturbed boundary value problem. Appl. Math. Comput. **181**, 432–439 (2006)
9. Khan, A., Khandelwal, P.: Non-polynomial sextic spline solution of singularly perturbed boundary-value problems. Int. J. Comput. Math. **19**(5), 1122–1135 (2014)
10. Kumar, D., Yadaw, A.S., Kadalbajoo, M.K.: A parameter-uniform method for two parameter singularly perturbed boundary value problems via asymptotic expansion. Appl. Math. Inf. Sci. **7**(4), 1525–1532 (2013)
11. LinB, T., Roos, H.G.: Analysis of a finite-difference scheme for a singularly perturbed problem with two small parameters. J. Math. Anal. Appl. **289**, 355–366 (2004)
12. Lin, B., Li, K., Cheng, Z.: B-spline solution of a singularly perturbed boundary value problem arising in biology. Chaos Solitons Fractals **42**, 2934–2948 (2009)
13. Miller, J.J.H., O'Riordan, E., Shishkin, G.I.: Fitted Numerical Methods for Singular Perturbation Problem. World Scientific, Singapore (1996)
14. O'Malley Jr., R.E.: Two parameter singular perturbation problems for second order equations. J. Math. Mech. **16**, 1143–1164 (1967)
15. O'Malley Jr., R.E.: Introduction to Singular Perturbations. Academic Press, New York (1974)
16. Pandit, S., Kumar, M.: Haar wavelet approach for numerical solution of two parameters singularly perturbed boundary value problems. Appl. Math. Inf. Sci. **8**(6), 2965–2974 (2014)
17. Rashidinia, J., Ghasemi, M., Mahmoodi, Z.: Spline approach to the solution of a singularly perturbed boundary value problems. Appl. Math. Comput. **189**, 72–78 (2007)
18. Valarmathi, S., Ramanujam, N.: Computational methods for solving two-parameter singularly perturbed boundary value problems for second-order ordinary differential equations. Appl. Math. Comput. **136**, 415–441 (2003)
19. Zahra, W.K., El Mhlawy, A.M.: Numerical solution of two-parameter singularly perturbed boundary value problems via exponential spline. J. King Saud Univ. **25**, 201–208 (2013)
20. Brdar, M., Zarin, H.: On a graded meshes for a two-parameter singularly perturbed problem. Appl. Math. Comput. **282**, 97–107 (2016)

# 14

# Shorter proofs of some recent even-tupled coincidence theorems for weak contractions in ordered metric spaces

**Anupam Sharma · Mohammad Imdad · Aftab Alam**

**Abstract** In this paper, we prove some recent even coincidence theorems due to Imdad et al. (Bull Math Anal Appl 5(4): 19-39, 2013) using a method of reduction from the respective coincidence theorems for mappings with one variable in ordered complete metric spaces. Our technique of proof is different, slightly simpler, shorter and more effective than the ones used in Imdad et al.

**Keywords** Partially ordered set · Compatible mapping · Mixed $g$-monotone property · $n$-tupled coincidence point · $n$-tupled fixed point

**Mathematics Subject Classification** 54H10 · 54H25

## Introduction

The investigation of fixed points in ordered metric spaces is a relatively new development which appears to have its origin in the paper of Ran and Reurings [30] which was well complimented by Nieto and López [25]. Ran and Reurings' fixed point theorem extended and refined by many authors, (for details see [8, 12, 24–27, 37]).

The concept of coupled fixed point was introduced by Guo and Lakshmikantham [11]. In [5], Bhaskar and Lakshmikantham introduced the notion of mixed monotone property for a mapping $F: X^2 \to X$ and proved some coupled fixed point theorems for weakly linear contractions enjoying mixed monotone property in ordered complete metric spaces. In this continuation, Lakshmikantham and Ćirić [22] generalized these results for nonlinear contraction mappings by introducing two ideas namely: coupled coincidence point and mixed $g$-monotone property. In an attempt to extend the definition from $X^2$ to $X^3$, Berinde and Borcut [4] introduced the concept of tripled fixed point and utilize the same to prove some tripled fixed point theorems. After that, Karapınar [16] introduced the quadrupled fixed point to prove some quadrupled fixed point theorems for nonlinear contraction mappings satisfying mixed $g$-monotone property (for more details see [17, 18]). Recently, Samet and Vetro [32] extended the idea of coupled as well as quadrupled fixed point to higher dimensions by introducing the notion of fixed point of $n$-order (or $n$-tupled fixed point, where $n \in \mathbb{N}$ and $n \geq 3$) and presented some $n$-tupled fixed point results in complete metric spaces, using a new concept of $f$-invariant set. Here it can be pointed out that the notion of tripled fixed point due to Berinde and Borcut [4] is different from the one defined by Samet and Vetro [32] for $n = 3$ in the case of ordered metric spaces in order to keep the mixed monotone property working. Recently, Imdad et al. [13] extended the idea of mixed $g$-monotone property to the mapping $F: X^n \to X$ (where $n$ is even natural number)and proved an even-tupled coincidence point theorem for nonlinear contraction mappings satisfying mixed $g$-monotone property. Basically their results are true for only even $n$ but not for odd ones (for details see [15]). Further, Imdad et al. [14] proved some even-tupled coincidence theorems under nonlinear weak contractions due to Choudhury et al. [9].

---

A. Sharma (✉) · M. Imdad · A. Alam
Department of Mathematics, Aligarh Muslim University,
Aligarh 202 002, India
e-mail: annusharma241@gmail.com

M. Imdad
e-mail: mhimdad@yahoo.co.in

A. Alam
e-mail: aafu.amu@gmail.com

Very recently, Samet et al. [34] have shown that the coupled (analogously $n$-tupled) fixed results can be more easily obtained using well-known fixed point theorems on ordered metric spaces (see also [10, 28, 29]). This technique of proof is different, slightly simpler, shorter and more effective than classical technique. In this paper, we prove the main results of Imdad et al. [14] following the techniques of Samet et al. [34].

## Preliminaries

With a view to make, our presentation self-contained, we collect some basic definitions and needed results which will be used frequently in the text later.

**Definition 2.1** Let $X$ be a non-empty set. A relation '$\preceq$' on $X$ is said to be a partial order if the following properties are satisfied:

(i) reflexive: $x \preceq x$ for all $x \in X$,
(ii) anti-symmetric: $x \preceq y$ and $y \preceq x$ imply $x = y$,
(iii) transitive: $x \preceq y$ and $y \preceq z$ imply $x \preceq z$ for all $x, y, z \in X$.

A non-empty set $X$ together with a partial order '$\preceq$' is said to be an ordered set and we denote it by $(X, \preceq)$.

**Definition 2.2** Let $(X, \preceq)$ be an ordered set. Any two elements $x$ and $y$ are said to be comparable elements in $X$ if either $x \preceq y$ or $y \preceq x$.

**Definition 2.3** ([27]) A triplet $(X, d, \preceq)$ is called an ordered metric space if $(X, d)$ is a metric space and $(X, \preceq)$ is an ordered set. Moreover, if $d$ is a complete metric on $X$, then we say that $(X, d, \preceq)$ is an ordered complete metric space.

Recently, Kutbi et al. [21] introduced the concept of regular map.

**Definition 2.4** ([21]) An ordered metric space $(X, d, \preceq)$ is said to be nondecreasing regular (resp. nonincreasing regular) if it satisfies the following property: if $\{x_m\}$ is a nondecreasing (resp. nonincreasing) sequence and $x_m \to x$, then $x_m \preceq x$ (resp. $x \preceq x_m$) $\forall m \in \mathbb{N} \cup \{0\}$.

**Definition 2.5** ([21]) An ordered metric space $(X, d, \preceq)$ is said to be regular if it is both nondecreasing regular and nonincreasing regular.

**Definition 2.6** Let $(X, d, \preceq)$ be an ordered metric space and $g : X \to X$ be a mapping. Then $X$ is said to be nondecreasing $g$-regular (resp. nonincreasing $g$-regular) if it satisfies the following property: if $\{x_m\}$ is a nondecreasing (resp. nonincreasing) sequence and $x_m \to x$, then $gx_m \preceq gx$ (resp. $gx \preceq gx_m$) $\forall m \in \mathbb{N} \cup \{0\}$.

**Definition 2.7** An ordered metric space $(X, d, \preceq)$ is said to be $g$-regular if it is both nondecreasing $g$-regular and nonincreasing $g$-regular.

Notice that, on setting $g = I$ (identity mapping on $X$), Definitions 2.6 and 2.7 reduce to Definitions 2.4 and 2.5 respectively.

Throughout the paper, $n$ stands for a general even natural number. Let us denote by $X^n$ the product space $X \times X \times \ldots \times X$ of $n$ identical copies of $X$.

**Definition 2.8** ([13]) Let $(X, \preceq)$ be an ordered set and $F : X^n \to X$ and $g : X \to X$ be two mappings. Then $F$ is said to have the mixed $g$-monotone property if $F$ is $g$-nondecreasing in its odd position arguments and $g$-nonincreasing in its even position arguments, that is, for $x^1, x^2, x^3, ..., x^n \in X$,

if

for all $x_1^1, x_2^1 \in X$, $gx_1^1 \preceq gx_2^1 \Rightarrow F(x_1^1, x^2, x^3, ..., x^n) \preceq F(x_2^1, x^2, x^3, ..., x^n)$

for all $x_1^2, x_2^2 \in X$, $gx_1^2 \preceq gx_2^2 \Rightarrow F(x^1, x_2^2, x^3, ..., x^n) \preceq F(x^1, x_1^2, x^3, ..., x^n)$

for all $x_1^3, x_2^3 \in X$, $gx_1^3 \preceq gx_2^3 \Rightarrow F(x^1, x^2, x_1^3, ..., x^n) \preceq F(x^1, x^2, x_2^3, ..., x^n)$

$\vdots$

for all $x_1^n, x_2^n \in X$, $gx_1^n \preceq gx_2^n \Rightarrow F(x^1, x^2, x^3, ..., x_2^n) \preceq F(x^1, x^2, x^3, ..., x_1^n)$.

For $g = I$ (identity mapping), Definition 2.8 reduces to mixed monotone property (for details see [13]).

**Definition 2.9** ([32]) An element $(x^1, x^2, \ldots, x^n) \in X^n$ is called an $n$-tupled fixed point of the mapping $F : X^n \to X$ if

$$\begin{cases} F(x^1, x^2, x^3, \ldots, x^n) = x^1 \\ F(x^2, x^3, \ldots, x^n, x^1) = x^2 \\ F(x^3, \ldots, x^n, x^1, x^2) = x^3 \\ \vdots \\ F(x^n, x^1, x^2, \ldots, x^{n-1}) = x^n. \end{cases}$$

**Definition 2.10** ([13]) An element $(x^1, x^2, \ldots, x^n) \in X^n$ is called an $n$-tupled coincidence point of mappings $F : X^n \to X$ and $g : X \to X$ if

$$\begin{cases} F(x^1, x^2, x^3, \ldots, x^n) = g(x^1) \\ F(x^2, x^3, \ldots, x^n, x^1) = g(x^2) \\ F(x^3, \ldots, x^n, x^1, x^2) = g(x^3) \\ \vdots \\ F(x^n, x^1, x^2, \ldots, x^{n-1}) = g(x^n). \end{cases}$$

*Remark 2.1* For $n = 2$, Definitions 2.9 and 2.10 yield the definitions of coupled fixed point and coupled coincidence point respectively while on the other hand, for $n = 4$ these definitions yield the definitions of quadrupled fixed point and quadrupled coincidence point respectively.

**Definition 2.11** An element $(x^1, x^2, \ldots, x^n) \in X^n$ is called an $n$-tupled common fixed point of mappings $F : X^n \to X$ and $g : X \to X$ if

$$\begin{cases} F(x^1, x^2, x^3, \ldots, x^n) = g(x^1) = x^1 \\ F(x^2, x^3, \ldots, x^n, x^1) = g(x^2) = x^2 \\ F(x^3, \ldots, x^n, x^1, x^2) = g(x^3) = x^3 \\ \vdots \\ F(x^n, x^1, x^2, \ldots, x^{n-1}) = g(x^n) = x^n. \end{cases}$$

**Definition 2.12** ([14]) Let $X$ be a non-empty set. Then the mappings $F : X^n \to X$ and $g : X \to X$ are said to be compatible if

$$\begin{cases} \lim_{m\to\infty} d(g(F(x_m^1, x_m^2, \ldots, x_m^n)), F(gx_m^1, gx_m^2, \ldots, gx_m^n)) = 0 \\ \lim_{m\to\infty} d(g(F(x_m^2, \ldots, x_m^n, x_m^1)), F(gx_m^2, \ldots, gx_m^n, x_m^1)) = 0 \\ \vdots \\ \lim_{m\to\infty} d(g(F(x_m^n, x_m^1, \ldots, x_m^{n-1})), F(gx_m^n, gx_m^1, \ldots, gx_m^{n-1})) = 0, \end{cases}$$

where $\{x_m^1\}, \{x_m^2\}, \ldots, \{x_m^n\}$ are sequences in $X$ such that

$$\begin{cases} \lim_{m\to\infty} F(x_m^1, x_m^2, \ldots, x_m^n) = \lim_{m\to\infty} g(x_m^1) = x^1 \\ \lim_{m\to\infty} F(x_m^2, \ldots, x_m^n, x_m^1) = \lim_{m\to\infty} g(x_m^2) = x^2 \\ \vdots \\ \lim_{m\to\infty} F(x_m^n, x_m^1, \ldots, x_m^{n-1}) = \lim_{m\to\infty} g(x_m^n) = x^n, \end{cases}$$

for some $x^1, x^2, \ldots, x^n \in X$ are satisfied.

The following families of control functions are indicated in Choudhury et al. [9].

$\Im = \{\zeta \ [,\infty) \to [,\infty) \quad \zeta$ is continuous and $\zeta(t) = 0$ if and only if $t = 0\}$
$\Omega := \{\varphi : [0,\infty) \to [0,\infty) :$
$\varphi$ is continuous and monotone nondecreasing and $\varphi(t) = 0$ if and only if $t = 0\}$

Notice that members of $\Omega$ are called altering distance functions (cf. [20]).

Now, we state the main result of Imdad et al. [14], which is indeed $n$-tupled extension of that of Choudhury et al. [9].

**Theorem 2.1** *Let $(X, d, \preceq)$ be an ordered complete metric space and $F : X^n \to X$ and $g : X \to X$ be two mappings. Suppose that the following conditions are satisfied:*

(i) $F(X^n) \subseteq g(X)$,
(ii) *F and g are compatible,*
(iii) *F has the mixed g-monotone property,*
(iv) *g is continuous,*
(v) *either F is continuous or X is g-regular,*
(vi) *there exist $x_0^1, x_0^2, x_0^3, \ldots, x_0^n \in X$ such that*

$$\begin{cases} gx_0^1 \preceq F(x_0^1, x_0^2, x_0^3, \ldots, x_0^n) \\ F(x_0^2, x_0^3, \ldots, x_0^n, x_0^1) \preceq gx_0^2 \\ gx_0^3 \preceq F(x_0^3, \ldots, x_0^n, x_0^1, x_0^2) \\ \vdots \\ F(x_0^n, x_0^1, x_0^2, \ldots, x_0^{n-1}) \preceq gx_0^n, \end{cases}$$

(vii) *there exist $\varphi \in \Omega$ and $\zeta \in \Im$ such that*

$$\varphi(d(FU, FV)) \leq \varphi(\max_{1\leq i\leq n} d(gx^i, gy^i)) - \zeta(\max_{1\leq i\leq n} d(gx^i, gy^i)),$$

*for all $U = (x^1, x^2, \ldots, x^n)$, $V = (y^1, y^2, \ldots, y^n) \in X^n$ with $gy^1 \preceq gx^1, gx^2 \preceq gy^2, gy^3 \preceq gx^3, \ldots, gx^n \preceq gy^n$. Then F and g have an n-tupled coincidence point.*

## Main results

Let $(X, \preceq)$ be an ordered set. Define the following partial order $\sqsubseteq$ on the product space $X^n$, for $U = (x^1, x^2, \ldots, x^n)$, $V = (y^1, y^2, \ldots, y^n) \in X^n$

$$U \sqsubseteq V \Leftrightarrow x^1 \preceq y^1,\ y^2 \preceq x^2,\ x^3 \preceq y^3, \ldots, y^n \preceq x^n.$$

Let $(X, d)$ be a metric space. Define the following metric $\tilde{D}$ on the product space $X^n$, for $U = (x^1, x^2, \ldots, x^n)$, $V = (y^1, y^2, \ldots, y^n) \in X^n$,

$$\tilde{D}(U, V) = \max_{1\leq i\leq n} d(x^i, y^i).$$

The proofs of the following lemmas follow immediately. We note the same idea here, but in the case of coupled and tripled fixed point theorems, we have been first used in ([3, 28, 33]).

**Lemma 3.1** *Let $(X, d, \preceq)$ be an ordered complete metric space. Then $(X^n, \tilde{D}, \sqsubseteq)$ is an ordered complete metric space.*

**Lemma 3.2** *Let $(X, d, \preceq)$ be an ordered metric space and $F : X^n \to X$ and $g : X \to X$ be two mappings. Define mappings $T_F : X^n \to X^n$ and $T_g : X^n \to X^n$ by*

$$T_F(x^1, x^2, \ldots, x^n) = (F(x^1, x^2, \ldots, x^n), F(x^2, \ldots, x^n, x^1), \ldots, F(x^n, x^1, \ldots, x^{n-1}))$$

*and $T_g(x^1, x^2, \ldots, x^n) = (gx^1, gx^2, \ldots, gx^n)$. Then the following hold:*

(1) *If $F$ has the mixed $g$-monotone property, then $T_F$ is monotone $T_g$-nondecreasing with respect to $\sqsubseteq$.*
(2) *If $F$ and $g$ are compatible, then $T_F$ and $T_g$ are compatible.*
(3) *If $g$ is continuous, then $T_g$ is continuous.*
(4) *If $F$ is continuous, then $T_F$ is continuous.*
(5) *If $(X, d, \preceq)$ is $g$-regular, then $(X^n, \tilde{D}, \sqsubseteq)$ is nondecreasing $g$-regular.*
(6) *A point $(x^1, x^2, \ldots, x^n) \in X^n$ is an $n$-tupled coincidence point of $F$ and $g$ iff $(x^1, x^2, \ldots, x^n)$ is a coincidence point of $T_F$ and $T_g$.*

The following lemma is crucial for our main result.

**Lemma 3.3** *Let $(X, d, \preceq)$ be an ordered complete metric space and $f$ and $g$ be two self-mappings on $X$. Suppose that the following conditions are satisfied:*

(i) $f(X) \subseteq g(X)$,
(ii) *$f$ is monotone $g$-nondecreasing,*
(iii) *$f$ and $g$ are compatible,*
(iv) *$g$ is continuous,*
(v) *either $f$ is continuous or $X$ is nondecreasing $g$-regular,*
(vi) *there exists $x_0 \in X$ such that $g(x_0) \preceq f(x_0)$,*
(vii) *there exist $\varphi \in \Omega$ and $\zeta \in \Im$ such that for all $x, y \in X$,*

$$\varphi(d(f(x), f(y))) \leq \varphi(d(g(x), g(y))) - \zeta(d(g(x), g(y))), \text{ with } g(x) \preceq g(y). \tag{3.1}$$

*Then $f$ and $g$ have a coincidence point.*

*Proof* In view of assumption (vi), if $g(x_0) = f(x_0)$, then $x_0$ is a coincidence point of $f$ and $g$ and hence proof is finished. On the other hand if $g(x_0) \neq f(x_0)$, then we have $g(x_0) \prec f(x_0)$. So according to assumption (i), that is, $f(X) \subseteq g(X)$, we can choose $x_1 \in X$ such that $g(x_1) = f(x_0)$. Again from $f(X) \subseteq g(X)$, we can choose $x_2 \in X$ such that $g(x_2) = f(x_1)$. Continuing this process, we define a sequence $\{x_m\} \subset X$ of joint iterates such that

$$g(x_{m+1}) = f(x_m) \quad \forall m \in \mathbb{N} \cup \{0\}. \tag{3.2}$$

Now, we assert that $\{g(x_m)\}$ is a nondecreasing sequence, that is

$$g(x_m) \preceq g(x_{m+1}) \quad \forall m \in \mathbb{N} \cup \{0\}. \tag{3.3}$$

We prove this fact by mathematical induction. On using (3.2) for $m = 0$ and assumption (vi), we have

$$g(x_0) \preceq f(x_0) = g(x_1).$$

Thus, (3.3) holds for $m = 0$. Suppose that (3.3) holds for $m = r > 0$, that is,

$$g(x_r) \preceq g(x_{r+1}). \tag{3.4}$$

Then we have to show that (3.3) holds for $m = r + 1$. To accomplish this we use (3.2), (3.4) and assumption (ii) so that

$$g(x_{r+1}) = f(x_r) \preceq f(x_{r+1}) = g(x_{r+2}).$$

Thus, by induction, (3.3) holds for all $m \in \mathbb{N} \cup \{0\}$.

If $g(x_m) = g(x_{m+1})$ for some $m \in \mathbb{N}$, then using (3.2), we have $g(x_m) = f(x_m)$, that is, $x_m$ is a coincidence point of $f$ and $g$ and hence proof is finished. On the other hand if $g(x_m) \neq g(x_{m+1})$ for each $m \in \mathbb{N} \cup \{0\}$, we can define a sequence

$$\delta_m := d(g(x_m), g(x_{m+1})), \quad m \in \mathbb{N} \cup \{0\}. \tag{3.5}$$

On using (3.2), (3.3), (3.5) and assumption (vii), we obtain

$$\begin{aligned} \varphi(\delta_{m+1}) &= \varphi(d(g(x_{m+1}), g(x_{m+2}))) \\ &= \varphi(d(f(x_m), f(x_{m+1}))) \\ &\leq \varphi(d(g(x_m), g(x_{m+1}))) - \zeta(d(g(x_m), g(x_{m+1}))) \\ &= \varphi(\delta_m) - \zeta(\delta_m). \end{aligned} \tag{3.6}$$

On using the property of $\varphi$, we have $\varphi(\delta_{m+1}) \leq \varphi(\delta_m)$, which implies that $\delta_{m+1} \leq \delta_m$. Therefore, $\{\delta_m\}$ is a monotone decreasing sequence of nonnegative real numbers. Hence there exists $\delta \geq 0$ such that $\delta_m \to \delta$ as $m \to \infty$. Taking limit as $m \to \infty$ in (3.6) and using the continuities of $\varphi$ and $\zeta$, we have $\varphi(\delta) \leq \varphi(\delta) - \zeta(\delta)$, which is a contradiction under $r = 0$. Therefore,

$$\lim_{m \to \infty} \delta_m = \lim_{m \to \infty} d(g(x_m), g(x_{m+1})) = 0. \tag{3.7}$$

Now, we show that $\{g(x_m)\}$ is a Cauchy sequence. On the contrary, suppose that $\{g(x_m)\}$ is not a Cauchy sequence. Then, there exists an $\epsilon > 0$ and sequences of positive integers $\{m(k)\}$ and $\{t(k)\}$ such that for all positive integers $k$, $t(k) > m(k) > k$, such that

$$\eta_k = d(g(x_{m(k)}), g(x_{t(k)})) \geq \epsilon, \text{ and } d(g(x_{m(k)}), g(x_{t(k)-1})) < \epsilon.$$

Now,

$$\begin{aligned} \epsilon \leq \eta_k &= d(g(x_{m(k)}), g(x_{t(k)})) \\ &\leq d(g(x_{m(k)}), g(x_{t(k)-1})) + d(g(x_{t(k)-1}), g(x_{t(k)})) \\ &< \epsilon + \delta_{t(k)-1} \end{aligned}$$

that is,

$$\epsilon \leq \eta_k < \epsilon + \delta_{t(k)-1}.$$

Letting $k \to \infty$ in above inequality and using (3.7), we get

$$\lim_{k \to \infty} \eta_k = \epsilon. \tag{3.8}$$

Again,

$$\begin{aligned}\eta_{k+1} &= d(g(x_{m(k)+1}), g(x_{t(k)+1})) \\ &\le d(g(x_{m(k)+1}), g(x_{m(k)})) + d(g(x_{m(k)}), g(x_{t(k)})) \\ &\quad + d(g(x_{t(k)}), g(x_{t(k)+1})) \\ &< \delta_{m(k)+1} + \eta_k + \delta_{t(k)+1} \\ &\Rightarrow \eta_{k+1} < \delta_{m(k)+1} + \eta_k + \delta_{t(k)+1}.\end{aligned}$$

Letting $k \to \infty$ in above inequality and using (3.7) and (3.8), we get

$$\lim_{k\to\infty} \eta_{k+1} = \epsilon. \tag{3.9}$$

Since $t(k) > m(k)$, hence by (3.3), we get $g(x_{m(k)}) \le g(x_{t(k)})$. Therefore, owing to (3.1) and assumption (vii), we get

$$\begin{aligned}\varphi(\eta_{k+1}) &= \varphi(d(g(x_{m(k)+1}), g(x_{t(k)+1}))) \\ &= \varphi(d(f(x_{m(k)}), f(x_{t(k)}))) \\ &\le \varphi(d(g(x_{m(k)}), g(x_{t(k)}))) - \zeta(d(g(x_{m(k)}), g(x_{t(k)}))) \\ &= \varphi(\eta_k) - \zeta(\eta_k)\end{aligned}$$

that is,

$$\varphi(\eta_{k+1}) \le \varphi(\eta_k) - \zeta(\eta_k).$$

Letting $k \to \infty$ in above inequality and using (3.8), (3.9) and continuities of $\varphi$ and $\zeta$, we get

$$\varphi(\epsilon) \le \varphi(\epsilon) - \zeta(\epsilon)$$

which is a contradiction by virtue of property of $\zeta$. Therefore, the sequence $\{g(x_m)\}$ is Cauchy. From the completeness of $X$, there exists $x \in X$ such that

$$\lim_{m\to\infty} f(x_m) = \lim_{m\to\infty} g(x_m) = x. \tag{3.10}$$

Since $F$ and $g$ are compatible, we have from (3.10),

$$\lim_{m\to\infty} d(f(gx_m), g(fx_m)) = 0. \tag{3.11}$$

Now, we use assumption (v). Firstly, we assume that $f$ is continuous. Then for all $m \in \mathbb{N} \cup \{0\}$, we have

$$d(g(x), f(gx_m)) \le d(g(x), g(fx_m)) + d(g(fx_m), f(gx_m)).$$

Taking $k \to \infty$ in above inequality and using (3.10), (3.11) and continuities of $f$ and $g$, we get $d(g(x), f(x)) = 0$, that is, $g(x) = f(x)$. Hence, the element $x \in X$ is a coincidence point of $f$ and $g$. Next, we suppose that $X$ is nondecreasing $g$-regular. From (3.3) and (3.10), we get

$$g(gx_m) \preceq g(x). \tag{3.12}$$

Since $f$ and $g$ are compatible and $g$ is continuous by (3.10) and (3.11), we have

$$\lim_{m\to\infty} g(gx_m) = g(x) = \lim_{m\to\infty} g(fx_m) = \lim_{m\to\infty} f(gx_m). \tag{3.13}$$

Now, using triangle inequality, we have

$$\begin{aligned}d(f(x), g(x)) &\le d(f(x), g(gx_{m+1})) + d(g(gx_{m+1}), g(x)) \\ &= d(f(x), g(fx_m)) + d(g(gx_{m+1}), g(x)).\end{aligned}$$

Taking $k \to \infty$ in above inequality and using (3.13), we have

$$\begin{aligned}d(f(x), g(x)) &\le \lim_{m\to\infty} d(f(x), g(fx_m)) + \lim_{m\to\infty} d(g(gx_{m+1}), g(x)) \\ &= \lim_{m\to\infty} d(f(x), f(gx_m)).\end{aligned}$$

Since $\varphi$ is continuous and monotone nondecreasing, from the above inequality we have

$$\begin{aligned}\varphi(d(f(x), g(x))) &\le \varphi(\lim_{m\to\infty} d(f(x), f(gx_m))) \\ &= \lim_{m\to\infty} \varphi(d(f(x), f(gx_m))).\end{aligned}$$

By (3.12) and assumption (vii), we get

$$\begin{aligned}\varphi(d(f(x), g(x))) &\le \lim_{m\to\infty} \varphi(d(f(x), f(gx_m))) \\ &= \lim_{m\to\infty} \varphi(d(g(x), g(gx_m))) \\ &\quad + \lim_{m\to\infty} \zeta(d(g(x), g(gx_m))).\end{aligned}$$

Using (3.13) and the properties of $\varphi$ and $\zeta$, we have $\varphi(d(f(x), g(x))) = 0$, which implies that $d(f(x), g(x)) = 0$, that is, $g(x) = f(x)$. Hence, $x \in X$ is a coincidence point of $f$ and $g$.

**Lemma 3.4** *In addition to the hypotheses of Lemma 3.3, suppose that for real $x, y \in X$ there exists, $z \in X$ such that $f(z)$ is comparable to $f(x)$ and $f(y)$. Then $f$ and $g$ have a unique common fixed point.*

*Proof* The set of coincidence points of $f$ and $g$ is non-empty due to Lemma 3.3. Assume now, $x$ and $y$ are two coincidence points of $f$ and $g$, that is,

$$f(x) = g(x) \text{ and } f(y) = g(y).$$

Now we will show that $g(x) = g(y)$. By assumption, there exists $z \in X$ such that $f(z)$ is comparable to $f(x)$ and $f(y)$. Put $z_0^1 = z$ and choose $z_1 \in X$ such that $g(z_1) = f(z_0)$. Further define sequence $\{g(z_m)\}$ such that $g(z_{m+1}) = f(z_m)$. Further set $x_0 = x$ and $y_0 = y$. In the same way, define the sequences $\{g(x_m)\}$ and $\{g(y_m)\}$. Then, it is easy to show that

$$g(x_{m+1}) = f(x_m) \text{ and } g(y_{m+1}) = f(y_m).$$

Since $f(x) = g(x_1) = g(x)$ and $f(z) = g(z_1)$ are comparable, we have

$$g(x) \preceq g(z_1).$$

It is easy to show that $g(x)$ and $g(z_m)$ are comparable, that is, for all $m \in \mathbb{N}$,

$g(x) \preceq g(z_m)$.

Thus from (3.1) we have

$$\varphi(d(g(x),g(z_{m+1})))=\varphi(d(f(x),f(z_m))) \leq \varphi(d(g(x),g(z_m)))-\zeta(d(g(x),g(z_m))).$$

Let $R_m = d(g(x),g(z_{m+1}))$. Then

$$\varphi(R_m) \leq \varphi(R_{m-1}) - \zeta(R_{m-1}). \tag{3.14}$$

Using the property of $\varphi$, we have $\varphi(R_m) \leq \varphi(R_{m-1})$, which implies that $R_m \leq R_{m-1}$ (by the property of $\varphi$). Therefore $\{R_m\}$ is a monotone decreasing sequence of nonnegative real numbers. Hence, there exists $r \geq 0$ such that $R_m \to r$ as $m \to \infty$. Taking the limit as $m \to \infty$ in (3.14) and using the continuities of $\varphi$ and $\zeta$, we have $\varphi(r) \leq \varphi(r) - \zeta(r)$, which is a contradiction unless $r = 0$. Therefore $R_m \to 0$ as $m \to \infty$, that is,

$$\lim_{m\to\infty} d(g(x), g(z_{m+1})) = 0.$$

Similarly we can prove that

$$\lim_{m\to\infty} d(g(y), g(z_{m+1})) = 0.$$

Therefore by triangle inequality

$$d(g(x), g(y)) \leq d(g(x), g(z_{m+1})) + d(g(z_{m+1}), g(y)) \to 0 \text{ as } m \to \infty.$$

Hence

$$g(x) = g(y). \tag{3.15}$$

Since $g(x) = f(x)$ and $f$ and $g$ are compatible, we have $gg(x) = f(gx)$. Write $g(x) = a$, then we have

$$g(a) = f(a). \tag{3.16}$$

Thus $a$ is the coincidence point of $f$ and $g$. Then owing to (3.15) with $y = a$, it follows that $g(x) = g(a)$, that is,

$$g(a) = a. \tag{3.17}$$

Using (3.16) and (3.17), we have $a = g(a) = f(a)$. Thus $a$ is the common fixed point of $f$ and $g$. To prove the uniqueness, assume that $b$ is another common fixed point of $f$ and $g$. Then by (3.15), we have
$b = g(b) = g(a) = a$.

This completes the proof of Lemma.

**Theorem 3.1** *Theorem* 2.1 *is obtained using Lemmas* 3.1, 3.2 *and* 3.3.

*Proof* Consider the product space $Y = X^n$ equipped with the metric $\tilde{D}$ [given by (B)] and the partial order $\sqsubseteq$ [given by (A)]. Then by Lemma 3.1, $(Y, \tilde{D}, \sqsubseteq)$ is an ordered complete metric space. Also $F$ and $g$ induce mappings $T_F : Y \to Y$ and $T_g : Y \to Y$ (defined in Lemma 3.2). Clearly,

- (i) implies that $T_F(Y) \subseteq T_g(Y)$,
- (ii) implies that $T_F$ is monotone $T_g$-nondecreasing (by item (1) of Lemma 3.2),
- (iii) implies that $T_F$ and $T_g$ are compatible (by item (2) of Lemma 3.2),
- (iv) implies that $T_g$ is continuous (by item (3) of Lemma 3.2),
- (v) implies that either $T_F$ is continuous [by item (4) of Lemma 3.2] or $(Y, \tilde{D}, \sqsubseteq)$ is nondecreasing $g$-regular [by item (5) of Lemma 3.2],
- (vi) is equivalent to the condition: there exists $U_0 = (x_0^1, x_0^2, \ldots, x_0^n) \in Y$ such that $T_g(U_0) \subseteq T_F(U_0)$.

Now, in view of (vii), for given $U, V \in Y$ such that $T_g(U) \sqsubseteq T_g(V)$ implies that

$$(gx^1, gx^2, \ldots, gx^n) \sqsubseteq (gy^1, gy^2, \ldots, gy^n).$$

It follows that for odd $i$,

$$(gx^i, gx^{i+1}, \ldots, gx^n, gx^1, gx^2, \ldots, gx^{i-1}) \sqsubseteq (gy^i, gy^{i+1}, \ldots, gy^n, gy^1, gy^2, \ldots, gy^{i-1}), \tag{3.18}$$

and for even $i$,

$$(gy^i, gy^{i+1}, \ldots, gy^n, gy^1, gy^2, \ldots, gy^{i-1}) \sqsubseteq (gx^i, gx^{i+1}, \ldots, gx^n, gx^1, gx^2, \ldots, gx^{i-1}). \tag{3.19}$$

If $i$ is odd, then using (3.18) and (vii), we get

$$\begin{aligned} &d(F(x^i,x^{i+1},\ldots,x^n,x^1,x^2,\ldots,x^{i-1}),F(y^i,y^{i+1},\ldots,y^n,y^1,y^2,\ldots,y^{i-1}))\\ &\leq \varphi(\max\{d(gx^i,gy^i),d(gx^{i+1},gy^{i+1}),\ldots,d(gx^n,gy^n),d(gx^1,gy^1),\\ &d(gx^2,gy^2),\ldots,d(gx^{i-1},gy^{i-1})\}) - \zeta(\max\{d(gx^i,gy^i),d(gx^{i+1},gy^{i+1}),\ldots,\\ &d(gx^n,gy^n),d(gx^1,gy^1),d(gx^2,gy^2),\ldots,d(gx^{i-1},gy^{i-1})\})\\ &= \varphi(\max_{1\leq i\leq n} d(gx^i,gy^i)) - \zeta(\max_{1\leq i\leq n} d(gx^i,gy^i)). \end{aligned}$$

If $i$ is even, then using (3.19) and (vii), we get

$$\begin{aligned} &d(F(x^i,x^{i+1},\ldots,x^n,x^1,x^2,\ldots,x^{i-1}),F(y^i,y^{i+1},\ldots,y^n,y^1,y^2,\ldots,y^{i-1}))\\ &= d(F(y^i,y^{i+1},\ldots,y^n,y^1,y^2,\ldots,y^{i-1}),F(x^i,x^{i+1},\ldots,x^n,x^1,x^2,\ldots,x^{i-1}))\\ &\leq \varphi(\max\{d(gy^i,gx^i),d(gy^{i+1},gx^{i+1}),\ldots,d(gy^n,gx^n),d(gy^1,gx^1),\\ &d(gy^2,gx^2),\ldots,d(gy^{i-1},gx^{i-1})\}) - \zeta(\max\{d(gy^i,gx^i),d(gy^{i+1},gx^{i+1}),\ldots,\\ &d(gy^n,gx^n),d(gy^1,gx^1),d(gy^2,gx^2),\ldots,d(gy^{i-1},gx^{i-1})\})\\ &= \varphi(\max_{1\leq i\leq n} d(gx^i,gy^i)) - \zeta(\max_{1\leq i\leq n} d(gx^i,gy^i)). \end{aligned}$$

Hence, in both the cases, for each $i (1 \leq i \leq n)$, we have

$$d(F(x^i, x^{i+1}, \ldots, x^n, x^1, x^2, \ldots, x^{i-1}), F(y^i, y^{i+1}, \ldots, y^n, y^1, y^2, \ldots, y^{i-1})) \leq \varphi(\max_{1\leq i\leq n} d(gx^i, gy^i)) - \zeta(\max_{1\leq i\leq n} d(gx^i, gy^i)). \tag{3.20}$$

Hence using (3.20), we have

$$
\begin{aligned}
&\tilde{D}(T_F(U),T_F(V)) \\
&= \max_{1\le i\le n} d(F(x^i,x^{i+1},\ldots,x^n,x^1,x^2,\ldots,x^{i-1}), \\
&\quad F(y^i,y^{i+1},\ldots,y^n,y^1,y^2,\ldots,y^{i-1})) \\
&\le \max_{1\le i\le n} [\varphi(\max_{1\le i\le n} d(gx^i,gy^i)) - \zeta(\max_{1\le i\le n} d(gx^i,gy^i))] \\
&= \varphi(\max_{1\le i\le n} d(gx^i,gy^i)) - \zeta(\max_{1\le i\le n} d(gx^i,gy^i)) \\
&= \varphi(\tilde{D}(T_g(U),T_g(V)) - \zeta(\tilde{D}(T_g(U),T_g(V)).
\end{aligned}
$$

Thus all conditions of Lemma 3.3 are satisfied for ordered complete metric space $(Y,\tilde{D},\sqsubseteq)$ and mappings $T_F: Y\to Y$ and $T_g: Y\to Y$. Therefore, $T_F$ and $T_g$ have a coincidence point in $Y = X^n$. According to item (6) of Lemma 3.2, the mappings $F$ and $g$ have an $n$-tupled coincidence point.

**Corollary 3.1** *Let $(X,d,\preceq)$ be an ordered complete metric space and $F: X^n \to X$ be a mapping. Suppose that the following conditions are satisfied:*

(i) *$F$ has the mixed monotone property,*
(ii) *either $F$ is continuous or $X$ is regular,*
(iii) *there exist $x_0^1,x_0^2,x_0^3,\ldots,x_0^n \in X$ such that*

$$
\begin{cases}
x_0^1 \preceq F(x_0^1,x_0^2,x_0^3,\ldots,x_0^n) \\
F(x_0^2,x_0^3,\ldots,x_0^n,x_0^1) \preceq x_0^2 \\
x_0^3 \preceq F(x_0^3,\ldots,x_0^n,x_0^1,x_0^2) \\
\vdots \\
F(x_0^n,x_0^1,x_0^2,\ldots,x_0^{n-1}) \preceq x_0^n,
\end{cases}
$$

(iv) *there exist $\varphi\in\Omega$ and $\zeta\in\Im$ such that*

$$\varphi(d(FU,FV)) \le \varphi(\max_{1\le i\le n} d(x^i,y^i)) - \zeta(\max_{1\le i\le n} d(x^i,y^i)),$$

*for all $U=(x^1,x^2,\ldots,x^n)$, $V=(y^1,y^2,\ldots,y^n)\in X^n$ with $x^1\preceq y^1, y^2\preceq x^2, x^3\preceq y^3,\ldots,y^n\preceq x^n$.*

*Then $F$ has an n-tupled fixed point.*

*Proof* It is sufficient to take $g = I$ (identity mapping) in Theorem 3.1. □

**Corollary 3.2** *Corollary 3.1 remains true if condition (iv) is replaced by the following: (iv)' there exists $\zeta\in\Im$ such that*

$$d(FU,FV) \le \max_{1\le i\le n} d(x^i,y^i) - \zeta(\max_{1\le i\le n} d(x^i,y^i)),$$

*for all $U=(x^1,x^2,\ldots,x^n)$, $V=(y^1,y^2,\ldots,y^n)\in X^n$ with $x^1\preceq y^1, y^2\preceq x^2, x^3\preceq y^3,\ldots,y^n\preceq x^n$.*

*Proof* It is sufficient to take $\varphi$ and $g$ to be identity mappings in Theorem 3.1.

**Corollary 3.3** *Corollary 3.1 remains true if condition (iv) is replaced by the following:*

*(iv)" there exists $k\in(0,1)$ such that*

$$d(FU,FV) \le k \max_{1\le i\le n} d(x^i,y^i),$$

*for all $U=(x^1,x^2,\ldots,x^n)$, $V=(y^1,y^2,\ldots,y^n)\in X^n$ with $x^1\preceq y^1, y^2\preceq x^2, x^3\preceq y^3,\ldots,y^n\preceq x^n$.*

*Proof* It is sufficient to take $\varphi$ and $g$ to be identity mappings and $\zeta(t) = (1-k)t$, $k\in(0,1)$ in Theorem 3.1.

*Remark 3.1*

1. On setting $n = 2$ in Theorem 3.1, we get Theorem 3.1 of Choudhury et al. [9].
2. On setting $n = 2$ in Corollaries 3.1–3.3, we get Corollaries 3.2–3.4 of Choudhury et al. [9].
3. On setting $n = 4$ in Theorem 3.1 and Corollaries 3.1–3.3, we get their corresponding quadrupled fixed point results.

Now we shall prove the uniqueness of $n$-tupled fixed point.

**Theorem 3.2** *In addition to the hypotheses of Theorem 3.1, suppose that for real $(x^1,x^2,\ldots,x^n)$ and $(y^1,y^2,\ldots,y^n)\in X^n$ there exists, $(z^1,z^2,\ldots,z^n)\in X^n$ such that $(F(z^1,z^2,\ldots,z^n),F(z^2,\ldots,z^n,z^1),\ldots,F(z^n,z^1,\ldots,z^{n-1}))$ is comparable to $(F(x^1,x^2,\ldots,x^n),F(x^2,\ldots,x^n,x^1),\ldots,F(x^n,x^1,\ldots,x^{n-1}))$ and $(F(y^1,y^2,\ldots,y^n),F(y^2,\ldots,y^n,y^1),\ldots,F(y^n,y^1,\ldots,y^{n-1}))$. Then $F$ and $g$ have a unique n-tupled common fixed point.*

*Proof* Set $U=(x^1,x^2,\ldots,x^n)$, $V=(y^1,y^2,\ldots,y^n)$ and $W=(z^1,z^2,\ldots,z^n)$. Then we have

$$T_F(W)\sqsubseteq T_F(U) \text{ or } T_F(U)\sqsubseteq T_F(W)$$

and

$$T_F(W)\sqsubseteq T_F(V) \text{ or } T_F(V)\sqsubseteq T_F(W).$$

Hence using Lemma 3.4, $T_F$ and $T_g$ have a unique $n$-tupled common fixed point. □

*Remark 3.2* From Theorem 3.2, for $n = 2$, we can get unique coupled common fixed point theorem contained in Choudhury et al. [9].

## References


1. Al-Mezel, S. A., Alsulami, A. H., Karapınar, E. and Roldán, A.: Discussion on Multidimensional Coincidence Points via recent publications. Abstr. Appl. Anal. (2014) Art. ID 287492, 13 pages
2. Banach, S.: Sur les operations dans les ensembles abstraits et leur application aux quations intgrales. Fund. Math. **3**, 133–181 (1922)
3. Berinde, V.: Generalized coupled fixed point theorems for mixed monotone mappings in partially ordered metric spaces. Nonlinear Anal. **74**, 7347–7355 (2011)

4. Berinde, V., Borcut, M.: Tripled fixed point theorems for contractive type mappings partially ordered metric spaces. Nonlinear Anal. **75**(15), 4889–4897 (2011)
5. Bhaskar, T.G., Lakshmikantham, V.: Fixed points theorems in partially ordered metric spaces and applications. Nonlinear Anal. TMA **65**, 1379–1393 (2006)
6. Bhaskar, T.G., Lakshmikantham, V.: Fixed points theorems in partially ordered cone metric spaces and applications. Nonlinear Anal. **65**(7), 825–832 (2006)
7. Berzig, M., Samet, B.: An extension of coupled fixed point's concept in higher dimension and applications. Comput. Math. Appl. **63**, 1319–1334 (2012)
8. Ćirić, Lj. B., Cakić, M., Rajović and Ume, J. S.: Monotone generalized contractions in partially ordered metric spaces. Fixed Point Theory Appl. (2008) 11, Article ID 131294
9. Choudhury, B.S., Metiya, N., Kundu, A.: Coupled coincidence point theorems in ordered metric spaces. Ann. Univ. Ferrara **57**, 1–16 (2011)
10. Dalal, S., Khan, L.A., Masmali, I., Radenovic, S.: Some remarks on multidimensional fixed point theorems in partially ordered metric spaces. J. Adv. Math. **7**(1), 1084–1094 (2014)
11. Guo, D.J., Lakshmikantham, V.: Coupled fixed points of nonlinear operators with applications. Nonlinear Anal. **11**(5), 623–632 (1987)
12. Harjani, J., López, B., Sadarangani, K.: A fixed point theorem for weakly $C$-contractive mappings in ordered metric spaces. Comput. Math. Appl. **61**, 790–796 (2011)
13. Imdad, M., Soliman, A. H., Choudhury, B. S. and Das, P.: On $n$-tupled coincidence and common fixed points results in metric spaces. J. Oper. (2013) Article ID 532867, 9 pages
14. Imdad, M., Sharma, A., Rao, K.P.R.: $n$-tupled coincidence and common fixed point results for weakly contractive mappings in complete metric spaces. Bull. Math. Anal. Appl. **5**(4), 19–39 (2013)
15. Imdad, M., Alam, A., Soliman, A.H.: Remarks on a recent general even-tupled coincidence theorem. J. Adv. Math. **9**(1), 1787–1805 (2014)
16. Karapınar, E.: Quartet fixed point for nonlinear contraction. http://arxiv.org/abs/1106.5472
17. Karapınar, E., Berinde, V.: Quadruple fixed point theorems for nonlinear contractions in partially ordered metric spaces. Banach J. Math. Anal. **6**(1), 74–89 (2012)
18. Karapınar, E., Luong, N.V.: Quadruple fixed point theorems for nonlinear contractions. Comput. Math. Anal. **64**, 1839–1848 (2012)
19. Karapınar, E., Roldan, A., Martinez-Moreno, J. and Roldan, C.: Meir-Keeler type multidimensional fixed point theorems in partially ordered metric spaces. Abstr. Appl. Anal. (2013) (Article ID 406026)
20. Khan, M.S., Swaleh, M., Sessa, S.: Fixed point theorems by altering distance functions between the points. Bull. Aust. Math. Soc. **30**, 1–9 (1984)
21. Kutbi, M.A., Roldán, A., Sintunavarat, W., Moreno, J.M., Roldán, C.: $F$-closed sets and coupled fixed point theorems without the mixed monotone property. Fixed Point Theory Appl. **2013**, 330 (2013)
22. Lakshmikantham, V., Ćirić, L.B.: Coupled fixed point theorems for nonlinear contractions in partially ordered metric spaces. Nonlinear Anal. **70**, 4341–4349 (2009)
23. Luong, N.V., Thuan, N.X.: Coupled fixed point theorems in partially ordered metric spaces. Bull. Math. Anal. Appl. **2**(4), 16–24 (2010). ISSN
24. Nashine, H.K., Samet, B.: Fixed point results for mappings satisfying $(\psi, \phi)$-weakly contractive condition in partially ordered metric spaces. Nonlinear Anal. **74**, 2201–2209 (2011)
25. Nieto, J.J., López, R.R.: Contractive mapping theorems in partially ordered sets and applications to ordinary differential equations. Order **22**, 223–239 (2005)
26. Nieto, J.J., López, R.R.: Existence and uniqueness of fixed point in partially ordered sets and applications to ordinary differential equations. Acta Math. Sinica, Engl. Ser. **23**(12), 2205–2212 (2007)
27. O'Regan, D., Petrusel, A.: Fixed point theorems for generalized contractions in ordered metric spaces. J. Math. Anal. Appl. **341**, 1241–1252 (2008)
28. Radenovic, S.: Remarks on some coupled coincidence point in partially ordered metric spaces. Arab J. Math. Sci. **20**(1), 29–39 (2014)
29. Radenovic, S.: A note on tripled coincidence and tripled common fixed point theorems in partially ordered metric spaces. Appl. Math. Comput. **236**, 367–372 (2014)
30. Ran, A.C.M., Reurings, M.C.B.: A fixed point theorem in partially ordered sets and some applications to matrix equations. Proc. Am. Math. Soc. **132**, 1435–1443 (2004)
31. Roldán, A., Martínez-Moreno, J., Roldán, C.: Multidimensional fixed point theorems in partially ordered metric spaces. J. Math. Anal. Appl. **396**, 536–545 (2012)
32. Samet, B., Vetro, C.: Coupled fixed point, $f$-invariant set and fixed point of $N$-order. Ann. Funct. Anal. **1**(2), 4656–4662 (2010)
33. Samet, B., Vetro, C., Vetro, F.: From metric spaces to partial metric spaces. Fixed Point Theory Appl. **2013**, 5 (2013). doi:10.1186/1687-2012-2013-5
34. Samet, B., Karapınar, E., Aydi, H. and Rajic, V. C.: Discussion on some coupled fixed point theorems. Fixed Point Theory Appl. 2013:50, p. 12 (2013)
35. Sastry, K.P.R., Babu, G.V.R.: Some fixed point theorems by altering distances between the points. Ind. J. Pure Appl. Math. **30**(6), 641–647 (1999)
36. Shatanawi, W.: Fixed point theorems for nonlinear weakly $C$-contractive mappings in metric spaces. Math. Comput. Model. **54**, 2816–2826 (2011)
37. Shatanawi, W., Samet, B.: On $(\psi, \phi)$-weakly contractive condition in partially ordered metric spaces. Comput. Math. Appl. **62**, 3204–3214 (2011)

# 15

# Improved Jacobi matrix method for the numerical solution of Fredholm integro-differential-difference equations

M. Mustafa Bahşı[1] · Ayşe Kurt Bahşı[2] · Mehmet Çevik[3] · Mehmet Sezer[2]

**Abstract** This study is aimed to develop a new matrix method, which is used an alternative numerical method to the other method for the high-order linear Fredholm integro-differential-difference equation with variable coefficients. This matrix method is based on orthogonal Jacobi polynomials and using collocation points. The improved Jacobi polynomial solution is obtained by summing up the basic Jacobi polynomial solution and the error estimation function. By comparing the results, it is shown that the improved Jacobi polynomial solution gives better results than the direct Jacobi polynomial solution, and also, than some other known methods. The advantage of this method is that Jacobi polynomials comprise all of the Legendre, Chebyshev, and Gegenbauer polynomials and, therefore, is the comprehensive polynomial solution technique

**Keywords** Orthogonal Jacobi polynomials · Fredholm integro-differential-difference equation · Residual error technique · Matrix method

✉ M. Mustafa Bahşı
mustafa.bahsi@cbu.edu.tr

Ayşe Kurt Bahşı
ayse.kurt@cbu.edu.tr

Mehmet Çevik
mehmet.cevik@ikc.edu.tr

Mehmet Sezer
mehmet.sezer@cbu.edu.tr

1 Department of Mechanical Engineering, Celal Bayar University, Muradiye Campus, Manisa, Turkey

2 Department of Mathematics, Celal Bayar University, Muradiye Campus, Manisa, Turkey

3 Department of Mechanical Engineering, Izmir Katip Çelebi University, Çiğli Main Campus, Izmir, Turkey

## Introduction

### Orthogonal Jacobi polynomials

The systems of polynomials remain a very active research area in mathematics, physics, engineering and other applied sciences; and the orthogonal polynomials, among others, are definitely the most thoroughly studied and widely applied systems [1–3]. The three of these systems, namely, Hermite, Laguerre, and Jacobi, are called collectively the classical orthogonal polynomials [4]. There is excessive literature on these polynomials, and the most comprehensive single account of the classical polynomials is found in the classical treatise of Szegö [5].

Jacobi polynomials are the common set of orthogonal polynomials defined by the formula [4]

$$P_n^{(\alpha,\beta)}(x) = (n!)^{-1}(-2)^n(1-x)^{-\alpha}(1+x)^{-\beta}\frac{d^n}{dx^n} \times \left[(1+x)^{n+\beta}(1-x)^{n+\alpha}\right] \quad (1)$$

Here, $\alpha$ and $\beta$ are parameters that, for integrability purposes, are restricted to $\alpha > -1, \beta > -1$. However, many of the identities and other formal properties of these polynomials remain valid under the less restrictive condition that neither $\alpha$ is $\beta$ a negative integer. Among the many special cases, the following is the most important [4]

(a) The Legendre polynomials ($\alpha = \beta = 0$)
(b) The Chebyshev polynomials ($\alpha = \beta = -1/2$)
(c) The Gegenbouer (or ultraspherical) polynomials ($\alpha = \beta$)

The Jacobi polynomials $P_n^{(\alpha,\beta)}(x)$ are defined [6, 7] with respect to the weight function $\omega^{\alpha,\beta}(x) = (1-x)^{\alpha}(1+x)^{\beta}$

$(\alpha > -1, \beta > -1)$ on $(-1,1)$. It is proved that the Jacobi polynomials satisfy the following relation [7]:

$$P_n^{(\alpha,\beta)}(x) = \sum_{k=0}^{n} B_n^{(\alpha,\beta,n)}(x-1)^k; \quad \alpha,\beta > -1 \tag{2}$$

where

$$B_n^{(\alpha,\beta,n)} = 2^{-k}\binom{n+\alpha+\beta+k}{k}\binom{n+\alpha}{n-k}; \quad k = 0,1,2,\ldots,n \tag{3}$$

These polynomials play role in rotation matrices [8], in the trigonometric Reson–Morse potential [9], and the cases of a few exact solutions in quantum mechanics [10, 11].

In very recent years, several researchers developed new numerical algorithms for some problems using Jacobi polynomials. Eslahchi et al. [12] gave a numerical solution for some nonlinear ordinary differential equations using the spectral method. Bojdi et al. [13] proposed a Jacobi matrix method for differential-difference equations with variable coefficients. Kazem [14] used the Tau method for solving fractional-order differential equations by means of Jacobi polynomials.

Recently, Bharwy et al. [15–22] have used Jacobi polynomials both in operational matrix method and in spectral collocation method for solving some class of fractional differential equations; for instance, nonlinear sub-diffusion equations, delay fractional optimal control problems, time fractional Kdv equations, Caputo fractional diffusion-wave equations, fractional nonlinear cable equation, and fractional differential equations.

### Integro-differential-difference equations

Fredholm integro-differential-difference equations (FIDDEs) are encountered in many model problems in biology, physics, and engineering. Also, they have been investigated using different methods by scientists [23–28]. Various numerical schemes for solving a partial integro-differential equation are presented by Dehghan [29].

In this study, we generate a procedure to find a Jacobi polynomial solution for the $n$th order linear FIDDE with variable coefficients

$$\sum_{i=0}^{n} P_i(x)y^{(i)}(x) + \sum_{j=0}^{m} Q_j(x)y^{(j)}(x-\tau) = g(x) + \int_a^b K(x,t)y(t-\tau)\mathrm{d}t, \quad \tau \geq 0 \tag{4}$$

under mixed conditions

$$\sum_{i=0}^{n-1}\left[\alpha_{ki}y^{(i)}(a) + \beta_{ki}y^{(i)}(b) + \gamma_{ki}y^{(i)}(\eta)\right] = \mu_k, \quad k = 0,1,\ldots,n-1 \tag{5}$$

where $P_i(x)$, $Q_j(x)$, $K(x,t)$, and $g(x)$ are known functions and $\alpha_{ki}$, $\beta_{ki}$, $\gamma_{ki}$, and $\mu_k$ are appropriate constants, while $y(x)$ is the unknown function. Note that $a \leq \eta \leq b$ is a given point in the spatial domain of problem.

The main aim of our study, using orthogonal Jacobi polynomials, is to provide an approximate solution for the problem (4, 5), which is usually hard to find analytical solutions.

We assume a solution expressed as the truncated series of orthogonal Jacobi polynomials defined by

$$y(x) \cong y_N^{(\alpha,\beta)}(x) = \sum_{n=0}^{N} a_n P_n^{(\alpha,\beta)}(x) \tag{6}$$

where $P_n^{(\alpha,\beta)}(x)$, $n = 0,1,\ldots,N$ denote the orthogonal Jacobi polynomials defined by (2, 3); $N$ is chosen $N \geq n$ and $a_n, n = 0,1,\ldots,N$ are unknown coefficients to be determined. Note that $\alpha$ and $\beta$ are arbitrary parameters, such that $(\alpha > -1, \beta > -1)$.

### Fundamental matrix relations

We can transform the orthogonal Jacobi polynomials $P_n^{(\alpha,\beta)}(x)$ from algebraic form into matrix form as follow:

$$\mathbf{P}^{(\alpha,\beta)}(x) = \mathbf{X}(x)\mathbf{M}^{(\alpha,\beta)} \tag{7}$$

where

$$\mathbf{P}^{(\alpha,\beta)}(x) = \left[P_0^{(\alpha,\beta)}(x) \quad P_1^{(\alpha,\beta)}(x) \quad P_2^{(\alpha,\beta)}(x) \quad \ldots \quad P_N^{(\alpha,\beta)}(x)\right] \tag{8}$$

$$\mathbf{X}(x) = \left[1 \quad (x-1) \quad (x-1)^2 \quad \ldots \quad (x-1)^N\right] \tag{9}$$

and

$$\mathbf{M}^{(\alpha,\beta)} = \left[m_{ij}^{(\alpha,\beta)}\right], \quad 1 \leq i, \quad j \leq N+1 \tag{10}$$

such that

$$m_{ij}^{(\alpha,\beta)} = \begin{cases} 2^{1-i}\binom{\alpha+\beta+i-2+j}{i-1}\binom{\alpha+j-1}{j-i}, & i \leq j \\ 0, & i > j \end{cases}$$

We assume the solution $y(x)$, which is defined by the truncated orthogonal Jacobi series (6) in matrix form as follow

$$\left[y_N^{(\alpha,\beta)}(x)\right] = \mathbf{P}^{(\alpha,\beta)}(x)\mathbf{A} \tag{11}$$

where

$$\mathbf{A} = \left[a_0 \quad a_1 \quad \ldots \quad a_N\right]^{\mathrm{T}} \tag{12}$$

By substituting the matrix form of Jacobi polynomials (7) to (11) into (4, 5), we can obtain the fundamental matrix equation of approximate solution of unknown function as

$$\left[y_N^{(\alpha,\beta)}(x)\right] = \mathbf{X}(x)\mathbf{M}^{(\alpha,\beta)}\mathbf{A} \tag{13}$$

### Matrix representation of differential-difference part of problem

Differential-difference part of problem is $\sum_{i=0}^{n} P_i(x)y^{(i)}(x) + \sum_{j=0}^{m} Q_j(x)y^{(j)}(x-\tau)$.First, to explain the relation between the matrix form of the unknown function and the matrix form of its derivative $y^{(i)}(x)$, we introduce the relation between $\mathbf{X}(x)$ and its derivatives $\mathbf{X}^{(i)}(x)$ can be expressed as

$$\mathbf{X}^{(i)}(x) = \mathbf{X}(x)\mathbf{B}^i \tag{14}$$

where

$$\mathbf{B} = \begin{bmatrix} 0 & 1 & 0 & \cdots & 0 \\ 0 & 0 & 2 & & 0 \\ & \vdots & & \ddots & \vdots \\ 0 & 0 & 0 & \cdots & N \\ 0 & 0 & 0 & & N \end{bmatrix} \tag{15}$$

Then, using (13) and (14), we may write

$$\begin{aligned}\left[y^{(i)}(x)\right] \cong \left[\left(y_N^{(\alpha,\beta)}\right)^{(i)}(x)\right] &= \mathbf{X}^{(i)}(x)\mathbf{M}^{(\alpha,\beta)}\mathbf{A} \\ &= \mathbf{X}(x)\mathbf{B}^i\mathbf{M}^{(\alpha,\beta)}\mathbf{A}\end{aligned} \tag{16}$$

Similarly, the relation between the matrix form of unknown function and matrix form of its delay forms' derivatives $y^{(j)}(x-\tau)$ can be expressed as

$$\begin{aligned}y^{(j)}(x-\tau) &= \mathbf{X}^{(j)}(x-\tau)\mathbf{M}^{(\alpha,\beta)}\mathbf{A} \\ &= \mathbf{X}(x-\tau)\mathbf{B}^j\mathbf{M}^{(\alpha,\beta)}\mathbf{A} \\ &= \mathbf{X}(x)\mathbf{B}_\tau\mathbf{B}^j\mathbf{M}^{(\alpha,\beta)}\mathbf{A}\end{aligned} \tag{17}$$

where

$$\mathbf{B}_\tau = \begin{bmatrix} \binom{0}{0}(-\tau)^0 & \binom{1}{0}(-\tau)^1 & \binom{2}{0}(-\tau)^2 & \cdots & \binom{N}{0}(-\tau)^N \\ 0 & \binom{1}{1}(-\tau)^0 & \binom{2}{1}(-\tau)^1 & & \binom{N}{1}(-\tau)^{N-1} \\ & \vdots & & \ddots & \vdots \\ 0 & 0 & 0 & \cdots & \binom{N}{N}(-\tau)^0 \end{bmatrix} \tag{18}$$

Thus, it is seen that

$$y^{(j)}(x-\tau) = \mathbf{X}(x)\mathbf{B}_\tau\mathbf{B}^i\mathbf{M}^{(\alpha,\beta)}\mathbf{A} \tag{19}$$

Using (16) and (19), the matrix form of differential-difference part of Eq. (4) becomes

$$\begin{aligned}&\sum_{i=0}^{n} P_i(x)y^{(i)}(x) + \sum_{j=0}^{m} Q_j(x)y^{(j)}(x-\tau) \\ &\quad = \sum_{i=0}^{n} P_i(x)\mathbf{X}(x)\mathbf{B}^i\mathbf{M}^{(\alpha,\beta)}\mathbf{A} + \sum_{j=0}^{m} Q_j(x)\mathbf{X}(x)\mathbf{B}_\tau\mathbf{B}^j\mathbf{M}^{(\alpha,\beta)}\mathbf{A}\end{aligned} \tag{20}$$

### Matrix representation of integral part of problem

Fredholm integral part of problem is $\int_a^b K(x,t)y(t-\tau)\mathrm{d}t$, where $K(x,t)$ the kernel function of the Fredholm integral part of main problem is. This function can be written using the truncated Taylor Series [30] and the truncated orthogonal Jacobi series, respectively, as

$$K(x,t) = \sum_{m=0}^{N}\sum_{n=0}^{N} k_{mn}^T x^m t^n \tag{21}$$

and

$$K(x,t) = \sum_{m=0}^{N}\sum_{n=0}^{N} k_{mn}^J P_m^{(\alpha,\beta)}(x)P_n^{(\alpha,\beta)}(t) \tag{22}$$

where

$$k_{mn}^T = \frac{1}{m!n!}\frac{\partial^{m+n}K(0,0)}{\partial x^m \partial t^n}, \quad m,n = 0,1,\ldots,N$$

is the Taylor coefficient and $k_{mn}^J$ is the Jacobi coefficient. The expressions (21) and (22) can be written using matrix forms of the Jacobi polynomials, respectively, as

$$K(x,t) = \mathbf{X}(x)\mathbf{B}_{-1}\mathbf{K}_\mathrm{T}(\mathbf{X}(t)\mathbf{B}_{-1})^\mathrm{T}, \quad \mathbf{K}_\mathrm{T} = \left[k_{mn}^T\right] \tag{23}$$

$$K(x,t) = \mathbf{P}^{(\alpha,\beta)}(x)\mathbf{K}_\mathrm{J}\left(\mathbf{P}^{(\alpha,\beta)}(t)\right)^\mathrm{T}, \quad \mathbf{K}_\mathrm{J} = \left[k_{mn}^J\right] \tag{24}$$

The following relation can be obtained from Eqs. (7), (23), and (24),

$$\mathbf{X}(x)\mathbf{B}_{-1}\mathbf{K}_T(\mathbf{B}_{-1})^T\mathbf{X}^T(t) = \mathbf{P}^{(\alpha,\beta)}(x)\mathbf{K}_J\left(\mathbf{P}^{(\alpha,\beta)}(t)\right)^T$$
$$= \mathbf{X}(x)\mathbf{M}^{(\alpha,\beta)}\mathbf{K}_J\left(\mathbf{M}^{(\alpha,\beta)}\right)^T\mathbf{X}^T(t)$$
$$\Rightarrow \mathbf{B}_{-1}\mathbf{K}_T(\mathbf{B}_{-1})^T = \mathbf{M}^{(\alpha,\beta)}\mathbf{K}_J\left(\mathbf{M}^{(\alpha,\beta)}\right)^T$$
$$\Rightarrow \mathbf{K}_J = \left(\mathbf{M}^{(\alpha,\beta)}\right)^{-1}\mathbf{B}_{-1}\mathbf{K}_T(\mathbf{B}_{-1})^T\left(\left(\mathbf{M}^{(\alpha,\beta)}\right)^T\right)^{-1}$$

or

$$\mathbf{K}_T = (\mathbf{B}_{-1})^{-1}\mathbf{M}^{(\alpha,\beta)}\mathbf{K}_J\left(\mathbf{M}^{(\alpha,\beta)}\right)^T\left((\mathbf{B}_{-1})^T\right)^{-1} \quad (25)$$

By substituting the Eqs. (19) and (25) into $\int_a^b K(x,t)y(t-\tau)dt$, we derive the matrix relation

$$\int_a^b K(x,t)y(t-\tau)dt = \int_a^b \mathbf{P}^{(\alpha,\beta)}(x)\mathbf{K}_J\left(\mathbf{P}^{(\alpha,\beta)}(t)\right)^T \times \mathbf{X}(t)\mathbf{B}_\tau\mathbf{M}^{(\alpha,\beta)}\mathbf{A}dt$$
$$= \mathbf{P}^{(\alpha,\beta)}(x)\mathbf{K}_J\mathbf{Q}\mathbf{A} \quad (26)$$

such that

$$\mathbf{Q} = \int_a^b \left(\mathbf{P}^{(\alpha,\beta)}(t)\right)^T\mathbf{X}(t)\mathbf{B}_\tau\mathbf{M}^{(\alpha,\beta)}dt$$
$$= \int_a^b \left(\mathbf{M}^{(\alpha,\beta)}\right)^T\mathbf{X}^T(t)\mathbf{X}(t)\mathbf{B}_\tau\mathbf{M}^{(\alpha,\beta)}dt$$
$$= \left(\mathbf{M}^{(\alpha,\beta)}\right)^T\mathbf{H}\mathbf{B}_\tau\mathbf{M}^{(\alpha,\beta)}$$

where

$$\mathbf{H} = \int_a^b \mathbf{X}^T(t)\mathbf{X}(t)dt = [h_{ij}];$$
$$h_{ij} = \frac{1}{i-1+j}\left((b-1)^{i-1+j} - (a-1)^{i-1+j}\right),$$
$$i,j = 1,2,3\ldots,N+1$$

Finally, substituting the form (7) into expression (26) yields the matrix relation

$$\int_a^b K(x,t)y(t-\tau)dt = \mathbf{X}(x)\mathbf{M}^{(\alpha,\beta)}\mathbf{K}_J\mathbf{Q}\mathbf{A} \quad (27)$$

## Matrix representation of conditions

In this section, we write to the matrix form of mixed conditions of the problem given Eq. (5), using the matrix relation (16), as

$$\sum_{i=0}^{n-1}\left[\alpha_{ki}y^{(i)}(a) + \beta_{ki}y^{(i)}(b) + \gamma_{ki}y^{(i)}(\eta)\right]$$
$$= \sum_{i=0}^{n-1}[\alpha_{ki}\mathbf{X}(a) + \beta_{ki}\mathbf{X}(b) + \gamma_{ki}\mathbf{X}(\eta)]\mathbf{B}^i\mathbf{M}^{(\alpha,\beta)}\mathbf{A} = \mu_k,$$
$$k = 0,1,\ldots,n-1 \quad (28)$$

## Method of solution

We substitute obtained matrix relations in the previous subsections given in Eqs. (20) and (27) into fundamental problem to build the fundamental matrix equation of the problem. For this purpose, we can define collocation points as follow:

$$x_s = a + \frac{b-a}{N}s, \quad s = 0,1,2,\ldots,N$$

As can be observed, standard collocation points dividing the domain interval $[a,b]$ of the problem into $N$ equal parts are employed.

Accordingly, we obtain the system of matrix equations

$$\sum_{i=0}^{n} P_i(x_s)\mathbf{X}(x_s)\mathbf{B}^i\mathbf{M}^{(\alpha,\beta)}\mathbf{A} + \sum_{j=0}^{m} Q_j(x_s)\mathbf{X}(x_s)\mathbf{B}_\tau\mathbf{B}^j\mathbf{M}^{(\alpha,\beta)}\mathbf{A}$$
$$= g(x_s) + \mathbf{X}(x_s)\mathbf{M}^{(\alpha,\beta)}\mathbf{K}_J\mathbf{Q}\mathbf{A}$$

The fundamental matrix equation becomes

$$\left\{\sum_{i=0}^{n}\mathbf{P}_i\mathbf{X}\mathbf{B}^i\mathbf{M}^{(\alpha,\beta)} + \sum_{j=0}^{m}\mathbf{Q}_j\mathbf{X}\mathbf{B}_\tau\mathbf{B}^j\mathbf{M}^{(\alpha,\beta)} - \mathbf{X}\mathbf{M}^{(\alpha,\beta)}\mathbf{K}_J\mathbf{Q}\right\}\mathbf{A} = \mathbf{G} \quad (29)$$

where

$$\mathbf{P}_i = \begin{bmatrix} P_i(x_0) & 0 & \cdots & 0 \\ 0 & P_i(x_1) & & 0 \\ \vdots & & \ddots & \vdots \\ 0 & 0 & \cdots & P_i(x_N) \end{bmatrix}, \quad \mathbf{G} = \begin{bmatrix} g(x_0) \\ g(x_1) \\ \vdots \\ g(x_N) \end{bmatrix}$$

$$\mathbf{Q}_j = \begin{bmatrix} Q_j(x_0) & 0 & \cdots & 0 \\ 0 & Q_j(x_1) & & 0 \\ \vdots & & \ddots & \vdots \\ 0 & 0 & \cdots & Q_j(x_N) \end{bmatrix}, \quad \mathbf{X} = \begin{bmatrix} \mathbf{X}(x_0) \\ \mathbf{X}(x_1) \\ \vdots \\ \mathbf{X}(x_N) \end{bmatrix}$$

Equation (29), which is matrix representation of the Eq. (4), corresponds to a system of $N+1$ algebraic equations. This system indicates $N+1$ unknown coefficients, such that $a_0, a_1, a_2, \ldots, a_N$. Briefly, if we define

$$\mathbf{W} = \sum_{i=0}^{n}\mathbf{P}_i\mathbf{X}\mathbf{B}^i\mathbf{M}^{(\alpha,\beta)} + \sum_{j=0}^{m}\mathbf{Q}_j\mathbf{X}\mathbf{B}_\tau\mathbf{B}^j\mathbf{M}^{(\alpha,\beta)} - \mathbf{X}\mathbf{M}^{(\alpha,\beta)}\mathbf{K}_J\mathbf{Q}$$

under last definition **W** matrix Eq. (29) transforms into the augmented matrix form

$$[\mathbf{W};\mathbf{G}]. \tag{30}$$

Similarly, from (28), the matrix form of mixed conditions can be obtained briefly as

$$\mathbf{U_k}\mathbf{A} = \mu_k \text{ or } [\mathbf{U_k};\mu_k], \quad k = 0,1,2,\ldots,n-1 \tag{31}$$

such that

$$\mathbf{U_k} = \sum_{i=0}^{n-1}[\alpha_{ki}\mathbf{X}(a) + \beta_{ki}\mathbf{X}(b) + \gamma_{ki}\mathbf{X}(\eta)]\mathbf{B}^i\mathbf{M}^{(\alpha,\beta)}$$

Consequently, to find the Jacobi polynomial solution of Eq. (4) under the mixed conditions (5), we replace the row matrix (31) by last $n$ rows of the augmented matrix (30), which yields the new matrix equation form written as follow

$$[\tilde{\mathbf{W}};\tilde{\mathbf{G}}]. \tag{32}$$

If rank $\tilde{\mathbf{W}} = \text{rank}[\tilde{\mathbf{W}};\tilde{\mathbf{G}}] = N+1$, then we can find the matrix of unknown coefficient of Jacobi series via $\mathbf{A} = (\tilde{\mathbf{W}})^{-1}\tilde{\mathbf{G}}$. Note that the matrix **A** (thereby, the coefficients $a_0, a_1, a_2, \ldots, a_N$) is uniquely determined [22]. Equation (4) has also a unique solution under the conditions (5). Thus, we get the Jacobi polynomial solution for arbitrary parameters $\alpha$ and $\beta$:

$$y(x) \cong y_N^{(\alpha,\beta)}(x) = \sum_{n=0}^{N} a_n P_n^{(\alpha,\beta)}(x)$$

## Error analysis

In this part of study, it is given to a useful error estimation procedure for orthogonal Jacobi polynomial solution of the problem. Also, this procedure is used to obtain the improved solution of the problem (4, 5) according to the direct Jacobi polynomial solution. For this purpose, we use the residual correction technique [31, 32] and error estimation by the known Tau method [33, 34].

Recently, Yüzbaşı and Sezer [35] solved a class of the Lane–Emden equations using the improved BCM with residual error function. Yüzbaşı et al. [36] proposed an improved Legendre method for to obtain the approximate solutions of a class of the integro-differential equations. Wei and Chen [37] presented a numerical method called spectral methods for classes Volterra type integro-differential equations with weakly singular kernel and smooth solutions.

For the purpose of calculating the corrected solution, we now define the residual function using the Jacobi polynomial solution by obtained the our method as

$$\begin{aligned} R_N(x) = &\sum_{i=0}^{n} P_i(x)\left(y_N^{(\alpha,\beta)}\right)^{(i)}(x) \\ &+ \sum_{j=0}^{m} Q_j(x)\left(y_N^{(\alpha,\beta)}\right)^{(j)}(x-\tau) - g(x) \\ &- \int_a^b K(x,t)y_N^{(\alpha,\beta)}(t-\tau)dt \end{aligned} \tag{33}$$

where $y_N^{(\alpha,\beta)}(x)$ is the approximate solution of Eqs. (4, 5) for arbitrary parameters $\alpha$ and $\beta$. Hence, $y_N^{(\alpha,\beta)}(x)$ satisfies the problem

$$\begin{aligned} &\sum_{i=0}^{n} P_i(x)\left(y_N^{(\alpha,\beta)}\right)^{(i)}(x) + \sum_{j=0}^{m} Q_j(x)\left(y_N^{(\alpha,\beta)}\right)^{(j)}(x-\tau) \\ &\quad - \int_a^b K(x,t)y_N^{(\alpha,\beta)}(t-\tau)dt = g(x) + R_N(x) \\ &\sum_{i=0}^{n-1}\left[\alpha_{ki}\left(y_N^{(\alpha,\beta)}\right)^{(i)}(a) + \beta_{ki}\left(y_N^{(\alpha,\beta)}\right)^{(i)}(b) + \gamma_{ki}\left(y_N^{(\alpha,\beta)}\right)^{(i)}(\eta)\right] \\ &\quad = \mu_k, \quad k = 0,1,\ldots,n-1 \end{aligned} \tag{34}$$

The error function $e_N(x)$ can also be defined as

$$e_N^{(\alpha,\beta)}(x) = y(x) - y_N^{(\alpha,\beta)}(x) \tag{35}$$

where $y(x)$ is the exact solution of the Eqs. (4, 5). Substituting (35) into (4, 5) and also using (33) and (34), we derive the error differential equation with homogenous conditions:

$$\begin{aligned} &\sum_{i=0}^{n} P_i(x)\left(e_N^{(\alpha,\beta)}\right)^{(i)}(x) + \sum_{j=0}^{m} Q_j(x)\left(e_N^{(\alpha,\beta)}\right)^{(j)}(x-\tau) \\ &\quad - \int_a^b K(x,t)e_N^{(\alpha,\beta)}(t-\tau)dt = -R_N(x) \\ &\sum_{i=0}^{n-1}\left[\alpha_{ki}\left(e_N^{(\alpha,\beta)}\right)^{(i)}(a) + \beta_{ki}\left(e_N^{(\alpha,\beta)}\right)^{(i)}(b) + \gamma_{ki}\left(e_N^{(\alpha,\beta)}\right)^{(i)}(\eta)\right] = 0 \end{aligned} \tag{36}$$

By solving the problem (36) using the present method given in the previous section, we get the error estimation function $e_{N,M}^{(\alpha,\beta)}(x)$ to $e_N^{(\alpha,\beta)}(x)$. Note that $M$ must be bigger than $N$ and error estimation is found using the residual function $R_N(x)$. Consequently, by means of the orthogonal Jacobi polynomials $y_N^{(\alpha,\beta)}(x)$ and $e_{N,M}^{(\alpha,\beta)}(x)$, we obtain the corrected Jacobi solution

$$y_{N,M}^{(\alpha,\beta)}(x) = y_N^{(\alpha,\beta)}(x) + e_{N,M}^{(\alpha,\beta)}(x) \tag{37}$$

Finally, we construct the Jacobi error function $e_N^{(\alpha,\beta)}(x)$ and the corrected Jacobi error function $E_{N,M}^{(\alpha,\beta)}(x)$

$$e_N^{(\alpha,\beta)}(x) = y(x) - y_N^{(\alpha,\beta)}(x), \tag{38}$$

$$E_{N,M}^{(\alpha,\beta)}(x) = y(x) - y_{N,M}(x). \tag{39}$$

## Illustrative examples

We apply the Jacobi matrix method to four examples via the symbolic computation program Maple [38]. In these examples, the term $\left|e_N^{(\alpha,\beta)}(x)\right|$ represent absolute error function and also $\left|E_{N,M}^{(\alpha,\beta)}(x)\right|$ represent the absolute error function of the corrected Jacobi polynomial solution.

*Example 1* As the first example, we consider the FIDDE [23, 24]

$$y'(x) + xy'(x-1) - y(x) + y(x-1) = g(x) + \int_{-1}^{1} K(x,t)y(t-1)\mathrm{d}t \tag{40}$$

with initial condition

$$y(1) - 2y(0) + y(-1) = 0$$

Here, $m = 1, \tau = 1, P_1(x) = 1, P_0(x) = -1, g(x) = x - 2, Q_1(x) = x, Q_0(x) = 1, K(x,t) = x + t, a = -1, b = 1, \eta = 0$, $a_{00} = 1, \beta_{00} = -2, \gamma_{00} = 1, \mu_0 = 0$. We assume that the Eq. (40) has a Jacobi polynomial solution in the following form,

$$y(x) = a_0 P_0^{(\alpha,\beta)}(x) + a_1 P_1^{(\alpha,\beta)}(x) + a_2 P_2^{(\alpha,\beta)}(x)$$

where $N = 2$ and $(\alpha,\beta) = (0.5, -0.5)$, which are chosen arbitrary; then, according to (8)

$$\mathbf{P}^{(\alpha,\beta)}(x) = \begin{bmatrix} P_0^{(\alpha,\beta)}(x) & P_1^{(\alpha,\beta)}(x) & P_2^{(\alpha,\beta)}(x) \end{bmatrix} = \begin{bmatrix} 1 & \frac{1}{2} + x & -\frac{15}{8} + \frac{15}{4}x + \frac{3(x-1)^2}{2} \end{bmatrix}.$$

The collocation points are computed as

$$\{x_0 = -1, x_1 = 0, x_2 = 1\}$$

and from Eq. (30), the matrix equation of the Eq. (40) is

$$\{\mathbf{P}_1\mathbf{XBM} + \mathbf{P}_0\mathbf{XM} + \mathbf{Q}_1\mathbf{XB}_1\mathbf{BM} - \mathbf{Q}_0\mathbf{XB}_1\mathbf{M} - \mathbf{XMK}_J\mathbf{Q}\} \boldsymbol{A} = \boldsymbol{G}$$

where

$$\mathbf{P}_1 = \mathbf{Q}_1 = \begin{bmatrix} 1 & 0 & 0 \\ 0 & 1 & 0 \\ 0 & 0 & 1 \end{bmatrix},$$

$$\mathbf{P}_0 = \begin{bmatrix} -1 & 0 & 0 \\ 0 & -1 & 0 \\ 0 & 0 & -1 \end{bmatrix},$$

$$\mathbf{Q}_1 = \begin{bmatrix} -1 & 0 & 0 \\ 0 & 0 & 0 \\ 0 & 0 & 1 \end{bmatrix},$$

$$\mathbf{X} = \begin{bmatrix} 1 & -2 & 4 \\ 1 & -1 & 1 \\ 1 & 0 & 0 \end{bmatrix}, \quad \mathbf{B} = \begin{bmatrix} 0 & 1 & 0 \\ 0 & 0 & 2 \\ 0 & 0 & 0 \end{bmatrix},$$

$$\mathbf{B}_1 = \begin{bmatrix} 1 & -1 & 0 \\ 0 & 1 & -2 \\ 0 & 0 & 0 \end{bmatrix}, \quad \mathbf{G} = \begin{bmatrix} -3 \\ -2 \\ -1 \end{bmatrix}$$

$$\mathbf{M} = \begin{bmatrix} 1 & \frac{3}{2} & \frac{15}{8} \\ 0 & 1 & \frac{15}{4} \\ 0 & 0 & \frac{3}{2} \end{bmatrix}, \quad \mathbf{K}_J = \begin{bmatrix} -1 & 1 & 0 \\ 1 & 0 & 0 \\ 0 & 0 & 0 \end{bmatrix},$$

$$\mathbf{Q} = \begin{bmatrix} 2 & -1 & \frac{7}{4} \\ 1 & \frac{1}{6} & \frac{-5}{8} \\ \frac{1}{4} & \frac{3}{8} & \frac{-81}{160} \end{bmatrix}$$

Hence, we obtain the matrix $\mathbf{W}$ as follows

$$\mathbf{W} = \begin{bmatrix} 2 & \frac{-8}{3} & 10 \\ 0 & \frac{-2}{3} & 3 \\ -2 & \frac{4}{3} & 2 \end{bmatrix}$$

Using Eq. (31), we can write the matrix equation of the condition of the problem as

$$[\mathbf{U};\lambda] = [0 \quad 0 \quad 3; \quad 0]$$

Consequently, to find the Jacobi polynomial solution of the problem under the mixed conditions, we replace the row matrix $[\mathbf{U};\lambda]$ by last row of the matrix $[\mathbf{W};\mathbf{G}]$ and obtain the matrix $[\tilde{\mathbf{W}};\tilde{\mathbf{G}}]$ as

$$[\tilde{\mathbf{W}};\tilde{\mathbf{G}}] = \begin{bmatrix} 2 & \frac{-8}{3} & 10; & -3 \\ 0 & \frac{-2}{3} & 3; & -2 \\ 0 & 0 & 3; & 0 \end{bmatrix}.$$

Solving the new augmented matrix $[\tilde{\mathbf{W}};\tilde{\mathbf{G}}]$, we obtain the Jacobi polynomial coefficient matrix

$$\mathbf{A} = \begin{bmatrix} \frac{5}{2} & 3 & 0 \end{bmatrix}^{\mathrm{T}}.$$

From Eq. (11), the Jacobi polynomial solution of the problem is $y_2^{(0.5,-0.5)}(x) = 3x+4$, which is the exact solution of the problem. Furthermore, we can obtain the exact solution of the problem for any value of $N$ and corresponding suitable values of $(\alpha, \beta)$.

*Example 2* We consider a third-order FIDDE with variable coefficients

$$y'''(x) + y''(x-1) - xy'(x) - xy(x-1) = g(x) + \int_{-1}^{1} y(t-1)\mathrm{d}t \tag{41}$$

with the initial conditions$y(0) = 0, y'(0) = 1, y''(0) = 0$. where $g(x) = -(x+1)(\sin(x-1)+\cos(x)) - \cos(2) + 1$. The exact solution of problem is $y(x) = \sin(x)$.

After several trials, it has been determined that $(\alpha, \beta) = (-0.4, 0.5)$ gives the most accurate result; therefore, the approximate solution of the third-order FIDDE has been derived by employing these values, as $y_6^{(-0.4,0.5)}(x) = 0.299\,3926085 + 0.5377164192x - 0.4041110104(x-1)^2 - 0.68\,37651331e - 1(x-1)^3 + 0.4006530189e - 1(x-1)^4 + 0,2888171229e - 2(x-1)^5 - 0.8352418987e - 3(x-1)^6$ and the estimated error function is $e_{6,7}^{(-0.4,0.5)}(x) = 0.58\,5645023e - 3 + 0.460274411e - 2x - 0.1546818100\,e - 1(x-1)^2 - 0.2185948038e - 1(x-1)^3 - 0.47\,16294\,762e - 2(x-1)^4 + 0.2241198695e - 2(x-1)^5 - 0.167994\,29\,91\,e - 3(x-1)^6 - 0.1485433260e - 3(x-1)^7$. Then, we calculate the corrected solution function simply as the sum of the approximate solution and the estimated error function

$$y_{6,7}^{(-0.4,0.5)}(x) = y_6^{(-0.4,0.5)}(x) + e_{6,7}^{(-0.4,0.5)}(x)$$

Table 1 shows the relative absolute error function $\left|e_6^{(-0.4,0.5)}(x)\right|$ and the corrected absolute error function $\left|E_{6,7}^{(-0.4,0.5)}(x)\right|$ for this example.

Now, we determine the maximum error for $y_N^{(\alpha,\beta)}(x)$ as,

$$E_N^{(\alpha,\beta)} = y_N^{(\alpha,\beta)}(x) - y(x)_\infty = \max\left\{\left|y_N^{(\alpha,\beta)}(x) - y(x)\right|, \quad a \le x \le b\right\}$$

The maximum errors $E_N^{(\alpha,\beta)}$ for different values of $N$ are given in Table 2, and it is seen that the error decreases continually as $N$ increases.

The maximum error for the corrected Jacobi polynomial solution (37) is calculated in a similar way,

$$E_{N,M}^{(\alpha,\beta)} = y_{N,M}^{(\alpha,\beta)}(x) - y(x)_\infty = \max\left\{\left|y_{N,M}^{(\alpha,\beta)}(x) - y(x)\right|, \quad a \le x \le b\right\}$$

**Table 1** Relative absolute error function $\left|e_6^{(-0.4,0.5)}(x)\right|$ and the corrected absolute error function $\left|E_{6,7}^{(-0.4,0.5)}(x)\right|$ for Example 2

| $x_i$ | Present method | | Taylor method [23] | | Tau method [24] | |
|---|---|---|---|---|---|---|
| | $N=6$ | $N,M=6,7$ | $N=6$ | $N=7$ | $N=6$ | $N=7$ |
| −1.0 | 2.88e−02 | 1.04e−03 | 8.58e−02 | 6.03e−02 | 3.84e−02 | 5.05e−03 |
| −0.8 | 1.36e−02 | 5.72e−04 | 3.93e−02 | 2.28e−02 | 1.83e−02 | 2.38e−03 |
| −0.6 | 5.22e−03 | 2.53e−04 | 1.50e−02 | 6.63e−03 | 7.00e−03 | 9.14e−04 |
| −0.4 | 1.38e−03 | 7.70e−05 | 4.12e−03 | 1.20e−03 | 1.86e−03 | 2.42e−04 |
| −0.2 | 1.52e−04 | 9.68e−06 | 4.85e−04 | 6.90e−05 | 2.04e−04 | 2.65e−05 |
| 0.0 | 0 | 0 | 0 | 0 | 0 | 0 |
| 0.2 | 1.10e−04 | 9.22e−06 | 4.59e−04 | 5.30e−05 | 1.48e−04 | 1.91e−05 |
| 0.4 | 7.20e−04 | 6.97e−05 | 3.69e−03 | 8.09e−04 | 9.67e−04 | 1.25e−04 |
| 0.6 | 1.90e−03 | 2.15e−04 | 1.28e−02 | 3.82e−03 | 2.55e−03 | 3.30e−04 |
| 0.8 | 3.34e−03 | 4.71e−04 | 3.17e−02 | 1.14e−02 | 4.44e−03 | 5.78e−04 |
| 1.0 | 4.36e−03 | 8.26e−04 | 6.57e−02 | 2.73e−02 | 5.76e−03 | 7.53e−04 |

**Table 2** Maximum error ($E_N^{(-0.4,0.5)}$) for Example 2

| $N$ | 4 | 6 | 8 | 10 | 12 |
|---|---|---|---|---|---|
| $E_N^{(\alpha,\beta)}$ | 1.6321e−01 | 2.8883e−02 | 6.2300e−03 | 2.9748e−03 | 2.9748e−04 |

and the results are shown in Table 3 for miscellaneous values of $N$, $M$. The decrease in maximum error, as $M$ increases, is indisputable.

Finally, the third-order FIDDE has also been solved using Legendre, Gegenbauer (also Chebyshev), and Jacobi polynomials, for comparison purposes. The maximum error values are given in Table 4, and it is seen that Jacobi-based solution gives slightly better results.

**Table 3** Maximum error ($E_{N,M}^{(-0.4,0.5)}$) for Example 2

| $N, M$ | 6, 7 | 6, 8 | 10, 11 | 10, 15 |
|---|---|---|---|---|
| $E_{N,M}^{(\alpha,\beta)}$ | 1.0480$e$−03 | 6.7241$e$−04 | 3.5602$e$−04 | 1.7160$e$−05 |

**Table 4** Comparison of different polynomial bases for maximum error values for Example 2

| | $(\alpha,\beta)$ | $E_4^{(\alpha,\beta)}$ | $E_6^{(\alpha,\beta)}$ |
|---|---|---|---|
| Legendre base | (0, 0) | 0.1632109624 | 0.02888311052 |
| Gegenbauer base | (−0.5, −0.5) | 0.1632109623 | 0.0288831090 |
| Jacobi base | (−0.4, 0.5) | 0.1632109622 | 0.0288831072 |

*Example 3* The third example is a second-order FIDE [24–26] with variable coefficients

$$y''(x) + 4xy'(x) = \frac{-8x^4}{(x^2+1)^3} - 2\int_0^1 \frac{t^2+1}{(x^2+1)^2} y(t)dt, \quad (42)$$

$0 \le x \le 1$

under the boundary conditions

$y(0) = 1, y(1) = 1/2$.

Here, $P_2(x) = 1, P_1(x) = 4x, P_0(x) = 0, Q_0(x) = 0, g(x) = -8x^4/(x^2+1)^3, \alpha = 0, \beta = 0, K(x,t) = -2(t^2+1)/(x^2+1)^2, \tau = 0, a = 0$.

The exact solution of this problem is $(x^2+1)^{-1}$.

Figure 1 shows a comparison of the Jacobi polynomial solution $y_N^{(0,0)}(x)$, and the corrected Jacobi polynomial solution is $y_{N,M}^{(0,0)}(x)$, for $(N, M) = (5, 6)$ and $(\alpha = \beta = 0)$, with the exact solution $y(x)$. It is apparently seen that the corrected Jacobi polynomial solution almost coincides with the exact solution.

Table 5 and Fig. 2 show a comparison of the absolute error with the corrected absolute errors, for $N = 5, 8$ and

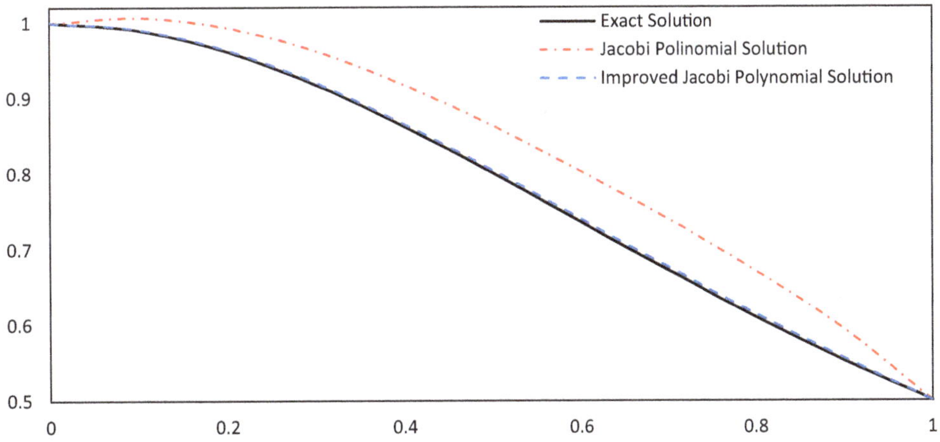


**Fig. 1** Comparison of the Jacobi polynomial solution $y_5^{(\alpha,\beta)}(x)$ and the corrected Jacobi polynomial solution $y_{5,6}^{(\alpha,\beta)}(x)$ with the exact solution for Example 3

**Table 5** Comparison of the absolute error with the corrected absolute errors for Example 3

| $x_i$ | Absolute error | | Corrected absolute errors | | | |
|---|---|---|---|---|---|---|
| | $\lvert e_5^{(0,0)}(x)\rvert$ | $\lvert e_8^{(0,0)}(x)\rvert$ | $\lvert E_{5,6}^{(0,0)}(x)\rvert$ | $\lvert E_{5,7}^{(0,0)}(x)\rvert$ | $\lvert E_{5,7,8}^{(0,0)}(x)\rvert$ | $\lvert E_{5,7,9}^{(0,0)}(x)\rvert$ |
| 0.125 | 2.097$e$−2 | 1.087$e$−2 | 1.298$e$−3 | 2.627$e$−4 | 2.723$e$−4 | 6.754$e$−5 |
| 0.250 | 3.883$e$−2 | 2.010$e$−2 | 2.347$e$−3 | 4.903$e$−4 | 5.024$e$−4 | 1.252$e$−4 |
| 0.375 | 5.295$e$−2 | 2.737$e$−2 | 3.197$e$−3 | 6.574$e$−4 | 6.821$e$−4 | 1.703$e$−4 |
| 0.500 | 6.296$e$−2 | 3.264$e$−2 | 3.799$e$−3 | 7.817$e$−4 | 8.114$e$−4 | 2.027$e$−4 |
| 0.625 | 6.815$e$−2 | 3.596$e$−2 | 4.173$e$−3 | 8.862$e$−4 | 8.957$e$−4 | 2.239$e$−4 |
| 0.750 | 6.581$e$−2 | 3.656$e$−2 | 4.257$e$−3 | 1.128$e$−3 | 9.439$e$−4 | 2.345$e$−4 |
| 0.875 | 4.838$e$−2 | 2.992$e$−2 | 3.403$e$−3 | 1.501$e$−3 | 9.147$e$−4 | 2.260$e$−4 |

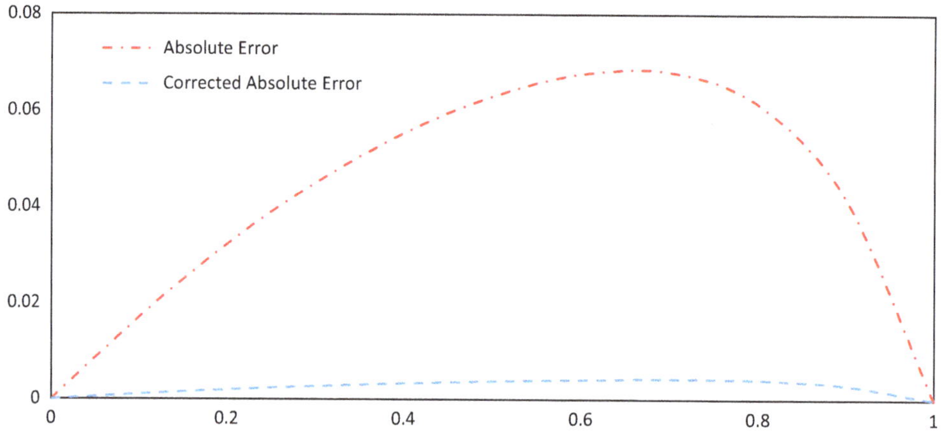


**Fig. 2** Comparison of the absolute error functions of the Jacobi polynomial solution $\left|e_5^{(\alpha,\beta)}(x)\right|$ with the corrected Jacobi polynomial solution $\left|E_{5,6}^{(\alpha,\beta)}(x)\right|$ for Example 3

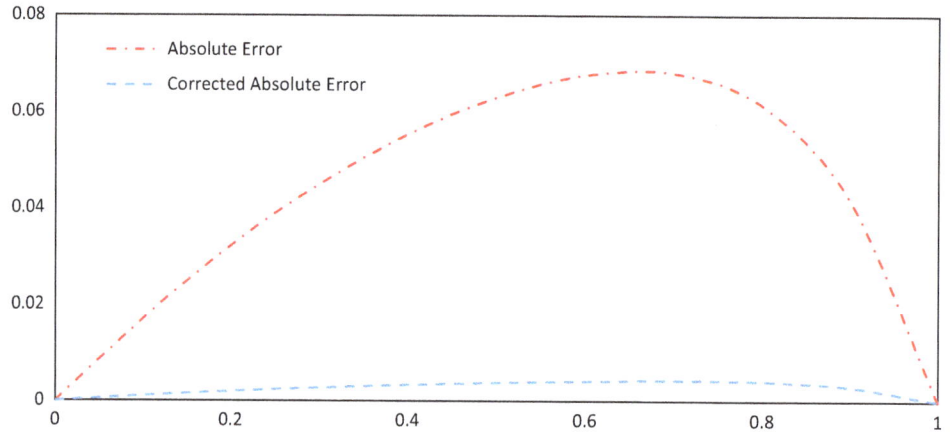

**Table 6** Comparison of the absolute errors of Jacobi polynomial solution, improved Jacobi polynomial solution, wavelet collocation, wavelet Galerkin, and ChFD methods for Example 4

| $x_i$ | Present method | | Wavelet collocation [26] | Wavelet Galerkin [26] | ChFD [17] |
|---|---|---|---|---|---|
| | $\left|e_7^{(.4,.3)}(x)\right|$ | $\left|E_{7,8}^{(.4,.3)}(x)\right|$ | | | $N=7$ |
| 0.125 | 8.5$e$−10 | 1.1$e$−11 | 2.6$e$−02 | 2.7$e$−04 | 1.8$e$−10 |
| 0.250 | 6.3$e$−10 | 5.0$e$−11 | 1.3$e$−02 | 3.0$e$−05 | 4.4$e$−10 |
| 0.375 | 1.0$e$−09 | 5.8$e$−11 | 9.3$e$−03 | 2.6$e$−04 | 1.4$e$−09 |
| 0.500 | 1.3$e$−09 | 7.6$e$−11 | 5.1$e$−03 | 4.3$e$−04 | 2.4$e$−10 |
| 0.625 | 1.6$e$−09 | 7.5$e$−11 | 2.5$e$−03 | 5.6$e$−04 | 1.7$e$−09 |
| 0.750 | 4.0$e$−09 | 1.1$e$−10 | 1.0$e$−03 | 6.5$e$−04 | 7.7$e$−10 |
| 0.875 | 3.8$e$−09 | 1.5$e$−10 | 2.6$e$−04 | 7.2$e$−04 | 1.3$e$−09 |

$M = 6, 7, 8, 9$. The parameters are taken as $(\alpha = \beta = 0)$. The corrected absolute errors are corrected, once more and the last two columns show these values. It is noticed that sequential corrections tend to decrease the absolute error.

*Example 4* [25, 26] The last example is the second-order Fredholm integro-differential equation

$$x^2y'' + 50xy' - 35y = \frac{1-e^{x+1}}{x+1} + (x^2 + 50x - 35)e^x + \int_0^1 e^{xt}y(t)dt$$

with conditions

$$y(0) = 1,\ y(1) = e$$

The exact solution of problem is

Taking $(\alpha = 0.4, \beta = 0.5)$, the absolute errors of Jacobi polynomial solution for $N = 7$ and the absolute errors of the improved Jacobi polynomial solution for $N = 7, M = 8$ are compared with those of the wavelet Galerkin, the wavelet collocation, and the Chebyshev finite difference (ChFD) methods [25, 26], in Table 6. Considering the errors of the different methods, it is observed that the smallest errors are obtained using the improved Jacobi polynomial solution.

*Example 5* Consider the first-order linear FIDDE [39]

$$\int_0^x \frac{y(t)}{(x-t)^{\frac{1}{2}}}dt = \frac{4}{105}x^{3/2}(24 - x^2), \quad 0 \le x \le 1.$$

We assume that the problem has a Jacobi polynomial solution in the form

$$y(x) = a_0P_0^{(\alpha,\beta)}(x) + a_1P_1^{(\alpha,\beta)}(x) + a_2P_2^{(\alpha,\beta)}(x) + a_3P_3^{(\alpha,\beta)}(x)$$

where $N = 3$ and $(\alpha, \beta) = (0.2, -0.3)$, which are chosen arbitrary. Using the mentioned methods, the Jacobi polynomial solution of the problem is obtained by $y_3^{(0.2,-0.3)}(x) = x - x^3$, which is the exact solution of the problem [39]. Furthermore, we can obtain the exact solution of the problem for any value of $N \ge 3$ and corresponding suitable values of $(\alpha, \beta)$.

## Conclusions

A new matrix method based on Jacobi polynomials and collocation points has been introduced to solve high-order linear FIDDE with variable coefficients. Jacobi polynomials are the common set of orthogonal polynomials, which are the most extensively studied and widely applied systems. The solution of the FIDDE is expressed as a truncated series of orthogonal Jacobi polynomials, which is then transformed from algebraic form into matrix form. The problem and the mixed conditions are also represented in matrix form. Finally, the solution is obtained as a truncated Jacobi series written in matrix form using collocation points. A new error estimation procedure for polynomial solution and a technique to find a high accuracy solution are developed.

Most of the previous studies dealt with solutions using Legendre, Chebyshev, and Gegenbauer polynomials. In this study, however, we have proposed a Jacobi polynomial solution that comprises all of these polynomial solutions.

The new Jacobi matrix method has been applied to four illustrative examples. It is well seen from these examples that the method yields either the exact solution or a high accuracy approximate solution for delay integro-differential equation problems. The accuracy of the approximate solution can be increased using the proposed error analysis technique depending on residual function.

## References


1. Pcoolen-Schrijner, P., Van Doorn, E.A.: Analysis of random walks using orthogonal polynomials. J. Comput. Appl. Math. **99**, 387–399 (1998)
2. Fischer, B., Prestin, J.: Wavelets based on orthogonal polynomials. Math. Comp. **66**, 1593–1618 (1997)
3. El-Mikkawy, M.E.A., Cheon, G.S.: Combinatorial and hypergeometric identities via the Legendre polynomials—a computational approach. Appl. Math. Comput. **166**, 181–195 (2005)
4. Chihara, T.S.: An Introduction to Orthogonal Polynomials. Gordon and Breach, Philadelphia (1978)
5. Szegö, G.: Orthogonal Polynomials, vol. 23. Amer Mathema Soci, Colloquium Publication, New York (1939)
6. Grümbaum, F.A.: Matrix valued Jacobi polynomials. Bull. Des. Sci. Math. **127**, 207–214 (2003)
7. Eslahchi, M.R., Dehghan, M.: Application of Taylor series in obtaining the orthogonal operational matrix. Comput. Math Appl. **61**, 2596–2604 (2011)
8. Rose, M.E.: Elementary Theory of Angular Momentum. Wiley, Oxford (1957)
9. Bijker, R., et al.: Latin-American School of Physics: XXXV ELAF: Supersymmetry and Its Applications in Physics. AIP Conf. Series **744** (2004)
10. Weber, H.J.: A simple approach to Jacobi polynomials: Integ. Transf. Spec. Funct. **18**, 217–221 (2007)
11. De, R., Dutt, R., Sukhatme, U.: Mapping of shape invariant potentials under point canonical transformations. J. Phys. A Math. Genet. **25**, 843–850 (1992)
12. Eslahchi, M.R., Dehghan, M., Ahmadi-Asl, S.: The general Jacobi matrix method for solving some nonlinear ordinary differential equations. Appl. Math. Model. **36**, 3387–3398 (2012)
13. Kalateh Bojdi, Z., Ahmadi-Asl, S., Aminataei, A.: The general shifted Jacobi matrix method for solving the general high order linear differential-difference equations with variable coefficients. J. Math. Res. Appl. **1**, 10–23 (2013)
14. Kazem, S.: An integral operational matrix based on Jacobi polynomials for solving fractional-order differential equations. Appl. Math. Model. **37**, 1126–1136 (2013)
15. Bhrawy, A.H., Taha, M., José, A.T.M.: A review of operational matrices and spectral techniques for fractional calculus. Nonlinear Dyn. **81**(3), 1023–1052 (2015)
16. Bhrawy, A.H: A Jacobi spectral collocation method for solving multi-dimensional nonlinear fractional sub-diffusion equations. Num. Algorithms 1–23 (2015). doi:10.1007/s11075-015-0087-2
17. Bhrawy, A.H., Alofi, A.S.: The operational matrix of fractional integration for shifted Chebyshev polynomials. Appl. Math. Lett. **26**(1), 25–31 (2013)
18. Bhrawy, A.H., Ezz-Eldien, S.S.: A new Legendre operational technique for delay fractional optimal control problems. Calcolo 1–23 (2015). doi:10.1007/s10092-015-0160-1
19. Bhrawy, A.H, et al.: A numerical technique based on the shifted Legendre polynomials for solving the time-fractional coupled KdV equations. Calcolo 1–17 (2015). doi:10.1007/s10092-014-0132-x
20. Bhrawy, A.H., Zaky, M.A., Van Gorder, R.A.: A space-time Legendre spectral tau method for the two-sided space-time Caputo fractional diffusion-wave equation. Numer. Algorithms **71**(1), 151–180 (2016)
21. Bhrawy, A.H., Zaky, M.A.: Numerical simulation for two-dimensional variable-order fractional nonlinear cable equation. Nonlinear Dyn. **80**(1-2), 101–116 (2015)
22. Doha, E.H., Bhrawy, A.H., Ezz-Eldien, S.S.: A new Jacobi operational matrix: an application for solving fractional differential equations. Appl. Math. Model. **36**(10), 4931–4943 (2012)
23. Gülsu, M., Sezer, M.: Approximations to the solution of linear Fredholm integro differential-difference equation of high order. J. Franklin Inst. **343**, 720–737 (2006)
24. Saadatmandi, A., Dehghan, M.: Numerical solution of the higher-order linear Fredholm integro-differential-difference equation with variable coefficients. Comput. Math Appl. **59**, 2996–3004 (2010)
25. Dehghan, M., Saadatmandi, A.: Chebyshev finite difference method for Fredholm integro-differential equations. Int. J. Comput. Math. **85**, 123–130 (2008)
26. Behirly, S.H., Hasnish, H.: Wavelet methods for the numerical solution of Fredholm integro-differential equations. Int. J. Appl. Math. **11**, 27–36 (2002)
27. Şahin, N., Yüzbaşi, Ş., Sezer, M.: A Bessel polynomial approach for solving general linear Fredholm integro-differential-difference equations. Int. J. Comput. Math. **88**, 3093–3111 (2011)
28. Kurt, A., Yalçinbaş, S., Sezer, M.: Fibonacci collocation method for solving high-order linear Fredholm integro-differential-difference equations. Int. J. Math. Math. Sci.1–9 (2013). doi:10.1155/2013/486013
29. Dehghan, M.: Solution of a partial integro-differential equation arising from viscoelasticity. Int. J. Comput. Math. **83**, 123–129 (2006)
30. Kurt, N., Sezer, M.: Polynomial solution of high-order linear Fredholm integro-differential equations with constant coefficients. J. Franklin Inst. **345**, 839–850 (2008)
31. Oliveira, F.A.: Collocation and residual correction. Numer. Math. **36**, 27–31 (1980)
32. Çelik, İ.: Collocation method and residual correction using Chebyshev series. Appl. Math. Comput. **174**, 910–920 (2006)
33. Pour-Mahmoud, J., Rahimi-Ardabili, M.Y., Shahmorad, S.: Numerical solution of the system of Fredholm integro-differential equations by the Tau method. Appl. Math. Comput. **168**, 465–478 (2005)

34. Shahmorad, S.: Numerical solution of the general form linear Fredholm-Volterra integro-differential equations by the Tau method with an error estimation. Appl. Math. Compt. **167**, 1418–1429 (2005)
35. Yüzbaşi, Ş., Sezer, M.: An improved Bessel collocation method with a residual error function to solve a class of Lane-Emden differential equations. Math. Comput. Model. **57**, 1298–1311 (2013)
36. Yüzbaşi, Ş., Sezer, M., Kemanci, B.: Numerical solutions of integro-differential equations and application of a population model with an improved Legendre method. Appl. Math. Model. **37**, 2086–2101 (2013)
37. Wei, Y., Chen, Y.: Convergence analysis of the spectral methods for weakly singular Volterra integro-differential equations with smooth solutions. Adv. Appl. Math. Mech. **4**, 1–20 (2012)
38. Maple 11 User Manual, Waterloo Maple Inc. http://www.maplesoft.com/view.aspx?sl=5883 (2007). Accessed 1 Jun 2016
39. Abdelkawy, M.A., Mohamed, A., Ezz-Eldien, S.S., Ahmad, Z.M.A.: A Jacobi spectral collocation scheme for solving Abel's integral equations. Progr. Fract. Differ. Appl. **1**(3), 1–14 (2015)

16

# Asymptotic equilibrium of integro-differential equations with infinite delay

Le Anh Minh[1] · Dang Dinh Chau[2]

**Abstract** The asymptotic equilibrium problems of ordinary differential equations in a Banach space have been considered by several authors. In this paper, we investigate the asymptotic equilibrium of the integro-differential equations with infinite delay in a Hilbert space.

**Keywords** Asymptotic equilibrium · Integro-differential equations · Infinite delay

## Introduction

The asymptotic equilibrium problems of ordinary differential equations in a Banach space have been considered by several authors, Mitchell and Mitchell [3], Bay et al. [1], but the results for the asymptotic equilibrium of integro-differential equations with infinite delay still is not presented. In this paper, we extend the results in [1] to a class of integro-differential equations with infinite delay in a Hilbert space $H$ which has the following form:

$$\begin{cases} \dfrac{\mathrm{d}x(t)}{\mathrm{d}t} = A(t)\left(x(t) + \displaystyle\int_{-\infty}^{t} k(t-\theta)x(\theta)\mathrm{d}\theta\right), & t \geqslant 0, \\ x(t) = \varphi(t), & t \leqslant 0 \end{cases} \tag{1}$$

where $A(t) : H \to H$, $\varphi$ in the phase space $\mathscr{B}$, and $x_t$ is defined as

$$x_t(\theta) = x(t+\theta), \quad -\infty < \theta \leqslant 0.$$

## Preliminaries

We assume that the phase space $(\mathscr{B}, ||.||_{\mathscr{B}})$ is a seminormed linear space of functions mapping $(-\infty, 0]$ into $H$ satisfying the following fundamental axioms (we refer reader to [2])

(A$_1$) For $a > 0$, if $x$ is a function mapping $(-\infty, a]$ into $H$, such that $x \in \mathscr{B}$ and $x$ is continuous on [0, $a$], then for every $t \in [0, a]$ the following conditions hold:

(i) $x_t$ belongs to $\mathscr{B}$;
(ii) $||x(t)|| \leqslant G||x_t||_{\mathscr{B}}$;
(iii) $||x_t||_{\mathscr{B}} \leqslant K(t)\sup_{s\in[0,t]}||x(s)|| + M(t)||x_0||_{\mathscr{B}}$

where $G$ is a possitive constant, $K, M : [0, \infty) \to [0, \infty)$, $K$ is continuous, $M$ is locally bounded, and they are independent of $x$.

(A$_2$) For the function $x$ in ($A_1$), $x_t$ is a $\mathscr{B}$-valued continuous function for $t$ in [0, $a$].

(A$_3$) The space $\mathscr{B}$ is complete.

✉ Le Anh Minh
leanhminh@hdu.edu.vn

Dang Dinh Chau
chaudida@gmail.com

1 Department of Mathematical Analysis, Hong Duc University, Thanh Hóa, Vietnam

2 Department of Mathematics, Hanoi University of Science, VNU, Hanoi, Vietnam

*Example 1*

(i) Let $BC$ be the space of all bounded continuous functions from $(-\infty, 0]$ to $H$, we define $C^0 := \{\varphi \in BC : \lim_{\theta \to -\infty} \varphi(\theta) = 0\}$ and $C^\infty := \{\varphi \in BC : \lim_{\theta \to -\infty} \varphi(\theta) \text{ exists in } H\}$ endowed with the norm

$$||\varphi||_{\mathscr{B}} = \sup_{\theta \in (-\infty, 0]} ||\varphi(\theta)||$$

then $C^0, C^\infty$ satisfies ($A_1$)–($A_3$). However, $BC$ satisfies ($A_1$) and ($A_3$), but ($A_2$) is not satisfied.

(ii) For any real constant $\gamma$, we define the functional spaces $C_\gamma$ by

$$C_\gamma = \left\{\varphi \in C((-\infty, 0], X) : \lim_{\theta \to -\infty} e^{\gamma\theta}\varphi(\theta) \text{ exists in } H\right\}$$

endowed with the norm

$$||\varphi||_{\mathscr{B}} = \sup_{\theta \in (-\infty, 0]} e^{\gamma\theta}||\varphi(\theta)||.$$

Then conditions ($A_1$)–($A_3$) are satisfied in $C_\gamma$.

*Remark 1* In this paper, we use the following acceptable hypotheses on $K(t)$, $M(t)$ in $(A_1)$(iii) which were introduced by Hale and Kato [2] to estimate solutions as $t \to \infty$,

($\gamma_1$) $K = K(t)$ is a constant for all $t \geqslant 0$;
($\gamma_2$) $M(t) \leqslant M$ for all $t \geqslant 0$ and some $M$.

*Example 2* For the functional space $C_\gamma$ in Example 1, the hypotheses ($\gamma_1$) and ($\gamma_2$) are satisfied if $\gamma \geqslant 0$.

**Definition 1** Equation (1) has an asymptotic equilibrium if every solution of it has a finite limit at infinity and, for every $h_0 \in H$, there exists a solution $x(t)$ of it such that $x(t) \to h_0$ as $t \to \infty$.

## Main results

Now, we consider the asymptotic equilibrium of Eq. (1) which satisfies the following assumptions:

($M_1$) $A(t)$ is a strongly continuous bounded linear operator for each $t \in \mathbb{R}^+$;

($M_2$) $A(t)$ is a self-adjoint operator for each $t \in \mathbb{R}^+$;

($M_3$) $k$ satisfies

$$\int_0^{+\infty} |k(\theta)| d\theta = L < +\infty;$$

and

($M_4$) There exists a constant $T > 0$ such that

$$\sup_{h \in S(0,1)} \int_T^\infty ||A(t)h|| dt < q < \frac{1}{\kappa}, \tag{2}$$

herein $S(0, 1)$ is a unit ball in $H$, $\kappa = L(K + M) + 1$, where $K$, $M$, $L$ are given in ($\gamma_1$), ($\gamma_2$) and ($M_3$).

**Theorem 1** *If* ($M_1$), ($M_2$), ($M_3$) *and* ($M_4$) *are satisfied, then Eq.* (1) *has an asymptotic equilibrium.*

*Proof* We shall begin with showing that all solutions of (1) has a finite limit at infinity. Indeed, Eq. (1) may be rewritten as

$$\frac{dx(t)}{dt} = A(t)\left(x(t) + \int_{-\infty}^{0} k(-\theta)x_t(\theta)d\theta\right),$$

then for $t \geqslant s \geqslant T$ we have

$$x(t) = x(s) + \int_s^t A(\tau)\left(x(\tau) + \int_{-\infty}^{0} k(-\theta)x_\tau(\theta)d\theta\right)d\tau$$

and

$$\begin{aligned} &||x(t)|| \\ &= \sup_{h \in S(0,1)} \left|\left\langle x(s) + \int_s^t A(\tau)\left(x(\tau) + \int_{-\infty}^{0} k(-\theta)x_\tau(\theta)d\theta\right)d\tau, h\right\rangle\right| \\ &\leqslant ||x(s)|| + \sup_{h \in S(0,1)} \int_s^t \left|\left\langle x(\tau) + \int_{-\infty}^{0} k(-\theta)x_\tau(\theta)d\theta, A(\tau)h\right\rangle\right| d\tau \\ &\leqslant ||x(s)|| + q\left((LK+1)\sup_{\xi \in [0,t]} ||x(\xi)|| + LM||\varphi||_{\mathscr{B}}\right) \end{aligned} \tag{3}$$

implies

$$|||x(t)||| \leqslant ||x(s)|| + q\big((LK + 1)|||x(t)||| + LM||\varphi||_{\mathscr{B}}\big)$$

or

$$|||x(t)||| \leqslant \frac{||x(s)|| + qLM||\varphi||_{\mathscr{B}}}{1 - q(LK + 1)} \tag{4}$$

where

$$|||x(t)||| = \sup_{0 \leqslant \xi \leqslant t} ||x(\xi)||.$$

Now, we conclude that $x(t)$ is bounded since

$$0<q<\frac{1}{\kappa}=\frac{1}{L(K+M)+1}<\frac{1}{LK+1}\Rightarrow q(LK+1)<1$$

and by (4).

Putting

$$M^*=\sup_{t\in\mathbb{R}}||x(t)||,$$

we have

$$||x(t)-x(s)||=\sup_{h\in S(0,1)}|<x(t)-x(s),h>|$$

$$\leqslant\sup_{h\in S(0,1)}\int_s^t\left|<A(\tau)\left(x(\tau)+\int_{-\infty}^0 k(-\theta)x_\tau(\theta)\mathrm{d}\theta\right),h>\right|\mathrm{d}\tau,$$

$$\leqslant[M^*(LK+1)+LM||\varphi||_{\mathscr{B}}]\sup_{h\in S(0,1)}\int_s^t||A(\tau)h||d\tau\to 0$$

as $t\geqslant s\to+\infty$. That means all solutions of (1) have a finite limit at infinity. To complete the proof, it remains to show that for any $h_0\in H$, there exists a solution $x(t)$ of (1) such that

$$\lim_{t\to+\infty}x(t)=h_0.$$

Indeed, let $h_0$ be an arbitrary fixed element of $H$; we choose the initial function $\varphi$ belongs to $\mathscr{B}$ such that $\varphi(0)=h_0$ and $||\varphi||_{\mathscr{B}}\leqslant||h_0||$ and consider the functional

$$g_1(t,h)=\langle h_0,h\rangle-\int_t^\infty\left\langle A(\tau)\left(h_0+\int_{-\infty}^\tau k(\tau-\theta)x_0(\theta)\mathrm{d}\theta\right),h\right\rangle\mathrm{d}\tau$$

We have

$$|g_1(t,h)|\leqslant||h_0||||h||+\int_t^{+\infty}||x_0(\tau)+\int_{-\infty}^\tau k(\tau-\theta)x_0(\theta)\mathrm{d}\theta||||A(\tau)h||\mathrm{d}\tau.$$

Since $x_0(\tau)\equiv h_0$, then

$$|g_1(t,h)|\leqslant||h_0||(||h||+q\kappa).$$

It follows from Riesz representation theorem that there exists an element $x_1(t)$ in $H$, such that

$$g_1(t,h)=\langle x_1(t),h\rangle$$

and

$$||x_1(t)||\leqslant||h_0||(1+q\kappa).$$

Now, we consider the functional

$$g_2(t,h)=\langle h_0,h\rangle-\int_t^{+\infty}\left\langle A(\tau)\left(x_1(t)+\int_{-\infty}^\tau k(\tau-\theta)x_1(\theta)\mathrm{d}\theta\right),h\right\rangle\mathrm{d}\tau.$$

By an argument analogous to the previous one, we get

$$|g_2(t,h)|\leqslant||h_0||[||h||+q\kappa+(q\kappa)^2]$$

and there exists an element $x_2(t)$ in $H$, such that

$$g_2(t,h)=\langle x_2(t),h\rangle$$

with

$$||x_2(t)||\leqslant||h_0||(1+q\kappa+(q\kappa)^2).$$

Continuing this process, we obtain the linear continuous functional

$$g_n(t,h)=\langle h_0,h\rangle-\int_t^{+\infty}\left\langle A(\tau)\left(x_{n-1}(t)+\int_{-\infty}^\tau k(\tau-\theta)x_{n-1}(\theta)\mathrm{d}\theta\right),h\right\rangle\mathrm{d}\tau \tag{5}$$

and $x_n(t)\in H$ such that

$$g_n(t,h)=\langle x_n(t),h\rangle$$

satisfies the following estimate

$$||x_n(t)||\leqslant(1+q\kappa+(q\kappa)^2+\cdots+(q\kappa)^n)||h_0||\leqslant\frac{||h_0||}{1-q\kappa}.$$

Futhermore,

$$||x_n(t)-x_{n-1}(t)||\leqslant||h_0||(q\kappa)^n.$$

This inequality shows that $\{x_n(t)\}$ is uniformly convergent on $[T,+\infty)$ since $q\kappa<1$. Put

$$x(t)=\lim_{n\to+\infty}x_n(t).$$

In (5), let $n\to+\infty$, we have

$$\langle x(t),h\rangle=\langle h_0,h\rangle-\int_t^{+\infty}\left\langle A(\tau)\left(x(t)+\int_{-\infty}^\tau k(\tau-\theta)x(\theta)\mathrm{d}\theta\right),h\right\rangle\mathrm{d}\tau \tag{6}$$

and since

$$|\langle x_n(t), h_0\rangle| < \int\limits_T^{+\infty} ||x_{n-1}(\tau)$$

$$+ \int\limits_{-\infty}^{\tau} k(\tau - \theta)x_{n-1}(\theta)\mathrm{d}\theta|| \, ||A(\tau)h||\mathrm{d}\tau$$

or

$$|\langle x_n(t), h_0\rangle| \leqslant \frac{||h_0||q}{1 - q\kappa},$$

we have $x_n(t) \to h_0$ as $q \to 0$, which means that there exists a solution of (1) converging to $h_0$. The theorem is proved.

## References

1. Bay, N.S., Hoan, N.T., Man, N.M.: On the asymptotic equilibrium and asymptotic equivalence of differential equations in Banach spaces. Ukr. Math. J. **60**(5), 716–729 (2008)
2. Hale, J.K., Kato, J.: Phase space for retarded equations with infinite delay. Fukcialaj Ekvacioj **21**, 11–41 (1978)
3. Mitchell, A.R., Mitchell, R.W.: Asymptotic equilibrium of ordinary differential systems in a Banach space. Theory Comput. Syst. **9**(3), 308–314 (1975)

# 17

# Homotopy analysis for the influence of Navier slip flow in a vertical channel with cross diffusion effects

K. Kaladhar[1] · E. Komuraiah[1]

**Abstract** This research is to examine the laminar, incompressible free convective Navier slip flow between vertical plates with cross diffusion effects. This investigation includes the first order chemical reaction also. The resulting equations with boundary conditions are reduced into dimensionless form using suitable transformations. Homotopy analysis method has been applied to solve the system. The influence of emerging parameters on fluid flow quantities have been presented graphically. In addition, the nature of physical quantities are shown in tabular form.

**Keywords** Natural convection · Soret effect · Dufour effect · Chemical reaction · Navier slip · HAM

## Introduction

Combined heat and mass transfer in free convection flow between vertical parallel plates has significant importance in many applications. Kairi and Murthy [1] presented the applications and past theoretical investigations on free convection flow in a channel. Fattahi et al. [2] applied the lattice Boltzmann technique to study the nature of free convection flow of nano fluids. Huelsz and Rechtman [3] also applied the lattice Boltzmann technique for heat transfer in an inclined square cavity due to natural convection. Most recently, Terekhov et al. [4] studied the laminar free convection heat transfer between vertical isothermal plates.

✉ K. Kaladhar
kaladhar@nitpy.ac.in

1 Department of Mathematics, National Institute of Technology Puducherry, Karaikal 609605, India

Generally, the no-slip condition has been accepted for the fluid over a solid surface boundary condition. At the solid boundary, the general boundary conditions for slip flow has been proposed by Navier [5]. Eegunjobi and Makinde [6] presented the early literature and application of the slip flow in various geometries. Recently, Rundora and Makinde [7] analyzed the influence of Navier slip and variable viscosity through a porous medium with asymmetric convective boundary conditions. Most recently, Ng [8] investigated the nature of starting flow in channels with boundary slip.

The Soret and Dufour effects are encountered in the areas of chemical engineering, geosciences, etc. The applications and early literature can be seen in [9]. Recently, Srinivasacharya et al. [10] studied the mixed convection flow along a vertical wavy surface with the effects of cross diffusion and variable properties in a porous medium. Most recently, Umar et al. [11] presented the effects of cross diffusions and chemical reaction flow in converging and diverging channels.

Most of the chemical reactions comprise both homogeneous and heterogeneous reactions. The homogeneous reaction takes place in bulk of the fluid, while heterogeneous reaction occurs on some catalytic surfaces. Generally, the interaction between the homogeneous and heterogeneous reactions are very complex and is involved in the production and consumption of reactant species at various rates, both on the catalytic surfaces and within the fluid. The applications and early literature has been reported by Kothandapani and Prakash [12]. Most recently, RamReddy and Pradeepa [13] used the spectral quasi-linearization method to study the effect of homogeneous–heterogeneous reactions on non-linear convection flow of micropolar fluid in a porous medium with convective boundary condition.

In this paper, the natural convection flow in a vertical channel saturated with Navier slip condition has been investigated in the presence of cross diffusions and the rate of chemical reaction. The survey clearly shows that the combined effects with slip flow condition through a vertical channel has not been presented elsewhere. In view of applications and importance, the authors are motivated to take this problem. Homotopy analysis method (HAM) [14–23] has been used for the solution of the present problem. Convergence of the resulting series solution is explained. Then the discussion with respect to the pertinent parameters of the present study on the solutions of velocity, temperature and concentration components.

## Mathematical modelling

We consider two-dimensional free convection fluid flow with Navier slip boundary in a vertical channel. Flow configuration is presented in Fig. 1. All the fluid properties are considered to be constant with the exception that the density in the buoyancy term of the balance of momentum equation. Since the plates are infinitely long, all physical variables depend on $y$ only. The governing two-dimensional steady flow equations can be put into the form:

$$v_y = 0 \Rightarrow v = v_0 = \text{constant} \quad (1)$$

$$v_0\rho\frac{\partial u}{\partial y} = \rho g(\beta_T(T - T_1) + \beta_C(C - C_1)) + \mu\frac{\partial^2 u}{\partial y^2} \quad (2)$$

$$\rho C_P v_0\frac{\partial T}{\partial y} = K_f\frac{\partial^2 T}{\partial y^2} + \mu\left(\frac{\partial u}{\partial y}\right)^2 + \frac{DK_T}{C_S C_P}\frac{\partial^2 C}{\partial y^2} \quad (3)$$

$$v_0\frac{\partial C}{\partial y} = D\frac{\partial^2 C}{\partial y^2} + \frac{DK_T}{T_m}\frac{\partial^2 T}{\partial y^2} - K_1(C - C_1) \quad (4)$$

with

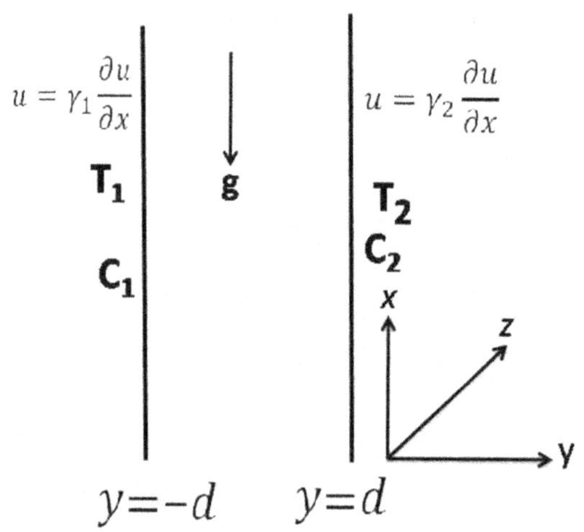


**Fig. 1** Physical model and coordinate system.

$$u = \gamma_1\frac{\partial u}{\partial y}, \quad T = T_1, \quad C = C_1 \quad \text{at} \quad y = -d \quad (5a)$$

$$u = \gamma_2\frac{\partial u}{\partial y}, \quad T = T_2, \quad C = C_2 \quad \text{at} \quad y = d \quad (5b)$$

where the velocity components in $x$- and $y$- are $u$, $v$, respectively, $g$ is the acceleration due to gravity, $\rho$ is the density, $C_p$ is the specific heat, $\mu$ is the coefficient of viscosity, $\beta_T$ is the coefficient of thermal expansion, $\beta_C$ is the coefficient of solutal expansion, $\alpha$ is the thermal diffusivity, $D$ is the mass diffusivity, $K_T$ is the coefficient of thermal conductivity, $C_S$ is the concentration susceptibility, $T_m$ is the mean fluid temperature, $K_f$ is the thermal diffusion ratio, $K_1$ is the rate of chemical reaction, $\gamma_1$ and $\gamma_2$ are the slip coefficients at both the plates.

Introducing the following dimensionless variables

$$\eta = \frac{y}{d}, u = \frac{vGr}{d}f, \theta = \frac{T - T_1}{T_2 - T_1}, \phi = \frac{C - C_1}{C_2 - C_1} \quad (6)$$

in Eqs. (2)–(4), we obtain the governing dimensionless equations as

$$f'' + \theta + N\phi - Ref' = 0 \quad (7)$$

$$\theta'' - RePr\theta' + BrGr^2(f')^2 + PrD_f\phi'' = 0 \quad (8)$$

$$\phi'' - ReSc\phi' + ScSr\theta'' - KSc\phi = 0 \quad (9)$$

with

$$\begin{aligned} &\eta = -1 : f - \beta_1 f' = \theta = \phi = 0 \\ &\eta = 1 : f - \beta_2 f' = 0 \quad \text{and} \quad \theta = \phi = 1 \end{aligned} \quad (10)$$

where the primes represents differentiation with respect to $\eta$, $N = \frac{\beta_C(C_2 - C_1)}{\beta_T(T_2 - T_1)}$ is the buoyancy parameter, $Re = \frac{\rho v_0 d}{\mu}$ is the Reynolds number, $Sc = \frac{v}{D}$ is the Schmidth number, $Gr = \frac{g_a\beta_T(T_2 - T_1)d^3}{v^2}$ is the Grashof number, $Pr = \frac{\mu C_P}{K_f}$ is the Prandtl number, $Br = \frac{\mu v^2}{K_f d^2(T_2 - T_1)}$ is the Brinkman number, $K = \frac{K_1 d^2}{v}$ is the chemical reaction parameter, $S_r = \frac{DK_T(T_2 - T_1)}{vT_m(C_2 - C_1)}$ is the thermo-diffusion parameter and $D_f = \frac{DK_T(C_2 - C_1)}{vC_S C_P(T_2 - T_1)}$ is the Dufour number. $\beta_1 = \frac{\gamma_1}{d}$, $\beta_2 = \frac{\gamma_2}{d}$ are the slip parameters.

The skin friction coefficients($C_f$), heat ($Nu$) and mass ($Sh$) fluxes at the vertical walls can be given by

$$\begin{aligned} &\frac{Re^2}{Gr}C_{f1} = 2f'(-1), \frac{Re^2}{Gr}C_{f2} = 2f'(1) \\ &Nu_{1,2} = -\theta'(\eta)|_{\eta=-1,1}, Sh_{1,2} = -\phi'(\eta)|_{\eta=-1,1} \end{aligned} \quad (11)$$

where

$$C_f = \frac{\tau_w}{\rho v_0^2},\ Nu = \frac{q_w d}{K_f(T_2 - T_1)},\ Sh = \frac{q_m d}{D(C_2 - C_1)} \tag{12}$$

and

$$\tau_w = \mu \frac{\partial u}{\partial y}\bigg|_{y=\pm d};\ q_w = -K_f \frac{\partial T}{\partial y}\bigg]_{y=\pm d} \\ q_m = -D\frac{\partial C}{\partial y}\bigg]_{y=\pm d} \tag{13}$$

Effect of the various parameters involved in the investigation on physical coefficients are discussed in the following section.

## Homotopy solution

For HAM solutions, we choose the initial approximations of $f(\eta)$, $\theta(\eta)$ and $\phi(\eta)$ as follows:

$$f_0(\eta) = 0, \quad \theta_0(\eta) = \frac{1+\eta}{2}, \quad \phi_0(\eta) = \frac{1+\eta}{2} \tag{14}$$

with the auxiliary linear operator

$$L = \frac{\partial^2}{\partial \eta^2} \quad \text{such that} \quad L_1(c_1\eta + c_2) = 0 \tag{15}$$

in which $c_1$ and $c_2$ are the arbitrary constants. $h_1$, $h_2$ and $h_3$ (the convergence control parameters) are introduced in zeroth-order deformations as

$$(1-p)L[f(\eta;p) - f_0(\eta)] = ph_1N_1[f(\eta;p)] \tag{16}$$

$$(1-p)L[\theta(\eta;p) - \theta_0(\eta)] = ph_2N_2[\theta(\eta;p)] \tag{17}$$

$$(1-p)L[\phi(\eta;p) - \phi_0(\eta)] = ph_3N_3[\phi(\eta;p)] \tag{18}$$

subject to the boundary conditions

$$f(-1;p) - \beta_1 f'(-1;p) = 0, f(1;p) - \beta_2 f'(1;p) = 0 \\ \theta(-1;p) = 0, \theta(1;p) = 1, \phi(-1;p) = 0, \phi(1;p) = 1 \tag{19}$$

where $p \in [0,1]$ is the embedding parameter and the nonlinear operators $N_1$, $N_2$ and $N_3$ are defined as:

$$N_1[f(\eta,p),\theta(\eta,p),\phi(\eta,p)] = f''(\eta,p) + \theta(\eta,p) \\ + N\phi(\eta,p) - Ref'(\eta,p) \tag{20}$$

$$N_2[f(\eta,p),\theta(\eta,p),\phi(\eta,p)] = \theta''(\eta,p) - RePr\theta'(\eta,p) \\ + BrGr^2(f'(\eta,p))^2 \\ + PrD_f\phi''(\eta,p) \tag{21}$$

$$N_3[f(\eta,p),\theta(\eta,p),\phi(\eta,p)] = \phi''(\eta,p) - ReSc\phi'(\eta,p) \\ + ScSr\theta''(\eta,p) - KSc\phi(\eta,p) \tag{22}$$

For $p = 0$, we have the initial guess approximations

$$f(\eta;0) = f_0(\eta),\ \theta(\eta;0) = \theta_0(\eta),\ \phi(\eta;0) = \phi_0(\eta) \tag{23}$$

When $p = 1$, Eqs. (16)–(18) are same as (7)–(9), respectively, therefore, at $p = 1$, we get the final solutions

$$f(\eta;1) = f(\eta),\ \theta(\eta;1) = \theta(\eta),\ \phi(\eta;1) = \phi(\eta) \tag{24}$$

Hence, the process of giving an increment to $p$ from 0 to 1 is the process of $f(\eta;p)$ varying continuously from the initial guess $f_0(\eta)$ to the final solution $f(\eta)$ (similar for $\theta(\eta,p)$ and $\phi(\eta,p)$). This kind of continuous variation is called deformation in topology so that we call system of Eqs. (16)–(19), the zeroth-order deformation equation. Next, the *mth*-order deformation equations follow as

$$L_1[f_m(\eta) - \chi_m f_{m-1}(\eta)] = h_1 R_m^f(\eta) \tag{25}$$

$$L[\theta_m(\eta) - \chi_m \theta_{m-1}(\eta)] = h_2 R_m^\theta(\eta) \tag{26}$$

$$L[\phi_m(\eta) - \chi_m \phi_{m-1}(\eta)] = h_3 R_m^\phi(\eta) \tag{27}$$

with the boundary conditions

$$f_m(-1) = 0, \quad f_m(1) = 0, \quad \theta_m(-1) = 0 \\ \theta_m(1) = 0, \quad \phi_m(-1) = 0, \quad \phi_m(1) = 0, \tag{28}$$

where

$$R_m^f(\eta) = f'' + \theta + N\phi - Ref' \tag{29}$$

$$R_m^\theta(\eta) = \theta'' - RePr\theta' + BrGr^2 \sum_{n=0}^{m-1} f'_{m-1-n} f'_n + PrD_f\phi'' \tag{30}$$

$$R_m^\theta(\eta) = \phi'' - ReSc\phi' + ScSr\theta'' - KSc\phi \tag{31}$$

for $m$ being integer

$$\chi_m = 0 \quad \text{for} \quad m \leq 1 \\ = 1 \quad \text{for} \quad m > 1 \tag{32}$$

The initial guess approximations $f_0(\eta)$, $\theta_0(\eta)$ and $\phi_0(\eta)$, the linear operators $L$ and the auxiliary parameters $h_1, h_2$ and $h_3$ are assumed to be selected such that Eqs. (16)–(19) have solution at each point $p \in [0,1]$ and also with the help of Taylor's series and due to Eq. (23); $f(\eta;p)$, $\theta(\eta;p)$ and $\phi(\eta;p)$ can be expressed as

$$f(\eta;p) = f_0(\eta) + \sum_{m=1}^{\infty} f_m(\eta) p^m \tag{33}$$

$$\theta(\eta;p) = \theta_0(\eta) + \sum_{m=1}^{\infty} \theta_m(\eta)p^m \quad (34)$$

$$\phi(\eta;p) = \phi_0(\eta) + \sum_{m=1}^{\infty} \phi_m(\eta)p^m \quad (35)$$

in which $h_1$, $h_2$ and $h_3$ are chosen in such a way that the series (33)–(35) are convergent [15] at $p = 1$. Therefore, we have from (24) that

$$f(\eta) = f_0(\eta) + \sum_{m=1}^{\infty} f_m(\eta) \quad (36)$$

$$\theta(\eta) = \theta_0(\eta) + \sum_{m=1}^{\infty} \theta_m(\eta) \quad (37)$$

$$\phi(\eta) = \phi_0(\eta) + \sum_{m=1}^{\infty} \phi_m(\eta) \quad (38)$$

for which we presume that the initial guesses to $f$, $\theta$ and $\phi$, the auxiliary linear operators $L$ and the non-zero auxiliary parameters $h_1$, $h_2$ and $h_3$ are so properly selected that the deformation $f(\eta,p)$, $\theta(\eta,p)$ and $\phi(\eta,p)$ are smooth enough and their *mth*-order derivatives with respect to $p$ in Eqs. (36)–(38) exist and are given, respectively, by $f_m(\eta) = \frac{1}{m!}\frac{\partial^m f(\eta;p)}{\partial p^m}\Big|_{p=0}$, $\theta_m(\eta) = \frac{1}{m!}\frac{\partial^m \theta(\eta;p)}{\partial p^m}\Big|_{p=0}$ and $\phi_m(\eta) = \frac{1}{m!}\frac{\partial^m \phi(\eta;p)}{\partial p^m}\Big|_{p=0}$. It is clear that the convergence of Taylor series at $p = 1$ is a prior assumption, whose justification is provided via a theorem [25], so that the system in (36)–(38) holds true. The formulae in (36)–(38) provide us with a direct relationship between the initial guesses and the exact solutions. All the effects of interaction of the magnetic field as well as of the heat transfer, Hall and Ion effects and couple stress flow field can be studied from the exact formulas (36)–(38). Moreover, a special emphasize should be placed here that the *mth*-order deformation system (25)–(28) is a linear differential equation system with the auxiliary linear operators $L$ whose fundamental solution is known.

## Convergence of the HAM solution

The expressions for $f$, $\theta$ and $\phi$ contain the auxiliary parameters $h_1, h_2$ and $h_3$. As pointed out by Liao [14], the convergence and the rate of approximation for the HAM solution strongly depend on the values of auxiliary parameter $h$. For this purpose, $h$-curves are plotted to choose $h_1, h_2$ and $h_3$ in such a manner that the solutions (33)–(35) ensure convergence [14]. Here to see the admissible values of $h_1, h_2$ and $h_3$, the $h$-curves are plotted for 15th-order of approximation in Figs. 2, 3 and 4 by

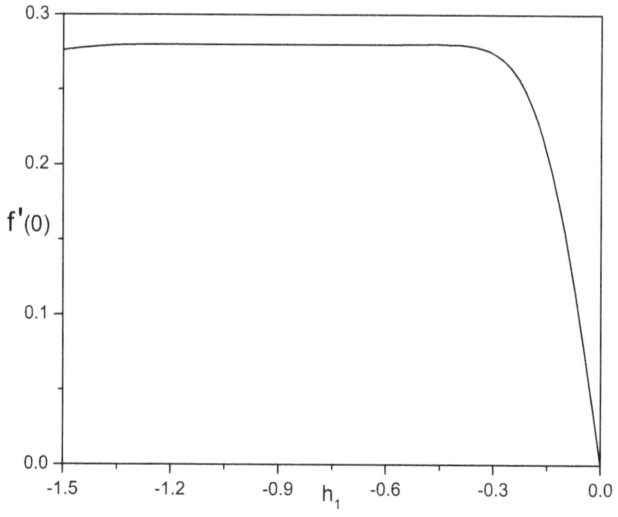


**Fig. 2** The $h$ curve of $f(\eta)$

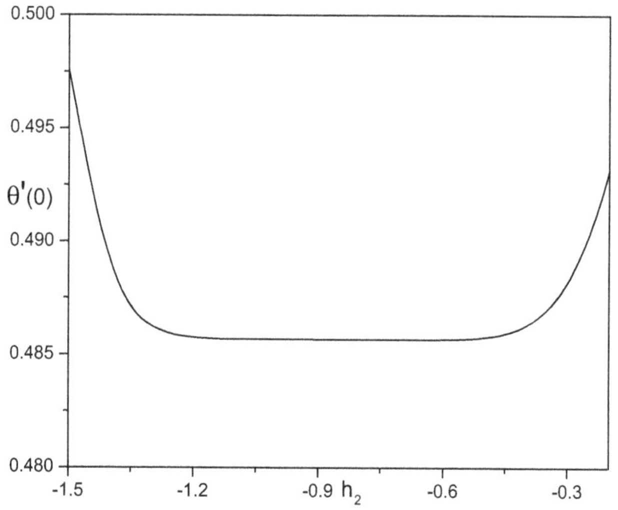


**Fig. 3** The $h$ curve of $\theta(\eta)$

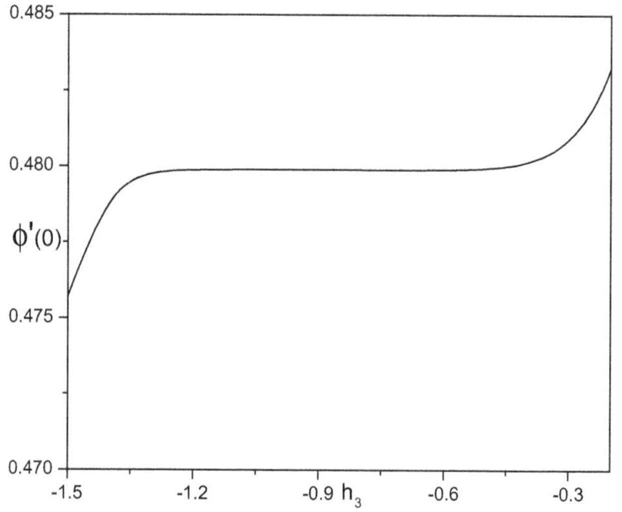


**Fig. 4** The $h$ curve of $\phi(\eta)$

taking the values of the parameters $N = 2, Re = 2, Pr = 0.71, Sc = 0.22, Br = 0.5, Gr = 0.5, Sr = 0.5, Df = 0.5, K = 1.0$. It is clearly noted from Fig. 2 that the range for the admissible values of $h_1$ is $-1.25 < h_1 < -0.6$. From Fig. 3, it can be seen that the $h$-curve has a parallel line segment that corresponds to a region $-1.15 < h_2 < -0.65$. Figure 4 depicts that the admissible value of $h_3$ are $-1.2 < h_3 < -0.6$. A wide valid zone is evident in these figures ensuring convergence of the series. To choose optimal value of auxiliary parameters, the average residual errors (see Ref. [15] for more details) are defined as

$$E_{f,m} = \frac{1}{2K}\sum_{i=-K}^{K}\left(N_1\left[\sum_{j=0}^{m} f_j(i\Delta t)\right]\right)^2 \quad (39)$$

$$E_{\theta,m} = \frac{1}{2K}\sum_{i=-K}^{K}\left(N_2\left[\sum_{j=0}^{m} \theta_j(i\Delta t)\right]\right)^2 \quad (40)$$

$$E_{\phi,m} = \frac{1}{2K}\sum_{i=-K}^{K}\left(N_3\left[\sum_{j=0}^{m} \phi_j(i\Delta t)\right]\right)^2 \quad (41)$$

where $\Delta t = 1/K$ and $K = 5$. At different order of approximations ($m$), minimum of average residual errors are shown in Tables 1, 2 and 3. It is clear from Table 1 that the average residual error for $f$ is minimum at $h_1 = -0.93$. It can be seen from Table 2 that the minimum of average residual error for $\theta$ attains at $h_2 = -0.93$. Table 3 depicts that at $h_3 = -0.93$, $E_\phi$ attains minimum. Therefore, the optimum values of convergence control parameters are taken as $h_1 = -0.93$, $h_2 = -0.93$, $h_3 = -0.93$.

To see the accuracy of the solutions, the residual errors are defined for the system as

$$\mathrm{RE}_f = f_n''(\eta) + \theta_n(\eta) + N\phi_n(\eta) - Ref_n'(\eta) \quad (42)$$

**Table 1** Optimal value of $h_1$ at different order of approximations

| Order | Optimal of $h_1$ | Minimum of $E_m$ |
|---|---|---|
| 10 | −0.93 | $1.30 \times 10^{-10}$ |
| 15 | −0.93 | $5.45 \times 10^{-16}$ |
| 20 | −0.93 | $1.36 \times 10^{-18}$ |

**Table 2** Optimal value of $h_2$ at different order of approximations

| Order | Optimal of $h_2$ | Minimum of $E_m$ |
|---|---|---|
| 10 | −0.94 | $2.01 \times 10^{-10}$ |
| 15 | −0.93 | $4.33 \times 10^{-14}$ |
| 20 | −0.93 | $1.44 \times 10^{-16}$ |

**Table 3** Optimal value of $h_3$ at different order of approximations

| Order | Optimal of $h_3$ | Minimum of $E_m$ |
|---|---|---|
| 10 | −0.93 | $1.50 \times 10^{-11}$ |
| 15 | −0.93 | $5.63 \times 10^{-15}$ |
| 20 | −0.93 | $2.10 \times 10^{-17}$ |

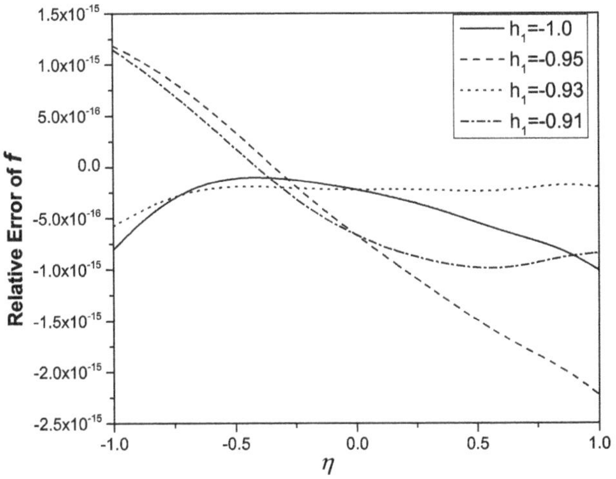


**Fig. 5** Residual error of $f(\eta)$ when $\beta_h = 2, \beta_i = 2, \alpha = 0.5, Ha = 5$

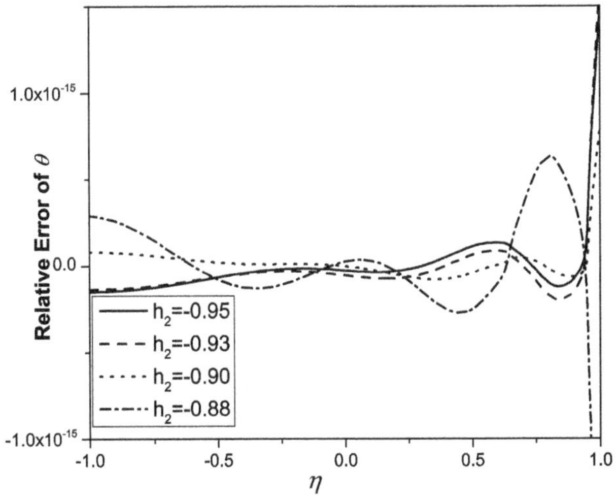


**Fig. 6** Residual error of $\theta(\eta)$ when $\beta_h = 2, \beta_i = 2, \alpha = 0.5, Ha = 5$

$$\mathrm{RE}_\theta = \theta_n''(\eta) - RePr\theta_n'(\eta) + BrGr^2(f_n'(\eta))^2 + PrD_f\phi_n''(\eta) \quad (43)$$

$$\mathrm{RE}_\theta = \phi_n''(\eta) - ReSc\phi_n'(\eta) + ScSr\theta_n''(\eta) - KSc\phi_n(\eta) \quad (44)$$

where $f_n(\eta)$, $\theta_n(\eta)$ and $\phi_n(\eta)$ are the HAM solutions for $f(\eta)$, $\theta(\eta)$ and $\phi(\eta)$, respectively. For optimality of the

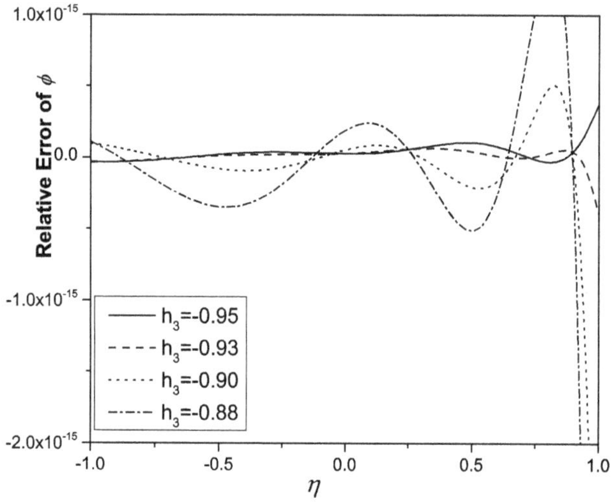


**Fig. 7** Residual error of $\phi(\eta)$ when $\beta_h = 2, \beta_i = 2, \alpha = 0.5, Ha = 5$

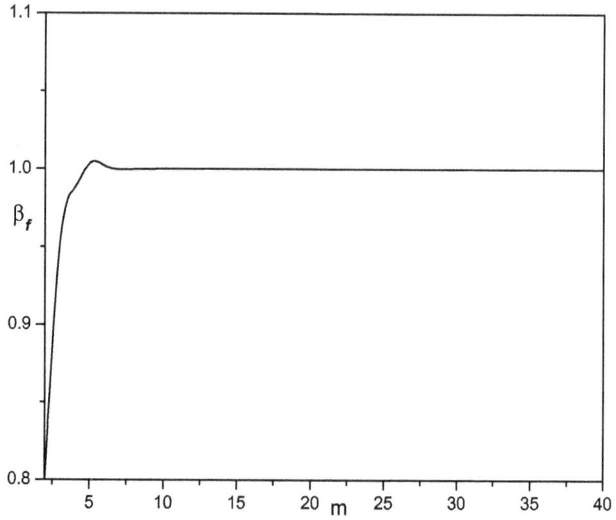


**Fig. 8** The ratio $\beta_f$ from the theorem to reveal the convergence of the HAM solutions

convergence control parameters, residual error [24] for different values of $h$ in the convergence region displayed in Figs. 5, 6 and 7. We examine that $h_1 = -0.93$, $h_2 = -0.93$, $h_3 = -0.93$ gives a better solution. Table 4 establishes the convergence of the obtained series solution. It is found from the above observations that the series given by (33)–(35) converge in the whole region of $\eta$ when $h_1 = -0.93$, $h_2 = -0.93$, $h_3 = -0.93$

To pursue the convergence of the HAM solutions to the exact ones, the graphs for the ratio (following the recent work of [25])

$$\beta_f = \left|\frac{f_m(h)}{f_{m-1}(h)}\right|, \beta_\theta = \left|\frac{\theta_m(h)}{\theta_{m-1}(h)}\right|, \beta_\phi = \left|\frac{\phi_m(h)}{\phi_{m-1}(h)}\right| \tag{45}$$

against the number of terms $m$ in the homotopy series is presented in Figs. 8, 9 and 10. Figures strongly indicate that a finite limit of $\beta$ will be attained in the limit of $m \to \infty$, which will remain less than or equals to unity. The velocity and temperature solutions seem to converge in an oscillatory manner requiring more terms in the homotopy series. Thus, the convergence to the exact solution is assured by the HAM method.

## Discussion of results

In the absence of Brinkman ($Br = 0$) number, the system of Eqs. (7)–(9) reduces to the linear system of equations for which exact solution exist. Comparison between analytical solution and HAM solution is shown in Table 5. The comparisons are found to be in a very good agreement. Therefore, the HAM code can be used with great confidence to study the problem considered in this paper.

The profiles of velocity ($f(\eta)$), temperature ($\theta(\eta)$) and concentration ($\phi(\eta)$) are computed and presented through plots in Figs. 11, 12, 13, 14, 15, 16, 17, 18, 19, 20 and 21 with different values of $\beta_1, \beta_2, Sr, D_f, K$. Computations were carried out by fixing the parameters $Pr = 0.71, N = 2, Br = 0.5, Da = 0.5, Re = 2, Sc = 0.22, Gr = 0.5$ to analyze the effects of the emerging parameters $\beta_1$, $\beta_2$, $Sr$, $D_f$ and $K$.

Figure 11 presents the influence of the slip flow parameter $\beta_1$ on $f(\eta)$. It is clear that increase in the parameter $\beta_1$ leads to increase the flow velocity $f(\eta)$ at the injection wall and there is slight reverse trend can be found at the terminal wall. Figure 12 shows that the effect of $\beta_2$

**Table 4** Convergence of HAM solutions for different order of approximations

| Order | $f(0)$ | $\theta(0)$ | $\phi(0)$ |
|---|---|---|---|
| 4 | 0.4910169474433473893 | 0.387920461619234213657136 | 0.428633469889216020427361 |
| 8 | 0.4947894544255234236 | 0.378481745766847297706702 | 0.424271364200980527767338 |
| 12 | 0.4948135285227568114 | 0.378404924235389153211848 | 0.424337509824766393328840 |
| 16 | 0.4948137463089264307 | 0.378403051384050246849151 | 0.424339108852924445123161 |
| 20 | 0.4948137528096351068 | 0.378403073998202584356433 | 0.424339130178531704167978 |
| 24 | 0.4948137530282150863 | 0.378403078729212125985981 | 0.424339129132763810471734 |
| 28 | 0.4948137530306036798 | 0.378403079015738519907694 | 0.424339129022152340152007 |
| 30 | 0.4948137530306036798 | 0.378403079015738519907694 | 0.424339129022152340152007 |

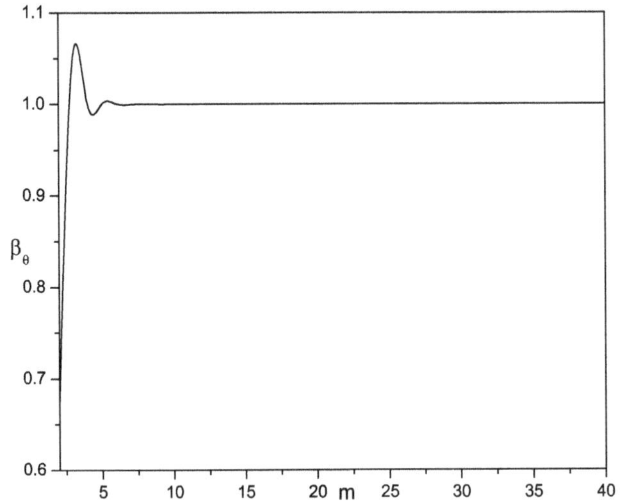


**Fig. 9** The ratio $\beta_\theta$ from the theorem to reveal the convergence of the HAM solutions

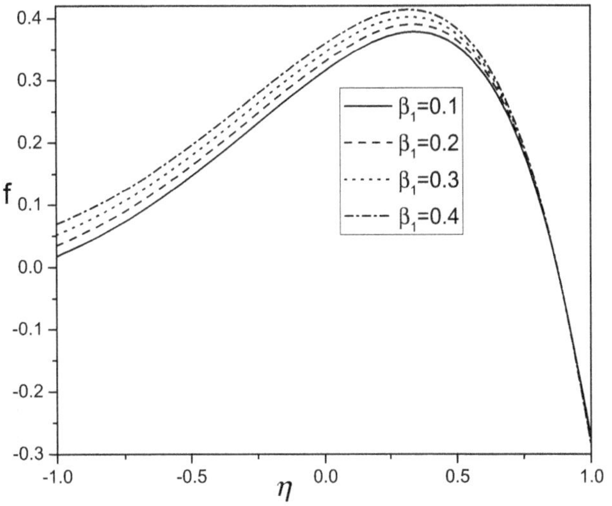


**Fig. 11** Variation $f(\eta)$ with $\beta_1$ when $\beta_2 = 0.1, Sr = 0.5, D_f = 0.5, K = 1.0$

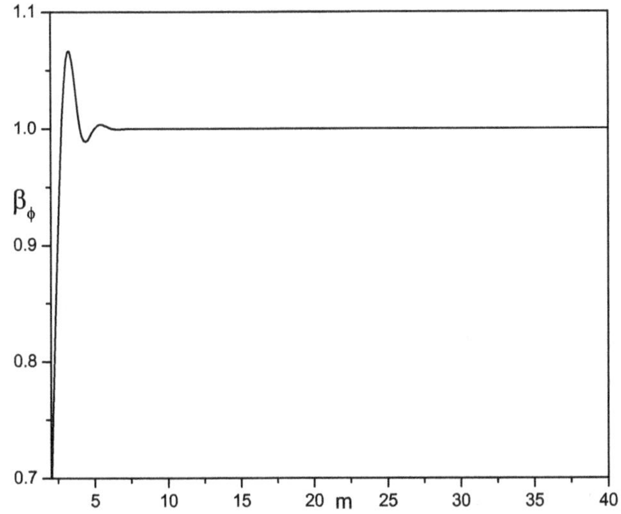


**Fig. 10** The ratio $\beta_\phi$ from the theorem to reveal the convergence of the HAM solutions

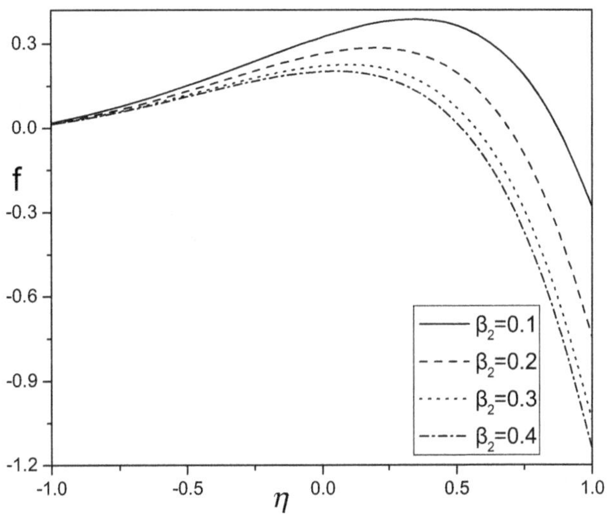


**Fig. 12** Variation $f(\eta)$ with $\beta_2$ when $\beta_1 = 0.1, Sr = 0.5, D_f = 0.5, K = 1.0$

on the flow velocity $f(\eta)$. It can be seen from this figure that as $\beta_2$ increases the flow velocity decreases negotiably at the initial wall but appreciable decrease in the velocity is noticed at the suction wall.

The effect of Soret parameter on velocity, temperature and concentration profiles are presented in Figs. 13, 14 and 15. It is clear from Fig. 13 that the higher values of Soret

**Table 5** Comparison of exact solution ($Br = 0$) with HAM solution

| $\eta$ | $f$ | | $\theta$ | | $\phi$ | |
|---|---|---|---|---|---|---|
| | Analytical | HAM | Analytical | HAM | Analytical | HAM |
| −1 | 0.026479 | 0.026479 | 0 | 0 | 0 | 0 |
| −0.6 | 0.148563 | 0.148563 | 0.112667 | 0.112667 | 0.173056 | 0.173056 |
| −0.2 | 0.275758 | 0.275758 | 0.257042 | 0.257042 | 0.359029 | 0.359029 |
| 0.2 | 0.345735 | 0.345735 | 0.443246 | 0.443246 | 0.558629 | 0.558629 |
| 0.6 | 0.254633 | 0.254633 | 0.684836 | 0.684836 | 0.772288 | 0.772288 |
| 1 | −0.16612 | −0.16612 | 1 | 1 | 1 | 1 |

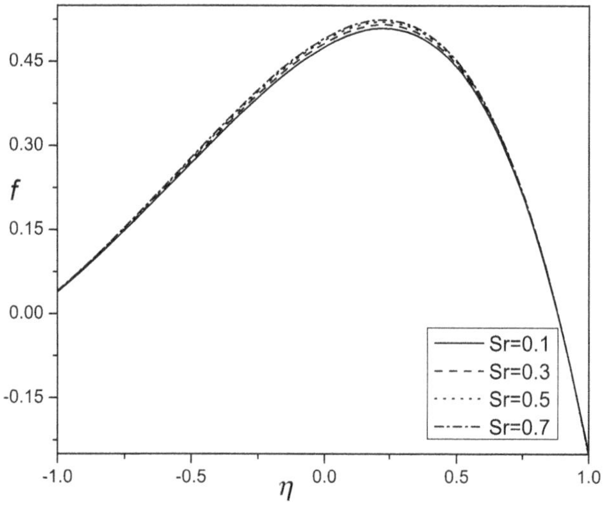


**Fig. 13** Soret effect on $f(\eta)$ when $\beta_1 = 0.1, \beta_2 = 0.1, D_f = 0.5, K = 1.0$

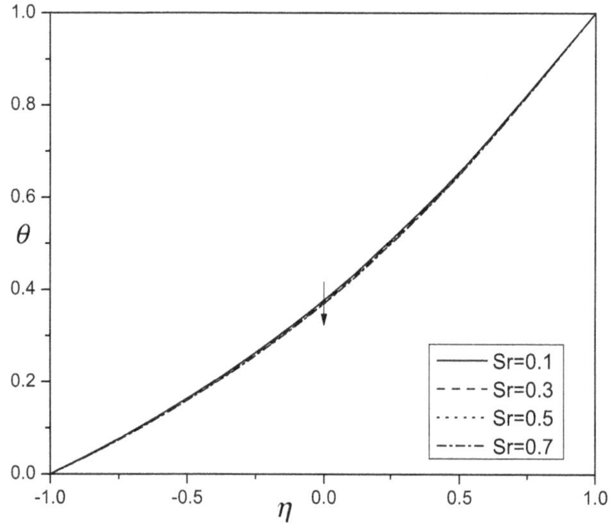


**Fig. 14** Soret effect on $\theta(\eta)$ when $\beta_1 = 0.1, \beta_2 = 0.1, D_f = 0.5, K = 1.0$

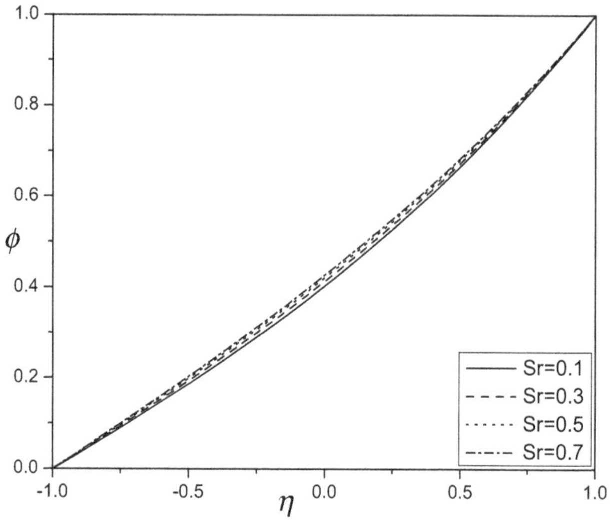


**Fig. 15** Soret effect on $\phi(\eta)$ when $\beta_1 = 0.1, \beta_2 = 0.1, D_f = 0.5, K = 1.0$

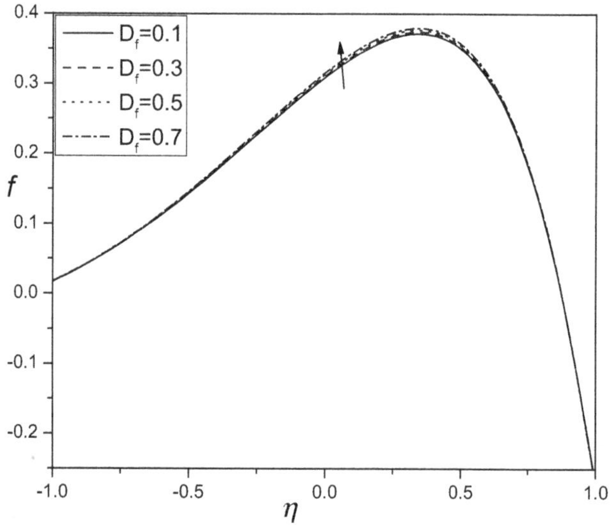


**Fig. 16** Dufour effect on $f(\eta)$ when $\beta_1 = 0.1, \beta_2 = 0.1, Sr = 0.5, K = 1.0$

number $Sr$ increases the velocity of the flow. Since the values of $S_r$ increases due to either an decrease in the temperature difference or increase in the concentration difference, with this lowest peak of the flow velocity compatible with the lowest Soret number. Figure 14 shows the influence of $Sr$ on $\theta$. It is observed that the temperature decreases as $Sr$ increases. The nature of $\phi$ with the influence of $Sr$ is shown in Fig. 15. It is noted that the dimensionless concentration increases as an increase in the Soret number. The results clearly exhibits that the flow field is significantly influenced by the Soret number.

The effect of Dufour number ($D_f$) on dimensionless velocity can be seen in Fig. 16. It is clear that the velocity of the fluid increases with an increase in diffusion thermo parameter. Since the values of $D_f$ increases due to either an increase in the temperature difference or decrease in the concentration difference, i.e., the lowest peak of the flow velocity compatible with the lowest Dufour number. The effect of diffusion thermo parameter on dimensionless temperature is presented in Fig. 17. It is noted that the temperature profile increases with an increase in Dufour number. Figure 18 demonstrates that the concentration of the fluid decreases as Dufour effect increases. This is because of temperature gradients contribution to diffusion of the species.

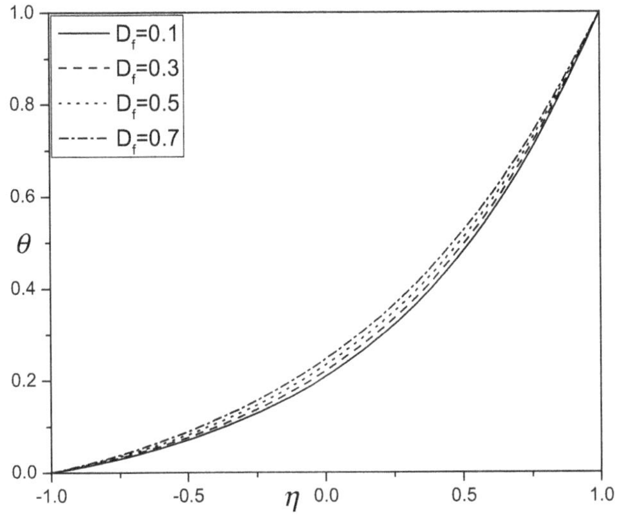


**Fig. 17** Dufour effect on $\theta(\eta)$ when $\beta_1 = 0.1, \beta_2 = 0.1, Sr = 0.5, K = 1.0$

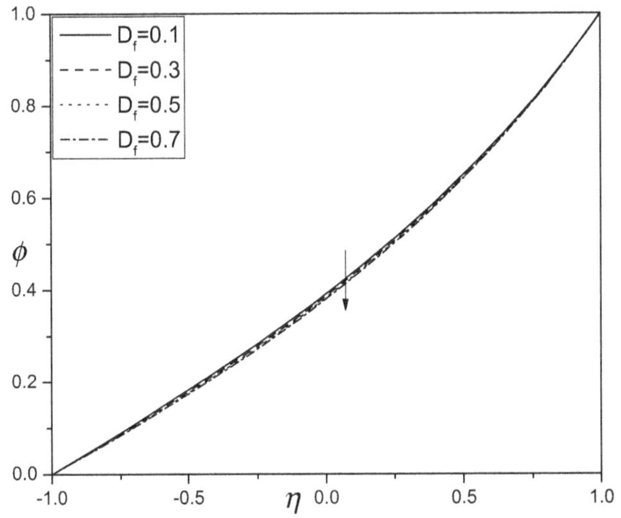


**Fig. 18** Dufour effect on $\phi(\eta)$ when $\beta_1 = 0.1, \beta_2 = 0.1, Sr = 0.5, K = 1.0$

The chemical reaction parameter $K$ effect on velocity, temperature and concentration profiles are shown in Figs. 19, 20 and 21. It is observed from these figures that the fluid flow velocity decreases as $K$ increases. The temperature increases and concentration of the fluid decreases when the chemical reaction parameter increases. Since the chemical molecular diffusivity drop down when chemical reaction is high, i.e., lesser diffusion. Hence, they are procured by the transfer of species. The higher $K$ will decrease the concentration species. Therefore, with the increase in $K$ suppresses the distribution of the concentration at all the points of the fluid flow. With this, it can be

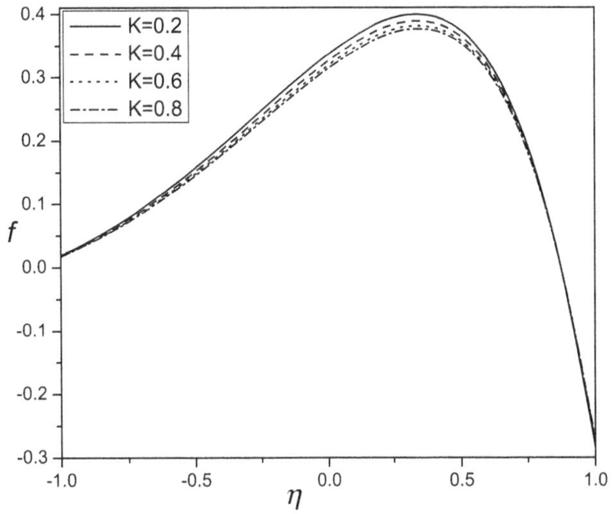


**Fig. 19** Chemical reaction effect on $f(\eta)$ when $\beta_1 = 0.1, \beta_2 = 0.1, Sr = 0.5, D_f = 0.5$

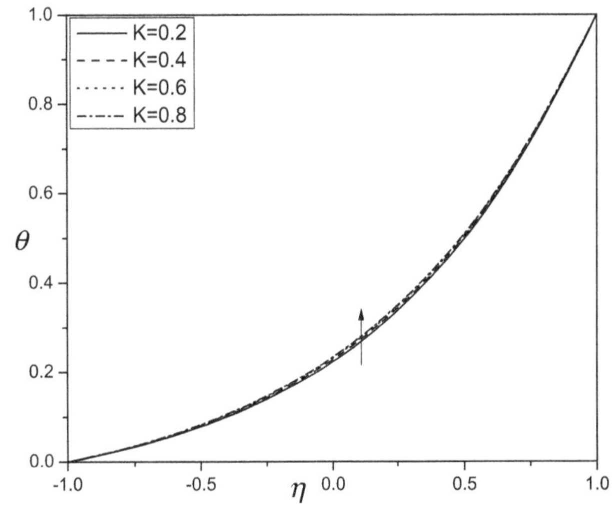


**Fig. 20** Chemical reaction effect on $\theta(\eta)$ when $\beta_1 = 0.1, \beta_2 = 0.1, Sr = 0.5, D_f = 0.5$

claimed that on the distribution of the concentration, massive effect is higher with heavier diffusing species.

Variation of slip parameters ($\beta_1$ and $\beta_2$), thermal diffusion parameter ($Sr$), diffusion thermo parameter ($D_f$), together with the chemical reaction parameter $K$ are shown in Table 6 by fixing the remaining parameters. It is seen that the friction factor decreases at both the walls with an increase in the slip parameters $\beta_1$ and $\beta_2$. It is clear that increase in $\beta_1$ leads to increase in heat transfer coefficient at the plates where as the mass transfer rate increases at the initial wall and decreases at the terminal wall. It can be observed that at the initial plate the heat transfer rate decreases and at the second plate it is found to be increasing while the reverse trend is perceived on mass transfer rate with an increase in $\beta_2$. It is observed from this table that the skin friction coefficient, heat transfer rate increases at the injection wall and decreases at the terminal wall with the increase of Soret parameter ($Sr$), where as mass transfer rate have reverse trend at both the plates. It is noticed that higher values of Dufour parameter ($D_f$) increases the friction factor and mass transfer rate at the initial plate and decreases at the suction wall, while heat transfer rate shows the reverse trend at two walls. The effect of chemical reaction parameter $K$ on skin friction, heat and mass transfer rates are presented in Table 6. It is

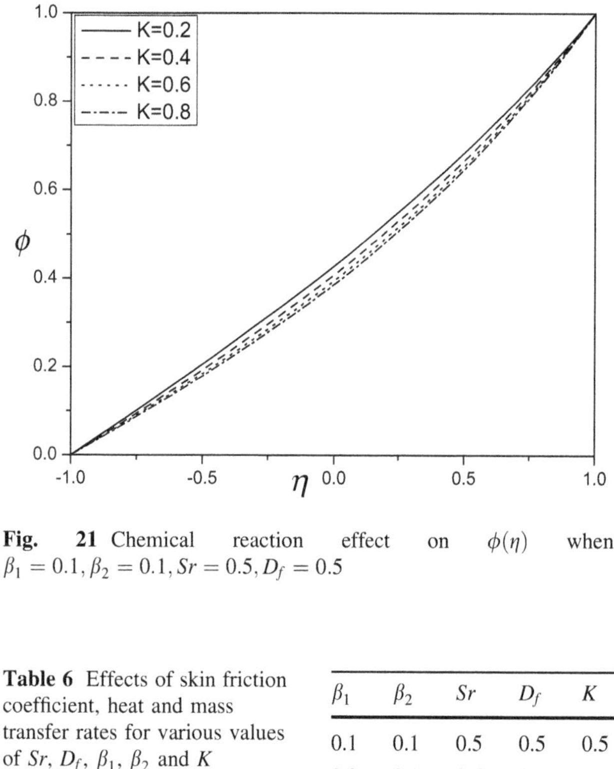


**Fig. 21** Chemical reaction effect on $\phi(\eta)$ when $\beta_1 = 0.1, \beta_2 = 0.1, Sr = 0.5, D_f = 0.5$

seen from the table that all the physical parameters are decreasing at the injection wall and those are increasing at the suction wall with the enhance in $K$. It is also seen from this table that the higher values of $K$ increases the mass transfer rate at $\eta = -1$ and decreases at $\eta = 1$. The influence of the emerging parameters are patently obvious from the Table 6, and hence are not discussed for conciseness.

## Conclusions

Two-dimensional natural convection flow in a vertical channel in the presence of chemical reaction and the cross diffusion effects with Navier slip. Homotopy analysis method has been employed to solve the final system of equations. The markable findings are summarized as:

- Flow velocity and the rate of mass transfer increases at the injection wall, decreases slightly at the suction wall and the rate of heat transfer increases at both the walls with an enhance in slip parameter $\beta_1$.
- As $\beta_2$ increases, the fluid flow velocity and friction factors decrease (slightly at the injection wall and marginally at the suction wall). Where as heat transfer rate disintegrates at the injection wall and increases at the end wall, while the opposite nature is noticed on the rate of mass transfer with an enhance in $\beta_2$.
- It is noticed that the velocity and concentration of the fluid increases and temperature profile decreases with an increase in Soret parameter. In addition, it is identified that the friction factor, Nusselt number increases at the injection plate and decreases at the end wall as $Sr$ increases, where as mass transfer rate shows the reverse trend at both the walls.

**Table 6** Effects of skin friction coefficient, heat and mass transfer rates for various values of $Sr$, $D_f$, $\beta_1$, $\beta_2$ and $K$

| $\beta_1$ | $\beta_2$ | $Sr$ | $D_f$ | $K$ | $f'(-1)$ | $f'(1)$ | $Nu_1$ | $Nu_1$ | $Sh_1$ | $Sh_2$ |
|---|---|---|---|---|---|---|---|---|---|---|
| 0.1 | 0.1 | 0.5 | 0.5 | 0.5 | 0.1829 | −2.7919 | −0.1154 | −1.2373 | −0.3733 | −0.7861 |
| 0.2 | 0.1 | 0.5 | 0.5 | 0.5 | 0.1820 | −2.8385 | −0.1153 | −1.2317 | −0.3732 | −0.7866 |
| 0.3 | 0.1 | 0.5 | 0.5 | 0.5 | 0.1811 | −2.8846 | −0.1152 | −1.2261 | −0.3731 | −0.7871 |
| 0.1 | 0.1 | 0.5 | 0.5 | 0.5 | 0.1829 | −2.7919 | −0.1154 | −1.2373 | −0.3733 | −0.7861 |
| 0.1 | 0.2 | 0.5 | 0.5 | 0.5 | 0.1651 | −3.7536 | −0.1162 | −1.0957 | −0.3709 | −0.7990 |
| 0.1 | 0.3 | 0.5 | 0.5 | 0.5 | 0.1287 | −5.7585 | −0.1252 | −0.6181 | −0.3600 | −0.8402 |
| 0.1 | 0.1 | 0.2 | 0.5 | 0.5 | 0.1762 | −2.7462 | −0.1191 | −1.2214 | −0.3476 | −0.8326 |
| 0.1 | 0.1 | 0.4 | 0.5 | 0.5 | 0.1806 | −2.7765 | −0.1166 | −1.2319 | −0.3646 | −0.8018 |
| 0.1 | 0.1 | 0.6 | 0.5 | 0.5 | 0.1852 | −2.8076 | −0.1140 | −1.2427 | −0.3819 | −0.7701 |
| 0.1 | 0.1 | 0.5 | 0.2 | 0.5 | 0.1817 | −2.7756 | −0.1028 | −1.3161 | −0.3792 | −0.7824 |
| 0.1 | 0.1 | 0.5 | 0.4 | 0.5 | 0.1825 | −2.7863 | −0.1110 | −1.2639 | −0.3753 | −0.7849 |
| 0.1 | 0.1 | 0.5 | 0.6 | 0.5 | 0.1833 | −2.7977 | −0.1198 | −1.2104 | −0.3711 | −0.7873 |
| 0.1 | 0.1 | 0.5 | 0.5 | 0.2 | 0.1882 | −2.8287 | −0.1127 | −1.2507 | −0.3924 | −0.7466 |
| 0.1 | 0.1 | 0.5 | 0.5 | 0.4 | 0.1846 | −2.8040 | −0.1145 | −1.2417 | −0.3795 | −0.7731 |
| 0.1 | 0.1 | 0.5 | 0.5 | 0.6 | 0.1812 | −2.7801 | −0.1161 | −1.2329 | −0.3671 | −0.7989 |

- Fluid flow velocity, $\theta$ increases and the dimensionless concentration profile decreases as $D_f$ increases. It is seen that as Dufour parameter increases, the skin friction and mass transfer rate at $\eta = -1$ and decreases at $\eta = 1$ wall, while the reverse trend is noticed in case of heat transfer rate at both the walls.
- As reaction parameter increases, the velocity and concentration of the fluid decreases and the temperature profile decreases. It is also observed that the values of $f'(0)$, Nusselt and Sherwood numbers decrease at the initial wall and those increase at the terminal plate with the increase in K.

## References


1. Kairi, R.R., Murthy, P.: Effect of viscous dissipation on natural convection heat and mass transfer from vertical cone in a non-Newtonian fluid saturated non-Darcy porous medium. Appl. Math. Comput. **217**(20), 8100–8114 (2011)
2. Fattahi, E., Farhadi, M., Sedighi, K., Nemati, H.: Lattice Boltzmann simulation of natural convection heat transfer in nanofluids. Int. J. Therm. Sci. **52**, 137–144 (2012)
3. Huelsz, G., Rechtman, R.: Heat transfer due to natural convection in an inclined square cavity using the lattice Boltzmann equation method. Int. J. Therm. Sci. **65**, 111–119 (2013)
4. Terekhov, V.I., Ekaid, A.L., Yassin, K.F.: Laminar free convection heat transfer between vertical isothermal plates. J. Eng. Thermophys. **25**(4), 509–519 (2016)
5. Navier, C.L.M.H.: Memoire sur les lois du mouvement des fluides. Mem. Acad. R. Sci. Paris **6**, 389–416 (1823)
6. Eegunjobi, A.S., Makinde, O.D.: Effects of the slip boundary condition on non-Newtonian flows in a channel. Entropy **14**(12), 1028–1044 (2012)
7. Rundora, L., Makinde, O.D.: Effects of Navier slip on unsteady flow of a reactive variable viscosity non-Newtonian fluid through a porous saturated medium with asymmetric convective boundary conditions. J. Hydrodyn. Ser. B. **27**(6), 934–944 (2015)
8. Ng, C.-O.: Starting flow in channels with boundary slip. Meccanica. doi:10.1007/s11012-016-0384-4
9. Mahmoud, M.A.A., Megahed, A.M.: Thermal radiation effect on mixed convection heat and mass transfer of a non-Newtonian fluid over a vertical surface embedded in a porous medium in the presence of thermal diffusion and diffusion-thermo effects. J. Appl. Mech. Tech. Phys. **54**(1), 90–99 (2013)
10. Srinivasacharya, D., Mallikarjuna, B., Bhuvanavijaya, R.: Soret and Dufour effects on mixed convection along a vertical wavy surface in a porous medium with variable properties. Ain Shams Eng. J. **6**(2), 553–564 (2015)
11. Umar, K., Naveed, A., Tauseef, S., Mohyud-Din.: Soret and Dufour effects on flow in converging and diverging channels with chemical reaction. Aerospace Sci. Technol. **49**, 135–143 (2016)
12. Kothandapani, M., Prakash, J.: Effects of thermal radiation and chemical reactions on peristaltic flow of a Newtonian nanofluid under inclined magnetic field in a generalized vertical channel using homotopy perturbation method. Asia-Pac. J. Chem. Eng. **10**(2), 259–272 (2015)
13. RamReddy, C., Pradeepa, T.: Spectral quasi-linearization method for homogeneous-heterogeneous reactions on nonlinear convection flow of micropolar fluid saturated porous medium with convective boundary condition. Open Eng. **6**(1), 106–119 (2016)
14. Liao, S.: Beyond perturbation. Introduction to homotopy analysis method. Chapman and Hall/CRC Press, Boca Raton (2003)
15. Liao, S.: An optimal homotopy-analysis approach for strongly nonlinear differential equations. Commun. Nonlinear Sci. Numer. Simul. **15**(8), 2003–2016 (2010)
16. Srinivasacharya, D., Kaladhar, K.: Mixed convection flow of couple stress fluid between parallel vertical plates with Hall and Ion-slip effects. Commun. Nonlinear Sci. Numer. Simul. **17**(6), 2447–2462 (2012)
17. Abbasbandy, S., Shivanian, E.: Solution of singular linear vibrational BVPs by the homotopy analysis method. J. Numer. Math. Stoch. **1**(1), 77–84 (2009)
18. Abbasbandy, S., Shivanian, E., Vajravelu, K.: Mathematical properties of h-curve in the frame work of the homotopy analysis method. Commun. Nonlinear Sci. Numer. Simul. **16**(11), 4268–4275 (2011)
19. Ellahi, R., Shivanian, E., Abbasbandy, S., Rahman, S.U., Hayat, T.: Analysis of steady flows in viscous fluid with heat/mass transfer and slip effects. Int. J. Heat Mass Transfer. **55**(23), 6384–6390 (2012)
20. Shaban, M., Shivanian, E., Abbasbandy, S.: Analyzing magneto-hydrodynamic squeezing flow between two parallel disks with suction or injection by a new hybrid method based on the Tau method and the homotopy analysis method. Eur. Phys. J. Plus. **128**(11), 1–10 (2013)
21. Shivanian, E., Abbasbandy, S.: Predictor homotopy analysis method: two points second order boundary value problems. Nonlinear Anal. Real World Appl. **15**, 89–99 (2014)
22. Ahmad Soltani, L., Shivanian, E., Ezzati, R.: Convection-radiation heat transfer in solar heat exchangers filled with a porous medium: exact and shooting homotopy analysis solution. Appl. Therm. Eng. **103**, 537–542 (2016)
23. Ellahi, R., Shivanian, E., Abbasbandy, S., Hayat, T.: Numerical study of magnetohydrodynamics generalized Couette flow of Eyring–Powell fluid with heat transfer and slip condition. Int. J. Numer. Methods Heat Fluid Flow **26**(5), 1433–1445 (2016)
24. Rashidi, M.M., Mohimanian pour, S.A., Abbasbandy, S.: Analytic approximate solutions for heat transfer of a micropolar fluid through a porous medium with radiation. Commun. Nonlinear Sci. Numer. Simul. **16**, 1874–1889 (2011)
25. Turkyilmazoglu, M.: Numerical and analytical solutions for the flow and heat transfer near the equator of an MHD boundary layer over a porous rotating sphere. Int. J. Therm. Sci. **50**, 831–842 (2011)

# 18

# Some inequalities for the $s$-Godunova–Levin type functions

**M. Emin Özdemir**

**Abstract** In this paper, first, we obtained two new $s$-Godunova–Levin type inequalities about "the mean value Theorem for integrals". Second, some inequalities were proved for mappings $q$th powers of first derivatives belonging to class $Q_s(I)$ using the Čebyšev's inequality, H ölder inequality, Power mean inequality, and some other classical inequalities. Finally, some error estimates for the Trapezoidal formula are also given.The results obtained are consistent with literature.

**Keywords** Hadamard's inequality · Differentiable mappings · $s$-Godunova–Levin type functions · Fractional integral · Power mean inequality

## Introduction

One of the most famous inequalities for convex functions is Hadamard's inequality. This double inequality is stated as follows (see for example [1, 2]): Let $f : I \subset \mathbb{R} \to \mathbb{R}$ be a convex function on the interval $I$ of real numbers and $a, b \in I$ with $a<b$. Then

$$f\left(\frac{a+b}{2}\right) \leq \frac{1}{b-a}\int_a^b f(x)\mathrm{d}x \leq \frac{f(a)+f(b)}{2}. \tag{1}$$

For several recent results concerning the inequality (1), we refer the interested reader to [1–5].

M. E. Özdemir (✉)
Department of Mathematics, K.K. Education Faculty, Atatürk University, Kampus, 25240 Erzurum, Turkey
e-mail: emos@atauni.edu.tr

**Definition 1** [6] We say that $f : I \to \mathbb{R}$ is a Godunova–Levin function or that f belongs to class Q(I) if f is non-negative and for all $x, y \in I$ and $t \in (0, 1)$ we have

$$f(tx+(1-t)y) \leq \frac{1}{t}f(x) + \frac{1}{1-t}f(y). \tag{2}$$

Some further properties of this class of functions can be found in [7–12]. Among others, it has been noted that non-negative monotone and non-negative convex functions belong to this class of functions. The above concept can be extended for functions $f : C \subseteq X \to [0, \infty)$ where C is a convex subset of the real or complex linear space X and the inequality (2) is satisfied for any vectors $x, y \in C$ and $t \in (0, 1)$. If the function $f : C \subseteq X \to \mathbb{R}$ is non-negative and convex, then it is of Godunova–Levin type.

**Definition 2** Let s be a real number, $s \in (0, 1]$. A function $f : [0, \infty) \to [0, \infty)$ is said to be s-convex (in the second sense)

$$f(tx+(1-t)y) \leq t^s f(x) + (1-t)^s f(y)$$

for all $x, y \in [0, \infty)$ and $t \in [0, 1]$.

For some properties of this class of functions see [13–17].

This concept can be extended for functions defined on convex subsets of linear spaces in the same way as above replacing the interval $I$ be the corresponding convex subset $C$ of the linear space $X$:

**Definition 3** [18] We say that the function $f : C \subset X \to [0, \infty)$ is of $s-$Godunova–Levin type, with $s \in [0, 1]$, if

$$f(tx+(1-t)y) \leq \frac{1}{t^s}f(x) + \frac{1}{(1-t)^s}f(y)$$

for all $t \in (0, 1)$ and $x, y \in C$.

We denote by $Q_s(C)$ the class of $s$-Godunova–Levin functions defined on $C$.

We observe that for $s=0$, we obtain the class of $p$-functions while for $s=1$ we obtain the class of Godunova–Levin. Thus,

$$P(C)=Q_0(C)\subseteq Q_{s_1}(C)\subseteq Q_{s_2}(C)\subseteq Q_1(C)=Q(C)$$

for $0\le s_1\le s_2\le 1$.

We recall the well-known Hölder's integral inequality which can be stated as follows, see [19].

**Theorem 1** *Let $p>1$ and $\frac{1}{p}+\frac{1}{q}=1$. If $f$ and $g$ are real functions defined on $[a,b]$ and if $|f|^p$ and $|g|^q$ are integrable functions on $[a,b]$, then*

$$\int_a^b |f(x)g(x)|\mathrm{d}x\le\left(\int_a^b |f(x)|^p\mathrm{d}x\right)^{\frac{1}{p}}\left(\int_a^b |g(x)|^q\mathrm{d}x\right)^{\frac{1}{q}}$$

*with equality holding if and only if $A|f(x)|^p=B|g(x)|^q$ almost everywhere, where $A$ and $B$ are constants.*

**Theorem 2** (Power Mean Inequality, see [20]) *Let $x=(x_i)$, $p=(p_i)$ be two positive n-tubles and let $r\in\mathbb{R}\cup\{+\infty,-\infty\}$, $i=1,2,\ldots,n$. Then, taking $p_n=\sum_{k=1}^{n}p_k$, the $r$ th power mean of $x$ with weights $p$ is defined by*

$$M_n^{[r]}=\begin{cases}\left(\dfrac{1}{p_n}\sum_{k=1}^{n}p_kx_k^r\right)^{\frac{1}{r}}, & r\ne+\infty,0,-\infty\\ \left(\prod_{k=1}^{n}x_k^{p_k}\right)^{\frac{1}{p_n}}, & r=0\\ \min(x_1,x_2,\ldots,x_n), & r=-\infty\\ \max(x_1,x_2,\ldots,x_n), & r=+\infty\end{cases}$$

*Note that if $-\infty\le r<s\le\infty$, then*

$$M_n^{[r]}\le M_n^{[s]}$$

*(see, e.g., [21]).*

**Theorem 3** [1] *Let $f\in Q(I)$, $a,b\in I$ with $a<b$ and $f\in L_1[a,b]$. Then one has the inequality*

$$f\left(\frac{a+b}{2}\right)\le\frac{4}{b-a}\int_a^b f(x)\mathrm{d}x. \tag{3}$$

**Theorem 4** [1] *Let $f\in P(I)$, $a,b\in I$ with $a<b$ and $f\in L_1[a,b]$. Then one has the inequality*

$$f\left(\frac{a+b}{2}\right)\le\frac{2}{b-a}\int_a^b f(x)\mathrm{d}x\le 2(f(a)+f(b)). \tag{4}$$

*Both inequalities are best possible.*

We need the following inequalities:

**Theorem 5** (see [21]) *Let $f,g:[a,b]\to\mathbb{R}$ be integrable functions, both increasing or both decreasing. Furthermore, let $p:[a,b]\to\mathbb{R}_+$ be an integrable function. Then*

$$\int_a^b p(x)f(x)\mathrm{d}x\int_a^b p(x)g(x)\mathrm{d}x\le\int_a^b p(x)\mathrm{d}x\int_a^b p(x)f(x)g(x)\mathrm{d}x. \tag{5}$$

*If one of the functions $f$ or $g$ is nonincreasing and the other nondecreasing then the inequality in (5) is reversed. Inequality (5) is known in the literature as Čebyšev's inequality and so are the following special cases of (5):*

$$\frac{1}{b-a}\int_a^b f(x)\mathrm{d}x\int_a^b g(x)\mathrm{d}x\le\int_a^b f(x)g(x)\mathrm{d}x$$

*and*

$$\int_0^1 f(x)\mathrm{d}x\int_0^1 g(x)\mathrm{d}x\le\int_0^1 f(x)g(x)\mathrm{d}x.$$

Now, we are giving some necessary definitions and mathematical preliminaries of fractional calculus theory which are used throughout this paper, see [22].

**Definition 4** Let $f\in L_1[a,b]$. The Riemann–Liouville integrals $J_{a^+}^{\alpha}f$ and $J_{b^-}^{\alpha}f$ of order $\alpha>0$ with $a\ge 0$ are defined by

$$J_{a^+}^{\alpha}f(x)=\frac{1}{\Gamma(\alpha)}\int_a^x (x-t)^{\alpha-1}f(t)\mathrm{d}t,\quad x>a$$

and

$$J_{b^-}^{\alpha}f(x)=\frac{1}{\Gamma(\alpha)}\int_x^b (t-x)^{\alpha-1}f(t)\mathrm{d}t,\quad x<b$$

respectively, where $\Gamma(\alpha)=\int_0^{\infty}e^{-u}u^{\alpha-1}\mathrm{d}u$. Here is $J_{a^+}^{0}f(x)=J_{b^-}^{0}f(x)=f(x)$.

In the case of $\alpha=1$, the fractional integral reduces to the classical integral.

For some recent results connected with fractional integral inequalities see [22–29].

In [30, Özdemir et al. proved the following result for fractional integrals.

**Lemma 1** *Let $f:I\subset\mathbb{R}\to\mathbb{R}$ be a differentiable mapping on $I$ with $a<r$, $a,r\in I$. If $f'\in L[a,r]$, then the following equality for fractional integrals holds:*

$$\frac{f(a)+f(r)}{2}-\frac{\Gamma(\alpha+1)}{2(r-a)^{\alpha}}\left[J_{r^-}^{\alpha}f(a)+J_{a^+}^{\alpha}f(r)\right]$$
$$=\frac{r-a}{2}\int_0^1[(1-t)^{\alpha}-t^{\alpha}]f'(r+(a-r)t)\mathrm{d}t.$$

The main purpose of this study is to obtain the inequalities for class of $s$-Godunova–Levin type functions.

## Main results

In [31], S. S. Dragomir has written the following inequalities (6) and (8) without proof.

**Theorem 6** *Let $f \in Q_s(C)$, with $a<b$ and $f \in L_1[a,b]$, $C=[a,b]$, $s \in [0,1]$. Then one has the inequalities*

$$f\left(\frac{a+b}{2}\right) \le \frac{2^{s+1}}{b-a}\int_a^b f(x)\mathrm{d}x \tag{6}$$

$$\frac{\Gamma((1+s)^2)}{\Gamma(2+2s)(b-a)}\int_a^b f(x)\mathrm{d}x \le \frac{f(a)+f(b)}{s+1} \tag{7}$$

*and*

$$\frac{1}{b-a}\int_a^b f(x)\mathrm{d}x \le \frac{f(a)+f(b)}{1-s}, \quad s\in[0,1) \tag{8}$$

*Proof* $f \in Q_s(C)$, we have for all $x,y \in C=[a,b]$ with $t=\frac{1}{2}$;

$$2^s(f(x)+f(y)) \ge f\left(\frac{x+y}{2}\right)$$

Now, if we choose $x=ta+(1-t)b$, $y=(1-t)a+tb$, we have

$$2^s(f(ta+(1-t)b)+f((1-t)a+tb)) \ge f\left(\frac{a+b}{2}\right).$$

By integrating, we have that

$$2^s\left[\int_0^1 f(ta+(1-t)b)\mathrm{d}t+\int_0^1 f((1-t)a+tb)\mathrm{d}t\right] \ge f\left(\frac{a+b}{2}\right) \tag{9}$$

On the other hand,

$$\int_0^1 f(ta+(1-t)b)\mathrm{d}t=\int_0^1 f((1-t)a+tb)\mathrm{d}t = \frac{1}{b-a}\int_0^1 f(x)\mathrm{d}x$$

we get the inequality (6) from (9).

For the proof of (7), if $f \in Q_s(C)$ for all $a,b \in C$ and $t\in(0,1)$, it yields

$$t^s(1-t)^s f(ta+(1-t)b) \le (1-t)^s f(a)+t^s f(b)$$

and

$$t^s(1-t)^s f((1-t)a+tb) \le t^s f(a)+(1-t)^s f(b).$$

By adding these inequalities and integrating over [0,1], we find that

$$\int_0^1 t^s(1-t)^s[f(ta+(1-t)b)+f((1-t)a+tb)]\mathrm{d}t \le \frac{2}{s+1}[f(a)+f(b)].$$

Now, by a simple computation, we have

$$\int_0^1 t^s(1-t)^s f(ta+(1-t)b)\mathrm{d}\text{and} \int_0^1 t^s(1-t)^s f((1-t)a+tb)\mathrm{d}t.$$

Let be $g(t)=t^s(1-t)^s$. We take symmetric of the functions $g(t)$ and $f$, respectively, $\frac{1}{2}$ and $\frac{a+b}{2}$. Also, let the functions $f$ and $g$ both be either increasing or decreasing. By applying Čebyš ev's inequality, we have

$$\int_0^1 t^s(1-t)^s f(ta+(1-t)b)\mathrm{d}t = \int_0^1 t^s(1-t)^s f((1-t)a+tb)\mathrm{d}t \ge \int_0^1 t^s(1-t)^s\mathrm{d}t\int_0^1 f((1-t)a+tb)\mathrm{d}t = \frac{\Gamma((1+s)^2)}{\Gamma(2+2s)(b-a)}\int_a^b f(x)\mathrm{d}x.$$

To obtain the inequality (8), as $f\in Q_s(C)$, we have

$$f(ta+(1-t)b) \le t^{-s}f(a)+(1-t)^{-s}f(b)$$

integrating this inequality on $[0,1]$, we get

$$\int_0^1 f(ta+(1-t)b)\mathrm{d}t \le f(a)\int_0^1 t^{-s}\mathrm{d}t+f(b)\int_0^1(1-t)^{-s}\mathrm{d}t = \frac{f(a)+f(b)}{1-s}, \quad (s\in[0,1))$$

As the change of variable $x=ta+(1-t)b$ gives us

$$\int_0^1 f(ta+(1-t)b)\mathrm{d}t = \frac{1}{b-a}\int_a^b f(x)\mathrm{d}x$$

which completes the proof of the inequality (8). □

**Theorem 7** *Combining the inequalities* (2.1) *and* (2.3) *under the conditions of Theorem* 6, *we get*

$$f\left(\frac{a+b}{2}\right) \le \frac{2^{s+1}}{b-a}\int_a^b f(x)\mathrm{d}x \le 2^{s+1}\frac{f(a)+f(b)}{1-s}.$$

*Remark 1* If we choose $s=1$ in (6) and (7), we obtain the inequality (3) and right hand side of (4), respectively.

**Theorem 8** *Let $f: I\subseteq \mathbb{R}\to[0,\infty)$ be a differentiable mapping on $I$, $a,r\in I$ and $a<r$. If $|f'|\in Q_\alpha(I)$ with $\alpha\in[0,1)$, $t\in(0,1)$, then the following inequality for fractional integrals holds:*

$$\left|\frac{f(a)+f(r)}{2}-\frac{\Gamma(\alpha+1)}{2(r-a)^\alpha}\left[J^\alpha_{r^-}f(a)+J^\alpha_{a^+}f(r)\right]\right| \le \frac{(r-a)}{2}\left\{\pi\alpha\operatorname{cosec}(\pi\alpha)[|f'(r)|+|f'(a)|]+2\beta_{\frac{1}{2}}(1-\alpha,1+\alpha)\right\} \tag{10}$$

*where* $\beta_x(y,z)=\int_0^x t^{y-1}(1-t)^{z-1}\mathrm{d}t,\ 0\le x\le 1$ *is incomplete Beta function.*

*Proof* Using the Lemma 1 and $|f'|\in Q_\alpha(I)$, it follows that

$$\left|\frac{f(a)+f(r)}{2}-\frac{\Gamma(\alpha+1)}{2(r-a)^\alpha}\left[J_{r^-}^\alpha f(a)+J_{a^+}^\alpha f(r)\right]\right|\le\frac{(r-a)}{2}\int_0^1|(1-t)^\alpha-t^\alpha||f'(r+(a-r)t)|\mathrm{d}t.$$

Since

$$|f'(r+(a-r)t)|=|f'(ta+(1-t)r)|\le\frac{1}{t^\alpha}|f'(a)|+\frac{1}{(1-t)^\alpha}|f'(r)|$$

$$\begin{aligned}&\left|\frac{f(a)+f(r)}{2}-\frac{\Gamma(\alpha+1)}{2(r-a)^\alpha}\left[J_{r^-}^\alpha f(a)+J_{a^+}^\alpha f(r)\right]\right|\\&\le\frac{(r-a)}{2}\int_0^1|(1-t)^\alpha-t^\alpha|\left[\frac{1}{t^\alpha}|f'(a)|+\frac{1}{(1-t)^\alpha}|f'(r)|\right]\mathrm{d}t\\&=\frac{(r-a)}{2}\left\{\int_0^{\frac12}[(1-t)^\alpha-t^\alpha]\left[\frac{1}{t^\alpha}|f'(a)|+\frac{1}{(1-t)^\alpha}|f'(r)|\right]\mathrm{d}t\right.\\&\left.+\int_{\frac12}^1[t^\alpha-(1-t)^\alpha]\left[\frac{1}{t^\alpha}|f'(a)|+\frac{1}{(1-t)^\alpha}|f'(r)|\right]\mathrm{d}t\right\}\end{aligned}$$

On the other hand, by a simple computation, we have

$$\int_0^{\frac12}\left[\frac{(1-t)^\alpha-t^\alpha}{t^\alpha}\right]\mathrm{d}t=\beta_{\frac12}(1-\alpha,1+\alpha)-\frac12$$

$$\int_0^{\frac12}\left[\frac{(1-t)^\alpha-t^\alpha}{(1-t)^\alpha}\right]\mathrm{d}t=\frac12-\beta_{\frac12}(\alpha+1,1-\alpha)$$

$$\int_{\frac12}^1\left[\frac{t^\alpha-(1-t)^\alpha}{t^\alpha}\right]\mathrm{d}t=\frac12-\pi\alpha\cos ec(\pi\alpha)+\beta_{\frac12}(\alpha+1,1-\alpha)$$

and

$$\int_{\frac12}^1\left[\frac{t^\alpha-(1-t)^\alpha}{(1-t)^\alpha}\right]\mathrm{d}t=\pi\alpha\cos ec(\pi\alpha)-\beta_{\frac12}(1+\alpha,1-\alpha)-\frac12.$$

Since, finally

$$|f'(r)|-|f'(a)|\le|f'(r)|+|f'(a)|.$$

Hence, we obtain the inequality (10). □

**Theorem 9** *Let* $f:I\subset\mathbb{R}\to[0,\infty)$ *be a differentiable mapping of I,* $a,r\in I$ *and* $a<r$. *If* $|f'|^q\in Q_s(I)$ *with* $\alpha\in[0,1]$, $t\in(0,1)$, *then the following inequality for fractional integrals holds:*

$$\left|\frac{f(a)+f(r)}{2}-\frac{\Gamma(\alpha+1)}{2(r-a)^\alpha}\left[J_{r^-}^\alpha f(a)+J_{a^+}^\alpha f(r)\right]\right|\le\frac{(r-a)}{2(\alpha p+1)^{\frac1p}(1-s)^{\frac1q}}\left(|f'(a)|^q+|f'(r)|^q\right)^{\frac1q}$$

*where* $\frac1p+\frac1q=1$, $s\in[0,1)$, $\Gamma(.)$ *is Gamma function.*

*Proof* From Lemma 1 and using Hölder inequality with properties of modulus, we have

$$\begin{aligned}&\left|\frac{f(a)+f(r)}{2}-\frac{\Gamma(\alpha+1)}{2(r-a)^\alpha}\left[J_{r^-}^\alpha f(a)+J_{a^+}^\alpha f(r)\right]\right|\\&\le\frac{(r-a)}{2}\int_0^1|(1-t)^\alpha-t^\alpha||f'(r+(a-r)t)|\mathrm{d}t\\&\le\frac{(r-a)}{2}\left(\int_0^1|(1-t)^\alpha-t^\alpha|^p\mathrm{d}t\right)^{\frac1p}\left(\int_0^1|f'(r+(a-r)t)|^q\mathrm{d}t\right)^{\frac1q}.\end{aligned}$$

We know that for $\alpha\in(0,1]$ and $\forall t_1,t_2\in(0,1)$,

$$|t_1^\alpha-t_2^\alpha|\le|t_1-t_2|^\alpha,$$

therefore,

$$\begin{aligned}\int_0^1|(1-t)^\alpha-t^\alpha|^p\mathrm{d}t&\le\int_0^1|1-2t|^{\alpha p}\mathrm{d}t\\&=\int_0^{\frac12}(1-2t)^{\alpha p}\mathrm{d}t+\int_{\frac12}^1(2t-1)^{\alpha p}\mathrm{d}t\\&=\frac{1}{\alpha p+1}.\end{aligned}$$

Since $|f'|\in Q_s(I)$, we obtain

$$|f'(r+(a-r)t)|=|f'(ta+(1-t)r)|\le\frac{1}{t^s}|f'(a)|+\frac{1}{(1-t)^s}|f'(r)|.$$

Hence, we get

$$\begin{aligned}&\left|\frac{f(a)+f(r)}{2}-\frac{\Gamma(\alpha+1)}{2(r-a)^\alpha}\left[J_{r^-}^\alpha f(a)+J_{a^+}^\alpha f(r)\right]\right|\\&\le\frac{(r-a)}{2(\alpha p+1)^{\frac1p}(1-s)^{\frac1q}}\left(|f'(a)|^q+|f'(r)|^q\right)^{\frac1q}\end{aligned}$$

which completes the proof. □

**Corollary 1** *In Theorem* 9, *if we choose* $\alpha=1$ *and* $s=0$, *then we have*

$$\begin{aligned}&\left|\frac{f(a)+f(r)}{2}-\frac{1}{(r-a)}\int_a^r f(t)\mathrm{d}t\right|\\&\le\frac{(r-a)}{2(p+1)^{\frac1p}(1-s)^{\frac1q}}\left(|f'(a)|^q+|f'(r)|^q\right)^{\frac1q}\\&\le\frac{(r-a)}{2}(|f'(a)|+|f'(r)|).\end{aligned}\tag{11}$$

*Proof* Let $a_1=|f'(a)|^q$, $b_1=|f'(r)|^q$, $0<\frac1q<1$ for $q>1$. Using the fact

$$\sum_{i=1}^n(a_i+b_i)^r\le\sum_{i=1}^n a_i^r+\sum_{i=1}^n b_i^r$$

for $a_1, a_2, \ldots, a_n \geq 0$, $b_1, b_2, \ldots, b_n \geq 0$, we obtain the inequality (11) and since $\frac{1}{2} \leq (\frac{1}{p+1})^{\frac{1}{p}} \leq 1$, $p \in (0,1)$. □

**Theorem 10** *Let $f : I \subset \mathbb{R} \to [0, \infty)$ be a differentiable mapping an $I$, $a, r \in I$, $a < r$ and $q \geq 1$. If $|f'|^q \in Q_\alpha(I)$ with $\alpha \in [0,1)$, $t \in (0,1)$, then the following inequality for fractional integrals hold:*

$$\left|\frac{f(a)+f(r)}{2} - \frac{\Gamma(\alpha+1)}{2(r-a)^\alpha}\left[J^\alpha_{r^-}f(a) + J^\alpha_{a^+}f(r)\right]\right| \leq \frac{(r-a)}{2^{\frac{1}{q}}(\alpha+1)^{1-\frac{1}{q}}}\left(1-\frac{1}{2^\alpha}\right)^{1-\frac{1}{q}}\left[(2\beta_{\frac{1}{2}}(1-\alpha, 1+\alpha) - \pi\alpha \operatorname{cosec}(\pi\alpha))\right]^{\frac{1}{q}} \times (|f'(a)| + |f'(r)|). \tag{12}$$

*Proof* From Lemma 1 and using the well-known power mean inequality, we have

$$\begin{aligned}
&\left|\frac{f(a)+f(r)}{2} - \frac{\Gamma(\alpha+1)}{2(r-a)^\alpha}\left[J^\alpha_{r^-}f(a) + J^\alpha_{a^+}f(r)\right]\right| \\
&\leq \frac{(r-a)}{2}\int_0^1 |(1-t)^\alpha - t^\alpha||f'(r+(a-r)t)|\mathrm{d}t \\
&\leq \frac{(r-a)}{2}\left(\int_0^1 |(1-t)^\alpha - t^\alpha|\mathrm{d}t\right)^{1-\frac{1}{q}} \\
&\quad \times \left(\int_0^1 |(1-t)^\alpha - t^\alpha||f'(r+(a-r)t)|^q \mathrm{d}t\right)^{\frac{1}{q}}.
\end{aligned}$$

On the other hand, we have

$$\begin{aligned}
\int_0^1 |(1-t)^\alpha - t^\alpha|\mathrm{d}t &= \int_0^{\frac{1}{2}} [(1-t)^\alpha - t^\alpha]\mathrm{d}t + \int_{\frac{1}{2}}^1 [t^\alpha - (1-t)^\alpha]\mathrm{d}t \\
&= \frac{2}{\alpha+1}\left(1-\frac{1}{2^\alpha}\right).
\end{aligned}$$

Since $|f'| \in Q_\alpha(I)$, we have

$$\begin{aligned}
|f'(r+(a-r)t)|^q = |f'(ta+(1-t)r)|^q &\leq \frac{1}{t^\alpha}|f'(a)|^q \\
&\quad + \frac{1}{(1-t)^\alpha}|f'(r)|^q
\end{aligned}$$

and

$$\begin{aligned}
&\int_0^1 |(1-t)^\alpha - t^\alpha||f'(r+(a-r)t)|^q \mathrm{d}t \\
&\leq \int_0^1 |(1-t)^\alpha - t^\alpha|\left[\frac{1}{t^\alpha}|f'(a)|^q + \frac{1}{(1-t)^\alpha}|f'(r)|^q\right]\mathrm{d}t \\
&\leq \int_0^{\frac{1}{2}} [(1-t)^\alpha - t^\alpha]\left[\frac{1}{t^\alpha}|f'(a)|^q + \frac{1}{(1-t)^\alpha}|f'(r)|^q\right]\mathrm{d}t \\
&\quad + \int_{\frac{1}{2}}^1 [t^\alpha - (1-t)^\alpha]\left[\frac{1}{t^\alpha}|f'(a)|^q + \frac{1}{(1-t)^\alpha}|f'(r)|^q\right]\mathrm{d}t.
\end{aligned}$$

and since $\sum_{i=1}^n (a_i + b_i)^r \leq \sum_{i=1}^n a_i^r + \sum_{i=1}^n b_i^r$, we obtain the required inequality (12). □

## Applications to numerical integration

We may not be given a formula for $f(x)$ as a function of $x$. For instance, $f(x)$ may be an unknown function whose values are at certain points of the interval $[a,b]$ . In this case, we investigate the problem of approximating the value of the integral $I = \int_a^b f(x)\mathrm{d}x$ using only the values of $f(x)$ at finitely many points of $[a,b]$. Obtaining such an approximation is called numerical integration. That is why, there are three methods for evaluating definite integrals numerically. One of them is Trapezoid Rule.

Let $d$ be a division of the interval $[a,r]$, i.e., $d : a = x_0 < x_1 < \cdots < x_{n-1} < x_n = r$, and consider the trapezoidal formula

$$T_n(f,d) = \sum_{i=0}^{n-1}\frac{f(x_i)+f(x_{i+1})}{2}(x_{i+1}-x_i).$$

So, the following approximation of the integral $\int_a^b f(x)\mathrm{d}x$ holds:

$$\int_a^b f(x)\mathrm{d}x \approxeq T_n(f,d) + E_n(f,d)$$

where the approximation error $E_n(f,d)$ of the integral $\int_a^b f(x)\mathrm{d}x$ by the trapezoidal formula $T_n(f,d)$ satisfies

$$|E_n(f,d)| \leq \frac{M}{12}\sum_{i=0}^{n-1}(x_{i+1}-x_i)^3.$$

We shall propose some new estimates of the remainder term $E_n(f,d)$.

**Proposition 1** *Let $f$ be a differentiable mapping on $I^\circ$, $a, r \in I^\circ$ with $a < r$. If $|f'|$ is $p$ -convex on $[a,r]$, then for every division $d$ of $[a,r]$, the following holds:*

$$\begin{aligned}
|E_n(f,d)| &\leq \frac{1}{2}\sum_{i=0}^{n-1}(x_{i+1}-x_i)^2(|f'(x_i)| + |f'(x_{i+1})|) \\
&\leq \sum_{i=0}^{n-1}(x_{i+1}-x_i)^2 \max\{|f'(a)|, |f'(r)|\}.
\end{aligned}$$

*Proof* Applying Corollary 1 on the subinterval $[x_i, x_{i+1}]$ $(i = 0, 1, 2, \ldots, n-1)$ of the division $d$, we get

$$\left|\frac{f(x_i)+f(x_{i+1})}{2}(x_{i+1}-x_i) - \int_{x_i}^{x_{i+1}} f(x)\mathrm{d}x\right| \leq \frac{(x_{i+1}-x_i)^2(|f'(x_i)| + |f'(x_{i+1})|)}{2}.$$

Summing over $i$ from 0 to $n-1$ on taking into account that $|f'|$ is $p$-convex, we deduce, by the triangle inequality that

$$\left|T_n(f,d)-\int_a^b f(x)\mathrm{d}x\right| \leq \frac{1}{2}\sum_{i=0}^{n-1}(x_{i+1}-x_i)^2(|f'(x_i)|+|f'(x_{i+1})|)$$
$$\leq \max\{|f'(x_i)|,|f'(x_{i+1})|\}\sum_{i=0}^{n-1}(x_{i+1}-x_i)^2$$
$$\leq \max\{|f'(a)|,|f'(r)|\}\sum_{i=0}^{n-1}(x_{i+1}-x_i)^2.$$

□

**Proposition 2** *Let $f$ be a differentiable mapping on $I^\circ \subset I$, $a, r \in I^\circ$ with $a<r$ and let $\frac{1}{p}+\frac{1}{q}=1$. If $|f'|^q \in Q_s(I^\circ)$ with $\alpha = 1$, $t \in (0,1)$. Then for every division of $[a,r]$, the following holds:*

$$|E_n(f,d)| \leq \frac{1}{2(p+1)^{\frac{1}{q}}(1-s)^{\frac{1}{q}}}\sum_{i=o}^{n-1}(x_{i+1}-x_i)^2\left[|f'(x_i)|^q+|f'(x_{i+1})|^q\right]^{\frac{1}{q}}$$
$$\leq \frac{\max\{|f'(a)|,|f'(r)|\}}{2(p+1)^{\frac{1}{q}}(1-s)^{\frac{1}{q}}}\sum_{i=o}^{n-1}(x_{i+1}-x_i)^2.$$

*Proof* If we apply the Theorem 9 for $\alpha = 1$, the proof is similar to that of Proposition 1. □

## References


1. Dragomir, S.S., Pearce, C.E.M.: Selected Topics on Hermite-Hadamard Inequalities and Applications. RGMIA Monographs. Victoria University, Melbourne (2000)
2. Pečarić, J.E., Proschan, F., Tong, Y.L.: Convex Functions. Partial Orderings and Statistical Applications. Academic Press, Boston (1992)
3. Kırmacı, U.S., Bakula, M.K., Özdemir M.E., Pe čarić, J.: Hadamard-type inequalities for s-convex functions. Appl. Math. and Comp. **193**, 26–35 (2007)
4. Özdemir, M.E., Kırmacı, U.S.: Two new theorem on mappings uniformly continuous and convex with applications to quadrature rules and means. Appl. Math. and Comp. **143**, 269–274 (2003)
5. Sarıkaya, M.Z., Set, E., Özdemir, M.E.: On some new inequalities of Hadamard type involving $h$-convex functions. Acta Math. Univ. Comenianae, vol. LXXIX, 2, pp. 265–272 (2010)
6. Godunova, E.K., Levin, V.I.: Inequalities for functions of a broad class that contains convex, monotone and some other forms of functions. In: Proceedings of (Russian) Numerical Mathematics and Mathematical Physics (Russian), vol. 166, pp. 138–142. Moskov. Gos. Ped. Inst., Moscow (1985)
7. Dragomir, S.S., Mond, B.: On Hadamard's inequality for a class of functions of Godunova and Levin. Indian J. Math. **39**(1), 1–9 (1997)
8. Dragomir, S.S., Pearce, C.E.M.: On Jensen's inequality for a class of functions of Godunova and Levin. Period. Math. Hungar. **33**(2), 93–100 (1996)
9. Dragomir, S.S., Pečarić, J.E., Persson, L.: Some inequalities of Hadamard type. Soochow J. Math. **21**(3), 335–341 (1995)
10. Mitrinović, D.S., Pečarić, J.E.: Note on a class of functions of Godunova and Levin. C. R. Math. Rep. Acad. Sci. Canada **12**(1), 33–36 (1990)
11. Pečarić, J.E., Dragomir, S.S.: A generalization of Hadamard's inequality for isotonic linear functionals. Radovi Mat. (Sarajevo) **7**, 103–107 (1991)
12. Radulescu, M., Radulescu, S., Alexandrescu, P.: On the Godunova–Levin–Schur class of functions. Math. Inequal. Appl. **12**(4), 853–862 (2009)
13. Sarıkaya, M.Z., Set, E., Özdemir, M.E.: On new inequalities of Simpson's type for s-convex functions. Comput. Math. Appl. **60**, 2191–2199 (2010)
14. Avcı, M., Kavurmacı, H., Özdemir, M.E.: New inequalities of Hermite–Hadamard type via s-convex functions in the second sense with applications. Appl. Math. Comput. **217**, 5171–5176 (2011)
15. Kirmaci, U.S., Bakula, M.K., Özdemir, M.E., Pečari ć, J.E.: Hadamard-type inequalities for s-convex functions. Appl. Math. Comput. **193**(1), 26–35 (2007)
16. Set, E., Özdemir, M.E., Sarıkaya, M.Z.: New inequalities of Ostrowski's type for s-convex functions in the second sense with applications. Facta Univ. Ser. Math. Inform. **27**(1), 67–82 (2012)
17. Dragomir, S.S., Fitzpatrick, S.: The Hadamard inequalities for s-convex functions in the second sense. Demonstr. Math. **32**(4), 687–696 (1999)
18. Dragomir, S.S.: Integral inequalities of Jensen type for $\lambda$-convex functions. In: Proceedings of RGMIA, Res. Rep. Coll. 17 (2014)
19. Mitrinović, D.S.: Analytic Inequalities. Springer, New York (1970)
20. Bullen, P.S., Mitrinović, D.S., Vasić, P.M.: Means and their inequalities. In: Proceedings of Mathematical and its Applications, D. Reidel, Dortdrecht (1988)
21. Mitrinović, D.S., Pečarić, J.E., Fink, A.M.: Mathematics and its Applications. Classical and new inequalities in analysis. Kluwer Academic Publishers, Dordrecht (1993)
22. Samko, S.G., Kilbas, A.A., Marichev, O.I.: Fractional Integrals and Derivatives Theory and Application. Gordan and Breach Science, New York (1993)
23. Belarbi, S., Dahmani, Z.: On some new fractional integral inequalities, J. Ineq. Pure Appl. Math. 10(3), Art. 86 (2009)
24. Dahmani, Z.: New inequalities in fractional integrals. Int. J. Nonlinear Sci. **9**(4), 493–497 (2010)
25. Dahmani, Z.: On Minkowski and Hermite–Hadamard integral inequalities via fractional integration. Ann. Funct. Anal. **1**(1), 51–58 (2010)
26. Dahmani, Z., Tabharit, L., Taf, S.: Some fractional integral inequalities. Nonl. Sci. Lett. A. **1**(2), 155–160 (2010)
27. Dahmani, Z., Tabharit, L., Taf, S.: New generalizations of Gruss inequality using Riemann–Liouville fractional integrals. Bull. Math. Anal. Appl. **2**(3), 93–99 (2010)
28. Özdemir, M.E., Kavurmacı, H., Avcı, M.: New inequalities of Ostrowski type for mappings whose derivatives are $(\alpha, m)$-convex via fractional integrals. In: Proceedings of RGMIA Research Report Collection, 15, Article 10, pp. 8 (2012)
29. Özdemir, M.E., Sarıkaya, M.Z., Yıldız, Ç.: New Generalizations of Ostrowski-Like Type Inequalities for Fractional Integrals, Submitted (2015)
30. Özdemir, M.E., Dragomir, S.S., Yıldız, Ç.: The Hadamard inequality for convex function via fractional integrals. Acta Math. Sci. **33B**(5), 1293–1299 (2013)
31. Dragomir, S.S.: Inequalities of Hermite–Hadamard type for $\lambda$-convex functions on linear spaces, preprint RGMIA Res. Rep. Coll. 17 (2014)

# 19

# Numerical simulation based on meshless technique to study the biological population model

**Saeid Abbasbandy**[1] · **Elyas Shivanian**[1]

**Abstract** A kind of spectral meshless radial point interpolation method is proposed to degenerate parabolic equations arising from the spatial diffusion of biological populations and satisfactory agreements is archived. This method is based on collocation methods with mesh-free techniques as a background. In contrast to the finite-element method and those meshless methods based on Galerkin weak form, such as element-free Galerkin, there is no integration tools in this approach. Furthermore, some numerical experiments are given to validate the accuracy of the results and stability of the present method.

**Keywords** Radial point interpolation · Spectral mesh-free method · Finite difference · Biological population equation

## Introduction

The biological population problems have attracted much attention and research recently [1, 2]. Biologists believe that dispersal or emigration play a key role in the regulation of population of some species. A persuasive example of this suggestion occurs in a paper by Carl [3], whose observations on a population of arctic ground squirrels showed that this species achieves population control by the dispersal of animals from densely populated areas of favorable habitat into unfavorable areas, where burrow sites are not available, and where they are subjected to intensive predation.

✉ Saeid Abbasbandy
abbasbandy@yahoo.com

[1] Department of Mathematics, Imam Khomeini International University, Ghazvin 34149-16818, Iran

Consider the following nonlinear biological population model

$$\frac{\partial u(\mathbf{x},t)}{\partial t} = \frac{\partial^2 u^2}{\partial x^2} + \frac{\partial^2 u^2}{\partial y^2} + \sigma(u), \quad \mathbf{x} = (x,y) \in \Omega \subseteq \mathbb{R}^2, \quad t \in [0,T], \tag{1}$$

with a given initial condition $u(\mathbf{x},0)$, where $u$ denotes the population density, and $\sigma$ represents the population supply due to births and deaths. We can consider a more general form for $\sigma$ as $hu^a(1-ru^b)$, in which $h,a,r,b \in \mathbb{R}$. The field $u(\mathbf{x},t)$ gives the number of individuals' per-unit volume at position $\mathbf{x}$ and time $t$, and it is integral over any subregion $\Omega$ to give the total population of $\Omega$ at time $t$. The field $\sigma(u)$ gives the rate at which individuals are supplied (per-unit volume) exactly at $\mathbf{x}$ through the births and deaths. The flow of population from point to point is then depicted by the diffusion velocity $v(\mathbf{x},t)$, which provides the average velocity of those individuals who occupies $\mathbf{x}$ at the time $t$. The fields $u$, $v$, and $\sigma$ should be consistent with the following law of population balance (for any regular subregion $\Omega$ for the defined region during all the time $t$):

$$\frac{\mathrm{d}}{\mathrm{dt}}\int_\Omega u\mathrm{dV} + \int_{\partial\Omega} \mathrm{u} \times \mathrm{v} \times \mathrm{ndV} = \int_\Omega \sigma \mathrm{dV}, \tag{2}$$

where $n$ is the outward unit normal vector to the boundary $\partial\Omega$ of $\Omega$. This equation means that the rate of change of population of $\Omega$ addition to the rate at which individuals left $\Omega$ across its boundary should be equal to the rate at which individuals are directly supplied to $\Omega$ [4, 5].

Several papers have considered the existence, uniqueness, and regularity of weak solutions [4–9] for general degenerate parabolic equations of Eq. (1). Numerical solutions of the biological population equation have seldom been explored and investigated, though some numerical struggles have been done in this field. Modeling of the biological population problem (1) has been explored using the element-free Kp-Ritz method in [1]. An improved element-free Galerkin method for numerical modeling of the biological population problem (1) has also been applied by Zhang et. al. [2]. Shakeri and Dehghan obtained numerical solution of the model using He's variational iteration method [10].

In general, the mesh-free methods can be grouped into two categories, first category uses the equation in weak form, for instance, element-free Galerkin method (EFG), improved element-free Galerkin method (IEFG), complex variable element-free Galerkin method (CVEFG), improved complex variable element-free Galerkin method (ICVEFG), meshless local Petrov–Galerkin method, and etc. [11–21], and the second category applies the strong form of the equation, for example, meshless collocation approach [22–28]. In addition, the meshless methods depend on how their shape or basis function constructed are different, and it might be based on interpolation or curve fitting and etc. [29–41].

In the current work, we testify spectral meshless radial point interpolation (SMRPI) method [42–44] on the problem (1) and then make simulations on two numerical experiments which leads to satisfactory results. A technique based on radial point interpolation is adopted to construct shape functions, also called basis functions, using radial basis functions. These shape functions have delta function property in the frame work of interpolation; therefore, they convince us to impose boundary conditions directly. The time derivatives are approximated by the finite-difference time-stepping method. This method is based on collocation methods with mesh-free techniques as a background. In contrast to those meshless methods based on weak form, there is no integration tools in this approach. Therefore, the computational complexity of SMRPI method seems to be of low order.

## The time discretization approximation

In the current work, let us to employ a time-stepping scheme to evaluate the time derivative. To this end, the following first-order finite-difference approximation for the time derivative operator is adopted:

$$\frac{\partial u(\mathbf{x},t)}{\partial t} \cong \frac{1}{\Delta t}(u^{k+1}(\mathbf{x}) - u^k(\mathbf{x})). \tag{3}$$

Moreover, we apply the following approximations using the Crank–Nicolson technique:

$$\frac{\partial^2 u^2}{\partial x^2} + \frac{\partial^2 u^2}{\partial y^2} \cong \frac{1}{2}\left(\frac{\partial^2 (u^{k+1})^2}{\partial x^2} + \frac{\partial^2 (u^{k})^2}{\partial x^2} + \frac{\partial^2 (u^{k+1})^2}{\partial y^2} + \frac{\partial^2 (u^{k})^2}{\partial y^2}\right), \tag{4}$$

$$\sigma(u(\mathbf{x},t)) \cong \frac{1}{2}\left(\sigma(u^{k+1}(\mathbf{x})) + \sigma(u^k(\mathbf{x}))\right), \tag{5}$$

where $u^k(\mathbf{x}) = u(\mathbf{x}, k\Delta t)$ and $\mathbf{x} = (x, y)$. Applying the above approximation and impose them to the original Eq. (1), we are conducted to the following time discretization equation:

$$u^{k+1} = u^k + \frac{\Delta t}{2}\left\{\frac{\partial^2 (u^{k+1})^2}{\partial x^2} + \frac{\partial^2 (u^{k})^2}{\partial x^2} + \frac{\partial^2 (u^{k+1})^2}{\partial y^2} + \frac{\partial^2 (u^{k})^2}{\partial y^2} + \sigma[u^{k+1}] + \sigma[u^k]\right\}. \tag{6}$$

## The basis functions in the frame of MLRPI

Consider a continuous function $u(\mathbf{x})$ defined in a domain $\Omega$, which is represented by a set of field nodes. The $u(\mathbf{x})$ at a point of interest $\mathbf{x}$ is approximated in the form of

$$u(\mathbf{x}) = \sum_{i=1}^{n} R_i(\mathbf{x}) a_i + \sum_{j=1}^{m} p_j(\mathbf{x}) b_j = \mathbf{R}^T(\mathbf{x})\mathbf{a} + \mathbf{P}^T(\mathbf{x})\mathbf{b}, \tag{7}$$

where $R_i(\mathbf{x})$ is a radial basis function (RBF), $n$ is the number of RBFs, $p_j(\mathbf{x})$ is monomial in the space coordinate $\mathbf{x}$, and $m$ is the number of polynomial basis functions. Coefficients $a_i$ and $b_j$ are unknown which should be determined. In the current work, we use the thin plate spline (TPS) as radial basis functions in Eq. (7) which is defined as follows:

$$R(\mathbf{x}) = r^{2m} \ln(r), \quad m = 1, 2, 3, \ldots \tag{8}$$

To determine $a_i$ and $b_j$ in Eq. (7), we consider a support domain, such as a disk or square, surrounding the point of interest $\mathbf{x}$ and use all nodes included in the support domain to form a system of equations based on Eq. (7). In this way, coefficients of $a_i$ and $b_j$ are obtained. Therefore, by the idea of interpolation, Eq. (7) is converted to the following form:

$$u(\mathbf{x}) = \Phi^T(\mathbf{x})\mathbf{U}_s = \sum_{i=1}^{n} \phi_i(\mathbf{x}) u_i, \tag{9}$$

where $\phi_i(\mathbf{x})$ are called the RPIM shape functions which have the Kronecker delta function property, that is

$$\phi_i(\mathbf{x}_j) = \begin{cases} 1, & i = j, \quad j = 1, 2, \ldots, n, \\ 0, & i \neq j, \quad i,j = 1, 2, \ldots, n. \end{cases} \tag{10}$$

This is because the RPIM shape functions are created to pass thorough nodal values. Moreover, the shape functions are the partitions of unity, that is

$$\sum_{i=1}^{n} \phi_i(\mathbf{x}) = 1, \tag{11}$$

for more details about RPIM shape functions and the way they are constructed, the readers are referred to see [42].

## Operational matrices of high-order derivatives

The essential tools of the current approach is operational matrices which is constructed and described briefly in this section. In fact, the operational matrices make the method easy to apply the high-order differential equations which are really difficult to handle by the majority of methods, especially those techniques based on weak forms. Consider the total $N$ nodes for covering the domain of the problem, i.e., $\overline{\Omega} = (\Omega \cup \partial\Omega)$. On the other hand, as we noted in the previous section, $n$ is depend on the point of interest $\mathbf{x}$ (so, we call it $n_{\mathbf{x}}$) in Eq. (9) which is the number of nodes included in support domain $\Omega_{\mathbf{x}}$ corresponding to the point of interest $\mathbf{x}$. Therefore, we have $n_{\mathbf{x}} \leq N$ and Eq. (9) could be reformulated as

$$u(\mathbf{x}) = \Phi^T(\mathbf{x})\mathbf{U}_s = \sum_{j=1}^{N} \phi_j(\mathbf{x})u_j. \tag{12}$$

In fact, there is one shape (basis) function $\phi_j(\mathbf{x})$, $j = 1, 2, 3, \ldots, N$ corresponding to the node $\mathbf{x}$, we define $\Omega_{\mathbf{x}}^c = \{\mathbf{x}_j : \mathbf{x}_j \notin \Omega_{\mathbf{x}}\}$, and then, it is straightforward, from the previous section, to conclude

$$\forall \mathbf{x}_j \in \Omega_{\mathbf{x}}^c : \phi_j(\mathbf{x}) = 0. \tag{13}$$

The derivatives of $u(\mathbf{x})$ are easily obtained as

$$\frac{\partial u(\mathbf{x})}{\partial x} = \sum_{j=1}^{N} \frac{\partial \phi_j(\mathbf{x})}{\partial x} u_j, \quad \frac{\partial u(\mathbf{x})}{\partial y} = \sum_{j=1}^{N} \frac{\partial \phi_j(\mathbf{x})}{\partial y} u_j, \tag{14}$$

and also high derivatives of $u(\mathbf{x})$ are clearly obtained as

$$\frac{\partial^s u(\mathbf{x})}{\partial x^s} = \sum_{j=1}^{N} \frac{\partial^s \phi_j(\mathbf{x})}{\partial x^s} u_j, \quad \frac{\partial^s u(\mathbf{x})}{\partial y^s} = \sum_{j=1}^{N} \frac{\partial^s \phi_j(\mathbf{x})}{\partial y^s} u_j, \tag{15}$$

where $\frac{\partial^s(\cdot)}{\partial x^s}$ and $\frac{\partial^s(\cdot)}{\partial y^s}$ are the $s$th derivative with respect to $x$ and $y$, respectively. By indicating $u_x^{(s)}(\cdot) = \frac{\partial^s(\cdot)}{\partial x^s}$ and $u_y^{(s)}(\cdot) = \frac{\partial^s(\cdot)}{\partial y^s}$, and substituting $\mathbf{x} = \mathbf{x}_i$ in Eq. (14), we can formulate the discrete differentiations process as a matrix-vector multiplications

$$\begin{pmatrix} u'_{x_1} \\ u'_{x_2} \\ \vdots \\ u'_{x_N} \end{pmatrix} = \begin{pmatrix} \frac{\partial \phi_1(\mathbf{x}_1)}{\partial x} & \frac{\partial \phi_2(\mathbf{x}_1)}{\partial x} & \cdots & \frac{\partial \phi_N(\mathbf{x}_1)}{\partial x} \\ \frac{\partial \phi_1(\mathbf{x}_2)}{\partial x} & \frac{\partial \phi_2(\mathbf{x}_2)}{\partial x} & \cdots & \frac{\partial \phi_N(\mathbf{x}_2)}{\partial x} \\ \vdots & \vdots & \ddots & \vdots \\ \frac{\partial \phi_1(\mathbf{x}_N)}{\partial x} & \frac{\partial \phi_2(\mathbf{x}_N)}{\partial x} & \cdots & \frac{\partial \phi_N(\mathbf{x}_N)}{\partial x} \end{pmatrix} \begin{pmatrix} u_1 \\ u_2 \\ \vdots \\ u_N \end{pmatrix}, \tag{16}$$

$$\begin{pmatrix} u'_{y_1} \\ u'_{y_2} \\ \vdots \\ u'_{y_N} \end{pmatrix} = \begin{pmatrix} \frac{\partial \phi_1(\mathbf{x}_1)}{\partial y} & \frac{\partial \phi_2(\mathbf{x}_1)}{\partial y} & \cdots & \frac{\partial \phi_N(\mathbf{x}_1)}{\partial y} \\ \frac{\partial \phi_1(\mathbf{x}_2)}{\partial y} & \frac{\partial \phi_2(\mathbf{x}_2)}{\partial y} & \cdots & \frac{\partial \phi_N(\mathbf{x}_2)}{\partial y} \\ \vdots & \vdots & \ddots & \vdots \\ \frac{\partial \phi_1(\mathbf{x}_N)}{\partial y} & \frac{\partial \phi_2(\mathbf{x}_N)}{\partial y} & \cdots & \frac{\partial \phi_N(\mathbf{x}_N)}{\partial y} \end{pmatrix} \begin{pmatrix} u_1 \\ u_2 \\ \vdots \\ u_N \end{pmatrix}, \tag{17}$$

we indicate above coefficients matrices as $\mathbf{D}x$ and $\mathbf{D}y$, respectively. In addition, clearly, we propose the following matrix-vector multiplications for high-order derivatives

$$U_x^{(s)} = \mathbf{D}^{(s)}xU, \quad U_y^{(s)} = \mathbf{D}^{(s)}yU, \tag{18}$$

where

$$U_x^{(s)} = \left(u_{x_1}^{(s)}, u_{x_2}^{(s)}, \ldots, u_{x_N}^{(s)}\right)^T, \tag{19}$$

$$U_y^{(s)} = \left(u_{y_1}^{(s)}, u_{y_2}^{(s)}, \ldots, u_{y_N}^{(s)}\right)^T, \tag{20}$$

$$\mathbf{D}^{(s)}x_{ij} = \frac{\partial^s \phi_j(\mathbf{x}_i)}{\partial x^s}, \tag{21}$$

$$\mathbf{D}^{(s)}y_{ij} = \frac{\partial^s \phi_j(\mathbf{x}_i)}{\partial y^s}, \tag{22}$$

$$U = (u_1, u_2, \ldots, u_N)^T. \tag{23}$$

## Discretization and numerical implementation for the method

Equation (6) can be rewritten as

$$\begin{aligned} u^{k+1} = u^k + \Delta t \left\{ u^{k+1}\frac{\partial^2 u^{k+1}}{\partial x^2} + \left(\frac{\partial u^{k+1}}{\partial x}\right)^2 + u^k\frac{\partial^2 u^k}{\partial x^2} + \left(\frac{\partial u^k}{\partial x}\right)^2 \right\} \\ + \Delta t \left\{ u^{k+1}\frac{\partial^2 u^{k+1}}{\partial y^2} + \left(\frac{\partial u^{k+1}}{\partial y}\right)^2 + u^k\frac{\partial^2 u^k}{\partial y^2} + \left(\frac{\partial u^k}{\partial y}\right)^2 \right\} \\ + \frac{\Delta t}{2}\{\sigma[u^{k+1}] + \sigma[u^k]\}. \end{aligned} \tag{24}$$

To overcome nonlinearity, suppose $u^{k+1} \approxeq u^k$ in the right-hand side of the above equation. This is possible if the time step $\Delta t$ be sufficiently small, therefore, Eq. (24) can be converted to

$$\begin{aligned} u^{k+1} = u^k + 2\Delta t \left\{ u^k\frac{\partial^2 u^k}{\partial x^2} + \left(\frac{\partial u^k}{\partial x}\right)^2 + u^k\frac{\partial^2 u^k}{\partial y^2} + \left(\frac{\partial u^k}{\partial y}\right)^2 \right\} \\ + \Delta t\sigma[u^k]. \end{aligned} \tag{25}$$

Now, consider $N$ scattered nodes on the boundary and domain of the problem (1), i.e., $\overline{\Omega} = (\Omega \cup \partial\Omega)$. Assuming that $u(\mathbf{x}_i, k\Delta t),\ i = 1, 2, \ldots, N$ are known, our purpose is to compute $u(\mathbf{x}_i, (k+1)\Delta t),\ i = 1, 2, \ldots, N$. For nodes which are included in the inside of the domain, i.e., $\mathbf{x}_i \in \Omega$, to obtain the discrete system of algebraic equations, let us substitute approximate formulas (12) and (14), (15) into equation (25), then

$$\begin{aligned} u^{k+1} &= u^k + 2\Delta t\left\{\left(\sum_{j=1}^{N}\phi_j(\mathbf{x})u_j^k\right)\left(\sum_{j=1}^{N}\frac{\partial^2\phi_j(\mathbf{x})}{\partial x^2}u_j^k\right)\right. \\ &\quad\left.+\left(\sum_{j=1}^{N}\frac{\partial\phi_j(\mathbf{x})}{\partial x}u_j^k\right)^2\right\} \\ &\quad+2\Delta t\left\{\left(\sum_{j=1}^{N}\phi_j(\mathbf{x})u_j^k\right)\left(\sum_{j=1}^{N}\frac{\partial^2\phi_j(\mathbf{x})}{\partial y^2}u_j^k\right)\right. \\ &\quad\left.+\left(\sum_{j=1}^{N}\frac{\partial\phi_j(\mathbf{x})}{\partial y}u_j^k\right)^2\right\} \\ &\quad+\Delta t\sigma\left[\left(\sum_{j=1}^{N}\phi_j(\mathbf{x})u_j^k\right)\right]. \end{aligned} \tag{26}$$

Now, by setting $\mathbf{x} = \mathbf{x}_i,\ i = 1, 2, 3, \ldots, N_\Omega$ ($N_\Omega$ is the number of nodes in $\Omega$) in the above equation and then applying notations (19)–(23), we obtain

$$\begin{aligned} u_i^{k+1} &= u_i^k + 2\Delta t\left\{u_i\left(\sum_{j=1}^{N}\mathbf{D}^{(2)}x_{ij}u_j^k\right)+\left(\sum_{j=1}^{N}\mathbf{D}x_{ij}u_j^k\right)^2\right\} \\ &\quad+2\Delta t\left\{u_i\left(\sum_{j=1}^{N}\mathbf{D}^{(2)}y_{ij}u_j^k\right)+\left(\sum_{j=1}^{N}\mathbf{D}y_{ij}u_j^k\right)^2\right\} \\ &\quad+\Delta t\sigma[u_i]. \end{aligned} \tag{27}$$

## Enforcement of boundary conditions

There are several techniques to enforcing essential boundary conditions in meshless methods, such as the use of penalty methods, Lagrange multipliers and modified variational principles, etc. In the current work, the essential boundary conditions are imposed directly.

### Simply-supported boundary conditions

In the case of simply-supported boundary conditions, we have

$$u(\mathbf{x}, t) = g(\mathbf{x}, t), \quad x \in \partial\Omega, \quad 0 \le t \le T. \tag{28}$$

For nodes which are located on the boundary $\partial\Omega$, we set as

$$u^{k+1}(\mathbf{x}_i) = g(\mathbf{x}_i, (k+1)\Delta t), \quad i = 1, 2, \ldots, N_{\partial\Omega}, \tag{29}$$

where $N_{\partial\Omega}$ is the number of nodes located on $\partial\Omega$.

### Clamped boundary conditions

In the case of clamped boundary conditions, it is usually included the following types of boundary conditions:

$$\frac{\partial u(\mathbf{x}, t)}{\partial n} = g(\mathbf{x}, t), \quad x \in \partial\Omega, \quad 0 \le t \le T. \tag{30}$$

Therefore, in this case, for nodes which are located on the boundary $\partial\Omega$, we set as

$$\begin{aligned} &\mathbf{n}_1(\mathbf{x}_i)\frac{\partial u}{\partial x}(\mathbf{x}_i, k+1\Delta t) + \mathbf{n}_2(\mathbf{x}_i)\frac{\partial u}{\partial y}(\mathbf{x}_i, (k+1)\Delta t) \\ &\quad= g(\mathbf{x}_i, (k+1)\Delta t), \end{aligned} \tag{31}$$

where $\mathbf{n} = \mathbf{n}_1(\mathbf{x}_i)i + \mathbf{n}_2(\mathbf{x}_i)j$ is the outward unit normal to the boundary $\partial\Omega$ at $\mathbf{x}_i \in \partial\Omega$. Then, we have the following equations:

$$\begin{aligned} &\mathbf{n}_1(\mathbf{x}_i)\sum_{j=1}^{N}\mathbf{D}x_{ij}u_j^{k+1} + \mathbf{n}_2(\mathbf{x}_i)\sum_{j=1}^{N}\mathbf{D}y_{ij}u_j^{k+1} \\ &\quad= g(\mathbf{x}_i, (k+1)\Delta t), \quad i = 1, 2, \ldots, N_{\partial\Omega}, \end{aligned} \tag{32}$$

where $N_{\partial\Omega}$ is the number of nodes on $\partial\Omega$.

## Numerical simulations

In this section, we aim to demonstrate that the SMRPI approach has a wider applications by testifying two examples of the type in Eq. (1). To show the accuracy and convergence of the method, maximum absolute error $\varepsilon_\infty$ defined by

$$\begin{aligned} \varepsilon_\infty(u) &= \|u_{\text{exact}} - u_{\text{approx}}\|_\infty \\ &= \{|u_{\text{exact}}(\mathbf{x}_i, t) - u_{\text{approx}}(\mathbf{x}_i, t)| : i = 1, 2, \ldots, N\}, \end{aligned} \tag{33}$$

is used, where $\{u_{\text{approx}}, p_{\text{approx}}\}$ are exact and numerical SMRPI solutions, respectively. In implementing the SMRPI method, we consider support domain as a disk with the radius $r_s = 4.2h$, where $h = 0.1$ is the distance length between two nodes in $x$- or $y$-directions.

*Example 1* Considering the following biological population equation:

$$\begin{aligned} &\frac{\partial u(\mathbf{x}, t)}{\partial t} = \frac{\partial^2 u^2}{\partial x^2} + \frac{\partial^2 u^2}{\partial y^2} + hu^a(1 - ru^b), \\ &\mathbf{x} = (x, y) \in [0, 1]^2, \quad t \in [0, T], \end{aligned} \tag{34}$$

with the initial condition

$$u(\mathbf{x}, 0) = \sqrt{xy}, \quad \mathbf{x} \in [0, 1]^2, \tag{35}$$

with $h = \frac{1}{5}$, $a = 1$, and $r = 0$, and the exact solution is

$$u(\mathbf{x}, t) = e^{\frac{t}{5}}\sqrt{xy}, \quad \mathbf{x} \in [0, 1]^2. \tag{36}$$

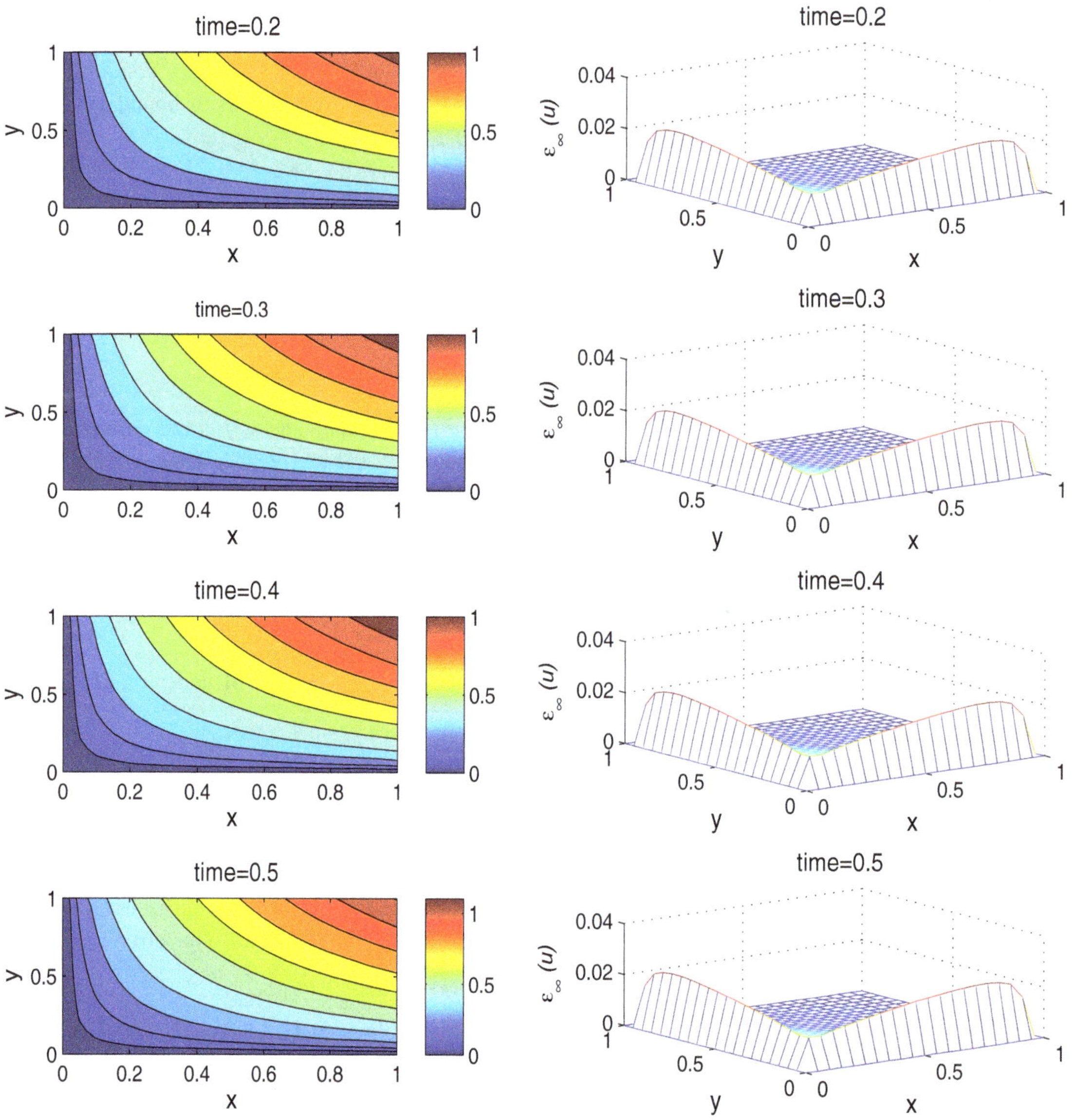


**Fig. 1** SMRPI simulations and the absolute errors for Example 1

In this problem, we set $\Delta t = 0.00001$. Figure 1 shows SMRPI simulations and the absolute errors in some levels of specific values of the time. Figure 2 compares the exact and approximate SMRPI solutions.

*Example 2* Considering the following biological population equation:

$$\frac{\partial u(\mathbf{x},t)}{\partial t} = \frac{\partial^2 u^2}{\partial x^2} + \frac{\partial^2 u^2}{\partial y^2} + hu^a(1 - ru^b), \qquad (37)$$
$$\mathbf{x} = (x, y) \in [0,1]^2, \quad t \in [0, T],$$

with the initial condition

$$u(\mathbf{x},0) = e^{\frac{x+y}{3}}, \quad \mathbf{x} \in [0,1]^2, \qquad (38)$$

with $h = -1$, $a = 1$, $r = -\frac{8}{9}$, and $b = 1$, and the exact solution is

$$u(\mathbf{x},t) = e^{\frac{1}{3}(x+y)-t}, \quad \mathbf{x} \in [0,1]^2. \qquad (39)$$

In this problem, we set $\Delta t = 0.00001$ as well. Figure 3 shows SMRPI simulations and the absolute errors in some levels of specific values of the time. Figure 4 compares the exact and approximate SMRPI solutions. As it is seen, the SMRPI and the exact solutions are not distinguishable, while we have adopted a very simple idea to overcome the

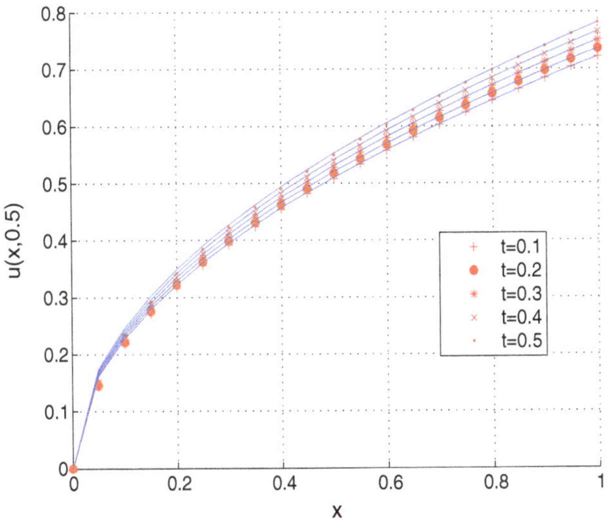


**Fig. 2** Comparison of SMRPI and exact solutions for Example 1

nonlinearity, as it has been pointed out in Sect. "Discretization and numerical implementation for the method".

## Conclusions

In this paper, the biological population equation has been investigated using the spectral meshless radial point interpolation method. The shape (basis) functions constructed by radial point interpolation augmented to monomials have been employed to approximate the 2D displacement field. A system of nonlinear discrete equations is obtained through application of the SMRPI. The nonlinear equation system is solved by iteration with a very simple scheme. A mesh-free method does not require a mesh to discretize the domain of the problem under consideration, and the approximate solution is constructed entirely based on a set

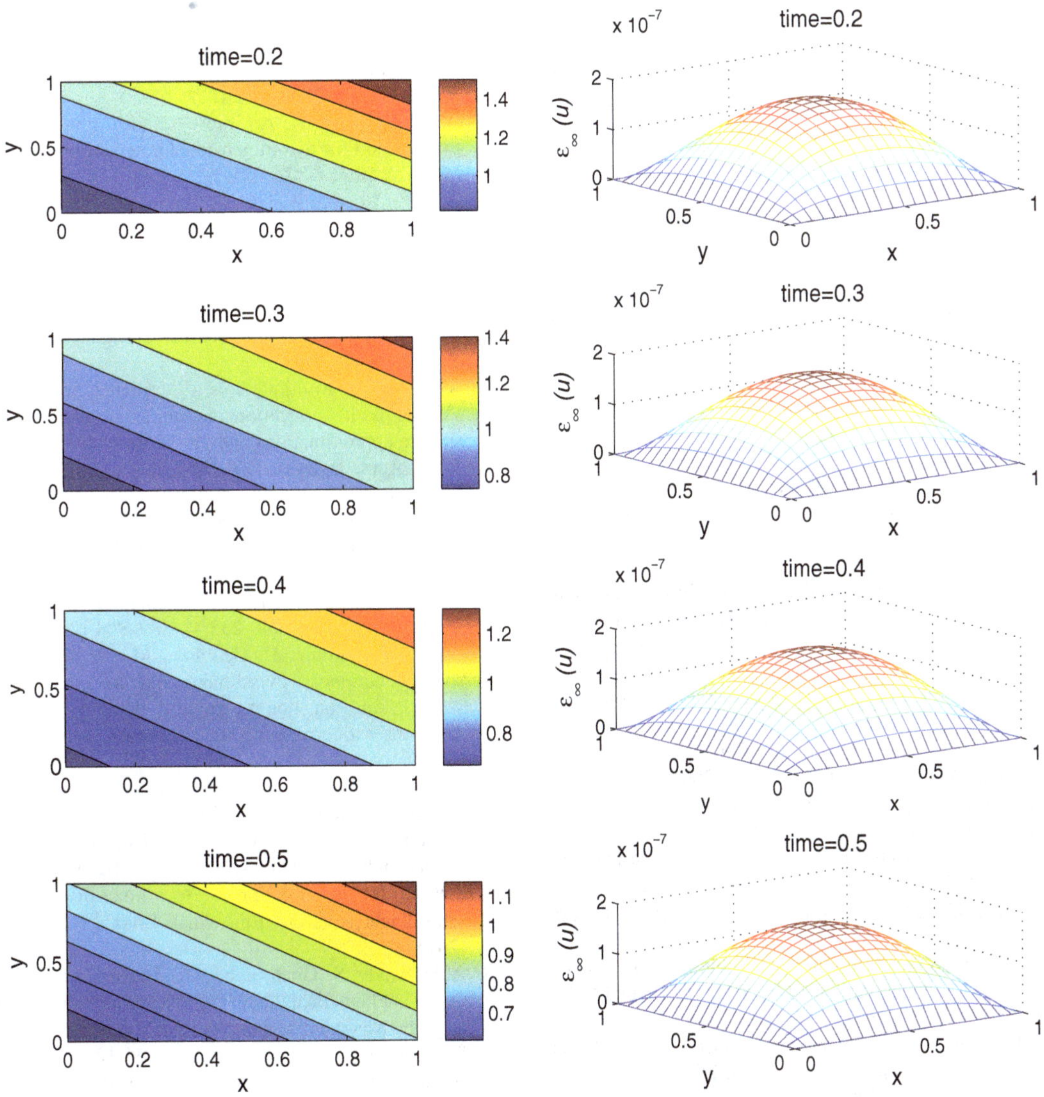


**Fig. 3** SMRPI simulations and the absolute errors for Example 2

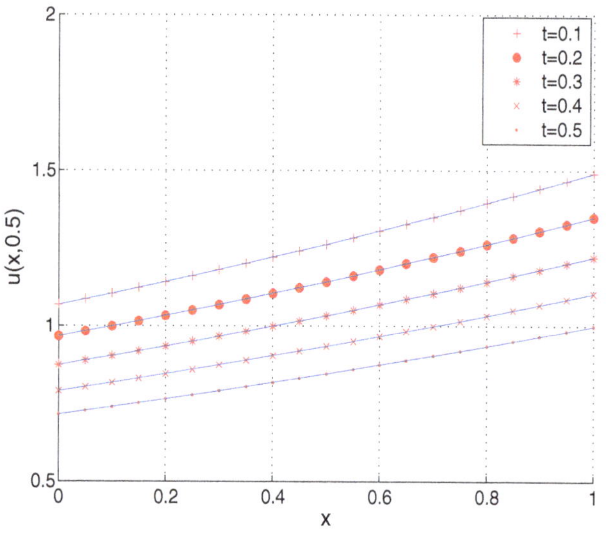


**Fig. 4** Comparison of SMRPI and exact solutions for Example 2

of scattered nodes. In SMRPI technique, in contrast to those meshless methods based on weak form, there is no integration tools in this approach. Therefore, the computational complexity of SMRPI method seems to be of low order. The numerical results are in excellent agreement with exact solutions.

**Acknowledgments** The authors acknowledge financial support from the Imam Khomeini International University project IKIU-11657. The authors would like to thank anonymous referees for valuable suggestions.

## References

1. Cheng, R., Zhang, L., Liew, K.: Modeling of biological population problems using the element-free kp-ritz method. Appl. Math. Comput. **227**, 274–290 (2014)
2. Zhang, L., Deng, Y., Liew, K.: An improved element-free galerkin method for numerical modeling of the biological population problems. Eng. Anal. Bound. Elem. **40**, 181–188 (2014)
3. Carl, E.A.: Population control in arctic ground squirrels. Ecology. **52**, 395–413 (1971)
4. Gurney, W., Nisbet, R.: The regulation of inhomogeneous populations. J. Theor. Biol. **52**(2), 441–457 (1975)
5. Lu, Y.-G.: Hölder estimates of solutions of biological population equations. Appl. Math. Lett. **13**(6), 123–126 (2000)
6. Gurtin, M.E., MacCamy, R.C.: On the diffusion of biological populations. Math. Biosci. **33**(1–2), 35–49 (1977)
7. Aronson, D.G.: The Porous Medium Equation. In: Fasano, A., Primicerio, M. (eds.) Nonlinear Diffusion Problems, Lecture Notes in Math, vol. 1224, pp. 12–46. Springer-Verlag, New York (1986)
8. Jäger, W., Lu, Y.: Global regularity of solution for general degenerate parabolic equations in 1-d. J. Diff. Equ. **140**(2), 365–377 (1997)
9. Giuggioli, L., Kenkre, V.: Analytic solutions of a nonlinear convective equation in population dynamics. Phys. D Nonlinear Phenom. **183**(3), 245–259 (2003)
10. Shakeri, F., Dehghan, M.: Numerical solution of a biological population model using hes variational iteration method. Comput. Math. Appl. **54**(7), 1197–1209 (2007)
11. Belytschko, T., Lu, Y.Y., Gu, L.: Element-free Galerkin methods. Int. J. Numer. Methods Eng. **37**(2), 229–256 (1994)
12. Belytschko, T., Lu, Y.Y., Gu, L.: Element free Galerkin methods for static and dynamic fracture. Int. J. Solids Struct. **32**, 2547–2570 (1995)
13. Duan, Y., Tan, Y.: A meshless Galerkin method for Dirichlet problems using radial basis functions. J. Comput. Appl. Math. **196**(2), 394–401 (2006)
14. Assari, P., Adibi, H., Dehghan, M.: A meshless discrete Galerkin (MDG) method for the numerical solution of integral equations with logarithmic kernels. J. Comput. Appl. Math. **267**, 160–181 (2014)
15. Dehghan, M., Salehi, R.: A meshless local Petrov-Galerkin method for the time-dependent Maxwell equations. J. Comput. Appl. Math. **268**, 93–110 (2014)
16. Mirzaei, D., Dehghan, M.: Meshless local Petrov-Galerkin (MLPG) approximation to the two dimensional sine-Gordon equation. J. Comput. Appl. Math. **233**(10), 2737–2754 (2010)
17. Fili, A., Naji, A., Duan, Y.: Coupling three-field formulation and meshless mixed Galerkin methods using radial basis functions. J. Comput. Appl. Math. **234**(8), 2456–2468 (2010)
18. Peng, M., Li, D., Cheng, Y.: The complex variable element-free Galerkin (CVEFG) method for elasto-plasticity problems. Eng. Struct. **33**(1), 127–135 (2011)
19. Dai, B., Cheng, Y.: An improved local boundary integral equation method for two-dimensional potential problems. Int. J. Appl. Mech. **2**(2), 421–436 (2010)
20. Bai, F., Li, D., Wang, J., Cheng, Y.: An improved complex variable element-free Galerkin method for two-dimensional elasticity problems. Chin. Phys. B **21**(2), 020204 (2012)
21. Atluri, S., Shen, S.: The meshless local Petrov-Galerkin (MLPG) method: a simple and less costly alternative to the finite element and boundary element methods. Comput. Model. Eng. Sci. **3**, 11–51 (2002)
22. Kansa, E.: Multiquadrics-a scattered data approximation scheme with applications to computational fluid-dynamics. I. surface approximations and partial derivative estimates. Comput. Math. Appl. **19**(8–9), 127–145 (1990)
23. Dehghan, M., Shokri, A.: A numerical method for solution of the two dimensional sine-Gordon equation using the radial basis functions. Math. Comput. Simul. **79**, 700–715 (2008)
24. Jakobsson, S., Andersson, B., Edelvik, F.: Rational radial basis function interpolation with applications to antenna design. J. Comput. Appl. Math. **233**(4), 889–904 (2009)
25. Assari, P., Adibi, H., Dehghan, M.: A meshless method for solving nonlinear two-dimensional integral equations of the second kind on non-rectangular domains using radial basis functions with error analysis. J. Comput. Appl. Math. **239**, 72–92 (2013)
26. Zhang, Y., Tan, Y.: Meshless schemes for unsteady Navier-Stokes equations in vorticity formulation using radial basis functions. J. Comput. Appl. Math. **192**(2), 328–338 (2006)
27. Abbasbandy, S., Ghehsareh, H.R., Hashim, I.: A meshfree method for the solution of two-dimensional cubic nonlinear schrödinger equation. Eng. Anal. Bound. Elem. **37**(6), 885–898 (2013)
28. Abbasbandy, S., Ghehsareh, H.R., Hashim, I.: Numerical analysis of a mathematical model for capillary formation in tumor angiogenesis using a meshfree method based on the radial basis function. Eng. Anal. Bound. Elem. **36**(12), 1811–1818 (2012)
29. Abbasbandy, S., Shirzadi, A.: A meshless method for two-dimensional diffusion equation with an integral condition. Eng. Anal. Bound. Elem. **34**(12), 1031–1037 (2010)

30. Shirzadi, A., Sladek, V., Sladek, J.: A local integral equation formulation to solve coupled nonlinear reaction-diffusion equations by using moving least square approximation. Eng. Anal. Bound. Elem. **37**, 8–14 (2013)
31. Dehghan, M., Ghesmati, A.: Numerical simulation of two-dimensional sine-Gordon solitons via a local weak meshless technique based on the radial point interpolation method (RPIM). Comput. Phys. Commun. **181**, 772–786 (2010)
32. Tadeu, A., Chen, C., Antonio, J., Simoes, N.: A boundary meshless method for solving heat transfer problems using the Fourier transform. Adv. Appl. Math. Mech. **3**, 572–585 (2011)
33. Shivanian, E.: Analysis of meshless local radial point interpolation (MLRPI) on a nonlinear partial integro-differential equation arising in population dynamics. Eng. Anal. Bound. Elem. **37**, 1693–1702 (2013)
34. Shivanian, E.: Analysis of meshless local and spectral meshless radial point interpolation (MLRPI and SMRPI) on 3-D nonlinear wave equations. Ocean Eng. **89**, 173–188 (2014)
35. Shivanian, E.: Meshless local Petrov-Galerkin (MLPG) method for three-dimensional nonlinear wave equations via moving least squares approximation. Eng. Anal. Bound. Elem. **50**, 249–257 (2015)
36. Shivanian, E., Khodabandehlo, H.: Meshless local radial point interpolation (MLRPI) on the telegraph equation with purely integral conditions. Eur. Phys. J. Plus **129**, 241–251 (2014)
37. Shivanian, E., Abbasbandy, S., Alhuthali, M.S., Alsulami, H.H.: Local integration of 2-d fractional telegraph equation via moving least squares approximation. Eng. Anal. Bound. Elem. **56**, 98–105 (2015)
38. Hosseini, V.R., Shivanian, E., Chen, W.: Local integration of 2-D fractional telegraph equation via local radial point interpolant approximation. Eur. Phys. J. Plus **130**(33), 1–21 (2015)
39. Aslefallah, M., Shivanian, E.: Nonlinear fractional integro-differential reaction-diffusion equation via radial basis functions. Eur. Phys. J. Plus **130**(47), 1–9 (2015)
40. Shivanian, E.: On the convergence analysis, stability, and implementation of meshless local radial point interpolation on a class of three-dimensional wave equations. Int. J. Numer. Methods Eng. **105**(2), 83–110 (2016)
41. Shivanian, E., Rahimi, A. Hosseini, M.: Meshless local radial point interpolation to three-dimensional wave equation with Neumann's boundary conditions. Int. J. Comput. Math. (2015). doi:10.1080/00207160.2015.1085032
42. Shivanian, E.: Spectral meshless radial point interpolation (SMRPI) method to two-dimensional fractional telegraph equation. Math. Methods Appl. Sci. **39**(7), 1820–1835 (2016)
43. Shivanian, E.: A new spectral meshless radial point interpolation (SMRPI) method: A well-behaved alternative to the meshless weak forms. Eng. Anal. Bound. Elem. **54**, 1–12 (2015)
44. Fatahi, H., Saberi-Nadjafi, J., Shivanian, E.: A new spectral meshless radial point interpolation (smrpi) method for the two-dimensional fredholm integral equations on general domains with error analysis. J. Comput. Appl. Math. **294**, 196–209 (2016)

# 20

# Parameter estimation of some Kumaraswamy-G type distributions

Güvenç Arslan[1] · Sevgi Yurt Oncel[1]

**Abstract** Since Kum-G distributions have additional two parameters, the estimation of parameters becomes an interesting problem by itself. In this study, we consider parameter estimation of Kum-Weibull, Kum-Pareto and Kum-Power distributions by using the maximum likelihood and the maximum spacing methods. These three distributions are important in reliability and other applications. The Kum-Pareto and Kum-Power distributions have parameter-dependent boundaries, which makes the estimation of parameters more interesting. We performed simulations for each of these considered distributions by using the R software for estimating parameters using the maximum likelihood and the maximum spacing method. In addition, an application of these distribution families to real data for modeling wind speed in a particular location in Turkey is discussed.

**Keywords** Kumaraswamy distribution · Maximum likelihood · Maximum spacing · Parameter estimation · Simulation

## Introduction

In 1980, Kumaraswamy [11] introduced a new distribution with applications in hydrology. The cumulative distribution function (cdf) of this new distribution is given by

$$F(x) = 1 - (1 - x^a)^b, \quad 0 < x < 1, \tag{1}$$

where $a > 0$ and $b > 0$. Jones [10] discussed properties of the Kumaraswamy distribution and its similarities with the beta distribution.

In recent years one can find many papers which generalize this distribution by replacing $x$ with some known distribution such as normal, Weibull, Pareto, and others (see, for example [2, 9, 12]). Based on the Kumaraswamy distribution Cordeiro and Castro [6] introduced a new generalized family of distributions, denoted in this paper by Kum-G, and discussed its basic statistical properties and application to a real data set.

It can be seen that in recent years many authors study applications and parameter estimation of special Kum-G distributions. For example, Cordeiro et al. [9] investigate the Kum-Weibull model and its application to failure data. Tamandi and Nadarajah [16] discuss parameter estimation of the Kum-Weibull, Kum-Normal and Kum-Inverse Gaussian families.

Since Kum-G distributions have additional two parameters, the estimation of parameters becomes an interesting problem by itself. The maximum likelihood method (ML) is one of the preferred methods for estimating the parameters in Kum-G distributions. Tamandi and Nadarajah [16] consider also the maximum spacing method (MSP) and compare it with the maximum likelihood (ML) method for estimating the parameters in some of the Kum-G distributions.

It is known that in situations like mixtures of distributions and distributions with a parameter-dependent lower bound, where the ML estimator leads to inconsistent estimators, the MSP estimator is consistent; see [13]. Motivated by this fact it is natural to consider the MSP estimator

✉ Güvenç Arslan
guvenc.arslan@gmail.com

Sevgi Yurt Oncel
syoncel@gmail.com

1 Department of Statistics, Kırıkkale University, Yahşihan, 71450 Kırıkkale, Turkey

in parameter estimation for the Kum-Pareto and Kum-Power distributions.

In this study, we consider parameter estimation of the Kum-Weibull [6], Kum-Pareto [2] and Kum-Power [12] families of distributions by using the ML and MSP methods. Although one may find some studies for the Kum-Weibull and Kum-Pareto distributions, there is only one paper dealing with the Kum-Power family of distributions. We performed simulations for each of the considered family of distributions. For calculations we used the R software [14]. In particular, for estimating parameters in the simulations the *optim* function in R was applied with the Nelder–Mead method. The parameter estimates for the Weibull distribution were obtained by applying the *fitdistr* method in R.

It can be seen from the literature that wind speed can be modeled by various distributions such as Weibull, Rayleigh, gamma, lognormal, beta, Burr, and inverse Gaussian distributions, among others [17]. For example, Chang [3] compared the performance of six numerical methods in estimating Weibull parameters for wind energy application. He concludes that the maximum likelihood, modified maximum likelihood and moment methods show better performance in simulation tests. In this study, we consider modeling wind speed by using the following generalized families of distributions: Kum-Weibull and Kum-Power. We note here that, for example, the Kum-Weibull family of distributions includes the Weibull and Rayleigh distributions. It is expected that the flexibility of the two additional two parameters in the Kum-G family of distributions will improve the modeling results. The parameter estimates for the real data were obtained by applying the *ga* method [15], which is a genetic algorithm method implemented in R.

## Kumaraswamy distributions considered

Cordeiro and Castro [6] introduced a new generalized family of distributions by replacing $x$ with a continuous base line distribution $G(x)$ in Kumaraswamy's distribution:

$$F(x) = 1 - \{1 - G^a(x)\}^b, \tag{2}$$

$$f(x) = ab\, g(x)\, G^{a-1}(x)(1 - G^a(x))^{b-1}, \tag{3}$$

where $g$ is the probability density function (pdf) of $G$ and $a > 0$, $b > 0$.

The cdf of the Kum-Weibull distribution is given by

$$F(x) = 1 - \left(1 - \left[1 - e^{-(\lambda x)^c}\right]^a\right)^b, \tag{4}$$

where $a > 0$, $b > 0$, $\lambda > 0$ and $c > 0$. We will denote this distribution by Kum-W$(a, b, \lambda, c)$. Some special cases of the Kum-W$(a, b, \lambda, c)$ are given in Table 1 [6]. Figure 1 shows some special cases of the density function for this family.

**Table 1** Some Kum-W special cases

| Distribution | $\lambda$ | c | a | b |
|---|---|---|---|---|
| Kum-exponential | | 1 | | |
| Kum-Rayleigh | | 2 | | |
| Exponentiated-Weibull | | | | 1 |
| Exponentiated-Rayleigh | | 2 | | 1 |
| Exponentiated-exponential | | 1 | | 1 |
| Weibull | | | 1 | 1 |
| Rayleigh | | 2 | 1 | 1 |
| Exponential | | 1 | 1 | 1 |

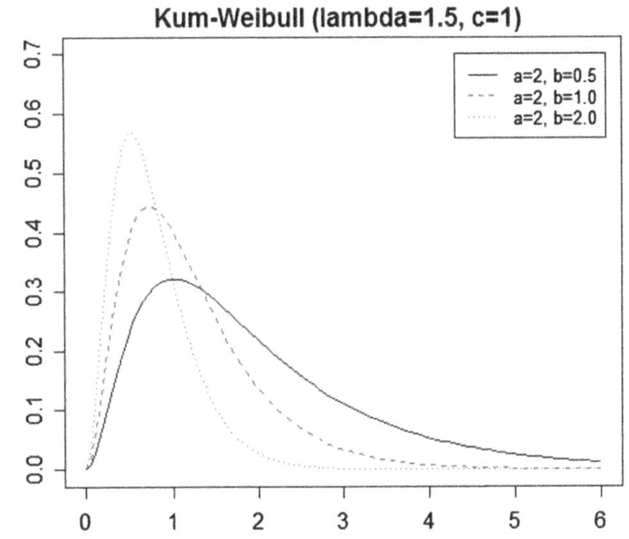


**Fig. 1** Some Kum-Weibull distributions

The cdf of the Kum-Pareto distribution is

$$F(x) = 1 - \left(1 - \left[1 - \left(\frac{\beta}{x}\right)^k\right]^a\right)^b, \tag{5}$$

where $a > 0$, $b > 0$, $\beta > 0$ and $k > 0$. We will denote this distribution by Kum-Par$(a, b, \beta, k)$. Figure 2 shows some special cases of Kum-Pareto density functions.

The cdf of the Kum-Power distribution is given by

$$F(x) = 1 - \left(1 - \left[\left(\frac{x}{\beta}\right)^\alpha\right]^a\right)^b \tag{6}$$

where $a > 0$, $b > 0$, $\alpha > 0$ and $\beta > 0$. We will denote this distribution by Kum-Pow$(a, b, \alpha, \beta)$. Figure 3 shows some special cases of Kum-Power density functions.

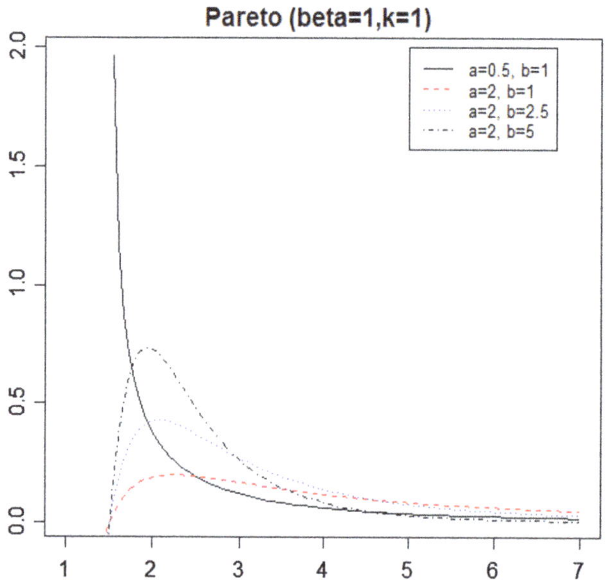


**Fig. 2** Some Kum-Pareto distributions

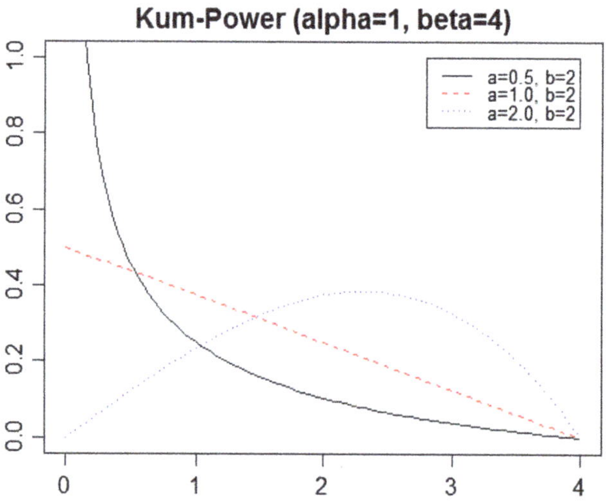


**Fig. 3** Some Kum-Power distributions

## Parameter estimation

The ML method is one of the most widely used parameter estimation methods in statistics. On the other hand, it is known that ML estimation may lead to inconsistent estimation results, especially in parameter-dependent boundary situations. Ranneby [13] showed that in such cases, the maximum spacing method is more reliable than the ML method. Ekström [7, 8], on the other hand, showed that the MSP estimators may give better results than ML estimators for small samples. Also, Cheng [4] showed that in unbounded likelihood problems such as estimation of three-parameters in the Weibull distribution, the MSP estimation method produces consistent and asymptotically efficient estimators. Recently, Tamandi and Nadarajah [16] investigated parameter estimation of some Kum-G distributions by using ML and MSP methods.

In this paper, we consider parameter estimation of the Kum-Weibull, Kum-Pareto and Kum-Power distributions by using ML and MSP methods. We note that in Kum-Par as well as Kum-Pow distributions parameter-dependent boundaries exist. Therefore, we hope that this study will contribute to parameter estimation in Kum-G distributions.

Since by definition of the Kum-G distributions two additional shape parameters are introduced to the family of $G(x, \boldsymbol{\theta})$ distributions, the estimation of parameters becomes an interesting problem. The additional two parameters $a$ and $b$ provide more flexibility in modeling and applications. On the other hand, it should be noted that this flexibility also causes some major problems in parameter estimation. It can be seen that one of the main problems is that one may have to deal with quite different support sets of the distribution for different parameter values. Thus classical hill-climbing approaches such as Newton–Raphson and as well as methods such as Nelder–Mead may actually not give consistent (or any) results in Kum-G distributions.

### Maximum likelihood method

To obtain the ML and MSP formulations for Kum-G distributions suppose that $X_1, X_2, \ldots, X_n$ is a random sample from some Kum-G distribution $G(x, \boldsymbol{\theta})$ with pdf given by (3) and baseline pdf $g$. Also suppose that $g$ is parameterized by a vector $\boldsymbol{\theta}$ of length $p$. The log-likelihood function of $a$, $b$ and $\boldsymbol{\theta}$ is

$$\begin{aligned} l(a, b, \boldsymbol{\theta}) = n \log a + n \log b + \sum_{i=1}^{n} \log g(x_i, \boldsymbol{\theta}) \\ + (a-1) \sum_{i=1}^{n} \log G(x_i, \boldsymbol{\theta}) \\ + (b-1) \sum_{i=1}^{n} \log[1 - G^a(x_i, \boldsymbol{\theta})] \end{aligned} \quad (7)$$

The ML estimates of the parameters can be found by solving the following equations simultaneously:

$$\frac{n}{a} + \sum_{i=1}^{n} \log G(x_i, \boldsymbol{\theta}) - (b-1) \sum_{i=1}^{n} \frac{G^a(x_i, \boldsymbol{\theta}) \log G(x_i, \boldsymbol{\theta})}{1 - G^a(x_i, \boldsymbol{\theta})} = 0,$$

$$\frac{n}{b} + \sum_{i=1}^{n} \log[1 - G^a(x_i, \boldsymbol{\theta})] = 0,$$

and

$$\begin{aligned} \sum_{i=1}^{n} \frac{1}{g(x_i, \boldsymbol{\theta})} \frac{\partial g(x_i, \boldsymbol{\theta})}{\partial \theta_k} - (a-1) \sum_{i=1}^{n} \frac{1}{G(x_i, \boldsymbol{\theta})} \frac{\partial G(x_i, \boldsymbol{\theta})}{\partial \theta_k} \\ - a(b-1) \sum_{i=1}^{n} \frac{G^{a-1}(x_i, \boldsymbol{\theta})}{1 - G^a(x_i, \boldsymbol{\theta})} \frac{\partial G(x_i, \boldsymbol{\theta})}{\partial \theta_k} = 0. \end{aligned} \quad (8)$$

It should be noted that in order to find numerical solutions by using the above formulae, one has to calculate among other functions, $g(x, \boldsymbol{\theta})$ for different parameter vectors $\boldsymbol{\theta}$, which may stop the iterations of the algorithm because $g(x, \boldsymbol{\theta})$ may not be defined for the corresponding $\boldsymbol{\theta}$ vector. Since in (8) the first term includes the reciprocal of $g(x, \boldsymbol{\theta})$ some algorithms may not converge or even work in this case. By considering how the MSP method (see Eq. (11)) is obtained, one may observe that this type of problem is less likely to occur in MSP.

## Maximum spacing method

The MSP method was introduced by Cheng [4] as an alternative to the ML method. Ranneby [13] derived the MSP method from an approximation of the Kullback–Leibler divergence (KLD). Cheng [4] showed that in unbounded likelihood problems such as estimation of three-parameter gamma, lognormal or Weibull distributions, the MSP estimation method produces consistent and asymptotically efficient estimators. In situations like mixtures of distributions and distributions with a parameter-dependent lower bound, where the MLE leads to inconsistent estimators, the MSP estimator is consistent; see [13]. Even in other situations, Ekström [8] showed that the MSP estimators have better properties than ML estimators for small samples. Ekström [8] showed that MSP estimators are L1-consistent for any unimodal pdf without any additional conditions. According to [13], the MSP method works better than the ML method for multivariate data too. MSP estimators have all the nice properties of ML estimators such as consistency, asymptotic normality, efficiency and invariance under one-to-one transformations. For a detailed survey of the MSP method, the reader is referred to [8]. On the other hand, MSP estimators have some disadvantages too. First of all, they are sensitive to closely spaced observations, and especially ties. They are also sensitive to secondary clustering: one example is when a set of observations is thought to come from a single normal distribution, but in fact comes from a mixture of normals with different means [5].

Let $x_1, x_2, \ldots, x_n$ be a random sample from a population with cdf $F(x, \boldsymbol{\theta})$ and let $f(x, \boldsymbol{\theta})$ denote the corresponding pdf. The Kullback–Leibler divergence between $F(x, \boldsymbol{\theta})$ and $F(x, \boldsymbol{\theta}_0)$ is given by

$$H(F_{\boldsymbol{\theta}}, F_{\boldsymbol{\theta}_0}) = \int f(x, \boldsymbol{\theta}_0) \log \left| \frac{f(x, \boldsymbol{\theta}_0)}{f(x, \boldsymbol{\theta})} \right| \mathrm{d}x$$

The KLD can be approximated by estimating $H(F_{\boldsymbol{\theta}}, F_{\boldsymbol{\theta}_0})$ by

$$\frac{1}{n} \sum_{i=1}^{n} \log \left| \frac{f(x_i, \boldsymbol{\theta}_0)}{f(x_i, \boldsymbol{\theta})} \right| \tag{9}$$

Minimizing (9) with respect to $\boldsymbol{\theta}$, the estimator of $\boldsymbol{\theta}_0$ can be found, which is actually the well-known MLE. It should be noted that for some continuous distributions, $\log f(x_i)$, $i = 1, \ldots, n$ may not be bounded from above. Ranneby [13], therefore, suggested another approximation of the KLD, namely

$$\frac{1}{n+1} \sum_{i=1}^{n+1} \log \left| \frac{F(x_{(i)}, \boldsymbol{\theta}_0) - F(x_{(i-1)}, \boldsymbol{\theta}_0)}{F(x_{(i)}, \boldsymbol{\theta}) - F(x_{(i-1)}, \boldsymbol{\theta})} \right|, \tag{10}$$

where $x_{(i-1)} \leq x_{(i-1)} \leq \cdots \leq x_{(i-1)}$ are the order statistics and $F(x_{(0)}, \boldsymbol{\theta}) \equiv 0$, $F(x_{(n+1)}, \boldsymbol{\theta}) \equiv 1$. $F(x_{(i)}, \boldsymbol{\theta}_0) - F(x_{(i-1)}, \boldsymbol{\theta})$ are the first-order spacings of $F(x_{(0)}, \boldsymbol{\theta}_0), \ldots, F(x_{(n+1)}, \boldsymbol{\theta})$.

By minimizing (10) the MSP estimator of $\boldsymbol{\theta}_0$ is obtained. Minimizing (10) is equivalent to maximizing:

$$M(\boldsymbol{\theta}) = \sum_{i=1}^{n+1} \log \left[ F(x_{(i)}, \boldsymbol{\theta}) - F(x_{(i-1)}, \boldsymbol{\theta}) \right], \tag{11}$$

where $\boldsymbol{\theta}$ is an unknown parameter. Therefore, the MSP estimator can obtained by maximizing $M(\boldsymbol{\theta})$ with respect to $\boldsymbol{\theta}$.

Consider estimation of some Kum-G distribution with baseline distribution $G$ by the MSP method. Suppose that $x_{(1)}, x_{(2)}, \ldots, x_{(n)}$ is an ordered sample and $x_{(0)} = 0, x_{(n+1)} = \infty$. These values for $x_{(0)}$ and $x_{(n+1)}$ assume that the support for $G$ is the positive real line. If the support for $G$ is different, then $x_{(0)}$ and $x_{(n+1)}$ can be chosen accordingly. Substituting (2) into (11) we obtain

$$M(a, b, \boldsymbol{\theta}) = \sum_{i=1}^{n+1} \log \left\{ \left(1 - G^a(x_{(i-1)}, \boldsymbol{\theta})\right)^b - \left(1 - G^a(x_{(i)}, \boldsymbol{\theta})\right)^b \right\}. \tag{12}$$

To find the ML estimates of the parameters, the simultaneous solutions of the equations obtained by taking partial derivatives with respect to the parameters $a$, $b$ and $\boldsymbol{\theta}$ have to be found. It should be noted that, in general, no analytical solution exists for these equations. Therefore, numerical methods need to be applied in order to find the corresponding parameter estimates.

## Simulation results

Simulation is a powerful tool that is used in many areas of science. For example, some recent simulation studies can be found in [1, 18]. Abbasbandy and Shivanian [1] used numerical simulation based on meshless technique to study the biological population model. Vajargah and Shoghi [18] used quasi-Monte Carlo method in prediction of total index of stock market and value at risk. To assess the performance of the ML and MSP estimators we conducted a small size simulation study for the Kum-W, Kum-Par and Kum-Pow distributions. It should be noted that these three Kum-G distributions have different characteristics and are also important in reliability problems and applications. The Kum-Par and Kum-Pow distributions both have parameter-dependent boundaries, which may have important implications in parameter estimation. We used 1000 runs in each simulation to compare estimation results for the estimators. In this study, we selected a sample size of $n = 25$.

In order to include the effect of initial values in the estimates, we used randomly generated starting values as follows. Let Kum-G$(a, b, \theta_1, \theta_2)$ be one of the considered Kum-G distributions, where $G$ is one of the Weibull, Pareto or Power distributions and $\theta_1$ and $\theta_2$ are the corresponding parameters. We generated random variates from Kum-G$(a_0, b_0, {\theta_1}^0, {\theta_2}^0)$ and as starting values the following values were used:

$$a_0 + u_1,\ u_1 \in U\left(-\frac{a_0}{2}, \frac{a_0}{2}\right)$$

$$b_0 + u_2,\ u_2 \in U\left(-\frac{b_0}{2}, \frac{b_0}{2}\right)$$

$${\theta_1}^0 + u_3,\ u_3 \in U\left(-\frac{{\theta_1}^0}{2}, \frac{{\theta_1}^0}{2}\right)$$

$${\theta_2}^0 + u_4,\ u_4 \in U\left(-\frac{{\theta_2}^0}{2}, \frac{{\theta_2}^0}{2}\right)$$

Table 2 shows that, in general, MSP estimates have smaller bias and MSEs. When $a$ is considerably larger than $b$ significant differences between the two estimates are observed. Also for $a = 10$ we observed some convergence problems related to the initial parameters in the ML method. Therefore, only 1000 iterations were conducted in the simulations. We note that this problem did not occur in the MSP method.

When $a < b$ (that is for heavy-tailed) and for fixed $a$ with increasing $b$ Table 3 shows that the MSEs for MSP are smaller then for MLE. On the other hand, when $a$ is considerably larger than $b$ significant differences between the two estimates are observed. In the remaining cases no significant differences are observed.

From Table 4 it can be observed that MLE, in general, outperforms MSP estimates. This can be explained by the fact that for the Kum-Pow distribution closely spaced observations are much more likely to occur. It is known that MSP is sensitive to closely spaced observations.

It should be noted that estimating all four parameters in the Kum-G families of distributions may result in inconsistent estimates. In addition, it can be observed that the estimates are highly dependent on the initial values which may also lead to convergence problems. For this reason when applying these families of distributions to real data, we preferred to use genetic algorithms for estimating the parameters.

## Application to real data

Wind energy is an important alternative to conventional energy resources. Therefore, one may find many studies related to modeling wind characteristics such as wind

**Table 2** Bias and MSE for sample size $n = 25$ (1000 runs)

| Weibull | $a$ | $b$ | $\hat{\lambda}$ | MSE($\hat{\lambda}$) | $\hat{c}$ | MSE($\hat{c}$) | $\hat{a}$ | MSE($\hat{a}$) | $\hat{b}$ | MSE($\hat{b}$) |
|---|---|---|---|---|---|---|---|---|---|---|
| MLE | 0.5 | 0.5 | 0.0622 | 0.0377 | -0.0134 | 0.0639 | −0.0319 | 0.0855 | 0.0587 | 0.0958 |
| MSP | | | −0.0057 | 0.0865 | 0.0284 | 0.0969 | 0.0334 | 0.0283 | 0.0427 | 0.0298 |
| MLE | 0.5 | 1.0 | 0.1059 | 0.0412 | 0.0578 | 0.0306 | −0.4037 | 0.2029 | 0.3809 | 0.1666 |
| MSP | | | 0.0123 | 0.0829 | 0.0123 | 0.0924 | 0.0547 | 0.0327 | 0.0544 | 0.1003 |
| MLE | 0.5 | 2.5 | 0.1357 | 0.0542 | 0.0764 | 0.0330 | −0.3916 | 0.1640 | 0.6279 | 0.4740 |
| MSP | | | 0.0658 | 0.0992 | 0.0161 | 0.0791 | 0.1449 | 0.0722 | 0.0484 | 0.5480 |
| MLE | 2 | 0.5 | 0.0016 | 0.0203 | 0.0113 | 0.0238 | 0.1850 | 0.135 | 0.0708 | 0.0089 |
| MSP | | | 0.0273 | 0.0843 | 0.0850 | 0.1250 | 0.0536 | 0.389 | 0.0200 | 0.0232 |
| MLE | 2 | 1.0 | −0.0891 | 0.1176 | 0.1390 | 0.1206 | 0.1730 | 0.240 | 0.1142 | 0.1038 |
| MSP | | | 0.0400 | 0.0824 | 0.1140 | 0.1460 | 0.0968 | 0.365 | −0.0017 | 0.0838 |
| MLE | 2 | 2.5 | 0.2786 | 0.3257 | 0.3100 | 0.2091 | −0.2550 | 0.702 | 0.4131 | 0.3369 |
| MSP | | | 0.1288 | 0.1416 | 0.1160 | 0.1370 | 0.0431 | 0.349 | 0.0315 | 0.5240 |
| MLE | 10 | 0.5 | 0.3982 | 0.3028 | −0.926 | 1.230 | 0.6820 | 2.65 | 0.5470 | 0.3457 |
| MSP | | | 0.2270 | 0.3490 | 0.598 | 0.599 | 0.0335 | 8.01 | 0.0084 | 0.0210 |
| MLE | 10 | 1.0 | 0.0042 | 0.0425 | 0.579 | 0.677 | 0.4750 | 2.30 | 0.1280 | 0.0246 |
| MSP | | | 0.1640 | 0.2290 | 0.668 | 0.775 | 0.2492 | 8.41 | 0.0121 | 0.0857 |
| MLE | 10 | 2.5 | 0.0485 | 0.0800 | 0.672 | 0.725 | 0.3150 | 2.61 | 0.3200 | 0.1363 |
| MSP | | | 0.2970 | 0.3330 | 0.565 | 0.634 | 0.2283 | 8.68 | −0.0023 | 0.4780 |

**Table 3** Bias and MSE for sample size $n = 25$ (1000 runs)

| Pareto | $a$ | $b$ | $\hat{\beta}$ | MSE($\hat{\beta}$) | $\hat{k}$ | MSE($\hat{k}$) | $\hat{a}$ | MSE($\hat{a}$) | $\hat{b}$ | MSE($\hat{b}$) |
|---|---|---|---|---|---|---|---|---|---|---|
| MLE | 0.5 | 0.5 | −0.396 | 0.199 | 0.0081 | 0.0856 | 0.0284 | 0.0312 | 0.0546 | 0.0411 |
| MSP | | | −0.412 | 0.204 | 0.0220 | 0.0883 | 0.0284 | 0.0264 | 0.0993 | 0.0394 |
| MLE | 0.5 | 1.0 | −0.425 | 0.218 | 0.0120 | 0.0914 | 0.0231 | 0.0273 | 0.1191 | 0.1239 |
| MSP | | | −0.418 | 0.204 | 0.0297 | 0.0814 | 0.0261 | 0.0250 | 0.0803 | 0.1005 |
| MLE | 0.5 | 2.5 | −0.597 | 0.418 | 0.1606 | 0.1176 | −0.0356 | 0.0404 | 0.2870 | 0.6307 |
| MSP | | | −0.560 | 0.350 | 0.2024 | 0.1297 | 0.0633 | 0.0371 | 0.0982 | 0.5622 |
| MLE | 2 | 0.5 | −0.314 | 0.155 | −0.0032 | 0.0891 | 0.1058 | 0.351 | 0.0322 | 0.0356 |
| MSP | | | −0.323 | 0.169 | 0.0071 | 0.0772 | −0.0416 | 0.335 | 0.1124 | 0.0476 |
| MLE | 2 | 1.0 | −0.296 | 0.151 | 0.0146 | 0.0807 | 0.0303 | 0.338 | 0.0583 | 0.1124 |
| MSP | | | −0.316 | 0.167 | 0.0900 | 0.1103 | 0.0059 | 0.334 | 0.0396 | 0.0760 |
| MLE | 2 | 2.5 | −0.369 | 0.213 | 0.1837 | 0.1597 | 0.0447 | 0.350 | 0.0393 | 0.5444 |
| MSP | | | −0.383 | 0.219 | 0.1971 | 0.1631 | 0.0200 | 0.342 | 0.0402 | 0.5619 |
| MLE | 10 | 0.5 | 0.429 | 0.423 | −0.0019 | 0.0892 | −0.1095 | 8.240 | 0.0871 | 0.1170 |
| MSP | | | −0.195 | 0.270 | 0.3750 | 0.4190 | 0.0129 | 8.260 | 0.4494 | 0.4548 |
| MLE | 10 | 1.0 | 0.437 | 0.463 | 0.0227 | 0.1005 | 0.0875 | 8.080 | 0.1025 | 0.1850 |
| MSP | | | −0.214 | 0.278 | 0.7960 | 1.0200 | −0.1422 | 8.610 | 0.0462 | 0.0912 |
| MLE | 10 | 2.5 | 0.207 | 0.419 | 0.3083 | 0.4534 | −0.1044 | 8.440 | −0.0056 | 0.5170 |
| MSP | | | −0.191 | 0.303 | 0.8350 | 1.0790 | 0.1349 | 8.220 | 0.0134 | 0.5236 |

**Table 4** Bias and MSE for sample size $n = 25$ (1000 runs)

| Power | $a$ | $b$ | $\hat{\alpha}$ | MSE($\hat{\alpha}$) | $\hat{\beta}$ | MSE($\hat{\beta}$) | $\hat{a}$ | MSE($\hat{a}$) | $\hat{b}$ | MSE($\hat{b}$) |
|---|---|---|---|---|---|---|---|---|---|---|
| MLE | 0.5 | 0.5 | 0.0063 | 0.1050 | 0.558 | 0.390 | 0.2170 | 0.163 | 0.1290 | 0.1460 |
| MSP | | | 0.7600 | 1.0700 | 0.530 | 0.366 | 0.3690 | 0.195 | 0.0600 | 0.0575 |
| MLE | 0.5 | 1.0 | 0.0600 | 0.208 | 0.605 | 0.476 | 0.1197 | 0.322 | 0.1380 | 0.3990 |
| MSP | | | 0.7610 | 1.100 | 0.485 | 0.317 | 0.3870 | 0.201 | 0.0421 | 0.1215 |
| MLE | 0.5 | 2.5 | 0.0691 | 0.299 | 0.548 | 0.472 | 0.0585 | 0.488 | 0.2000 | 0.9570 |
| MSP | | | 0.7140 | 1.010 | 0.485 | 0.318 | 0.3880 | 0.204 | 0.0423 | 0.5609 |
| MLE | 2 | 0.5 | 0.122 | 0.101 | 0.653 | 0.495 | 0.2350 | 0.426 | −0.1374 | 0.0731 |
| MSP | | | 0.814 | 1.180 | 0.884 | 0.876 | 0.0353 | 0.319 | 0.0020 | 0.0236 |
| MLE | 2 | 1.0 | 0.194 | 0.156 | 0.634 | 0.503 | 0.1590 | 0.566 | −0.1248 | 0.3205 |
| MSP | | | 0.878 | 1.240 | 0.655 | 0.515 | 0.0871 | 0.317 | 0.0229 | 0.1142 |
| MLE | 2 | 2.5 | 0.212 | 0.268 | 0.530 | 0.432 | 0.1180 | 0.594 | 0.0573 | 1.0092 |
| MSP | | | 0.976 | 1.430 | 0.497 | 0.336 | 0.1634 | 0.293 | 0.0949 | 0.5837 |
| MLE | 10 | 0.5 | 0.399 | 0.418 | 0.753 | 0.681 | −0.0829 | 8.470 | −0.0141 | 0.0232 |
| MSP | | | 0.876 | 1.260 | 1.400 | 2.140 | 0.0422 | 8.670 | −0.0067 | 0.0202 |
| MLE | 10 | 1.0 | 0.482 | 0.447 | 0.691 | 0.574 | 0.0283 | 7.980 | −0.0921 | 0.1209 |
| MSP | | | 0.885 | 1.250 | 1.320 | 1.930 | −0.1088 | 8.180 | 0.0081 | 0.0848 |
| MLE | 10 | 2.5 | 0.620 | 0.596 | 0.656 | 0.550 | 0.0286 | 8.270 | −0.1156 | 0.6505 |
| MSP | | | 1.040 | 1.510 | 1.180 | 1.610 | −0.1045 | 8.030 | −0.0001 | 0.5635 |

speed in order to estimate the potential for use in generating energy. It can be observed that distributions such as Weibull, Rayleigh, gamma, lognormal, beta, Burr, and inverse Gaussian distributions are used in modeling wind speed frequencies [17]. As noted before, the two additional parameters in the Kum-G distribution families may provide more flexibility in modeling. For example, the Kum-Weibull family of distributions include the Weibull and Rayleigh distributions as special cases. Motivated by this fact the Kum-Weibull, Kum-Pareto and Kum-Power families of distributions are applied to model wind speed frequencies for a particular location, Cide, in Turkey. The data represent daily average wind speed measurements at the given location for January 2016 and are obtained from the Turkish State Meteorological Service.

The results for the wind data are given in Table 5 and in Fig. 4. The parameter estimates for the Weibull distribution are obtained by applying the *fitdistr* method in R. The

**Table 5** Parameter estimates fitted to wind data

| Model | $\hat{a}$ | $\hat{b}$ | $\hat{\theta}_1^*$ | $\hat{\theta}_2^*$ | $\chi^2$ statistic | $p$ value |
|---|---|---|---|---|---|---|
| Weibull (MLE) | 1 | 1 | 4.366 | 2.774 | 0.09872 | 0.8943 |
| Kum-W (MSP) | 4.814 | 9.211 | 0.482 | 4.993 | 0.078944 | 0.982 |
| Kum-Pow (MSP) | 1.599 | 7.974 | 1.311 | 2.721 | 0.073136 | 0.9921 |

* For Weibull and Kum-W $\theta_1 = \lambda$, $\theta_2 = c$, for Kum-Pow $\theta_1 = \alpha$, $\theta_2 = \beta$

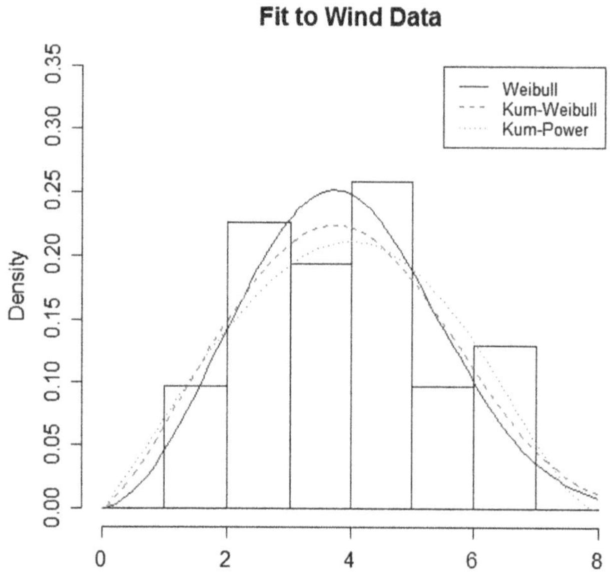


**Fig. 4** Parameter estimates for wind data

parameter estimates for the Kum-W and Kum-Pow distributions were obtained by applying the *ga* method ([15]), which is a genetic algorithm method implemented in R. Since in this particular application the Kum-Pareto families of distributions are not suited for the data we did not include the results for Kum-Pareto. On the other hand, due to convergence problems with ML estimation, only results for the MSP method with genetic algorithms are given. Table 5 clearly demonstrates that Kum-G families of distributions can be used as alternatives for classical distributions such as Weibull. Since many types of distributions, for example, are used in modeling wind characteristics it should be expected that Kum-G families of distributions may improve modeling results.

## Conclusion

Tamandi and Nadarajah [16] considered parameter estimation of Kum-Weibull, Kum-Normal and Kum-InverseNormal distributions. They stated that for these distributions, in general, the MSP method results in smaller bias and MSEs for small sample sizes. It should be noted that in these distributions no parameter-dependent boundaries exist, that is the domain of the random variable is independent of the parameters. In this study, we considered three Kum-G distributions, all with different characteristics. The Kum-Par and Kum-Pow distributions both have parameter-dependent bounds and may model different distributions. In addition, we applied these families of distributions to model real data for wind speed measurements.

The computations in the simulations and in application to real data have shown that the MSP method, in general, outperforms the ML method. Also, we have seen that in the ML method the initial values for parameters may cause the algorithms to stop before reaching any feasible parameter estimate. Thus, in general the ML approach is sensitive to initial values leading to convergence problems. In contrast, the MSP method, in general, seems to give more consistent results. Therefore, to model wind speed we have preferred to use genetic algorithms with the MSP approach in order to obtain parameter estimates for the Kum-W and Kum-Pow families of distributions.

## References

1. Abbasbandy, S., Shivanian, E.: Numerical solution based on meshless technique to study the biological population model. Math. Sci. **12**, 123–130 (2016)
2. Bourguignon, M., Silva, R.B., Zea, L.M., Cordeiro, G.M.: The Kumaraswamy Pareto distribution. J. Stat. Theory Appl. **12**, 129–144 (2013)
3. Chang, T.P.: Performance comparison of six methods in estimating Weibull parameters for wind energy application. Appl. Energy **88**, 272–282 (2011)
4. Cheng, R.C.H., Amin, N.: Estimating parameters in continuous univariate distributions with a shifted origin. J. R. Stat. Soc. B **45**, 394–403 (1980)
5. Cheng, R.C.H., Stephens, M.A.: A goodness-of-fit test using Moran's statistic with estimated parameters. Biometrika **76**, 386–392 (1989)
6. Cordeiro, G.M., Castro, M.: A new family of generalized distributions. J. Stat. Comput. Simul. **81**, 883–898 (2011)
7. Ekström, M.: Consistency of generalized maximum spacing estimates. Scand. J. Stat. **28**, 343–354 (2001)
8. Ekström, M.: Alternatives to maximum likelihood estimation based on spacing and the Kullback–Leibler divergence. J. Stat. Plan. Inference **138**, 1778–1791 (2008)
9. Gupta, R.C., Kannan, N., Raychaudhari, A.: The Kumaraswamy Weibull distribution with application to failure data. J. Frankl. Inst. **347**, 1399–1429 (1997)
10. Jones, M.C.: A beta-type distribution with some tractability advantages. Stat. Methodol. **12**, 70–81 (2009)

11. Kumaraswamy, P.: Generalized probability density-function for double-bounded random-processes. J. Hydrol. **462**, 79–88 (1980)
12. Oguntunde, P.E., Odetunmibi, O., Okagbue, H.I.: The Kumaraswamy-power distribution: a generalization of the power distribution. Int. J. Math. Anal. **9**, 637–645 (2015)
13. Ranneby, B.: The maximum spacing method. An estimation method related to the maximum likelihood method. Scand. J. Stat. **11**, 93–112 (1984)
14. R Core Team.: R: A language and environment for statistical computing, R Foundation for Statistical Computing, Vienna, Austria. 2016. https://www.R-project.org/
15. Scrucca, L. GA: A Package for Genetic Algorithms in R. J. Stat. Softw. **53**, 1–37 (2013). http://www.jstatsoft.org/v53/i04/
16. Tamandi, M., Nadarajah, S.: On the estimation of parameters of Kumaraswamy-G distributions. Commun. Stat. Simul. Comput. (2014). doi:10.1080/03610918.2014.957840
17. Vaishali, S., Gupta, S., Nema, R.: comparative analysis of wind speed probability distributions for wind power assessment of four sites. Turk. J. Electr. Eng. Comput. Sci. **24**, 4724–4735 (2016)
18. Vajargah, K.F., Shoghi, M.: Simulation of stochastic differential equation of geometric Brownian motion by quasi-Monte Carlo method and its application in prediction of total index of stock market and value at risk. Math. Sci. **9**, 115–125 (2015)

21

# A new approach for ranking of intuitionistic fuzzy numbers using a centroid concept

**K. Arun Prakash[1] · M. Suresh[1] · S. Vengataasalam[1]**

**Abstract** Ranking of intuitionistic fuzzy numbers is a difficult task. Many methods have been proposed for ranking of intuitionistic fuzzy numbers. In this paper we have ranked both trapezoidal intuitionistic fuzzy numbers and triangular intuitionistic fuzzy numbers using the centroid concept. Some of the properties of the ranking function have been studied. Also, comparative examples are given to show the effectiveness of the proposed method.

**Keywords** Intuitionistic fuzzy set · Trapezoidal intuitionistic fuzzy number · Triangular intuitionistic fuzzy number · Ranking of trapezoidal intuitionistic fuzzy number · Centroid of an intuitionistic fuzzy number

## Introduction

The fuzzy sets [29] were extended by Atanassov [4] to develop the intuitionistic fuzzy sets by including non-membership function which is useful to express vagueness more accurately as compared to fuzzy sets. Fuzzy numbers [1] are special kind of fuzzy sets which are of importance in solving fuzzy linear programming problems. An important issue in fuzzy set theory is ranking of uncertainty numbers. When numerical values are represented in uncertain nature termed as fuzzy numbers, a comparison of these numerical values is not easy. Several methods have been proposed in literature to rank fuzzy numbers. As a generalization of fuzzy numbers, an intuitionistic fuzzy number (IFN) seems to fit more suitably to describe uncertainty. After this, many research works have been carried out in defining and studying interesting properties of various types of intuitionistic fuzzy numbers. Recently, the research on IFN's has received high attention, since it is more suitable for solving intuitionistic fuzzy linear programming problems. Many ranking methods for ordering of IFNs have been introduced in the literature.

Grzegorzewski [7, 8] treated IFNs as two families of metrics and developed a ranking method for IFNs. Mitchell [11] proposed a ranking method to order triangular intuitionistic fuzzy numbers (TIFNs) by accepting a statistical viewpoint and interpreting each IFN as ensemble of ordinary fuzzy numbers. Ranking of TIFN on the basis of value index to ambiguity index is proposed by Li [9] and solved a multiattribute decision-making problem. Dubey et al. [6] extended the definitions given by Li [9] to the newly defined TIFNs. Thereafter, a ranking function was proposed to solve a class of linear programming problems. A ranking function based on score function was proposed and the same used to solve IFLP, in which the data parameters are TIFNs. In the past, Nayagam et al. [14] introduced TIFNs and proposed a method to rank them.

Nehi [15] proposed a new ranking method, in which the membership function and non-membership function of IFNs are treated as fuzzy quantities. In the same way, Li [10] defined the value and ambiguity index for TIFN which is similar to those of Delgado et al. [5]. These are then used to define the value index and the ambiguity index for TIFN. Using this concept, a method based on ratio ranking is developed for ranking TIFNs. The limitations and shortcomings of some of the existing ranking were overcome by a new ranking approach by modifying an existing ranking approach proposed for comparing IF numbers by Amit

✉ M. Suresh
mathssuresh84@gmail.com

1 Department of Mathematics, Kongu Engineering College, Erode, Tamil Nadu, India

Kumar [2]. With the help of the proposed ranking approach, a new method is proposed to find the optimal solution of unbalanced minimum cost flow (MCF) problems, in which all the parameters are represented by IF numbers. The values and ambiguities of the membership degree and the non-membership degree for trapezoidal intuitionistic fuzzy number are defined as well as the value index- and ambiguity index-based ranking approach given by Rezvani [18]. In 2011, Nayagam et al. [13] defined new intuitionistic fuzzy scoring method for the intuitionistic fuzzy number in which hesitation is greater than membership fuzzy number. Similarly, in the intuitionistic fuzzy number, the hesitation is less than the membership fuzzy number. This new method includes the concept of both membership and non-membership function of an intuitionistic fuzzy number. By this defined method, the problems involving hesitation can be easily studied. Also in that paper, the defined intuitionistic fuzzy scoring method was applied to the clustering problem.

In Peng et al. [17], defined the concepts of canonical intuitionistic fuzzy numbers and fuzzy cut sets, and the relation between generalized fuzzy numbers and canonical intuitionistic fuzzy numbers were studied. Next, the concept of center index and radius index of canonical intuitionistic fuzzy numbers, based on fuzzy cut sets are introduced, and the ranking index with the degree of optimism of the decision maker for canonical intuitionistic fuzzy numbers is defined. Then a new ranking method based on the ranking index is developed. The properties of values index and ambiguity index of TIFNs are studied and a compromise ratio ranking method for TIFNs is developed based on the value index and ambiguity index of TIFNs. Using the above ranking, MADM problems in which the ratings of alternatives on attributes are expressed as TIFNs are solved by the extended additive weighted method in [30] by Zhang et al. In [20], a novel approach for ranking triangular intuitionistic fuzzy numbers (TIFNs) is obtained by converting each TIFN to two related triangular fuzzy numbers (TFNs) based on their membership functions and non-membership functions. Then, a new defuzzification for the obtained TFNs using their values and ambiguities was suggested by Salahshour [20]. The average ranking index is introduced to find out order relations between two TIFNs by Seikh et al. in [23]. A method is described to approximate a TIFN to a nearly approximated interval number. Applying this result and using interval arithmetic, a bound unconstrained optimization problem is solved whose coefficients are fixed TIFNs.

Based on the possibility degree formula, Wei et al. [27] gave a possibility degree method to rank intuitionistic fuzzy numbers, which is used to rank the alternatives in multi-criteria decision-making problems. Cosine similarity measure method was extended for ranking alternatives, and then a practical example of the developed approach was given to select the investment alternatives by Ye [28].

The basic arithmetic operations of generalized triangular intuitionistic fuzzy numbers (GTIFNs) and the notion of (α,β)-cut sets were defined by Seikh et al. [24]. Also a nearest interval approximation method is described to approximate a GTIFN to a nearest interval number. Moreover, the average ranking index is introduced to find out order relations between two GTIFNs. Intuitionistic trapezoidal fuzzy weighted arithmetic averaging operator and weighted geometric averaging operator were proposed by Wang et al. in [25]. The expected values, score function, and the accuracy function of intuitionistic trapezoidal fuzzy numbers are also defined. By comparing the score function and the accuracy function values of integrated fuzzy numbers, a ranking of the whole alternative set was attained. Nagoorgani et al. [12] introduced a ranking technique for TIFN using α,β-cut, score function and accuracy function. The method is validated by applying the concept to solve the intuitionistic fuzzy variable linear programming problem.

In this paper, we introduce a new approach to rank intuitionistic fuzzy numbers which is easy to handle. The rest of the paper is organized as follows: Sect. 2 gives some basic definitions and notations of intuitionistic fuzzy sets and intuitionistic fuzzy numbers. In Sect. 3, a new approach for ranking trapezoidal and triangular intuitionistic fuzzy numbers is introduced and some properties of ranking functions are investigated. In Sect. 4, some numerical examples are illustrated to prove the advantage of the proposed paper. In Sect. 5 a comparative study is given. The last section gives the conclusion and future work.

## Preliminaries

This section introduces some definitions and basic concepts related to intuitionistic fuzzy sets and intuitionistic fuzzy numbers.

**Definition 2.1** [4] An IFS A in X is given by $A = \{(x, \mu_A(x), \nu_A(x), x \in X)\}$, where the functions $\mu_A(x) : X \to [0, 1]$ and $\nu_A(x) : X \to [0, 1]$ define, respectively, the degree of membership and degree of non-membership of the element $x \in X$ to the set A, which is a subset of X, and for every $x \in X, 0 \leq \mu_A(x) + \nu_A(x) \leq 1$. Obviously, every fuzzy set has the form $\{(x, \mu_A(x), \mu_{A^c}(x)), x \in X\}$.

For each IFS A in X, we will call $\Pi_A(x) = 1 - \mu(x) - \nu(x)$ the intuitionistic fuzzy index of $x$ in A. It is obvious that $0 \leq \Pi_A(x) \leq 1$, for all $x \in X$.

**Definition 2.2** [15] An intuitionistic fuzzy set (IFS) $A = \{(x, \mu_A(x), \nu_A(x), x \in X)\}$ is called IF-normal, if there exist at least two points $x_0.x_1 \in X$ such that $\mu_A(x_0) = 1$, $\nu_A(x_1) = 1$. It is easily seen that given intuitionistic fuzzy set A is IF-normal, if there is at least one point it surely belongs to A, and at least one point does not belong to A.

**Definition 2.3** [15] An intuitionistic fuzzy set (IFS) $A = \{(x, \mu_A(x), \nu_A(x), x \in X)\}$ of the real line is called IF-convex, if $\forall x_1, x_2 \in R, \forall \lambda \in [0,1]$

$$\mu_A(\lambda x_1 + (1-\lambda)x_2) \geq \mu_A(x_1)\Lambda\mu_A(x_2)$$
$$\nu_A(\lambda x_1 + (1-\lambda)x_2) \geq \nu_A(x_1)\Lambda\nu_A(x_2)$$

**Definition 2.4** [15] An IFS $A = \{(x, \mu_A(x), \nu_A(x),\ x \in X)\}$ of the real line is called an intuitionistic fuzzy number (IFN) if:

(i) A is IF-normal,
(ii) A is IF-convex,
(iii) $\mu_A$ is upper semi-continuous and $\nu_A$ is lower semi continuous,
(iv) *Supp* $A = \{(x \in X/\nu_A(x) < 1)\}$ is bounded.

**Definition 2.5** [15] A trapezoidal intuitionistic fuzzy number with parameters $b_1 \leq a_1 \leq b_2 \leq a_2 \leq a_3 \leq b_3 \leq a_4 \leq b_4$ is denoted by $A = (b_1, a_1, b_2, a_2, a_3, b_3, a_4, b_4)$. In this case, we have

$$\mu_A(x) = \begin{cases} 0; & x < a_1 \\ \dfrac{x - a_1}{a_2 - a_1}; & a_1 \leq x \leq a_2 \\ 1; & a_2 \leq x \leq a_3 \\ \dfrac{x - a_4}{a_3 - a_4}; & a_3 \leq x \leq a_4 \\ 0; & a_4 < x \end{cases}$$

and

$$\nu_A(x) = \begin{cases} 0; & x < b_1 \\ \dfrac{x - b_2}{b_1 - b_2}; & b_1 \leq x \leq b_2 \\ 0; & b_2 \leq x \leq b_3 \\ \dfrac{x - b_4}{b_3 - b_4}; & b_3 \leq x \leq b_4 \\ 1; & b_4 < x \end{cases}$$

In the above definition, if we let $b_2 = b_3$ (and hence $a_2 = a_3$), then we will get a triangular intuitionistic fuzzy number with parameters $b_1 \leq a_1 \leq (b_2 = a_2 = a_3 = b_3) \leq a_4 \leq b_4$ denoted by $A = (b_1, a_1, b_2, a_4, b_4)$.

**Definition 2.6** [16] An intuitionistic fuzzy number (IFN) A in R is said to be a symmetric trapezoidal intuitionistic fuzzy number if there exists real numbers $a_1, a_2, h, h'$ where $a_1 \leq a_2, h \leq h'$ and $h, h' > 0$ such that the membership and non-membership functions are as follows:

$$\mu_A(x) = \begin{cases} \dfrac{x - (a_1 - h)}{h}; & x \in [a_1 - h, a_1] \\ 1; & x \in [a_1, a_2] \\ \dfrac{a_2 + h - x}{h}; & x \in [a_2, a_2 + h] \\ 0; & \text{otherwise} \end{cases}$$

and

$$\nu_A(x) = \begin{cases} \dfrac{(a_1 - x)}{h'}; & x \in [a_1 - h', a_1] \\ 0; & x \in [a_1, a_2] \\ \dfrac{x - a_2}{h}; & x \in [a_2, a_2 + h'] \\ 1; & \text{otherwise} \end{cases},$$

where $A = [a_1, a_2, h, h; a_1, a_2, h', h']$.

## Centroid-based approach for ranking of intuitionistic fuzzy numbers

In this section, we derive the centroid point of the trapezoidal intuitionistic fuzzy number and triangular intuitionistic fuzzy numbers.

The method of ranking trapezoidal intuitionistic fuzzy numbers with centroid index uses the geometric center of a trapezoidal intuitionistic fuzzy number. The geometric center corresponds to $\tilde{x}(A)$ value on the horizontal axis and $\tilde{y}(A)$ value on the vertical axis.

Consider a trapezoidal intuitionistic fuzzy number of the form $A = (b_1, a_1, b_2, a_2, a_3, b_3, a_4, b_4)$, whose membership function can be defined as follows:

$$\mu_A = \begin{cases} 0, & x < a_1 \\ f_A^L(x), & a_1 \leq x \leq a_2 \\ 1, & a_2 \leq x \leq a_3 \\ f_A^R(x), & a_3 \leq x \leq a_4 \\ 0, & a_4 \leq x \end{cases},$$

and non-membership can generally be defined as

$$\nu_A = \begin{cases} 0, & x < b_1 \\ g_A^L(x), & b_1 \leq x \leq b_2 \\ 0, & b_2 \leq x \leq b_3 \\ g_A^R(x), & b_3 \leq x \leq b_4 \\ 1, & b_4 \leq x \end{cases},$$

where $f_A^L : R \to [0,1]$, $f_A^R : R \to [0,1]$, $g_A^L : R \to [0,1]$ and $g_A^R : R \to [0,1]$, called the sides of an intuitionistic fuzzy number, where $f_A^L$ and $g_A^R$ are non-decreasing and $f_A^R$, $g_A^L$ are non-increasing. Therefore, the inverse functions of. $f_A^L, f_A^R$, $g_A^L$ and $g_A^R$ exist which are also of the same nature. Let $h_A^L : [0,1] \to R$, $h_A^R : [0,1] \to R, k_A^L : [0,1] \to R$ and $k_A^L : [0,1] \to R$ be the inverse functions of $f_A^L$, $f_A^R$, $g_A^L$ and $g_A^R$, respectively. Then, $h_A^L$,$h_A^R$,$k_A^L$ and $k_A^L$ should be integrable on R. In the case of the above-defined trapezoidal intuitionistic fuzzy number, the above inverse functions can be analytically expressed as follows:

$$h_A^L\,(y) = a_1 + (a_2 - a_1)y, \quad 0 \le y \le 1,$$
$$h_A^R\,(y) = a_4 + (a_3 - a_4)y, \quad 0 \le y \le 1,$$
$$k_A^L\,(y) = b_2 + (b_1 - b_2)y, \quad 0 \le y \le 1,$$
$$k_A^R\,(y) = b_3 + (b_4 - b_3)y, \quad 0 \le y \le 1.$$

The diagrammatic representations are shown Figs. 1 and 2.

The centroid point $(\tilde{x}(A), \tilde{y}(A))$ of the trapezoidal intuitionistic fuzzy numbers $\tilde{A}$ is determined as follows:

$$\tilde{x}_\mu(A) = \frac{\int_{a_1}^{a_2} x f_A^L(x)dx + \int_{a_2}^{a_3} xdx + \int_{a_3}^{a_4} x f_A^R(x)dx}{\int_{a_1}^{a_2} f_A^L(x)dx + \int_{a_2}^{a_3} dx + \int_{a_3}^{a_4} f_A^R(x)dx},$$

$$= \frac{\int_{a_1}^{a_2} \frac{x^2 - xa_1}{a_2 - a_1} dx + \int_{a_2}^{a_3} xdx + \int_{a_3}^{a_4} \frac{x^2 - a_4 x}{a_3 - a_4} dx}{\int_{a_1}^{a_2} \frac{x - a_1}{a_2 - a_1} dx + \int_{a_2}^{a_3} dx + \int_{a_3}^{a_4} \frac{x - a_4}{a_3 - a_4} dx},$$

$$= \frac{\frac{1}{a_2 - a_1}\left[\frac{x^3}{3} - \frac{x^2 a_1}{2}\right]_{a_1}^{a_2} + \left[\frac{x^2}{2}\right]_{a_2}^{a_3} + \frac{1}{a_3 - a_4}\left[\frac{x^3}{3} - \frac{a_4 x^2}{2}\right]_{a_3}^{a_4}}{\frac{1}{a_2 - a_1}\left[\frac{x^2}{2} - a_1 x\right]_{a_1}^{a_2} + [x]_{a_2}^{a_3} + \frac{1}{a_3 - a_4}\left[\frac{x^2}{2} - x\right]_{a_3}^{a_4}},$$

$$= \frac{\left(\frac{a_2^2 + a_1 a_2 + a_1^2}{3}\right) - \frac{a_1}{2}[a_2 + a_1] + \left[\frac{a_3^2 - a_2^2}{2}\right] + \left[\frac{-a_4^2 - a_3 a_4 - a_3^2}{3}\right] - \frac{a_4}{2}[a_4 + a_3]}{\frac{a_2 + a_1}{2} - a_1 + a_3 - a_2 - \frac{a_4 - a_3}{2} + a_4},$$

$$\tilde{x}_\mu(A) = \frac{1}{3}\left[\frac{a_3^2 + a_4^2 - a_1^2 - a_2^2 - a_1 a_2 + a_3 a_4}{a_4 + a_3 - a_2 - a_1}\right].$$

Similarly,

$$\tilde{x}_\upsilon(A)7 = \frac{\int_{b_1}^{b_2} x\, g_A^L(x)dx + \int_{b_2}^{b_3} xdx + \int_{b_3}^{b_4} x g_A^R(x)dx}{\int_{b_1}^{b_2} g_A^L(x)dx + \int_{b_2}^{b_3} dx + \int_{b_3}^{b_4} g_A^R(x)dx}$$

$$= \frac{\int_{b_1}^{b_2} \frac{x^2 - xb_2}{b_1 - b_2} dx + \int_{b_2}^{b_3} xdx + \int_{b_3}^{b_4} \frac{x^2 - xb_3}{b_4 - b_3} dx}{\int_{b_1}^{b_2} \frac{x - b_2}{b_1 - b_2} dx + \int_{b_2}^{b_3} dx + \int_{b_3}^{b_4} \frac{x - b_3}{b_4 - b_3} dx},$$

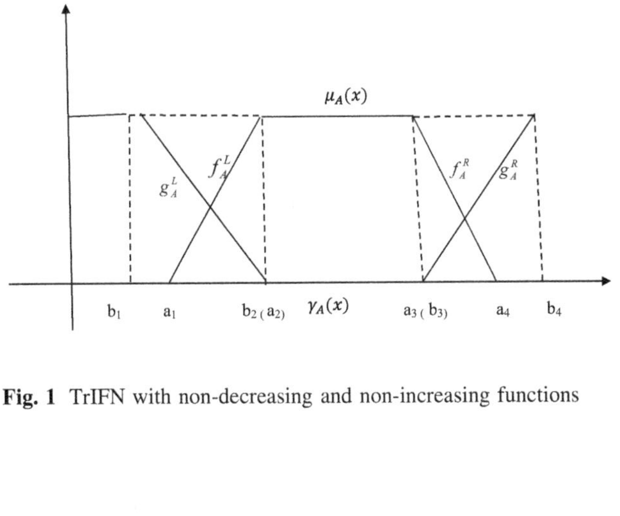


**Fig. 1** TrIFN with non-decreasing and non-increasing functions

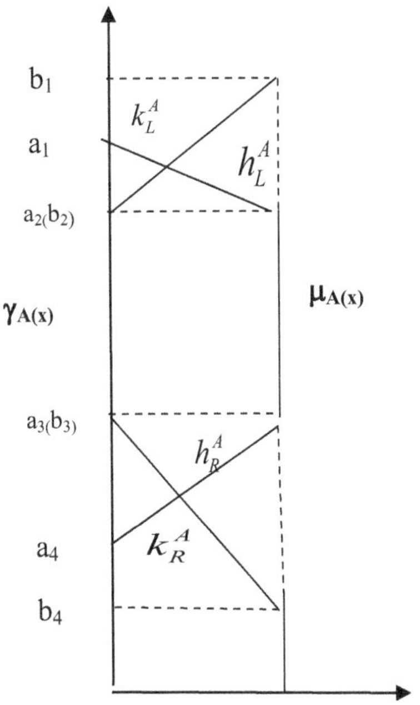


**Fig. 2** TrIFN with inverse non-decreasing and non-increasing functions

$$= \frac{\frac{1}{b_1 - b_2}\left[\frac{x^3}{3} - b_2\frac{x^2}{2}\right]_{b_1}^{b_2} + \left[\frac{x^2}{2}\right]_{b_2}^{b_3} + \frac{1}{b_4 - b_3}\left[\frac{x^3}{3} - b_3\frac{x^2}{2}\right]_{b_3}^{b_4}}{\frac{1}{b_1 - b_2}\left[\frac{x^2}{2} - b_2 x\right]_{b_1}^{b_2} + [x]_{b_2}^{b_3} + \frac{1}{b_4 - b_3}\left[\frac{x^2}{2} - b_3 x\right]_{b_3}^{b_4}},$$

$$\tilde{x}_\upsilon(A) = \frac{1}{3}\left[\frac{2b_4^2 - 2b_1^2 + 2b_2^2 + 2b_3^2 + b_1 b_2 - b_3 b_4}{b_3 + b_4 - b_1 - b_2}\right].$$

Next,

$$\tilde{y}_\mu(A) = \frac{\int_0^1 y h_A^L(y)dy - \int_0^1 y h_A^R(y)dy}{\int_0^1 h_A^L(y)dy - \int_0^1 h_A^R(y)dy},$$

$$\tilde{y}_\mu(A) = \frac{\int_0^1 (a_2 y^2 + a_1 y - a_1 y^2)dy - \int_0^1 (a_3 y^2 + a_4 y - a_4 y^2)dy}{\int_0^1 (a_2 y + a_1 - a_1 y)dy - \int_0^1 (a_3 y + a_4 - a_4 y)dy},$$

$$= \frac{\left(a_2\left[\frac{y^3}{3}\right] + a_1\left[\frac{y^2}{2}\right] - a_1\left[\frac{y^3}{3}\right]\right)_0^1 - \left(a_3\left[\frac{y^3}{3}\right] + a_4\left[\frac{y^2}{2}\right] - a_4\left[\frac{y^3}{3}\right]\right)_0^1}{\left(a_2\left[\frac{y^2}{2}\right] + a_1 y - a_1\left[\frac{y^2}{2}\right]\right)_0^1 - \left(a_3\left[\frac{y^2}{2}\right] + a_4 y - a_4\left[\frac{y^2}{2}\right]\right)_0^1},$$

$$= \frac{\left(\left[\frac{a_2}{3}\right] + \left[\frac{a_1}{2}\right] - \left[\frac{a_1}{3}\right]\right) - \left(\left[\frac{a_3}{3}\right] + \left[\frac{a_4}{2}\right] - \left[\frac{a_4}{3}\right]\right)}{\left(\left[\frac{a_2}{2}\right] + a_1 - \left[\frac{a_1}{2}\right]\right) - \left(\left[\frac{a_3}{2}\right] + a_4 - \left[\frac{a_4}{2}\right]\right)},$$

$$= \frac{1}{3}\left(\frac{2a_2 + 3a_1 - 2a_1 - 2a_3 - 3a_4 + 2a_4}{a_2 + 2a_1 - a_1 - a_3 - 2a_4 + a_4}\right),$$

$$\tilde{y}_\mu(A) = \frac{1}{3}\left(\frac{a_1 + 2a_2 - 2a_3 - a_4}{a_1 + a_2 - a_3 - a_4}\right),$$

$$\tilde{y}_v(A) = \frac{\int_0^1 y\, k_A^L(y)dy - \int_0^1 y k_A^R(y)dy}{\int_0^1 k_A^L(y)dy - \int_0^1 k_A^R(y)dy},$$

$$\tilde{y}_\upsilon(A) = \frac{\int_0^1 (b_1y^2 + b_2y - b_2y^2)dy - \int_0^1 (b_4y^2 + b_3y - b_3y^2)dy}{\int_0^1 (b_1y + b_2 - b_2y)dy - \int_0^1 (b_4y + b_3 - b_3y)dy},$$

$$= \frac{\left(b_1\left[\frac{y^3}{3}\right] + b_2\left[\frac{y^2}{2}\right] - b_2\left[\frac{y^3}{3}\right]\right)_0^1 - \left(b_4\left[\frac{y^3}{3}\right] + b_3\left[\frac{y^2}{2}\right] - b_3\left[\frac{y^3}{3}\right]\right)_0^1}{\left(b_1\left[\frac{y^2}{2}\right] + b_2y - b_2\left[\frac{y^2}{2}\right]\right)_0^1 - \left(b_4\left[\frac{y^2}{2}\right] + b_3y - b_3\left[\frac{y^2}{2}\right]\right)_0^1},$$

$$= \frac{\left(\left[\frac{b_1}{3}\right] + \left[\frac{b_2}{2}\right] - \left[\frac{b_2}{3}\right]\right) - \left(\left[\frac{b_4}{3}\right] + \left[\frac{b_3}{2}\right] - \left[\frac{b_3}{3}\right]\right)}{\left(\left[\frac{b_1}{2}\right] + b_2 - \left[\frac{b_2}{2}\right]\right) - \left(\left[\frac{b_4}{2}\right] + b_3 - \left[\frac{b_3}{2}\right]\right)},$$

$$= \frac{1}{3}\left(\frac{2b_1 + 3b_2 - 2b_2 - 2b_4 - 3b_3 + 2b_3}{b_1 + 2b_2 - b_2 - b_4 - 2b_3 + b_3}\right),$$

$$\tilde{y}_\upsilon(A) = \frac{1}{3}\left(\frac{2b_1 + b_2 - b_3 - 2b_4}{b_1 + b_2 - b_3 - b_4}\right).$$

Then, $(\tilde{x}_\mu(A), \tilde{y}_\mu(A)); (\tilde{x}_\upsilon(A), \tilde{y}_\upsilon(A))$ gives the centroid of the trapezoidal intuitionistic fuzzy number.

The above relations can be reduced to get the centroid point of the triangular intuitionistic fuzzy numbers, as they are a special case of trapezoidal intuitionistic fuzzy numbers with $b_2 = a_2 = a_3 = b_3$. Its centroid can be determined by

$$\tilde{x}_\mu(A) = \frac{1}{3}\left[\frac{a_4^2 - a_1^2 + a_2(a_4 - a_1)}{a_4 - a_1}\right] = \frac{a_1 + a_2 + a_4}{3};$$

$$\tilde{x}_\upsilon(A) = \frac{1}{3}\left[\frac{2(b_4^2 - b_1^2) + b_2(b_1 - b_4)}{b_1 - b_4}\right] = \frac{2b_1 - b_2 + 2b_4}{3};$$

$$\tilde{y}_\mu(A) = \frac{1}{3} \text{ and}$$

$$\tilde{y}_\upsilon(A) = \frac{2}{3}.$$

**Theorem 3.1** *Let* $A = (b_1, a_1, b_2, a_4, b_4)$ and $B = (b'_1, a'_1, b'_2, a'_4, b'_4)$ *be two triangular intuitionistic fuzzy numbers. Then the following equation is valid:* $\tilde{x}_\mu(A + B) = \tilde{x}_\mu(A) + \tilde{x}_\mu(B)$

*Proof* We have $A + B = (b_1 + b'_1, a_1 + a'_1, b_2 + b'_2, a_4 + a'_4, b_4 + b'_4)$.

Now,

$$\tilde{x}_\mu(A + B) = \frac{(a_1 + a'_1) + (a_2 + a'_2) + (a_4 + a'_4)}{3}$$
$$= \frac{a_1 + a_2 + a_4}{3} + \frac{a'_1 + a'_2 + a'_4}{3},$$
$$\tilde{x}_\mu(A + B) = \tilde{x}_\mu(A) + \tilde{x}_\mu(B).$$

**Theorem 3.2** *Let* $A = (b_1, a_1, b_2, a_4, b_4)$ and $B = (b'_1, a'_1, b'_2, a'_4, b'_4)$ *be two triangular intuitionistic fuzzy numbers. Then,* $\tilde{x}_v(A + B) = \tilde{x}_v(A) + \tilde{x}_v(B)$.

*Proof* We have $A + B = (b_1 + b'_1, a_1 + a'_1, b_2 + b'_2, a_4 + a'_4, b_4 + b'_4)$.

Now,

$$\tilde{x}_\upsilon(A + B) = \frac{2(b_1 + b'_1) - (b_2 + b'_2) + 2(b_4 + b'_4)}{3}$$
$$= \frac{2b_1 - b_2 + 2b_4}{3} + \frac{2b'_1 - b'_2 + 2b'_4}{3},$$
$$\tilde{x}_\upsilon(A + B) = \tilde{x}_\upsilon(A) + \tilde{x}_\upsilon(B).$$

**Theorem 3.3** *Let* $A = (b_1, a_1, b_2, a_4, b_4)$ and $B = (b'_1, a'_1, b'_2, a'_4, b'_4)$ *be two triangular intuitionistic fuzzy numbers. Then*

$$\tilde{x}(A + B) = \tilde{x}(A) + \tilde{x}(B) + \tilde{x}_v(A)*\tilde{x}_\mu(B) + \tilde{x}_\mu(A)*\tilde{x}_v(B)$$

*Proof* We have

$$\tilde{x}(A+B) = \tilde{x}_\mu(A+B) * \tilde{x}_\upsilon(A+B)$$
$$= [\tilde{x}_\mu(A) + \tilde{x}_\mu(B)] * [\tilde{x}_\upsilon(A) + \tilde{x}_\upsilon(B)]$$
$$= \left[\frac{a_1 + a_2 + a_4}{3} + \frac{a'_1 + a'_2 + a'_4}{3}\right] * \left[\frac{2b_1 - b_2 + 2b_4}{3} + \frac{2b'_1 - b'_2 + 2b'_4}{3}\right]$$
$$= \frac{\begin{array}{c}\times(2a_1b_1 + 2a_2b_1 + 2a_4b_1 - a_1b_2 - a_2b_2 - a_4b_2 + 2a_1b_4 + 2a_2b_4 + 2a_4b_4)\\ + (2a'_1b'_1 + 2a'_2b'_1 + 2a'_4b'_1 - a'_1b'_2 - a'_2b'_2 - a'_4b'_2 + 2a'_1b'_4 + 2a'_2b'_4 + 2a'_4b'_4)\\ + (a'_1 + a'_2 + a'_4) * (2b_1 - b_2 + 2b_4) + (a_1 + a_2 + a_4) * (2b'_1 - b'_2 + 2b'_4)\end{array}}{9},$$
$$\tilde{x}(A+B) = \tilde{x}(A) + \tilde{x}(B) + \tilde{x}_\upsilon(A)*\tilde{x}_\mu(B) + \tilde{x}_\mu(A)*\tilde{x}_\upsilon(B).$$

**Theorem 3.4** *Let* $A = (b_1, a_1, b_2, a_4, b_4)$ and $B = (b'_1, a'_1, b'_2, a'_4, b'_4)$ *be two triangular intuitionistic fuzzy numbers. Then the following equation is valid:* $\tilde{y}(A + B) = \tilde{y}(A) + \tilde{y}(B) + 2 * \tilde{y}(A)$.

*Proof* Consider

$$\tilde{y}(A + B) = \tilde{y}_\mu(A + B) * \tilde{y}_\upsilon(A + B)$$
$$= [\tilde{y}_\mu(A) + \tilde{y}_\mu(B)] * [\tilde{y}_\upsilon(A) + \tilde{y}_\upsilon(B)]$$
$$= \left[\frac{2}{3}\right] * \left[\frac{4}{3}\right] = \frac{8}{9}$$
$$= \frac{2}{9} + \frac{2}{9} + 2 * \frac{2}{9}$$
$$= \tilde{y}(A) + \tilde{y}(B) + 2 * \tilde{y}(A).$$

*Remark 3.5* The result of the above theorem can also be simplified as

$$\tilde{y}(A + B) = 4 * \tilde{y}(A) \text{ (or) } \tilde{y}(A + B) = 4 * \tilde{y}(B),$$

where $\tilde{y}(A) = \tilde{y}(B)$ in case of triangular intuitionistic fuzzy number.

**Theorem 3.6** *The centroids of the symmetric trapezoidal intuitionistic fuzzy number are given by the following relations:*

$$\tilde{x}_{\mu}\,(\mathrm{A}) = \frac{1}{2}\left[\frac{a_1 h + a_2 h + a_2^2 - a_1^2}{a_2 - a_1}\right],$$
$$\tilde{x}_{\upsilon}\,(\mathrm{A}) = \frac{a_1 + a_2}{2},$$
$$\tilde{y}_{\mu}\,(\mathrm{A}) = \frac{1}{3}\left[\frac{3a_1 - 3a_2 - 2h}{2a_1 - 2a_2 - 2h}\right],$$
$$\tilde{y}_{\upsilon}\,(\mathrm{A}) = \frac{1}{3}\left[\frac{3a_1 - 3a_2 - 4h'}{2a_1 - 2a_2 - 2h'}\right].$$

The proof of the result is similar to the one as derived earlier for trapezoidal intuitionistic fuzzy numbers.

**Definition 3.7** The ranking function of the trapezoidal (triangular) intuitionistic fuzzy number A is defined by

$$\Re(A) = \sqrt{\frac{1}{2}\left(\left[\tilde{x}_{\mu}(A) - \tilde{y}_{\mu}(A)\right]^2 + \left[\tilde{x}_{\upsilon}(A) - \tilde{y}_{\upsilon}(A)\right]^2\right)},$$

which is the Euclidean distance.

It is obvious that the proposed ranking function $\Re$ satisfies the properties A1;A2;A3;A4;A5 and A6 of [23]. We list these properties below for the completeness of the section. Let S be the set of fuzzy quantities, and M be an ordering approach.

A1: For an arbitrary finite subset A of S,$\tilde{a} \in A$, $\tilde{a} \succeq \tilde{a}$ by M on A.

A2: For an arbitrary finite subset A of S and $(\tilde{a}, \tilde{b}) \in A^2, \tilde{a} \succeq \tilde{b}$ and $\tilde{b} \succeq \tilde{a}$ by M on A, we should have $\tilde{a} \approx \tilde{b}$ by M on A.

A3: For an arbitrary finite subset A of S and $(\tilde{a}, \tilde{b}, \tilde{c}) \in A^3, \tilde{a} \succeq \tilde{b}$ and $\tilde{b} \succeq \tilde{c}$ by M on A, we should have $\tilde{a} \succeq \tilde{c}$ by M on A.

A4: For an arbitrary finite subset A of S and $(\tilde{a}, \tilde{b}) \in A^2, \inf supp(\tilde{a}) \succeq \sup supp(\tilde{b})$, we should have $\tilde{a} \succeq \tilde{b}$ by M on A.

A4′: For an arbitrary finite subset A of S and $(\tilde{a}, \tilde{b}) \in A^2, \inf supp(\tilde{a}) \succ \sup supp(\tilde{b})$, we should have $\tilde{a} \succ \tilde{b}$ by M on A.

A5: Let S and S′ be two arbitrary finite sets of fuzzy quantities in which M can be applied and $\tilde{a}$ and $\tilde{b}$ are in $S \cap S'$. We obtain the ranking order $\tilde{a} \succ \tilde{b}$ by M on S′ iff $\tilde{a} \succ \tilde{b}$ by M on S.

A6: Let $\tilde{a}, \tilde{b}, \tilde{a} + \tilde{c}, \tilde{b} + \tilde{c}$ be elements of S. If $\tilde{a} \geq \tilde{b}$ by M on $\{\tilde{a}, \tilde{b}\}$, then $\tilde{a} + \tilde{c} \succeq \tilde{b} + \tilde{c}$ by M on $\{\tilde{a} + \tilde{c}, \tilde{b} + \tilde{c}\}$.

A6′: Let $\tilde{a}, \tilde{b}, \tilde{a} + \tilde{c}, \tilde{b} + \tilde{c}$ be elements of S. If $\tilde{a} \succ \tilde{b}$ by M on $\{\tilde{a}, \tilde{b}\}$, then $\tilde{a} + \tilde{c} \succ \tilde{b} + \tilde{c}$ by M on $\{\tilde{a} + \tilde{c}, \tilde{b} + \tilde{c}\}$.

A7: Let $\tilde{a}, \tilde{b}, \tilde{a}\tilde{c}, \tilde{b}\tilde{c}$ be elements of S. If $\tilde{a} > \tilde{b}$ by M on $\{\tilde{a}, \tilde{b}\}$, then $\tilde{a}\tilde{c} \succeq \tilde{b}\tilde{c}$ by M on $\{\tilde{a}\tilde{c},\ \tilde{b}\tilde{c}\}$

It is easy to verify that $\Re(A)$ satisfies the axioms A1–A3,A5 and A7. We focus on verifying axioms A4 and A6.

**Theorem 3.8** *Let A and B be two TrIFNs; if $a_{A1} > a_{B4}$ and $b_{A1} > b_{B4}$, then $A \succ B$.*

*Proof* *It is known that*

$x_{\mu}(A) > a_{A1}$ and $y_{\mu}(A) > a_{A1}$,
$x_{\mu}(B) < a_{B4}$ and $y_{\mu}(A) < a_{B4}$,
from $a_{A1} > a_{B4}$,
$\Re(A) > \Re(B)$.
Similarly,
$x_{\upsilon}(A) > b_{A1}$ and $y_{\upsilon}(A) > b_{A1}$,
$x_{\upsilon}(B) < b_{B4}$ and $y_{\upsilon}(B) < b_{B4}$,
from $b_{A1} > b_{B4}$,
$\Re(A) > \Re(B)$.
Therefore, $\Re(A) > \Re(B) \Rightarrow A \succ B$.

**Theorem 3.9** *Let A and B be two TrIFNs, then* $\Re(A + C) > \Re(B + C) \Rightarrow A + C \succ B + C$.

*Proof*

$\Re(A + B) = \Re(A) + \Re(B)$,
similarly,
$\Re(B + C) = \Re(B) + \Re(C)$.
Therefore, if $A \succ B$,
$\Re(A + C) > \Re(B + C) \Rightarrow A + C \succ B + C$.

**Algorithm 3.10** The approach of ranking any two trapezoidal (triangular) intuitionistic fuzzy numbers:

$A = (b_1, a_1, b_2, a_2, a_3, b_3, a_4, b_4)$ and
$B = (b'_1, a'_1, b'_2, a'_2, a'_3, b'_3, a'_4, b'_4)$,

(respectively, $A = (b_1, a_1, b_2, a_4, b_4)$ and $B = (b'_1, a'_1, b'_2, a'_4, b'_4)$,) developed is summarized as follows:

**Step 1:** Compute $\tilde{x}_{\mu}(A), \tilde{x}_{v}(A), \tilde{y}_{\mu}(A)$ and $\tilde{y}_{v}(A)$.
**Step 2:** Compute $\tilde{x}_{\mu}(B), \tilde{x}_{v}(B), \tilde{y}_{\mu}(B)$ and $\tilde{y}_{v}(B)$.
**Step 3:** Evaluate $\Re(A)$.
**Step 4:** Evaluate $\Re(B)$
**Step 5:** Calculate $\Re(A)$ and $\Re$(B). Then,

(a) $A \prec B$ if and only if $\Re(\mathrm{A}) < \Re(\mathrm{B})$.
(b) $A \succ B$ if and only if $\Re(\mathrm{A}) > \Re(\mathrm{B})$.
(c) $A \approx B$ if and only if $\Re(\mathrm{A}) = \Re(\mathrm{B})$.

## Numerical examples

This section gives some numerical examples of the proposed method.

*Examples 4.1* Consider two TIFNs $A = (0.2, 0.4, 0.6, 0.8, 1.0, 1.2, 1.4, 1.6)$ and $B = (0.1, 0.3, 0.5, 0.7, 0.9, 1.1, 1.3, 1.5)$.

Then, using the proposed method we get

$$\tilde{x}_\mu(A) = 0.9; \tilde{x}_\upsilon(A) = 0.9; \tilde{y}_\mu(A) = 0.39; \tilde{y}_\upsilon(A) = 0.57,$$
$$\tilde{x}_\mu(B) = 0.8; \tilde{x}_\upsilon(B) = 0.8; \tilde{y}_\mu(B) = 0.39; \tilde{y}_\upsilon(B) = 0.57,$$
$$\Re(A) = \sqrt{\frac{1}{2}\left( [\tilde{x}_\mu(A) - \tilde{y}_\mu(A)]^2 + [\tilde{x}_\upsilon(A) - \tilde{y}_\upsilon(A)]^2 \right)} = 0.56,$$
$$\Re(B) = \sqrt{\frac{1}{2}\left( [\tilde{x}_\mu(B) - \tilde{y}_\mu(B)]^2 + [\tilde{x}_\upsilon(B) - \tilde{y}_\upsilon(B)]^2 \right)} = 0.44,$$
$$\Re(A) > \Re(B) \Rightarrow A \succ B.$$

*Example 4.2* Consider two triangular intuitionistic fuzzy numbers $A = (0.2,0.4,0.6,0.8,1)$ and $B = (0.1,0.3,0.5,0.7, 0.9)$, then

$$\tilde{x}_\mu(A) = \frac{a_1 + a_2 + a_4}{3} = 0.6;\ \tilde{x}_\mu(B) = 0.5,$$
$$\tilde{x}_\upsilon(A) = \frac{2b_1 - b_2 + 2b_4}{3} = 0.6; \tilde{x}_\upsilon(B) = 0.5,$$
$$\tilde{y}_\mu(A) = \frac{1}{3} = 0.33 = \tilde{y}_\mu(B),$$
$$\tilde{y}_\upsilon(A) = \frac{2}{3} = 0.67 = \tilde{y}_\upsilon(B),$$
$$\Re(A) = \sqrt{\frac{1}{2}\left( [\tilde{x}_\mu(A) - \tilde{y}_\mu(A)]^2 + [\tilde{x}_\upsilon(A) - \tilde{y}_\upsilon(A)]^2 \right)} = 0.27,$$
$$\Re(B) = \sqrt{\frac{1}{2}\left( [\tilde{x}_\mu(B) - \tilde{y}_\mu(B)]^2 + [\tilde{x}_\upsilon(B) - \tilde{y}_\upsilon(B)]^2 \right)} = 0.20,$$
$$\Re(A) > \Re(B) \Rightarrow A \succ B.$$

*Example 4.3* Consider two symmetric trapezoidal intuitionistic fuzzy numbers $A = (25,23,1,1;25,23,3,3)$ and $B = (5,7,2,2;5,7,4,4)$, then

$$\tilde{x}_\mu\ (\mathrm{A}) = \frac{1}{2}\left[\frac{a_1 h + a_2 h + a_2^2 - a_1^2}{a_2 - a_1}\right] = 12,$$
$$\tilde{x}_v\ (\mathrm{A}) = \frac{a_1 + a_2}{2} = 24,$$
$$\tilde{y}_\mu\ (\mathrm{A}) = \frac{1}{3}\left[\frac{3a_1 - 3a_2 - 2h}{2a_1 - 2a_2 - 2h}\right] = 0.67,$$
$$\tilde{y}_v\ (\mathrm{A}) = \frac{1}{3}\left[\frac{3a_1 - 3a_2 - 4h'}{2a_1 - 2a_2 - 2h'}\right] = 1.$$

$$\tilde{x}_\mu\ (\mathrm{B}) = \frac{1}{2}\left[\frac{a_1 h + a_2 h + a_2^2 - a_1^2}{a_2 - a_1}\right] = 12,$$
$$\tilde{x}_\upsilon\ (\mathrm{B}) = \frac{a_1 + a_2}{2} = 6,$$
$$\tilde{y}_\mu\ (\mathrm{B}) = \frac{1}{3}\left[\frac{3a_1 - 3a_2 - 2h}{2a_1 - 2a_2 - 2h}\right] = 0.42,$$
$$\tilde{y}_\upsilon\ (\mathrm{B}) = \frac{1}{3}\left[\frac{3a_1 - 3a_2 - 4h'}{2a_1 - 2a_2 - 2h'}\right] = 0.61,$$
$$\Re(A) = \sqrt{\frac{1}{2}\left( [\tilde{x}_\mu(A) - \tilde{y}_\mu(A)]^2 + [\tilde{x}_\upsilon(A) - \tilde{y}_\upsilon(A)]^2 \right)} = 18.24,$$
$$\Re(B) = \sqrt{\frac{1}{2}\left( [\tilde{x}_\mu(B) - \tilde{y}_\mu(B)]^2 + [\tilde{x}_\upsilon(B) - \tilde{y}_\upsilon(B)]^2 \right)} = 9.03.$$

## Comparative study

A comparative study between the proposed method and other existing methods in the literature is Table 1.

## Conclusion

In this paper, a new ranking method based on the centroid is proposed for TrIFNs and TIFNs. The comparative results are given for the justification of the proposed method with the existing ranking. The advantages of the proposed ranking method can be pointed out as follows:

**Table 1** Comparative Examples of the proposed ranking process with the existing ranking process

| S. no. | Examples | Wei's process | Wang and Zhong's process | Rezvani's process | Li's process | Dubey and Mehra's process | Satyajit's process | Sagaya's process | Proposed method |
|---|---|---|---|---|---|---|---|---|---|
| 1. | $A_1 = [(0.57,0.73,0.83);0.73,0.2]$<br>$A_2 = [(0.58,0.74,0.819);0.72,0.2]$ | $A_1 \approx A_2$ | $A_1 \succ A_2$ | $A_1 \prec A_2$ | $A_1 \succ A_2$ | $A_1 \succ A_2$ | $A_1 \succ A_2$ | $A_1 \approx A_2$ | $A_1 \succ A_2$ |
| 2. | $A_1 = [(-9,1.5,3);0.6,0.2]$<br>$A_2 = [(-9,1.5,3);0.7,0.3]$ | $A_1 \prec A_2$ | $A_1 \prec A_2$ | $A_1 \approx A_2$ | $A_1 \approx A_2$ | $A_1 \prec A_2$ | $A_1 \prec A_2$ | $A_1 \approx A_2$ | $A_1 \prec A_2$ |
| 3. | $A_1 = [(3,4,5);0.8,0.2]$<br>$A_2 = [(6,8,10);0.4,0.6]$ | $A_1 \approx A_2$ | $A_1 \succ A_2$ | $A_1 \prec A_2$ | $A_1 \approx A_2$ | $A_1 \approx A_2$ | $A_1 \prec A_2$ | $A_1 \prec A_2$ | $A_1 \succ A_2$ |
| 4. | $A_1 = [(1,2,3);0.6,0.4]$<br>$A_2 = [(2,4,6);0.3,0.7]$ | $A_1 \approx A_2$ | $A_1 \succ A_2$ | $A_1 \prec A_2$ | $A_1 \approx A_2$ | $A_1 \approx A_2$ | $A_1 \prec A_2$ | $A_1 \prec A_2$ | $A_1 \prec A_2$ |
| 5. | $A_1 = [(4,5.5,6,8);1,0]$<br>$A_2 = [(3.5,5,7,7.5);1,0]$ | $A_1 \succ A_2$ | $A_1 \prec A_2$ | $A_1 \approx A_2$ | $A_1 \succ A_2$ | $A_1 \succ A_2$ | $A_1 \succ A_2$ | $A_1 \prec A_2$ | $A_1 \succ A_2$ |
| 6. | $A_1 = [(0.55,0.75,0.8,0.9);0.5,0.5]$<br>$A_2 = [(0.5,0.7,0.85,0.95);0.5,0.5]$ | $A_1 \approx A_2$ | $A_1 \approx A_2$ | $A_1 \approx A_2$ | $A_1 \approx A_2$ | $A_1 \succ A_2$ | $A_1 \succ A_2$ | $A_1 \succ A_2$ | $A_1 \succ A_2$ |

1. The method proposed is time consuming compared to the existing methods.
2. The proposed ranking method can be applied for both TrIFNs and TIFNs, which reflects the uncertainty suitably.
3. The proposed ranking technique based on centroid is flexible to the researchers in the ranking index of their attitudinal analysis.

Using this ranking, many decision-making and optimization problems of uncertain nature can be solved. In future, we will use the proposed ranking technique to solve intuitionistic fuzzy linear programming and multi-criteria decision-making problems.

## References

1. Abbasbandy, S., Hajjari, T.: A new approach for ranking of trapezoidal fuzzy numbers. Comput. Math Appl. **57**, 413–419 (2009)
2. Kumar, A., Kaur, M.: A ranking approach for intuitionistic fuzzy numbers and its application. J. Appl. Res. Technol. **11**, 381–396 (2013)
3. Arun Prakash, K., Suresh, M., Vengataasalam, S.: An solution of intuitionistic fuzzy transportation problem. Asian J. Res. Soc. Sci. Human. **6**, 2650–2658 (2016)
4. Atanassov, K.T.: Intuitionistic fuzzy sets. Fuzzy Sets Syst. **20**, 87–96 (1986)
5. Delgado, M., Vila, M.A., Voxman, W.: On a canonical representation of fuzzy numbers. Fuzzy Sets Syst. **93**, 125–135 (1998)
6. Dubey, D., Mehra, A.: Linear programming with triangular intuitionistic fuzzy number, EUSFLAT-LFA, 563–569 (2011)
7. Grzegorzewski, P.: Distance and orderings in a family of intuitionistic fuzzy numbers. In proceedings of the third conference on fuzzy logic and technology (EUSFLAT 03), 223–227 (2003)
8. Grzegorzewski, P.: The hamming distance between intuitionistic fuzzy sets. In Proceedings of the IFSA 2003 World Congress
9. Li, D.F.: A ratio ranking method of triangular intuitionistic fuzzy numbers and its applications to MADM problems. Comput. Math Appl. **60**, 1557–1570 (2010)
10. Li, D.F., Nan J.X., Zhang M.J.: A ranking Method of triangular intuitionistic fuzzy numbers and applications to decision making, international Journal of Computational Intelligence Systems **3**, 522–530 (2010)
11. Mitchell, H.B.: Ranking - Intuitionistic Fuzzy numbers. Int. J. Uncert. Fuzz. Knowl. Based Syst. **12**, 377–386 (2004)
12. Nagoorgani, A., Ponnalagu, K.: A new approach on solving intuitionistic fuzzy linear programming problem. Appl. Math. Sci. **6**, 3467–3474 (2012)
13. Nayagam, V.L.G., Venkateshwari, G., Sivaraman, G.: Modified ranking of intuitionistic fuzzy numbers. Notes on Intuitionistic Fuzzy Sets **17**, 5–22 (2011)
14. Nayagam, V.L.G., Venkateshwari, G., Sivaraman, G.: Ranking of intuitionistic fuzzy numbers. Proceed. Int. Confer. Fuzzy Syst. 2008, Fuzz-IEEE 1971–1974 (2008)
15. Nehi, H.M.: A new ranking method for intuitionistic fuzzy numbers. Int. J. Fuzzy Syst. **12**, 80–86 (2010)
16. Parvathi, R., Malathi, C.: Arithmetic operations on symmetric trapezoidal intuitionistic fuzzy numbers. Int. J. Soft Comp. Eng. **2**, 268–273 (2012)
17. Peng, Z., Chen; A new method for ranking canonical intuitionistic fuzzy numbers. Proceed. Int. Confer. Inform. Eng. Appl. (IEA), 216, 609–616 (2013)
18. Rezvani, S.: Ranking method of trapezoidal intuitionistic fuzzy numbers. Ann. Fuzzy Math. Inform. **10**, 1–10 (2013)
19. Sagaya Roseline, S., Henry Amirtharaj, E.C.: A new method for ranking of intuitionistic fuzzy numbers. Indian J. Appl. Res. **3**, 1–2 (2013)
20. Salahshour, S., Shekari, G.A., Hakimzadeh, A.: A novel approach for ranking triangular intuitionistic fuzzy numbers. AWER Procedia Inform. Technol. Comp. Sci. **1**, 442–446 (2012)
21. Suresh, M., Vengataasalam, S., Arun Prakash, K.: Solving intuitionistic fuzzy linear programming by ranking function. J. Intell. Fuzzy Syst. **27**(2014), 3081–3087 (2014)
22. Suresh, M., Vengataasalam, S., Arun Prakash, K.: A method to solve fully intuitionistic fuzzy assignment problem. VSRD Int. J. Techn. Non-Techn. Res. **7**, 71–74 (2016)
23. Seikh, M.R., Nayak, P.K., Pal, M.: Generalized triangular fuzzy numbers in intuitionistic fuzzy environment. Int. J. Eng. Res. Dev. **5**, 8–13 (2012)
24. Seikh, M.R., Nayak, P.K., Pal, M.: Notes on triangular intuitionistic fuzzy numbers. Int. J. Math. Operat. Res. **5**, 446–465 (2013)
25. Wang, J., Zhang, Z.: Aggregation operators on Intuitionistic trapezoidal Fuzzy Numbers and its applications to Multi-Criteria Decision Making Problems. J. Syst. Eng. Elect. **20**, 321–326 (2009)
26. Wang, X., Kerre, E.E.: Reasonable properties for the ordering of fuzzy quantities (I). Fuzzy Sets Syst. **118**, 375–385 (2001)
27. Wei, C.P., Tang, X.: Possibility degree method for ranking intuitionistic fuzzy numbers, In the Proceed. Int. Confer. Web Intell. Intelligen Agent Technol. (2010)
28. Ye, J.: Multicriteria decision-making method based on a cosine similarity measure between trapezoidal fuzzy numbers. Int. J. Eng. Sci. Technol. **3**, 272–278 (2011)
29. Zadeh, A.: Fuzzy sets. Inf. Control **8**, 338–353 (1965)
30. Zhang, M. J., Nan, J. X.: A compromise ratio ranking method of triangular intuitionistic fuzzy numbers and its application to MADM problems. Iranian J. Fuzzy Syst. 10, 21–37 (2013)

# 22

# Universal approximation method for the solution of integral equations

**Mahmood Otadi[1] · Maryam Mosleh[1]**

**Abstract** In this paper, we present a novel approach for approximating Hammerstein–Volterra delay integral equations (HVDIEs) by applying the universal approximation method through an artificial intelligence utility in a simple way. In this paper, neural network model (NNM) is applied as universal approximators for any nonlinear continuous functions. Here, neural network is considered as a part of large field called neural computing or soft computing. With this capability, the solution of Hammerstein–Volterra delay integral equation can be approximated by the appropriate NNM within an arbitrary accuracy.

**Keywords** Hammerstein–Volterra delay integral equation · Universal approximation · Neural network model · Numerical method · Mathematical model in epidemiology

## Introduction

Proper design for engineering applications requires detailed information of the system-property distributions such as temperature, velocity, density, etc., in space and time domain. This information can be obtained by either experimental measurement or computational simulation. Although experimental measurement is reliable, it needs a lot of labor efforts and time. Therefore, the computational simulation has become a more and more popular method as a design tool, since it needs only a fast computer with a large memory [45].

The solutions of integral equations have a major role in the field of science and engineering. The theory and application of integral equation are an important subject within applied mathematics. Integral equations are used as mathematical models for many physical situations, and integral equations also occur as reformulations of other mathematical problems, such as partial differential equations and ordinary differential equations. A physical even can be modeled by the differential equation, an integral equation. Since few of these equations cannot be solved explicitly, it is often necessary to resort to numerical techniques which are appropriate combinations of numerical integration and interpolation [3, 4, 6, 7, 30]. There are several numerical methods for solving linear Volterra integral equation [16, 49] and system of nonlinear Volterra integral equations [9, 51]. Biazar [12] presented differential transform method for solving systems of integral equations. Kauthen in [25] used a collocation method to solve the Volterra–Fredholm integral equation numerically. Borzabadi and Fard in [14] obtained a numerical solution of nonlinear Fredholm integral equations of the second kind.

In recent years, meshless methods have been developed as alternative numerical approaches in effort to eliminate known shortcomings of the mesh-based methods [8]. The main advantage of these methods is to approximate the unknowns by a linear combination of shape functions. Shape functions are based on a set of nodes and a certain weight function with a local support associated with each of these nodes. Therefore, they can solve many engineering problems that are not suited to the conventional computational methods [17, 47, 53]. Bhrawy et al. [10] reported a new spectral collocation technique for solving second kind

✉ Mahmood Otadi
mahmoodotadi@yahoo.com

Maryam Mosleh
maryammosleh79@yahoo.com

[1] Department of Mathematics, Firoozkooh Branch, Islamic Azad University, Firoozkooh, Iran

Fredholm integral equations (FIEs). They developed a collocation scheme to approximate FIEs by means of the shifted Legendre–Gauss–Lobatto collocation (SL–GL–C) method. Then, they developed an efficient direct solver for solving numerically the high-order linear Fredholm integro-differential equations (FIDEs) with piecewise intervals under the initial-boundary conditions [11]. Maleknejad et al. [33] introduced an approach for obtaining the numerical solution of the nonlinear Volterra–Fredholm integro-differential (NVFID) equations using hybrid Legendre polynomials and block-pulse functions. These hybrid functions and their operational matrices are used for representing matrix form of these equations. Hashemizadeh and Rostami [20] obtained numerical solution of Hammerstein integral equations of mixed type by means of Sinc collocation method is presented. This proposed approximation reduces these kinds of nonlinear Hammerstein integral equations of mixed type to a nonlinear system of algebraic equations.

FNN systems are hybrid systems that combine the theories of fuzzy logic and neural networks. Designing the FNN system based on the input–output data is a very important problem. Several authors investigated FNN, to compute crisp and even fuzzy information with neural network (NN). There existed only a few approaches to learning algorithms for FNN when Ishibuchi et al. presented two NNs which can be trained with interval vectors and with vectors of fuzzy numbers. In both networks, Ishibuchi et al. used crisp weights. For these networks, they presented a backpropagation based on learning algorithm. In a later paper [23], Ishibuchi et al. developed an FNN with symmetric triangular fuzzy numbers as weights. For this NN, they evolved a learning algorithm in which the backpropagation algorithm is used to compute the new lower and upper limits of the support of the weights. The modal value of the new fuzzy weight is calculated as an average of the newly computed limits. Recently, FNN successfully was used for solving fuzzy polynomial equation and systems of fuzzy polynomials [1, 2], approximate fuzzy coefficients of fuzzy regression models [37, 38, 40, 41], approximate solution of fuzzy linear systems and fully fuzzy linear systems [42, 43], and fuzzy differential equations [34–36, 44].

In this work, we propose a new solution method for the approximate solution of integral equations using innovative mathematical tools and neural-like systems of computation. This hybrid method can result in improved numerical methods for solving integral equations. In this proposed method, NNM is applied as universal approximator. Neural computation research, together with related areas in approximation theory, has developed powerful methods for approximating continuous and integrable functions on compact subsets. In such schemes, function approximation capabilities critically depend on the activation function nature of the hidden layer. Ito and Saito [24] proved that if the activation function is continuous and nondecreasing sigmoidal function, then the interpolation can be made with inner weights.

## Neural network model

Artificial neural networks are an exciting form of artificial intelligence, which mimic the learning process of the human brain to extract patterns from historical data [2, 48]. For many years, this technology has been successfully applied to a wide variety of real-world applications [46]. Simple perceptrons need a teacher to tell the network what the desired output should be. These are supervised networks. In an unsupervised net, the network adapts purely in response to its inputs. Such networks can learn to pick out structure in their input. Figure 1 shows typical three-layered perceptron as a basic structural architecture with an input layer, a single hidden layer, and an output layer [19]. Here, the dimension of NNM is denoted by the number of neurons in each layer, that is $N \times M \times S$ NNM, where $N$, $M$, and $S$ are the number of neurons in the input layer, the hidden layer, and the output layer, respectively. The architecture of the model shows how NNM transforms the $N$ inputs $(t_1, \ldots, t_i, \ldots, t_N)$ into the $s$ outputs $(x_1, \ldots, x_k, \ldots, x_S)$ throughout the $m$ hidden neurons $(z_1, \ldots, z_j, \ldots, z_M)$, where the cycles represent the neurons in each layer. Let $b_j$ be the bias for neuron $z_j$, $c_k$ be the bias for neuron $x_k$, $w_{ji}$ be the weight connecting neuron $x_i$ to neuron $z_j$, and $w_{kj}$ be the weight connecting neuron $z_j$ to neuron $x_k$. Then, the output of NNM $f : \mathrm{R}^N \to \mathrm{R}^S$ can be determined as follows:

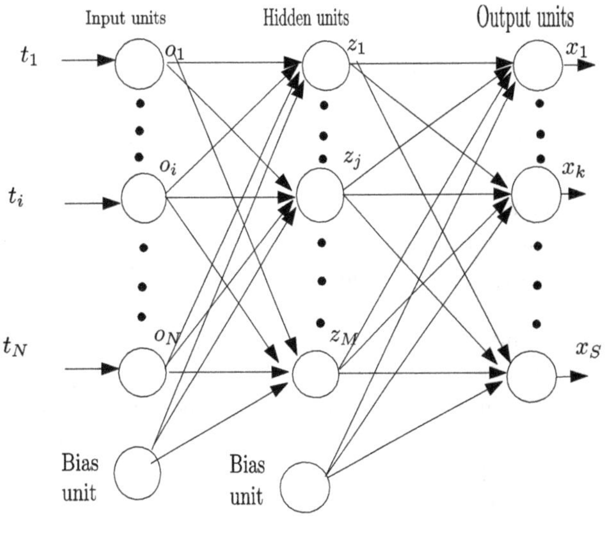


**Fig. 1** Multiple layer feed-forward NNM

$$x_k = f_k\left(\sum_{j=1}^{M} w_{kj} z_j + c_k\right), \tag{1}$$

with

$$z_j = f_j\left(\sum_{i=1}^{N} w_{ji} t_i + b_j\right), \tag{2}$$

where $f_k$ and $f_j$ are the activation functions which are linear, piecewise linear, hard limiter, unipolar sigmoidal, and bipolar sigmoidal functions. The usual choices of the activation function [21, 28, 45] are the unipolar sigmoidal of the form:

$$f(\theta) = \frac{1}{1 + \exp(-\theta)}. \tag{3}$$

Multi-layered perceptrons with more than three layers, which use more hidden layers [21, 26]. Multi-layered perceptrons correspond the input units to the output units by a specific nonlinear mapping [50]. The most important application of multi-layered perceptrons is their ability in function approximation [15]. From Kolmogorov existence theorem, we know that a three-layered perceptron with $n(2n + 1)$ nodes can compute any continuous function of $n$ variables [22, 31]. The accuracy of the approximation depends on the number of neurons in the hidden layer and does not depend on the number of the hidden layers [27].

## Hammerstein–Volterra delay integral equations

The present paper deals with the investigation of the Volterra integral equation having a constant delay:

$$x(t) = g(t) + \lambda \int_{t-\tau}^{t} H(t,s) \cdot x(s)\mathrm{d}s, \quad t \in [0,T], \tag{4}$$

with the initial condition

$$x(t) = \psi(t), \quad t \in [-\tau, 0], \tag{5}$$

where $\tau > 0$ and $T > 0$ are such that $T = p \bullet \tau$ for given $p \in \mathrm{N}$, and $g$ and $\psi$ are known functions and $H : [0,T] \times [-\tau, T] \to \mathrm{R}$ is a weight function. This equation is a mathematical model for the spread of certain infectious diseases with a contact rate that varies seasonally. Here, $x(t)$ is the proportion of infectives in the population at time $t$, $\tau > 0$ is the length of time in which an individual remains infectious, and $H(t, s).x(s)$ is the proportion of new infectives per unit time [39]. Throughout this paper, we always assume that the solution of (4) and (5) exists and is unique. Now, we study the NNM to approximate the solution of Eqs. (4) and (5).

The integral Eq. (4) contains a constant delay and its variant is the generalization of an epidemic model (see [18]), where $H(t, s) = P(t - s)$. First, we define the operator:

$$T_1(t, x(t)) = x(t) - g(t) - \lambda \int_{t-\tau}^{t} H(t,s) \cdot x(s)\mathrm{d}s, \tag{6}$$

$$T_2(t, x(t)) = x(t) - \psi(t). \tag{7}$$

To obtain approximate solution $x_M(t, P)$, we solve unconstrained optimization problem that is simpler to deal with; we define the trial function to be in the following form:

$$x_M(t, P) = N_M(t, P), \tag{8}$$

where the term in the right-hand side is a feed-forward neural network consisting of an input $t$ and $P$ is the vector containing all the adjustable parameters of NNM [32]. This NNM with some weights and biases is considered, and we train to compute the approximate solution of HVDIEs.

Substituting (8) into Eqs. (4) and (5), we can obtain the expression:

$$T_1(t, x_M(t)) = x_M(t) - g(t) - \lambda \int_{t-\tau}^{t} H(t,s).x_M(s)\mathrm{d}s, \quad t \in [0,T], \tag{9}$$

$$T_2(t, x_M(t)) = x_M(t) - \Phi(t), \quad t \in [-\tau, 0]. \tag{10}$$

For any $t \in [-\tau, 0], R_{M_2}(t) = T_2(t, x_M(t)) - T_2(t, x(t))$ and for any $t \in [0, T], R_{M_1}(t) = T_1(t, x_M(t)) - T_1(t, x(t))$ is called the remaining items of Eqs. (4) and (5), where $M$ is the numbers of neurons or hidden layers for NNM; also in addition, we have:

$$R_{M_1}(t) = (x_M(t) - x(t)) - \lambda \int_{t-\tau}^{t} H(t,s) \cdot (x_M(s) - x(s))\mathrm{d}s,$$

and

$$R_{M_2}(t) = x_M(t) - x(t).$$

*Remark 1* For any $t \in [-\tau, T]$, if $R_{M_1}(t) = 0$ and $R_{M_2}(t) = 0$, then $x(t) = x_M(t)$; if $\lim_{M\to\infty} R_{M_1}(t) = 0$ and $\lim_{M\to\infty} R_{M_2}(t) = 0$, then $\lim_{M\to\infty} x_M(t) = x(t)$.

*Remark 2* For any $t \in [-\tau, T]$, if $R_{M_1}(t) \equiv 0$ and $R_{M_2}(t) \equiv 0$, then $x_M(t)$ is an exact solution of Eqs. (4) and (5); if $\lim_{M\to\infty} R_{M_1}(t) = 0$ and $\lim_{M\to\infty} R_{M_2}(t) = 0$, then $x_M(t)$ converges to the exact solution of Eqs. (4) and (5).

*Remark 3* If $\lim_{M\to\infty} \frac{1}{2}\int_{-\tau}^{0} T_2^2(t, x_M(t))\mathrm{d}t + \frac{1}{2}\int_0^T T_1^2(t, x_M(t))\mathrm{d}t = 0$,, then the approximation solution $x_M(t)$ converges to the exact solution $x(t)$ of Eqs. (4) and (5).

To compute the integrals from Eqs. (4) and (5), we consider the uniform partition of $[-\tau, T]$:

$$\Delta : -\tau = t_0 < t_1 < \cdots < t_n = 0 < t_{n+1} < \cdots < t_q = T$$

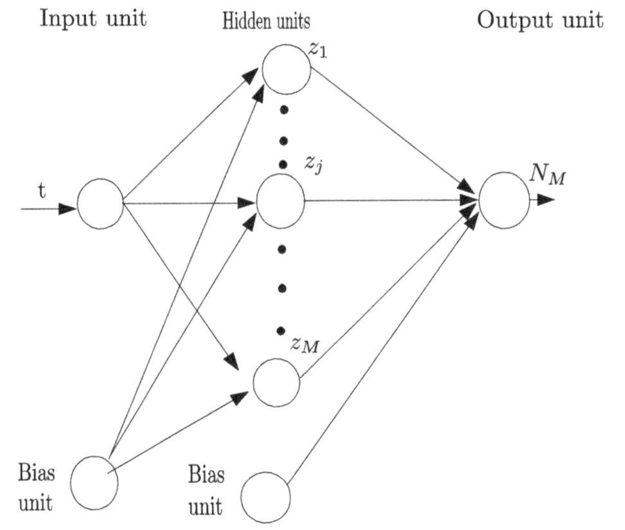


**Fig. 2** NNM with one input unit and one output unit

with $q = (p+1)n, t_i = t_{i-1} + \frac{\tau}{n} = -\tau + \frac{i\tau}{n}, i = \overline{1, q}$. On these knots, the terms of the sequence of successive approximations are the following:

$$x(t_i) = \psi(t_i), \quad i = \overline{0, n},$$

$$x(t_i) = g(t_i) + \lambda \int_{t_i-\tau}^{t_i} H(t_i, s) \cdot x(s)\mathrm{d}s, \quad i = \overline{n+1, q}. \tag{11}$$

The numerical calculation can be implemented to determine the integration of Eq. (11). Let us consider a three-layered NNM (see Fig. 2) with one unit entry *t*, one hidden layer consisting of *M* activation functions, and one unit output $N_M(t, P)$. In this paper, we use the unipolar sigmoidal activation function *f*(.). Here, the dimension of NNM is $1 \times M \times 1$.

Hence, in our proposed UAM, the solution of HVDIEs can be simply obtained with the algorithm of optimization; in addition, the adjustable parameters of NNM are systematically updated in such a way. Hence, the problem formulation can be expressed as the typical minimization problem:

$$\text{Minimize} \quad J(P) \tag{12}$$

where *P* is the vector containing all the adjustable parameters. For instance, we can use the penalty method for the minimization problem. Therefore, for solving HVDIEs by Eq. (12), we have:

$$\text{Minimize} \quad \sum_{g=1}^{G} \frac{1}{2} T_2^2(t_g, x_M(t_g)) + \frac{1}{2} T_1^2(t_g, x_M(t_g)), \tag{13}$$

where *G* is the total number of points chosen within the domain of $[-\tau, T]$ and *g* is the point index.

Below, we present the following algorithm that gives the approximate solution using NNM:

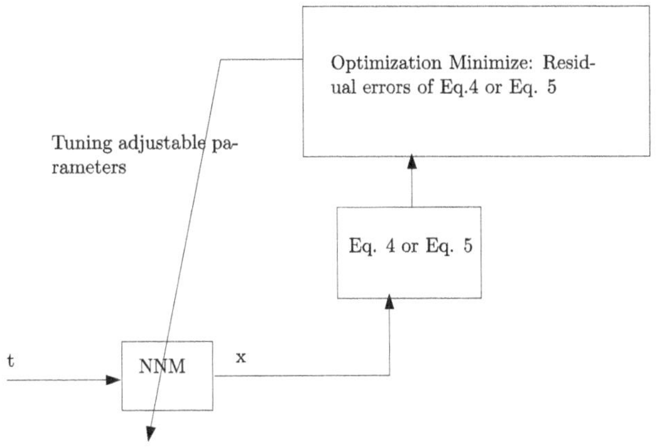


**Fig. 3** Diagram of proposed NNM

**Step 1** Choose the numbers of neurons and hidden layers for NNM as small as possible at the beginning.

**Step 2** Apply an optimization technique to determine the sub-optimal adjustable parameters of NNM in such a way that the residual errors are minimized.

**Step 3** If the residual errors are less than tolerance then stop.

If not, then try to increase the various numbers in Step 1 and go to Step 2.

Figure 3 shows the overall diagram of the proposed UAM in determining the solution of HVDIEs.

## Numerical examples

In this section, two examples are given to illustrate the technique proposed in this paper.

*Example 4.1* Consider the initial value problem:

$$x(t) = \begin{cases} e^{t-\tau} + \int_{t-\tau}^{t} x(s)\mathrm{d}s, & t \in [0, 0.5], \\ \psi(t), & t \in [-\tau, 0], \end{cases}$$

with $\tau = 0.5$, $T = 0.5$, and $\psi$:[ −0.5, 0] → R is defined by $\psi(t) = e^t$, $t \in [-\tau, 0]$. The exact solution is $x$:[ −0.5, 0.5] → $R$ given by $x(t) = e^t$, $t \in [-0.5, 0.5]$. Applying for $M = 9$, the above presented method, we obtain approximate solution with the error $3.21\mathrm{e}^{-13}$. The exact and obtained solutions of Hammerstein–Volterra delay integral equation in this example are shown in Fig. 4. We see that the approximate solution obtained by the NNM has good accuracy on the whole interval. In Table 1, we compare the error of the present method (Method 1), method 2 in [13], method 3 in [5], method 4 in [52], and method 5 in [39].

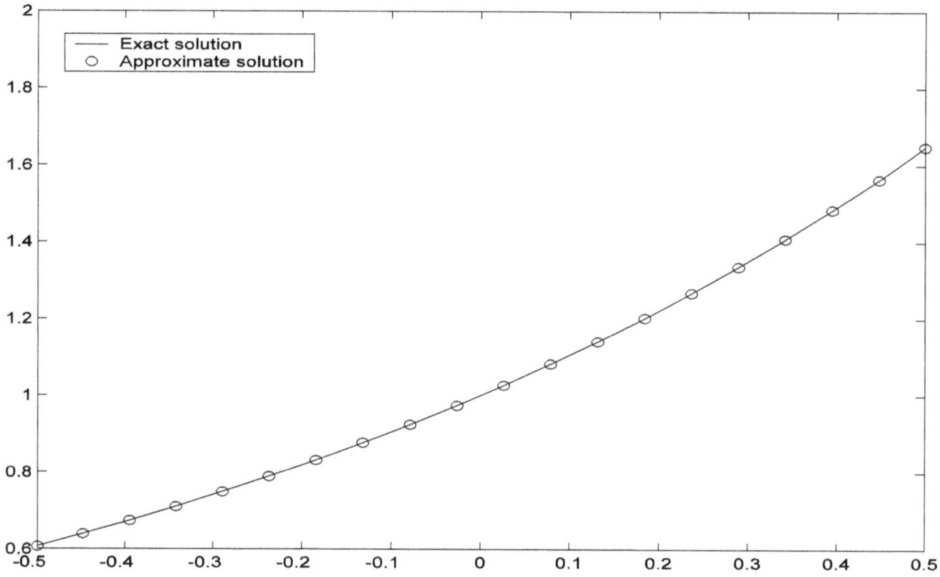


**Fig. 4** Compares the exact solution and obtained solution

**Table 1** Comparison of the errors for Example 4.1

| $M$ | Method 1 | Method 2 | Method 3 | Method 4 | Method 5 |
|---|---|---|---|---|---|
| 8 | $2.22e^{-11}$ | $3.54e^{-6}$ | $4.11e^{-8}$ | $5.31e^{-6}$ | $2.01e^{-9}$ |
| 9 | $3.21e^{-13}$ | $6.21e^{-7}$ | $3.30e^{-8}$ | $3.01e^{-6}$ | $4.71e^{-10}$ |

*Example 4.2* According to the epidemic model presented in [18], we consider that $x(t)$ be the proportion of infectious individuals at the moment $s$, $f(x(s))$ be the proportion of new infected cases on unit time, and $g(t)$ be the proportion of immigrants that still have the disease at the moment $t$. Considering $P(s)$ as the probability of having the infection for a time at least $s$ after infection, the spread of infection is governed by the integral equation:

$$x(t) = \begin{cases} g(t) + \int_{t-\tau}^{t} P(t-s) \cdot f(x(s)) \mathrm{d}s, & t \in [0, T], \\ \psi(t), & t \in [-\tau, 0]. \end{cases}$$

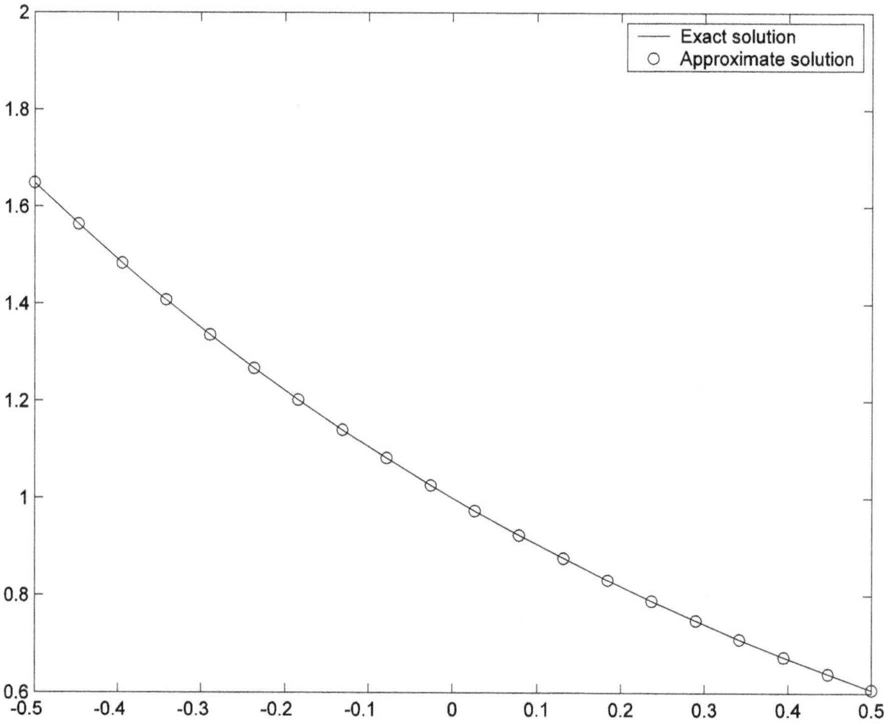


**Fig. 5** Compares the exact solution and obtained solution

Suppose that the proportion of new infected cases on unit time, $f(x(s))$ is proportional with $x(s)$, and $P$ is a positive decreasing function with $P(0) = 1$. Let $\tau = T = 0.5$. Since it is natural to suppose that the proportion of immigrants that still have the disease is decreasing in time according to the decisions of the authorities, we were shown to the following model:

$$x(t) = \begin{cases} g(t) + \int_{t-\tau}^{t} e^{-(t-s)} \cdot x(s)ds, & t \in [0, 0.5], \\ \psi(t), & t \in [-0.5, 0], \end{cases}$$

with $g(t) = 0.5e^{-t}$ and $\psi(t) = e^{-t}$. The exact solution is $x:[-0.5, 0.5] \to R$ given by $x(t) = e^{-t}$, $t \in [-0.5, 0.5]$. Applying, for $M = 10$, the above presented method, we obtain approximate solution with the error $4.23e^{-9}$. The exact and obtained solutions of Hammerstein–Volterra delay integral equation in this example are shown in Fig. 5.

## Conclusions

Solving Hammerstein–Volterra delay integral equations using the universal approximators, that is, NNM was presented in this paper. We proposed NNM approximation method based on unipolar sigmoidal functions. The reliability and efficiency of the proposed method are demonstrated on the numerical experiments. In addition, we can execute this method in a computer simply.

## References


1. Abbasbandy, S., Otadi, M.: Numerical solution of fuzzy polynomials by fuzzy neural network. Appl. Math. Comput. **181**, 1084–1089 (2006)
2. Abbasbandy, S., Otadi, M., Mosleh, M.: Numerical solution of a system of fuzzy polynomials by fuzzy neural network. Inform. Sci. **178**, 1948–1960 (2008)
3. Abbasbandy, S.: Application of He's homotopy perturbation method to functional integral equations. Chaos, Solitons Fractals **31**, 1243–1247 (2007)
4. Abbasbandy, S.: Numerical solutions of the integral equations: homotopy perturbation method and Adomian's decomposition method. Appl. Math. Comput. **173**, 493–500 (2006)
5. Avaji, M., Hafshejani, J.S., Dehcheshmeh, S.S., Ghahfarokhi, D.F.: Solution of delay Volterra integral equations using the vavariational iteration method. J. Appl. Sci. **12**, 196–200 (2012)
6. Babolian, E., Abbasbandy, S., Fattahzadeh, F.: A numerical method for solving a class of functional and two dimensional integral equations. Appl. Math. Comput. **198**, 35–43 (2008)
7. Baker, C.T.H.: A perspective on the numerical treatment of Volterra equations. J. Comput. Appl. Math. **125**, 217–249 (2000)
8. Belytschko, T., Krongauz, Y., Organ, D.: Meshless methods: an overview and recent developments. Comput. Methods Appl. Mech. Eng. **139**, 3–47 (1996)
9. Berenguer, M.I., Gamez, D., Garralda-Guillem, A.I., Galan, M.R., Serrano Perez, M.C.: Biorthogonal systems for solving Volterra integral equation systems of the second kind. J. Comput. Appl. Math. **235**, 1875–1883 (2011)
10. Bhrawy, A.H., Abdelkawy, M.A., Machado, J.T., Amin, A.Z.M.: Legendre–Gauss–Lobatto collocation method for solving multi-dimensional Fredholm integral equations. Comput. Math. Appl. (2016) (**in press**)
11. Bhrawy, A.H., Tohidi, E., Soleymani, F.: A new Bernoulli matrix method for solving high-order linear and nonlinear Fredholm integro-differential equations with piecewise intervals. Appl. Math. Comput. **219**, 482–497 (2012)
12. Biazar, J., Eslami, M., Islam, M.R.: Differential transform method for special systems of integral equations. J. King Saud Univ. Sci. **24**, 211–214 (2012)
13. Bica, A., Iancu, C.: A numerical method in terms of the third derivative for a delay integral equation from biomathematics. J. Inequal. Pure Appl. Math. **6**(42), 1–8 (2005)
14. Borzabadi, A.H., Fard, O.S.: A numerical scheme for a class of nonlinear Fredholm integral equations of the second kind. J. Comput. Appl. Math. **232**, 449–454 (2009)
15. Buckley, J.J., Hayashi, Y.: Can fuzzy neural nets approximate continuous fuzzy functions? Fuzzy Sets Syst. **61**, 43–51 (1994)
16. Chen, Y., Tang, T.: Spectral methods for weakly singular Volterra integral equations with smooth solutions. J. Comput. Appl. Math. **233**, 938–950 (2009)
17. Cheng, Y., Wang, J., Li, R.: The complex variable element-free Galerkin method for two-dimensional elastodynamics problems. Int. J. Appl. Mech. **4**, 1250042-1–125004223 (2012)
18. Cooke, K.L.: An epidemic equation with immigration. Math. Biosci. **29**, 135–158 (1979)
19. Hagan, M.T., Demuth, H.B., Beale, M.: Neural network design. PWS publishing company, Massachusetts (1996)
20. Hashemizadeh, E., Rostami, M.: Numerical solution of Hammerstein integral equations of mixed type using the Sinc-collocation method. J. Comput. Appl. Math. **279**, 31–39 (2015)
21. Haykin, S.: Neural Networks: A Comprehensive Foundation. Prentice Hall, New Jersey (1999)
22. Hornick, K., Stinchcombe, M., White, H.: Multilayer feedforward networks are universal approximators. Neural Netw. **2**, 359–366 (1989)
23. Ishibuchi, H., Kwon, K., Tanaka, H.: A learning algorithm of fuzzy neural networks with triangular fuzzy weights. Fuzzy Sets Syst. **71**, 277–293 (1995)
24. Ito, Y., Saito, K.: Superposition of linearly independent functions and finite mappings by neural networks. Math. Sci. **21**, 27–33 (1996)
25. Kauthen, J.P.: Continuous time collocation method for Volterra-Fredholm integral equations. Numer. Math. **56**, 409–424 (1989)
26. Khanna, T.: Foundations of neural networks. Addison-Wesly, Reading, MA (1990)
27. Lapedes, A., Farber, R.: How neural nets work. In: Anderson, D.Z. (ed.) Neural information processing systems, pp. 442–456. American Institute of Physics, New York (1988)
28. Leephakpreeda, T.: Novel determinatin of differential-equation solution: universal approximation method. J. Comput. Appl. Math. **146**, 443–457 (2002)
29. Liew, K.M., Cheng, Y., Kitipornchai, S.: Boundary element-free method (BEFM) and its application to two-dimensional elasticity problems. Int. J. Numer. Methods Eng. **65**, 1310–1332 (2006)
30. Linz, P.: Analytical and numerical methods for Volterra equations. SIAM, Philadelphia, PA (1985)
31. Lippmann, R.P.: An introduction to computing with neural nets. IEEE ASSP Mag. **4**, 4–22 (1987)
32. Malek, A., Beidokhti, R.S.: Numerical solution for high order differential equations using a hybrid neural network-Optimization method. Appl. Math. Comput. **183**, 260–271 (2006)
33. Maleknejad, K., Basirat, B., Hashemizadeh, E.: Hybrid Legendre polynomials and Block-Pulse functions approach for nonlinear

Volterra–Fredholm integro-differential equations. Comput. Math Appl. **61**, 2821–2828 (2011)
34. Mosleh, M.: Fuzzy neural network for solving a system of fuzzy differential equations. Appl. Soft Comput. **13**, 3597–3607 (2013)
35. Mosleh, M.: Numerical solution of fuzzy linear Fredholm integro-differential equation by fuzzy neural network. Iran. J. Fuzzy Syst. **11**, 91–112 (2014)
36. Mosleh, M., Otadi, M.: Simulation and evaluation of fuzzy differential equations by fuzzy neural network. Appl. Soft Comput. **12**, 2817–2827 (2012)
37. Mosleh, M., Otadi, M., Abbasbandy, S.: Fuzzy polynomial regression with fuzzy neural networks. Appl. Math. Model. **35**, 5400–5412 (2011)
38. Mosleh, M., Allahviranloo, T., Otadi, M.: Evaluation of fully fuzzy regression models by fuzzy neural network. Neural Comput Appl. **21**, 105–112 (2012)
39. Mosleh, M., Otadi, M.: Least squares approximation method for the solution of Hammerstein–Volterra delay integral equations. Appl. Math. Comput. **258**, 105–110 (2015)
40. Mosleh, M., Otadi, M., Abbasbandy, S.: Evaluation of fuzzy regression models by fuzzy neural network. J. Comput. Appl. Math. **234**, 825–834 (2010)
41. Otadi, M.: Fully fuzzy polynomial regression with fuzzy neural networks. Neurocomputing **142**, 486–493 (2014)
42. Otadi, M., Mosleh, M.: Simulation and evaluation of dual fully fuzzy linear systems by fuzzy neural network. Appl. Math. Model. **35**, 5026–5039 (2011)
43. Otadi, M., Mosleh, M., Abbasbandy, S.: Numerical solution of fully fuzzy linear systems by fuzzy neural network. Soft. Comput. **15**, 1513–1522 (2011)
44. Otadi, M., Mosleh, M.: Simulation and evaluation of interval-valued fuzzy linear Fredholm integral equations with interval-valued fuzzy neural network. Neurocomputing **205**, 519–528 (2016)
45. Otadi, M., Mosleh, M.: Numerical solution of quadratic Riccati differential equation by neural network. Math. Sci. **5**, 249–257 (2011)
46. Picton, P.: Neural networks, 2nd edn. Palgrave, Great Britain (2000)
47. Ren, H., Cheng, Y.: The interpolating element-free Galerkin (IEFG) method for two-dimensional potential problems. Eng. Anal. Boundary Elem. **36**, 873–880 (2012)
48. Schalkoff, R.J.: Artificial Neural Networks. McGraw-Hill, New York (1997)
49. Sorkun, H.H., Yalcinbas, S.: Approximate solutions of linear Volterra integral equation systems with variable coefficients. Appl. Math. Model. **34**, 3451–3464 (2010)
50. Stanley, J.: Introduction to Neural Networks, 3rd edn. Sierra Mardre (1990)
51. Wang, Q., Wang, K., Chen, S.H.: Least squares approximation method for the solution of Volterra-Fredholm integral equations. J. Comput. Appl. Math. **272**, 141–147 (2014)
52. Yalcinbas, S.: Taylor polynomial solutions of nonlinear Volterra-Fredholm integral equations. Appl. Math. Comput. **127**, 195–206 (2002)
53. Zheng, B., Dai, B.: A meshless local moving Kriging method for two-dimensional solids. Appl. Math. Comput. **218**, 563–573 (2011). **[199–249 (1975)]**

# 23

# Understanding image inpainting with the help of the Helmholtz equation

**Laurent Hoeltgen**[1]

**Abstract** Partial differential equations have recently been used for image compression purposes. One of the most successful frameworks solves the Laplace equation using a weighting scheme to determine the importance of individual pixels. We provide a physical interpretation of this approach in terms of the Helmholtz equation which explains its superiority. For better reconstruction quality, we subsequently formulate an optimisation task for the corresponding finite difference discretisation to maximise the influence of the physical traits of the Helmholtz equation. Our findings show that sharper contrasts and lower errors in the reconstruction are possible.

**Keywords** Laplace interpolation · Helmholtz equation · Image inpainting · Finite difference scheme

**Mathematics Subject Classification** 35Q99 · 49M25

## Introduction

The reconstruction of an image from a sparse subset of all pixels is known as inpainting [1]. For the application of image compression, the selection of the data has to be optimised. Let us emphasise that this is by no means a simple task. Selecting 5% of the pixels from a $256 \times 256$ pixel image offers more than $10^{5000}$ possible choices. In [2–7], corresponding strategies are devised via partial differential equations (PDEs). Related models for the inpainting step using the Allen–Cahn model have also been considered in [8], whereas the authors of [9] analysed the Cahn–Hilliard equation. Finally, a broader discussion on fluid dynamics for image reconstruction tasks has been discussed in [10]. The results from [2, 6] motivated the authors of [11] to suggest an optimal control-based approach with a relaxed formulation of the Laplace equation given for known data $f$ by

$$c(\mathbf{x})(u(\mathbf{x}) - f(\mathbf{x})) + (1 - c(\mathbf{x}))(-\Delta)u(\mathbf{x}) = 0 \tag{1}$$

and additional boundary conditions. The support of the function $c$ indicates that the data locations used for reconstruction should be minimised while preserving a good reconstruction quality. In [11], a local contrast enhancing effect was also observed if $c$ maps to values outside of $[0, 1]$. Based on [6], this finding was reinforced in [12] where an equivalence with a tuning of the data $f$ was proven. A concrete explanation for this behaviour was not given. However, the understanding of the influence of $c$ on the reconstruction $u$ is crucial for the understanding and improvement of current and future approaches to optimise the inpainting data.

**Our Contributions.** We show that (1) is related to the Helmholtz equation with a non-constant refraction index if $c(\mathbf{x}) > 1$ and deduce from this interpretation the benefits of non-binary-valued functions $c$ and thus the observations in [11].

In addition, we provide details on how to maximise the benefits gained from this insight. As discussed in [13], it is important to assert that solutions of (1) exist and are unique for each feasible choice of $c$, and 8 / 7 is stated as its upper bound. To improve this finding, we formulate the finite difference discretisation as an optimisation task. This allows us to specify larger feasible ranges for the values of

✉ Laurent Hoeltgen
hoeltgen@b-tu.de

1 Chair for Applied Mathematics, Brandenburg University of Technology Cottbus-Senftenberg, Platz der Deutschen Einheit 1, 03046 Cottbus, Germany

$c$ than in [5, 13]. We obtain more accurate reconstructions and a stronger contrast enhancing effect.

## Inpainting and the Helmholtz equation

Let us briefly recall the mechanisms of inpainting with homogeneous diffusion (IHD). Let $f : \Omega \to \mathbb{R}$ be a smooth function on some bounded and open domain $\Omega \subset \mathbb{R}^2$ with a sufficiently regular boundary $\partial\Omega$. Moreover, let us assume that there exists a closed non-empty set of known data $\Omega_K \subsetneq \Omega$ that we interpolate. IHD considers the following PDE with mixed boundary conditions:

$$\begin{aligned} -\Delta u &= 0 \text{ on } \Omega \setminus \Omega_K, \\ \text{with} \quad u = f \text{ on } \Omega_K, &\quad \text{and} \quad \partial_n u = 0 \text{ on } \partial\Omega \setminus \partial\Omega_K \end{aligned} \tag{2}$$

and where $\partial_n u$ denotes the derivative of $u$ in the outer normal direction. We refer to [14] for an extensive study on the existence and uniqueness of solutions.

Following [6], we introduce the confidence function $c : \Omega \to \mathbb{R}$ indicating the presence of data. We set $c(\mathbf{x})$ to 1 for $\mathbf{x} \in \Omega_K$ and 0 otherwise. Lets us rewrite (2) as

$$c(\mathbf{x})(u(\mathbf{x}) - f(\mathbf{x})) - (1 - c(\mathbf{x}))\Delta u(\mathbf{x}) = 0 \text{ on } \Omega \tag{3}$$

with Neumann boundary conditions along $\partial\Omega \setminus \partial\Omega_K$. As shown in [2, 6], the choice of $c$ has a substantial influence on the solution. Interestingly, (3) also makes sense when $c$ is not binary-valued but takes values in $\mathbb{R}$. This has been exploited in [11], where (3) is complemented by a convex energy to obtain a sparse and optimal support for $c$.

Let us now combine the idea of a non-binary-valued confidence function $c$ with the mixed boundary value problem given in (2). We consider:

$$\begin{aligned} c(\mathbf{x})(u(\mathbf{x}) - f(\mathbf{x})) - (1 - c(\mathbf{x}))\Delta u(\mathbf{x}) &= 0 \text{ on } \Omega \setminus \Omega_K, \\ \text{with} \quad u = f \text{ on } \Omega_K, \quad \text{and} \quad \partial_n u(\mathbf{x}) &= 0 \text{ on } \partial\Omega \setminus \partial\Omega_K. \end{aligned} \tag{4}$$

**Proposition 1** *If we define* $\Omega_K := \{x \in \Omega \mid c(x) \equiv 1\}$, *then* (4) *is equivalent to* (3) *for non-binary-valued c and equivalent to* (2) *for binary-valued c with range* $\{0, 1\}$.

The major difference between (4) and the previous formulations lies in the distinction between $c = 1$ and $c \neq 1$. We now proceed to the first important finding of this paper.

**Theorem 1** *If the confidence function c from* (3) *is continuous, then the inpainting equation from* (4) *corresponds to the Helmholtz equation in those regions where* $c(x) > 1$.

*Proof* Due to the intermediate value theorem and the fact that the level sets where $c \equiv 1$ form closed contours around the regions, where $c(x) > 1$, we can subdivide $\Omega \setminus \Omega_K$ into disjoint regions, where $c < 1$ and where $c > 1$. These regions are separated by $\Omega_K$ on which the solutions $u$ are enforced to coincide with $f$. Thus, the problem decouples and allows us to discuss these two cases independently. The case $0 \leqslant c(x) < 1$ has already been discussed in [13] and will not be investigated further in this paper. Inside those regions, where $c > 1$, we can divide (4) by $1 - c(\mathbf{x})$ and obtain the following formulation:

$$\begin{aligned} \Delta u(\mathbf{x}) + \eta^2(\mathbf{x})u(\mathbf{x}) &= g(\mathbf{x}) \text{on} \quad \Omega \setminus \Omega_K, \\ \text{with} \quad u = f \text{ on } \Omega_K \quad &\text{and} \quad \partial_n u(\mathbf{x}) = 0 \text{on } \partial\Omega \setminus \partial\Omega_K, \end{aligned} \tag{5}$$

where $\eta^2(\mathbf{x}) = \frac{c(\mathbf{x})}{c(\mathbf{x})-1}$ and $g(\mathbf{x}) = \eta^2(\mathbf{x})f(\mathbf{x})$. Equation (5) is the inhomogeneous Helmholtz equation with a refraction $\eta$ and mixed boundary conditions [15]. □

Theorem 1 gives us valuable insight into the properties of our inpainting equation. The Dirichlet data in $\Omega_K$ in (5) can be interpreted in physical terms as a radiation source. The solutions $u$ model the spread of this radiation inside $\Omega \setminus \Omega_K$ and the superposition of radiated waves causes the contrast enhancement. Thus, our observations provide a physically motivated explanation for the equivalence shown in [12] and the usage of mask values outside of the range $[0, 1]$. Furthermore, the largest refraction numbers $\eta$ are obtained for values of $c$ slightly above 1 ($\eta \to \infty$ for $c \searrow 1$). If $c \to \infty$, then $\eta \to 1$, and the sharpening effect vanishes. This explains why values for $c$ significantly larger than 1 are rarely observed in practice.

To get the best possible results for image reconstructions, it is essential to maximise the admissible range of $\eta$. At the same time, it should be asserted that the discrete setup is well posed. In [13], the author analysed (3) and provided bounds that guarantee that the discretised PDE in (3) has a unique solution. For standard finite difference schemes, these bounds are given by $0 \leqslant c_i \leqslant 8/7$ for all stencil points $c_i$ and $0 < c_i$ for at least one $c_i$.

## Improved schemes for the inpainting equation

We now follow the philosophy to optimise key features of our discrete operator for fixed grid parameters and maximise the feasible range for the refraction directly within the design process. For simplicity, we restrict ourselves to $3 \times 3$ stencils.

To pursue our goals, we need the 2D Taylor expansion. For a sufficiently smooth function $f : \mathbb{R}^2 \to \mathbb{R}$, the Taylor expansion of order $n$ around $\mathbf{x}_0$ is given by $f(\mathbf{x}) \approx \sum_{|\alpha| \leqslant n} \frac{D^\alpha f(\mathbf{x}_0)}{\alpha!}(\mathbf{x} - \mathbf{x}_0)^\alpha$, where we employ multi-index notation on $\alpha \in \mathbb{N}^2$ and where $D^\alpha f(\mathbf{x}_0)$ denotes the partial derivatives of $f$ evaluated at $\mathbf{x_0}$. Given a regular 2D grid with constant step sizes $h$ in each direction, we use the Taylor approximations in all 8 neighbouring pixels. Each Taylor expansion uses $\mathbf{x}_0 := (x_0, y_0)$ as centre. This yields the 9 combinations $f(x_0 + kh, y_0 + jh) \approx \sum_{|\alpha| \leqslant n} \frac{D^\alpha f(x_0, y_0)}{\alpha!} (kh, jh)^\alpha$ with $k$ and $j \in \{-1, 0, 1\}$ each. In a next step, we express the desired derivative $D^\alpha f(x_0, y_0)$ as a linear combination of all these positions in our stencil:

$$D^\alpha f(x_0, y_0) \approx \lambda_1 f(x_0 - h, y_0 - h) + \lambda_2 f(x_0 - h, y_0) + \ldots + \lambda_9 f(x_0 + h, y_0 + h). \tag{6}$$

Inserting corresponding Taylor expansions and performing a comparison of coefficients leads to a linear system of equations. Its solutions represent the coefficient vectors $\lambda$ for the stencil. Omitting, for clarity, the common argument $(x_0, y_0)$, we have

$$\begin{aligned} D^\alpha f = \sum_{j=1}^{9} \lambda_j f &+ \left( \sum_{i=1}^{3} \lambda_{6+i} - \lambda_i \right) h f_x + \\ &\left( \sum_{i=0}^{2} \lambda_{3(i+1)} - \lambda_{1+3i} \right) h f_y + \left( \sum_{i=1}^{3} \lambda_i + \lambda_{6+i} \right) \frac{h^2}{2} f_{xx} + \\ &\left( \sum_{i=0}^{1} \lambda_{7+2i} - \lambda_{1+2i} \right) h^2 f_{xy} + \left( \sum_{i=0}^{2} \lambda_{1+3i} + \lambda_{3(i+1)} \right) \frac{h^2}{2} f_{yy} \end{aligned} \tag{7}$$

up to $\mathcal{O}(h^3)$. Unfortunately, this approach does not lead, in general, to a square system matrix. The number of unknowns coincides with the number of stencil positions. On the other hand, if we perform a Taylor expansion up to order $n$, we obtain $n(n+1)/2$ equations. These numbers rarely match. Nevertheless, unless the equations contradict, it is still possible to determine a particular solution $\lambda = (\lambda_i)_{i=1}^{9}$ as well as a basis $\{v_1, v_2, \ldots\}$ for the nullspace of the matrix and express all solutions as $\lambda + \sum_j \beta_j v_j$.

By computing a particular solution of the linear system for an approximation of the second-order derivatives $f_{xx}$, we arrive at the parametric representations presented in Fig. 1. The stencil for $f_{yy}$ is obtained analogously and corresponds to transposing the stencil of $\partial_{xx} f$. The stencil for the Laplacian as in Fig. 1 is obtained by adding the stencils for $f_{xx}$ and $f_{yy}$.

Let us now compute the stencil entries for our inpainting task from (3) resp. (4). As discussed in [11], these equations can be discretised and written as follows:

$$\mathbf{A(c)u} := (\mathrm{diag}(\mathbf{c}) + (\mathbf{I} - \mathrm{diag}(\mathbf{c}))(-\mathbf{L}))\mathbf{u} = \mathrm{diag}(\mathbf{c})\mathbf{f} \tag{8}$$

where $\mathbf{c}$, $\mathbf{u}$, and $\mathbf{f}$ are the discretised variants of $c$, $u$, and $f$ respectively. The matrix $\mathbf{L}$ is the discrete analogue of the Laplace operator with incorporated boundary conditions, whereas $\mathbf{I}$ is the identity matrix. We assume that $\mathbf{L}$ is discretised with the stencil from Fig. 1. The system matrix of (8) is large and banded. A straightforward computation for non-boundary pixels $i$ shows that the stencil is given as in Fig. 2.

For $\beta = -1/2$, the corresponding stencils for the Laplacian as well as for the inpainting matrix perform an undesirable odd-even decoupling. As a remedy, values for $\beta$ should be chosen in the range $[-1, -1/2)$. If $\beta = -1$, we obtain the well-known standard five point stencil for the Laplacian.

Following [5, 13], we can use the stencil entries to obtain estimates on the $c_i$ for which invertibility of the system matrix is asserted below. This finding extends results in [5] with feasible range $\{0, 1\}$ and from [13] with feasible range $[0, 8/7]$.

$\partial_{xx}$ ○—●

| | | |
|---|---|---|
| $\beta + h^{-2}/2$ | $-2\beta$ | $\beta + h^{-2}/2$ |
| $-2\beta - h^{-2}$ | $4\beta$ | $-2\beta - h^{-2}$ |
| $\beta + h^{-2}/2$ | $-2\beta$ | $\beta + h^{-2}/2$ |

,

$\Delta$ ○—●

| | | |
|---|---|---|
| $\beta + h^{-2}$ | $-2\beta - h^{-2}$ | $\beta + h^{-2}$ |
| $-2\beta - h^{-2}$ | $4\beta$ | $-2\beta - h^{-2}$ |
| $\beta + h^{-2}$ | $-2\beta - h^{-2}$ | $\beta + h^{-2}$ |

**Fig. 1** Differential operators and corresponding stencils. The free parameter $\beta$ stems from the fact that the nullspace of the matrix is one dimension

**Theorem 2** *The inpainting matrix* $\mathbf{A}(\mathbf{c})$ *from* (8) *corresponding to the stencil from Fig. 2 is invertible when all* $c_i$ *lie in the range* $[0, 4/3]$ *for* $h = 1$ *and* $\beta = -1/2$. *This is the largest possible range for the Laplacian from Fig. 1. Furthermore, the lower bound* $c_i \geqslant 0$ *for the mask entries* $c_i$ *can only be asserted when* $\beta \in [-1, -1/2]$.

*Proof* We follow the ideas from [5, 13]. By applying Geršgorin's circle theorem at the rows of the matrix $\mathbf{A}(\mathbf{c})$, we obtain pointwise estimates for the position of its eigenvalues. We note that all non-zero entries in any of the

$\mathbf{A}(\mathbf{c})$ ○—●

| | | |
|---|---|---|
| $(\beta + h^{-2})(c_i - 1)$ | $(-2\beta - h^{-2})(c_i - 1)$ | $(\beta + h^{-2})(c_i - 1)$ |
| $(-2\beta - h^{-2})(c_i - 1)$ | $c_i + 4\beta(c_i - 1)$ | $(-2\beta - h^{-2})(c_i - 1)$ |
| $(\beta + h^{-2})(c_i \;\; 1)$ | $( \;\; 2\beta \;\; h^{-2})(c_i \;\; 1)$ | $(\beta + h^{-2})(c_i - 1)$ |

**Fig. 2** Stencil for the inpainting matrix $\mathbf{A}(\mathbf{c})$ from (8) for a non-boundary pixel at position $i$. These stencil entries correspond to the non-zero entries of $\mathbf{A}(\mathbf{c})$ in the $i$-th row. Along the image boundaries, the stencil has to be adapted to consider the boundary conditions

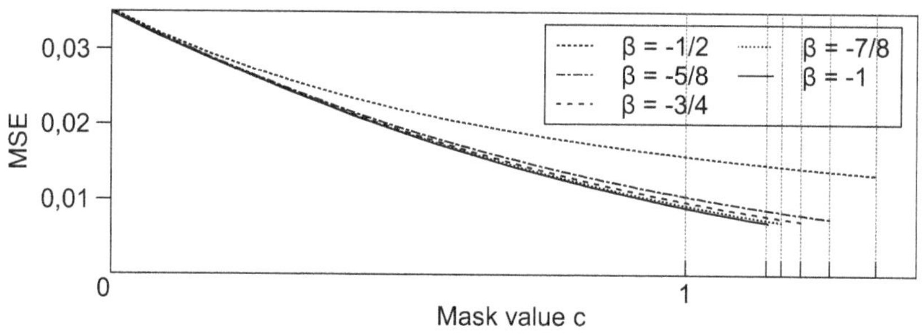


**Fig. 3** MSE for different stencil choices in function of the mask value. For each stencil, the MSE is monotonically decreasing. Most stencils yield similar errors. The huge gap in the error between the stencil for $\beta = -1/2$ and all the others is due to the odd-even decoupling of the corresponding stencil. The *vertical lines* indicate the maximal allowed value of $c$ for each case

rows of $\mathbf{A}(\mathbf{c})$ are given by the stencil entries from Fig. 2. To exclude 0 from the spectrum, we must solve

$$c_i + 4(c_i - 1) - 4\big(\big|(-2\beta - h^{-2})(c_i - 1)\big| + \big|(\beta + h^{-2})(c_i - 1)\big|\big) > 0. \tag{9}$$

For $h = 1$, a lengthy but simple computation shows that $0 < c_i < 8\beta/(1 + 8\beta)$ must hold whenever $\beta \leqslant -1/2$. Other values for $\beta$ yield ranges which do not include the interval (0, 1] and thus are of no interest to us. Maximising the upper bound is possible for $\beta = -1/2$ and yields the range (0, 4 / 3]. An identical analysis for the cases where a pixel is placed along a boundary or in a corner does not yield further restrictions. □

Let us now demonstrate the benefits of larger feasible ranges for the mask values $c$. Our objective in this is to improve image reconstructions by IHD. Thus, we consider for quantitative evaluation a small synthetic image and measure the reconstruction error in function of the mask values. Our test image consists of 25 × 25 pixels representing a sampled version of the function $x^2 + y^2$ over $[0, 1] \times [0, 1]$. Our mask contains 5 non-zero entries. In each corner, we fix the mask value to be 1. The remaining non-zero entry is placed at the centre of the image and we measure the mean-squared error (MSE) of the reconstruction as a function of the mask value at this position. We study this setup for different values $\beta$, as shown in Fig. 3.

In a second experiment, we show that our discretisations allow a contrast enhancing effect. To this end, we select a small 10 × 12 grey-scale image patch with pixel value 0.25 on the left half and 0.75 on the right half. Our masks consists of a small strip with 2 pixels width along the image edge. We set all non-zero mask values to the same value $8\beta/(1 + 8\beta)$, and let $\beta$ vary in the admissible range $[-1, -1/2]$. For each value of $\beta$, we consider the reconstruction at the two neighbouring pixels from each side of the image. Table 1 confirms the desired contrast improvements.

**Table 1** Possible contrast enhancement for different stencils. Values $\beta$ in $(-1, -1/2]$ correspond to our discretisations with enhanced contrast properties. The improved contrast is clearly visible for increasing $\beta$

| Parameter value $\beta$ | $-1$ | $-7/8$ | $-3/4$ | $-5/8$ | $-1/2$ |
|---|---|---|---|---|---|
| Maximal mask value for $c$ | 8/7 | 7/6 | 6/5 | 5/4 | 4/3 |
| Maximal difference | 2/3 | 7/10 | 3/4 | 5/6 | 1 |

## Conclusions

This paper shows the relation between the classic Helmholtz equation and the inpainting problem. Thus, we relate two up to now unconnected fields from science. In the future, we hope to develop better performing inpainting models based on this insight.

**Acknowledgements** The author thanks Michael Breuß for his support and many helpful comments.

**Compliance with ethical standards**

**Conflict of interest** The authors declare that they have no competing interests.

## References

1. Bertalmío, M., Sapiro, G., Caselles, V., Ballester, C.: Image inpainting. In: Proc. SIGGRAPH, pp. 417–424. ACM Press, New York (2000)
2. Belhachmi, Z., Bucur, D., Burgeth, B., Weickert, J.: How to choose interpolation data in images. SIAM J. Appl. Math. **70**(1), 333–352 (2009)
3. Chen, Y., Ranftl, R., Pock, T.: A bi-level view of inpainting-based image compression. Comput Res Repos (2014)
4. Galić, I., Weickert, J., Welk, M., Bruhn, A., Belyaev, A., Seidel, H.P.: Towards PDE-based image compression. In: LNCS 3752. Springer, Berlin, pp 37–48 (2005)
5. Mainberger, M., Bruhn, A., Weickert, J., Forchhammer, S.: Edge-based compression of cartoon-like images with homogeneous diffusion. Pattern Recogn. **44**(9), 1859–1873 (2011)
6. Mainberger, M., Hoffmann, S., Weickert, J., Tang, C.H., Johannsen, D., Neumann, F., Doerr, B.: Optimising spatial and

tonal data for homogeneous diffusion inpainting. In: LNCS 6667. Springer, Berlin, pp 26–37 (2012)
7. Schmaltz, C., Weickert, J., Bruhn, A.: Beating the quality of JPEG 2000 with anisotropic diffusion. In: LNCS 5748. Springer, Berlin, pp. 452–461 (2009)
8. Li, Y., Jeong, D., Choi, J., Lee, S., Kim, J.: Fast local image inpainting based on the allen-cahn model. Digital Signal Process. **37**, 65–74 (2015)
9. Bertozzi, A., Esedoglu, S., Gillette, A.: Inpainting of binary images using the cahn-hilliard equation. IEEE Trans. Image Proc. **16**(1), 285–291 (2007)
10. Bertalmio, M., Bertozzi, A., Sapiro, G.: Navier-stokes, fluid dynamics, and image and video inpainting. In: Computer Vision and Pattern Recognition, 2001. CVPR 2001. Proceedings of the 2001 IEEE Computer Society Conference on, vol. 1, pp. 355–362. IEEE (2001)
11. Hoeltgen, L., Setzer, S., Weickert, J.: An optimal control approach to find sparse data for Laplace interpolation. In: LNCS 8081. Springer, Berlin, pp. 151–164 (2013)
12. Hoeltgen, L., Weickert, J.: Why does non-binary mask optimisation work for diffusion-based image compression? In: LNCS 8932. Springer, Berlin, pp. 85–98 (2015)
13. Hoeltgen, L.: Optimal interpolation data for image reconstructions. Ph.D. thesis, Saarland University (2014)
14. Gilbarg, D., Trudinger, N.: Elliptic Partial Differential Equations of Second Order. Springer, Berlin (2001)
15. Colton, D., Kress, R.: Inverse Accoustic and Electromagnetic Scattering Theory, *Applied Mathematical Sciences*, vol. 93, 2nd edn. Springer (1998)

24

# Coupled coincidence point results in partially ordered generalized fuzzy metric spaces with applications to integral equations

**Binayak S. Choudhury**[1] · **Pradyut Das**[1]

**Abstract** Here we prove a coupled coincidence point theorem in $G$-fuzzy metric spaces for compatible mappings using *Hadžić* type $t$-norms which is characterized by the equi-continuity of its iterates. We apply our result toward obtaining a result in $G$-metric spaces. Two supporting examples are also given. Some existing results are extended by our theorem. We also apply our result to a problem of an integral equation. We further assume the $G$-metric space to be equipped with a partial ordering.

**Keywords** Partially ordered set · $G$-fuzzy metric space · $t$-Norm of *Hadžić* · Compatible mappings · Coupled coincidence point · Integral equation

## Introduction

The program of this work is to establish coupled coincidence point results in generalized fuzzy metric spaces which are actually fuzzy extensions of generalized metric spaces (abbreviated as $G$-metric spaces in the literatures). These spaces were introduced in the paper by Mustafa et al. [16, 17]. The generalization is effectuated by a non-negative real-valued mapping on $X^3$, where $X$ is a given non-empty set on which the generalized metric is defined. Fixed point results on this structure were proved in a good number of papers, as, for instance [1, 5, 8, 10, 19].

Fuzzy sets were introduced by Zadeh [23] which provided an approach to non-probabilistic uncertain situations. Several fuzzified versions of the exiting mathematical structures were introduced in the literatures, particularly the fuzzification of metric space followed through adoption of different approaches. The flexible structure of fuzzy ideas allow for adopting different approaches resulting into the definitions of fuzzy metric spaces which are not equivalent to each other. Here we work on the definition given by George et al. [9] in which the topology is a Hausdorff topology. The fuzzy fixed point theory has a commendable development in the context of this space. One of the possible causes for that is the nature of the topology which is Hausdorff. In the theory of fuzzy fixed points and related topics on the above mentioned space. Some important references, amongst others, are [6, 7, 11].

Putting together the concepts involved in the two above mentioned spaces, generalized fuzzy metric spaces were introduced by [22]. Works on fixed point theory on this space are obtainable in [13, 18, 22].

This paper aimed at establishing a new coupled fixed point theorem in $G$-fuzzy metric spaces with a partial order. For this purpose we prove a lemma which establishes a Cauchy criterion for two sequences simultaneously. Hadzic type $t$-norm is used in this paper which is characterized by the equi-continuity of iterates. It is used in the proof of a lemma. We apply the result in this space to obtain new coupled fixed point results in $G$-metric spaces. Finally, we have an application in which we establish an existing result of an integral equation. We also provide illustrations of our results.

✉ Pradyut Das
pradyutdas747@gmail.com

Binayak S. Choudhury
binayak12@yahoo.co.in

1 Department of Mathematics, Indian Institute of Engineering Science and Technology, Shibpur, Howrah, West Bengal, India

## Preliminaries

**Definition 2.1** [12] The mapping $*: [0,1] \times [0,1] \longrightarrow [0,1]$ is said to be a $t$-norm when the following hold:

1. $*$ is commutative as well as associative,
2. $1 * c_1 = c_1$ whenever $c_1 \in [0,1]$,
3. $c_1 * c_2 \leq c_3 * c_4$ whenever $c_1 \leq c_3$ and $c_2 \leq c_4$, for each $c_1, c_2, c_3, c_4 \in [0,1]$,
4. The operator $*$ is a continuous $t$-norm if $*$ is continuous.

Some illustrations of the above definition are given in [12].

**Definition 2.2** [22] The 3-tuple $(A, G, *)$ called $G$-fuzzy metric space if $A$ is any non-empty set, $*$ is a $t$-norm which is continuous and $G$ is a fuzzy membership function on $A^3 \times (0,\infty)$ which satisfies $z_1, z_2, z_3, z_4 \in A$ and $t_1, t_2 > 0$:

1. $G(z_1, z_1, z_2, t_1) > 0$,
2. $G(z_1, z_1, z_2, t_1) \geq G(z_1, z_2, z_3, t_1)$ with $z_2 \neq z_3$,
3. $G(z_1, z_2, z_3, t_1) = 1$ if and only if $z_1 = z_2 = z_3$,
4. $G(z_1, z_2, z_3, t_1) = G(p(z_1, z_2, z_3), t)$, where $p$ is a permutation function,
5. $G(z_1, z_4, z_4, t_1) * G(z_4, z_2, z_3, t_2) \leq G(z_1, z_2, z_3, t_1 + t_2)$ and
6. $G(z_1, z_2, z_3, .) : (0,\infty) \longrightarrow [0,1]$ is continuous.

**Example 2.3** [22] Let (*A*, *G*) be *G*-metric space. Let $c_1 * c_2 = c_1.c_2$ for all $c_1, c_2 \in [0,1]$. For each $t_1 > 0$, $z_1, z_2 \in A$, let

$$G(z_1, z_1, z_2, t_1) = \frac{t_1}{t_1 + G(z_1, z_1, z_2)}.$$

Then $(A, G, *)$ is a $G$-fuzzy metric space.

**Definition 2.4** [22] Let $(A, G, *)$ be a $G$-fuzzy metric space.

1. Any sequence $\{x_n\}$ in $A$ converges to a point $z \in A$ if $\lim_{n\to\infty} G(x_n, x_n, z, s) = 1$ for all $s > 0$.
2. Any sequence $\{x_n\}$ in $A$ is called a Cauchy sequence if corresponding to $0 < \varepsilon < 1$ and $s > 0$, there is a positive integer $n_0$ for which $G(x_n, x_n, x_m, s) > 1 - \varepsilon$ when $n, m \geq n_0$.
3. If every Cauchy sequence converges, then the space is complete.

**Lemma 2.5** [22] Let $(A, G, *)$ be a $G$-fuzzy metric space. Then $G(x_1, x_1, x_2, .)$ is nondecreasing for all $x_1, x_2 \in A$.

**Lemma 2.6** [22] $G$ is a continuous function on $A^3 \times (0,\infty)$.

Let $A$ be a set with a partial order $\preceq$ and $F$ be a function from $A$ to itself. Then $F$ is non-decreasing (non-increasing) whenever $x_1 \preceq x_2$ ($x_1 \succeq x_2$) and $x_1, x_2 \in A$ if $F(x_1) \preceq F(x_2)$ ($F(x_1) \succeq F(x_2)$) [3].

**Definition 2.7** [3] Let $(A, \preceq)$ be a set with a partial ordering $\preceq$ and $F : A^2 \to A$ be a function. The function $F$ has the mixed monotone property whenever for all $x_1, x_2 \in A$, $x_1 \preceq x_2$ implies $F(x_1, y) \preceq F(x_2, y)$, with fixed $y \in A$ and, for all $y_1, y_2 \in A$, $y_1 \preceq y_2$ implies $F(x, y_1) \succeq F(x, y_2)$, with fixed $x \in A$.

**Definition 2.8** [14] Let $(A, \preceq)$ be a partially ordered set and $F : A^2 \to A$ and $g : A \to A$ be two functions. The function $F$ has the mixed $g$-monotone property if for all $x_1, x_2 \in A$, $g(x_1) \preceq g(x_2)$ implies $F(x_1, y) \preceq F(x_2, y)$, with any $y \in A$ and, for all $y_1, y_2 \in A$, $g(y_1) \preceq g(y_2)$ implies $F(x, y_1) \succeq F(x, y_2)$, with fixed $x \in A$.

**Definition 2.9** [3] Let $A$ be any nonempty set. The ordered pair $(p, q) \in A \times A$ is a coupled fixed point of the function $F : A \times A \to A$ if

$F(p, q) = p$ as well as$F(q, p) = q$.

**Definition 2.10** [14] Let $A$ be any nonempty set. An element $(p, q) \in A^2$ is a coupled coincidence point of the functions $F : A \times A \to A$ and $g : A \to A$ if

$F(p, q) = g(p)$ and$F(q, p) = g(q)$.

**Definition 2.11** [14] Let $A$ be any non empty set. The functions $F : A \times A \to A$ and $g : A \to A$ are commuting if $(p, q) \in A^2$

$g(F(p, q)) = F(g(p), g(q))$.

**Definition 2.12** [20] Let (*A*, *G*) be a *G*-metric space. The pair (*g*, *F*) where $g : A \to A$ and $F : A \times A \to A$, is compatible if

$$\lim_{n\to\infty} G(g(F(p_n, q_n)), g(F(p_n, q_n)), F(g(p_n), g(q_n))) = 0$$

and

$$\lim_{n\to\infty} G(g(F(q_n, p_n)), g(F(q_n, p_n)), F(g(q_n), g(p_n))) = 0,$$

whenever $\{p_n\}$ and $\{q_n\}$ are sequences in $A$ such that $\lim_{n\to\infty} F(p_n, q_n) = \lim_{n\to\infty} g(p_n) = p$ and $\lim_{n\to\infty} F(q_n, p_n) = \lim_{n\to\infty} g(q_n) = q$ for some $p, q \in A$.

The intuitive idea is that the functions $F$ and $g$ are commuting in the limit in the situations where the functional values are the same in the limit.

**Definition 2.13** [13] Let $(A, G, *)$ be a $G$-fuzzy metric space. The pair (*F*, *g*) where $F : A \times A \to A$ and $g : A \to A$, are said to be compatible if for all $t > 0$

$$\lim_{n\to\infty} G(g(F(p_n, q_n)), g(F(p_n, q_n)), F(g(p_n), g(q_n), t) = 1$$

and

$$\lim_{n\to\infty} G(g(F(q_n,p_n)), g(F(q_n,p_n)), F(g(q_n), g(p_n), t) = 1,$$

whenever $\{p_n\}$ and $\{q_n\}$ are sequences in $A$ such that $\lim_{n\to\infty} F(p_n, q_n) = \lim_{n\to\infty} g(p_n) = p$ and $\lim_{n\to\infty} F(q_n, p_n) = \lim_{n\to\infty} g(q_n) = q$ for some $p, q \in A$.

**Lemma 2.14** Let $(A, G)$ be a $G$-metric space. If the pair $(F, g)$ where $F: A \times A \to A$ and $g: A \to A$ are compatible as per Definition 2.12, then he pair $(F, g)$ is also compatible as per Definition 2.13.

*Proof* As we have mentioned earlier, in the associated $G$-fuzzy metric space, for all $x, y \in A$, $t > 0$,

$$G(x,x,y,t) = \frac{t}{t + G(x,x,y)} \tag{2.1}$$

and $a * b$ = minimum $\{a, b\}$.

Let $\{p_n\}$ and $\{q_n\}$ be two sequences in $(X, G)$ such that $\lim_{n\to\infty} F(p_n, q_n) = \lim_{n\to\infty} g(p_n) = p$ and $\lim_{n\to\infty} F(q_n, p_n) = \lim_{n\to\infty} g(q_n) = q$. Then the same limits also hold in $(A, G, *)$.

Since $g$ and $F$ are compatible in $(A, G)$, we have

$$\lim_{n\to\infty} G(g(F(p_n,q_n)), g(F(p_n,q_n)), F(g(p_n), g(q_n))) = 0$$

and

$$\lim_{n\to\infty} G(g(F(q_n,p_n)), g(F(q_n,p_n)), F(g(q_n), g(p_n))) = 0,$$

Now from (2.1), we have for all $t > 0$

$$G(g(F(p_n,q_n)), g(F(p_n,q_n)), F(g(p_n), g(q_n)), t) = \frac{t}{t + G(g(F(p_n,q_n)), g(F(p_n,q_n)), F(g(p_n), g(q_n)))}$$

and

$$G(g(F(q_n,p_n)), g(F(q_n,p_n)), F(g(q_n), g(p_n)), t) = \frac{t}{t + G(g(F(q_n,p_n)), g(F(q_n,p_n)), F(g(q_n), g(p_n)))}.$$

Taking $n \to \infty$ in both the above equalities, for all $t > 0$, we have

$$\lim_{n\to\infty} G(g(F(p_n,q_n)), g(F(p_n,q_n)), F(g(p_n), g(q_n)), t) = 1$$

and

$$\lim_{n\to\infty} G(g(F(q_n,p_n)), g(F(q_n,p_n)), F(g(q_n), g(p_n)), t) = 1.$$

Therefore $(F, g)$ is compatible in $(A, G, *)$. □

Continuous *Hadžić* type $t$-norm is used in our theorem.

**Definition 2.15** [12]*Hadžić* type $t$-norms are $t$-norms such that $\{*^p\}_{p\geq 0}$ give by

$*^0(s) = 1$, $*^{p+1}(s) = *(*^p(s), s)$ for all $p \geq 0$, $0 < s < 1$, are equi-continuous at $s = 1$, which is that, for $\epsilon > 0$ there exists $a(\epsilon) \in (0, 1)$ for which

$1 \geq u > a(\epsilon) \Rightarrow *^p(u) > 1 - \epsilon$ for all $p \geq 0$.

Illustrations of the above $t$-norm type is given in [12].

**Lemma 2.16** Let $(A, G, *)$ be a $G$-fuzzy metric space having a $t$-norm of *Hadžić* type for which $G(x, x, y, u) \to 1$ as $u \to \infty$, $\{x_n\}$ and $\{y_n\}$ in $A$ satisfy, for all $n \geq 1$, $t > 0$,

$$G(x_n, x_n, x_{n+1}, t) * G(y_n, y_n, y_{n+1}, t) \geq G\left(x_{n-1}, x_{n-1}, x_n, \frac{t}{k}\right) * G\left(y_{n-1}, y_{n-1}, y_n, \frac{t}{k}\right) \tag{2.2}$$

with some $0 < k < 1$, then $\{x_n\}$ and $\{y_n\}$ are Cauchy sequences.

*Proof* We successively apply (2.2) to obtain for all $i \geq 1$ (integer) $t > 0$, $q \geq 0$,

$$G(x_{q+i}, x_{q+i}, x_{q+i+1}, t) * G(y_{q+i}, y_{q+i}, y_{q+i+1}, t) \geq G\left(x_q, x_q, x_{q+1}, \frac{t}{k^i}\right) * G\left(y_q, y_q, y_{q+1}, \frac{t}{k^i}\right) \tag{2.3}$$

Let $\epsilon > 0$ and $0 < \lambda < 1$ be given. Let $p$ be another integer such that $p > q$ be some other integer. Then

$$\epsilon = \epsilon\frac{(1-k)}{(1-k)} > \epsilon(1-k)(1 + k + \cdots + k^{p-q-1}).$$

Then, by Lemma 2.5, for all $p > q$, we obtain

$$G(x_q, x_q, x_p, \epsilon) * G(y_q, y_q, y_p, \epsilon) \geq G(x_q, x_q, x_p, \epsilon(1-k)(1 + k + \cdots + k^{p-q-1})) * G(y_q, y_q, y_p, \epsilon(1-k)(1 + k + \cdots + k^{p-q-1})),$$

or,

$$\begin{aligned} G(x_q, x_q, x_p, \epsilon) * G(y_q, y_q, y_p, \epsilon) &\geq G(x_q, x_q, x_{q+1}, \epsilon(1-k)) * G(x_{q+1}, x_{q+1}, x_{q+2}, \epsilon k(1-k)) * \cdots * \\ &\quad G(x_{p-1}, x_{p-1}, x_p, \epsilon k^{p-q-1}(1-k)) * G(y_q, y_q, y_{q+1}, \epsilon(1-k)) \\ &\quad * G(y_{q+1}, y_{q+1}, y_{q+2}, \epsilon k(1-k)) * \cdots * \\ &\quad G(y_{p-1}, y_{p-1}, y_p, \epsilon k^{p-q-1}(1-k)). \\ &= \{G(x_q, x_q, x_{q+1}, \epsilon(1-k)) * G(y_q, y_q, y_{q+1}, \epsilon(1-k))\} \\ &\quad * \{G(x_{q+1}, x_{q+1}, x_{q+2}, \epsilon k(1-k)) * \\ &\quad G(y_{q+1}, y_{q+1}, y_{q+2}, \epsilon k(1-k))\} * \cdots * \\ &\quad \{G(x_{p-1}, x_{p-1}, x_p, \epsilon k^{p-q-1}(1-k)) * \\ &\quad G(y_{p-1}, y_{p-1}, y_p, \epsilon k^{p-q-1}(1-k))\}. \end{aligned} \tag{2.4}$$

We put $t = (1-k)\epsilon k^i$ in (2.3); we get, for all $q \geq 0, i \geq 1$

$$G(x_{q+i}, x_{q+i}, x_{q+i+1}, (1-k)\epsilon k^i) * G(y_{q+i}, y_{q+i}, y_{q+i+1}, (1-k)\epsilon k^i) \geq G(x_q, x_q, x_{q+1}, (1-k)\epsilon) * G(y_q, y_q, y_{q+1}, (1-k)\epsilon).$$

From the above, and using (2.4), with $p > q$, we get

$$G(x_q,x_q,x_p,\epsilon) * G(y_q,y_q,y_p,\epsilon) \geq \{G(x_q,x_q,x_{q+1},\epsilon(1-k)) * G(y_q,y_q,y_{q+1},\epsilon(1-k))\} * \{G(x_{q+1},x_{q+1},x_{q+2},\epsilon(1-k)) * G(y_{q+1},y_{q+1},y_{q+2},\epsilon(1-k))\} * \cdots * \{G(x_{p-1},x_{p-1},x_p,\epsilon(1-k)) * G(y_{p-1},y_{p-1},y_p,\epsilon(1-k))\},$$

that is, $G(x_q,x_q,x_p,\epsilon) * G(y_q,y_q,y_p,\epsilon)$

$$\geq *^{(p-q)} \{G(x_q,x_q,x_{q+1},\epsilon(1-k)) * G(y_q,y_q,y_{q+1},\epsilon(1-k))\}. \tag{2.5}$$

By equi-continuity of $t$-norm at 1, there exists $\eta(\lambda) \in (0,1)$ such that for all $m > n$,

$$*^{(m-n)}(s) > 1 - \lambda, \tag{2.6}$$

whenever $1 \geq s > \eta(\lambda)$, where $0<\lambda<1$, as mentioned above, is given. Since $G(x_0, x_0, x_1, u) \to 1$ as $u \to \infty$, there exists $N(\epsilon, \lambda)$ such that

$$G(x_0, x_0, x_1, \frac{(1-k)\epsilon}{k^n}) * G(y_0, y_0, y_1, \frac{(1-k)\epsilon}{k^n}) > \eta(\lambda) \text{whenever} n \geq N(\epsilon, \lambda). \tag{2.7}$$

From (2.3) and (2.7), with $q = 0, i = n \geq N(\epsilon, \lambda)$ and $t = (1-k)\epsilon$, we have

$$G(x_n, x_n, x_{n+1}, (1-k)\epsilon) * G(y_n, y_n, y_{n+1}, (1-k)\epsilon) > \eta(\lambda).$$

Then, from (2.6), with $s = G(x_n, x_n, x_{n+1}, (1-k)\epsilon) * G(y_n, y_n, y_{n+1}, (1-k)\epsilon)$ and $m > n \geq N(\epsilon, \lambda)$, we have

$$*^{(m-n)} (G(x_n, x_n, x_{n+1}, (1-k)\epsilon) * G(y_n, y_n, y_{n+1}, (1-k)\epsilon)) > 1 - \lambda.$$

Then, by (2.5), for all $m > n \geq N(\epsilon, \lambda)$, we obtain

$$G(x_n, x_n, x_m, \epsilon) * G(y_n, y_n, y_m, \epsilon) > 1 - \lambda,$$

which implies that

$$G(x_n, x_n, x_m, \epsilon) > 1 - \lambda \text{and} G(y_n, y_n, y_m, \epsilon) > 1 - \lambda \text{for all} n, m \geq N(\epsilon, \lambda).$$

Again $\epsilon > 0$ and $\lambda$ are arbitrary with their range.

This proves that $\{x_n\}$ and $\{y_n\}$ are Cauchy sequences. □

**Note 2.17** Equi-continuous of the iterates is essential in the proof of the above lemma.

## Main results

**Theorem 3.1** Let $(A, G, *)$ be a complete $G$-fuzzy metric space having a $t$-norm whose *Hadžić* type is such that $G(x, y, z, s) \to 1$ as $s \to \infty$, for all $x, y, z \in A$. Let $\preceq$ be a partial ordering on $A$. Let $F : A \times A \to A$ and $g : A \to A$ be two functions of which $F$ has mixed $g$-monotone property and that the following is satisfied:

$$G(F(x,y),F(x,y),F(u,v),ks) * G(F(y,x),F(y,x),F(v,u),ks) \geq G(g(x),g(x),g(u),s) * G(g(y),g(y),g(v),s),$$

for all $x, y, u, v \in A$, $s > 0$ with $g(x) \preceq g(u)$ and $g(y) \succeq g(v)$, where $0<k<1$ and $F(A \times A) \subseteq g(A)$, $g$ is continuous and monotonic increasing, $(g, F)$ is a compatible pair. Suppose one of (a) and (b) holds:

(a) $F$ is continuous

(b)

$$1.\{z_p\} \to z \text{ is such that } z_p \preceq z_{p+1}, \text{for every } p \geq 0, \text{ then } z_p \preceq z \text{for every } p \geq 0, \tag{3.2}$$

$$2.\{z_p\} \to z \text{ is such that } z_p \succeq z_{p+1} \text{for every } p \geq 0, \text{then } z_p \succeq z \text{for every } p \geq 0 \tag{3.3}$$

If there are $x_0, y_0 \in A$ for which $g(x_0) \preceq F(x_0, y_0)$, $g(y_0) \succeq F(y_0, x_0)$, then we can find $x, y \in A$ for which $g(x) = F(x, y)$ and $g(y) = F(y, x)$.

*Proof* Let $x_0, y_0$ in A, for which $g(x_0) \preceq F(x_0, y_0)$, $g(y_0) \succeq F(y_0, x_0)$. We construct the sequences $\{x_p\}$ and $\{y_p\}$ in A, for all $p \geq 0$,

$$g(x_{p+1}) = F(x_p, y_p) \text{and} g(y_{p+1}) = F(y_p, x_p). \tag{3.4}$$

Then it follows that for all $p \geq 0$

$$g(x_p) \preceq g(x_{p+1}) \tag{3.5}$$

and

$$g(y_p) \succeq g(y_{p+1}). \tag{3.6}$$

Let for all $s > 0$, $p \geq 0$, due to (3.4), (3.5) and (3.6), from (3.1), for all $s > 0$, $k \geq 1$, we have

$$G(g(x_p), g(x_p), g(x_{p+1}), ks) * G(g(y_p), g(y_p), g(y_{p+1}), ks) = G(F(x_{p-1}, y_{p-1}), F(x_{p-1}, y_{p-1}), F(x_p, y_p), ks) * G(F(y_{p-1}, x_{p-1}), F(y_{p-1}, x_{p-1}), F(y_p, x_p), ks) \geq G(g(x_{p-1}), g(x_{p-1}), g(x_p), s) * G(g(y_{p-1}), g(y_{p-1}), g(y_p), s)$$

that is,

$$G(g(x_p), g(x_p), g(x_{p+1}), s) * G(g(y_p), g(y_p), g(y_{p+1}), s)$$
$$\geq G\Big(g(x_{p-1}), g(x_{p-1}), g(x_p), \frac{s}{k}\Big)$$
$$* G\Big(g(y_{p-1}), g(y_{p-1}), g(y_p), \frac{s}{k}\Big). \quad (3.7)$$

From (3.7), by the Lemma 2.16, we conclude that $\{g(x_p)\}$ and $\{g(y_p)\}$ are Cauchy sequences. Since $A$ is complete, there exist $x, y \in A$ such that

$$\lim_{p\to\infty} g(x_p) = x \text{ and } \lim_{n\to\infty} g(y_p) = y \quad (3.8)$$

Therefore,
$\lim_{p\to\infty} g(x_{p+1}) = \lim_{p\to\infty} F(x_p, y_p) = x$ and $\lim_{p\to\infty} g(y_{p+1}) = \lim_{p\to\infty} F(y_p, x_p) = y$.

Since $(g, F)$ is a compatible pair, using continuity of $g$ and Definition 2.13, we have

$$g(x) = \lim_{p\to\infty} g(g(x_{p+1})) = \lim_{p\to\infty} g(F(x_p, y_p))$$
$$= \lim_{p\to\infty} F(g(x_p), g(y_p)) \quad (3.9)$$

and

$$g(y) = \lim_{p\to\infty} g(g(y_{p+1})) = \lim_{p\to\infty} g(F(y_p, x_p))$$
$$= \lim_{p\to\infty} F(g(y_p), g(x_p)). \quad (3.10)$$

Now assume that(a) holds.

Then from (3.9), (3.10) and by using (3.8), we have

$$g(x) = \lim_{p\to\infty} g(F(x_p, y_p)) = \lim_{p\to\infty} F(g(x_p), g(y_p))$$
$$= F(\lim_{p\to\infty} g(x_p), \lim_{p\to\infty} g(y_p)) = F(x, y)$$

and

$$g(y) = \lim_{p\to\infty} g(F(y_p, x_p)) = \lim_{p\to\infty} F(g(y_p), g(x_p))$$
$$= F(\lim_{p\to\infty} g(y_p), \lim_{p\to\infty} g(x_p)) = F(y, x).$$

therefore $g(x) = F(x, y)$ and $g(y) = F(y, x)$.

Next we assume that (b) holds.

By (3.5), (3.6) and (3.8), it follows that, for all $n \geq 0$,

$$g(x_p) \preceq x \text{ and } g(y_p) \succeq y.$$

Since $g$ is monotonic increasing,

$$g(g(x_p)) \preceq g(x) \text{ and } g(g(y_p)) \succeq g(y). \quad (3.11)$$

Now, for all $s > 0$, $p \geq 0$, we have

$$G(F(x, y), F(x, y), g(F(x_p, y_p)), s) \geq G(F(x, y), F(x, y), g(g(x_{p+1})), ks) G(g(g(x_{p+1})), g(g(x_{p+1})), g(F(x_p, y_p)), (s - ks)).$$

Taking $p \to \infty$ on the both sides of the above inequality, for all $s > 0$,

$$\lim_{p\to\infty} G(F(x, y), F(x, y), g(F(x_p, y_p)), s)$$
$$\geq \lim_{p\to\infty} [G(F(x, y), F(x, y), g(g(x_{p+1})), ks) G(g(g(x_{p+1})), g(g(x_{p+1})), g(F(x_p, y_p)), (s - ks))],$$

that is, $G(F(x, y), F(x, y), g(x), s)$
$= \lim_{p\to\infty} [G(F(x, y), F(x, y), g(F(x_p, y_p)), ks)$
$* G(g(g(x_{p+1})), g(g(x_{p+1})), g(x), (s - ks))]$
$= G(F(x, y), F(x, y), \lim_{p\to\infty} g(F(x_p, y_p)), ks)$
$* G(\lim_{p\to\infty} g(g(x_{p+1})), \lim_{p\to\infty} g(g(x_{p+1})), g(x), (s - ks))$ (by lemma 2.6)
$= G(F(x, y), F(x, y), \lim_{p\to\infty} F(g(x_p), g(y_p)), ks)$
$* G(g(x), g(x), g(x), (s - ks)$ (by 3.9)
$= \lim_{p\to\infty} G(F(g(x_p), g(y_p)), F(x, y), F(x, y), ks)$
$* 1$ (by lemma 2.6)
$= \lim_{p\to\infty} G(F(g(x_p), g(y_p)), F(x, y), F(x, y), ks),$

that is, $G(F(x, y), F(x, y), g(x), s) \geq \lim_{p\to\infty} G(F(g(x_p), g(y_p)), F(x, y), F(x, y), ks)$. Similarly we obtain for all $s > 0$

$$G(F(y, x), F(y, x), g(y), s) \geq \lim_{p\to\infty} G(F(g(y_p), g(x_p)), F(y, x), F(y, x), ks). \quad (3.13)$$

From (3.12) and (3.13), using (3.1) and (3.11), for all $s > 0$, we have

$$G(F(x, y), F(x, y), g(x), s) * G(F(y, x), F(y, x), g(y), s)$$
$$\geq \lim_{p\to\infty} [G(F(g(x_p), g(y_p)), F(x, y), F(x, y), ks)$$
$$* G(F(g(y_p), g(x_p)), F(y, x), F(y, x), ks)]$$
$$\geq \lim_{p\to\infty} [G(g(g(x_p)), g(x), g(x), s) * G(g(g(y_p)), g(y), g(y), s)] (since * is continuous) = G(\lim_{p\to\infty} g(g(x_p)), g(x), g(x), s) * G(\lim_{p\to\infty} g(g(y_p)), g(y), g(y), s)$$
$$= G(g(x), g(x), g(x), s) * G(g(y), g(y), g(y), s) (\text{by}(3.9))$$
$$= 1 * 1 = 1,$$

that is,
$G(F(x, y), g(x), g(x), s) * G(F(y, x), g(y), g(y), s) \geq 1$, which implies that $g(x) = F(x, y)$ and $g(y) = F(y, x)$. Hence the proof.

**Corollary 3.2** Let $(A, G, *)$ be a complete $G$-fuzzy metric space having a $t$-norm which is *Hadžić* type for which $G(x, y, z, s) \to 1$ as $s \to \infty$, for all $x, y, z \in A$. Let $\preceq$ be a partial ordering on $A$. Let $F : A \times A \to A$ and $g : A \to A$ be two functions out of which $F$ has mixed $g$-monotone property and satisfies the following condition:

$$G(F(x,y),F(x,y),F(u,v),ks) \\ *G(F(y,x),F(y,x),F(v,u),ks) \geq G(g(x),g(x),g(u),s) \\ *G(g(y),g(y),g(v)s),$$

for all $x,y,u,v \in A$, $s>0$ with $g(x) \preceq g(u)$ and $g(y) \succeq g(v)$, where $0<k<1$ and $F(A \times A) \subseteq g(A)$, $g$ is continuous and monotonic increasing, ($g$, $F$) is a commuting pair. Suppose either

(a) $F$ is continuous or
(b) (3.2) and (3.3) hold.

If there are $x_0, y_0 \in A$ for which $g(x_0) \preceq F(x_0,y_0)$, $g(y_0) \succeq F(y_0,x_0)$, then we obtain $x,y \in A$ satisfying $g(x) = F(x,y)$ and $g(y) = F(y,x)$.

*Proof* As commuting pairs are also a compatible pairs, the result follows from Theorem 3.1.

Later, through an example, it is established that the Corollary 3.2 is actually contained within Theorem 3.1. □

**Corollary 3.3** Let $(A,G,*)$ be a complete $G$-fuzzy metric space having a $t$-norm which is *Hadžić* type for which $G(x,y,z,s) \to 1$ as $s \to \infty$, for all $x,y,z \in A$. Let $\preceq$ be a partial ordering on $A$. Let $F: A \times A \to A$ be a function for which $F$ has mixed mixed monotone property and satisfies the following condition:

$$G(F(x,y),F(x,y),F(u,v),ks) \\ *G(F(y,x),F(y,x),F(v,u),ks) \geq G(x,x,u,s) \\ *G(y,y,v,s),$$

for all $x,y,u,v \in A$, $s>0$ such that $x \preceq u$ and $y \succeq v$, and $0<k<1$ and $F(A \times A) \subseteq A$. Suppose either

(a) $F$ is continuous or
(b) (3.2) and (3.3) hold.

If there are $x_0, y_0 \in A$ for which $x_0 \preceq F(x_0,y_0)$, $y_0 \succeq F(y_0,x_0)$, then we obtain $x,y \in A$ such that $x = F(x,y)$ and $y = F(y,x)$

*Proof* With the assumption of $g = I$, the corollary follows by an application of Theorem 3.1. □

**Example 3.4** Let $(A,\preceq)$ be a partially ordered set with $A = [0,1]$ and the usual relation ordering $\leq$ on real numbers be the partial ordering $\preceq$. Let for all $s>0$, $p,q,z \in A$,

$$G(p,q,z,s) = e^{-\frac{|p-q|+|q-z|+|z-p|}{s}}.$$

Let $a*b = \min\{a,b\}$. Then $(A,G,*)$ is a complete $G$-fuzzy metric space such that $G(p,q,z,s) \to 1$ as $s \to \infty$, for all $p,q \in A$.

Let the mapping $g: A \to A$ be defined as

$$g(p) = \frac{5}{6}p^2 \text{ for all } p \in A$$

and the mapping $F: A \times A \to A$ be defined as

$$F(p,q) = \frac{p^2-q^2}{4}.$$

Then $F(A \times A) \subseteq g(A)$ and $F$ satisfies the mixed $g$-monotone property.

Let $\{t_n\}$ and $\{r_n\}$ be sequences in $A$ such that

$$\lim_{n\to\infty} F(t_n,r_n) = a, \quad \lim_{n\to\infty} g(t_n) = a, \lim_{n\to\infty} F(r_n,t_n) \\ = b \text{ and } \lim_{n\to\infty} g(r_n) = b.$$

Now, for all $n \geq 0$,

$$g(t_n) = \frac{5}{6}t_n^2, \; g(r_n) = \frac{5}{6}r_n^2, F(t_n,r_n) = \frac{t_n^2 - r_n^2}{4}$$

and

$$F(r_n,t_n) = \frac{r_n^2 - t_n^2}{4}.$$

Then necessarily $a = 0$ and $b = 0$.

It then follows from Lemma 2.6 that, for all $s>0$,

$$\lim_{n\to\infty} G(g(F(t_n,r_n)), g(F(t_n,r_n)), F(g(t_n),g(r_n),s) = 1$$

and

$$\lim_{n\to\infty} G(g(F(r_n,t_n)), g(F(r_n,t_n)), F(g(r_n),g(t_n)),s) = 1.$$

Therefore, ($g$, $F$) is compatible pair in $A$. Now we show that the inequality (3.1) holds.

$$2|F(p,q)-F(u,v)| \leq |g(p)-g(u)| + |g(q) \\ -g(v)|, p \geq u, q \leq v \tag{3.14}$$

and

$$2|F(q,p)-F(v,u)| \leq |g(q)-g(v)| + |g(p) \\ -g(u)|, p \geq u, q \leq v. \tag{3.15}$$

From (3.14), for all $s>0$ and $0<k<1$, we have

$$e^{-\frac{2|F(p,q)-F(u,v)|}{ks}} \geq e^{-\frac{[|g(p)-g(u)|+|g(q)-g(v)|]}{s}} \geq e^{-\frac{2|g(p)-g(u)|}{2s}}.e^{-\frac{2|g(q)-g(v)|}{2s}} \\ \geq \sqrt{e^{-\frac{2|g(p)-g(u)|}{s}}.e^{-\frac{2|g(q)-g(v)|}{s}}} \geq \min\left\{e^{-\frac{2|g(p)-g(u)|}{s}} e^{-\frac{2|g(q)-g(v)|}{s}}\right\} \\ e^{-\frac{2|F(p,q)-F(u,v)|}{ks}} \geq \min\{G(g(p),g(p),g(u),s), M(g(q), \\ g(q),g(v),s)\} \tag{3.16}$$

Similarly from (3.15), we get

$$e^{-\frac{2|F(q,p)-F(v,u)|}{ks}} \geq \min\{G(g(p),g(p),g(u),s), M(g(q), \\ g(q),g(v),s)\}. \tag{3.17}$$

From (3.16) and (3.17), we obtain

$\min\{G(F(p,q),F(p,q),F(u,v),ks),G(F(q,p),F(q,p),F(v,u),ks)\} \geq \min\{G(g(p),g(p),g(u),s),G(g(q),g(q),g(v),s)\}$, that is, $G(F(p,q),F(p,q),F(u,v),ks) * G(F(q,p),F(q,p),F(v,u),ks) \geq G(g(p),g(p),g(u),s) * G(g(q),g(q),g(v),s)$.

Hence (3.1) holds. Other cases can be similarly done. Then Theorem 3.1 is applicable. As can be seen by observation (0, 0) is the coupled coincidence point for the pair (*g*, *F*).

*Remark 3.5* (*g*, *F*) is not a commuting pair in Example 3.4 for which Corollary 3.2 is not applicable to this example. This establishes that Theorem 3.1 is actually more general than its Corollary 3.2.

## Results in *G*-metric spaces

Here we apply Theorem 3.1 to obtain a coupled coincidence point results in *G*-metric spaces. Several existing theorems are hereby extended.

**Theorem 4.1** Let *A* be a non empty set with a partial order $\preceq$ and *G* be a *G*- metric on *A* such that (*A*, *G*) is a complete *G*- metric space. Let $F: A \times A \to A$ and $g: A \to A$ be two mappings such that *F* has the mixed *g*-monotone property and satisfies the following condition:

$$\max\{G(F(p,q),F(p,q),F(u,v)),G(F(q,p),F(q,p),F(v,u))\} \leq \frac{k}{2}[G(g(p),g(p),g(u))+G(g(q),g(q),g(v))], \tag{4.1}$$

with $p,q,u,v \in A$ such that $g(p) \preceq g(u)$ and $g(q) \succeq g(v)$, and for some fixed $k \in (0,1)$. Suppose $F(A \times A) \subseteq g(A)$, (*g*, *F*) is compatible pair of functions in which we assume the continuity of *g*. Further, either

(a) *F* is continuous or
(b) (3.2) and (3.3) hold.

If there are $x_0, y_0 \in A$ such that $g(x_0) \preceq F(x_0,y_0)$, $g(y_0) \succeq F(y_0,x_0)$, then the pair (*g*, *F*) has a coupled coincidence point in A.

*Proof* For all $p,q,z \in A$ and $s > 0$, we define

$$G(p,q,z,s) = \frac{s}{s+G(p,q,z)}$$

and $a * b = \min\{a,b\}$. Then, $(A,G,*)$ is a complete *G*-fuzzy metric space. Further, from the form of *G*, $G(p,q,z,s) \to 1$ as $s \to \infty$, whenever $p,q,z \in A$. Using Lemma 2.14, we conclude that (*g*, *F*) is a compatible pair in this *G*-fuzzy metric space. Next we show that the inequality (4.1) implies (3.1). If otherwise, from (3.1), for some $s > 0$, $p,q,u,v \in A$ with $g(p) \preceq g(u)$ and $g(q) \succeq g(v)$, we have

$$\min\left\{\frac{s}{s+\frac{1}{k}G(F(p,q),F(p,q),F(u,v))},\frac{s}{s+\frac{1}{k}G(F(q,p),F(q,p),F(v,u))}\right\} < \min\left\{\frac{s}{s+G(g(p),g(p),g(u))},\frac{s}{s+G(g(q),g(q),g(v))}\right\},$$

Form the above inequality, we have either

$$\frac{s}{s+\frac{1}{k}G(F(p,q),F(p,q),F(u,v))} < \min\left\{\frac{s}{s+G(g(p),g(p),g(u))},\frac{s}{s+G(g(q),g(q),g(v))}\right\} \tag{4.2}$$

or

$$\frac{s}{s+\frac{1}{k}G(F(q,p),F(q,p),F(v,u))}\} < \min\left\{\frac{s}{s+G(g(p),g(p),g(u))},\frac{s}{s+G(g(q),g(q),g(v))}\right\}. \tag{4.3}$$

From (4.2), we have

$$s+\frac{1}{k}G(F(p,q),F(p,q),F(u,v)) > s + G(g(p),g(p),g(u)) \text{ and} s + \frac{1}{k}G(F(p,q),F(p,q),F(u,v)) > s + G(g(q),g(q),g(v)).$$

Combining the above two inequalities, we have that

$$G(F(p,q),F(p,q),F(u,v)) > \frac{k}{2}[G(g(p),g(p),g(u)) + G(g(q),g(q),g(v))]. \tag{4.4}$$

Similarly from (4.3), we have

$$G(F(q,p),F(q,p),F(v,u)) > \frac{k}{2}[G(g(q),g(q),g(v)) + G(g(p),g(p),g(u))]. \tag{4.5}$$

By (4.4) and (4.5), we have

$$\max\{G(F(p,q),F(p,q),F(u,v)),G(F(q,p),F(q,p),F(v,u))\} > \frac{k}{2}[G(g(p),g(p),g(u))+G(g(q),g(q),g(v))],$$

which contradicts (4.1). The proof is then completed by an application of Theorem 3.1. □

**Corollary 4.2** Let $(A,\preceq)$ be any non empty set with partial order $\preceq$ and *G* be a *G*-metric on *A* for which (*A*, *G*) is a complete *G*-metric space. Let $F: A \times A \to A$

and $g : A \to A$ be two mappings such that $F$ satisfies the property given in definition 2.8 and the following inequality:

$$[G(F(p,q),F(p,q),F(u,v)) + G(F(q,p),F(q,p),F(v,u))] \leq k[G(g(p),g(p),g(u)) + G(g(q),g(q),g(v))], \tag{4.6}$$

with $p,q,u,v \in A$ such that $g(p) \preceq g(u)$ and $g(q) \succeq g(v)$ and some fixed $k \in (0,1)$. Suppose $F(A \times A) \subseteq g(A)$, $(g, F)$ is compatible pair of functions in which we assume the continuity of $g$. Also suppose either

(a) $F$ is continuous or
(b) (3.2) and (3.3) hold.

If there are $x_0, y_0 \in A$ such that $g(x_0) \preceq F(x_0, y_0)$, $g(y_0) \succeq F(y_0, x_0)$, then the pair $(g, F)$ has a coupled coincidence point in A.

*Proof* Since $\frac{x+y}{2} \leq \max\{x,y\}$, the proof follows from Theorem 4.1. □

**Example 4.3** Let $A = [0,1]$ and with the natural ordering $\leq$ in $\mathbb{R}$ as the partial ordering $\preceq$. Let $p,q,z \in A, G(p,q,z) = |p-q| + |q-z| + |z-p|$. Then $(A, G)$ is a complete $G$-metric space. Let the mapping $g : A \to A$ be given by

$$g(p) = \frac{5}{6}p^2, p \in A$$

and the mapping $F : A^2 \to A$ be defined as

$$F(p,q) = \frac{p^2 - 2q^2}{4}.$$

Then $F(A \times A) \subseteq g(A)$ and $F$ satisfies the property in Definition 2.8. Let $\{t_n\}$ and $\{r_n\}$ be two sequences in $A$ for which

$$\lim_{n\to\infty} F(t_n, r_n) = a, \quad \lim_{n\to\infty} g(t_n) = a, \lim_{n\to\infty} F(r_n, t_n) = b \text{ and } \lim_{n\to\infty} g(r_n) = b.$$

Now, for all $n \geq 0$,

$$g(t_n) = \frac{5}{6}t_n^2, \; g(r_n) = \frac{5}{6}r_n^2, F(t_n, r_n) = \frac{t_n^2 - r_n^2}{4}$$

and

$$F(r_n, t_n) = \frac{r_n^2 - t_n^2}{4}.$$

Then necessarily $a = 0$ and $b = 0$. Applying Lemma 2.6, we have

$$\lim_{n\to\infty} G(g(F(t_n, r_n)), g(F(t_n, r_n)), F(g(t_n), g(r_n)) = 0$$

and

$$\lim_{n\to\infty} G(g(F(r_n, t_n)), g(F(r_n, t_n)), F(g(r_n), g(t_n))) = 0.$$

Therefore, the mappings $(g, F)$ are compatible pair in $A$. The mappings are not commuting. Now we show that the condition (4.6) holds.

$$2|F(p,q) - F(u,v)| \leq \frac{3}{10} \cdot 2|g(p) - g(u)| + \frac{3}{5} \cdot 2|g(q) - g(v)|, p \geq u, q \leq v \tag{4.7}$$

and

$$2|F(q,p) - F(v,u)| \leq \frac{3}{10} \cdot 2|g(q) - g(v)| + \frac{3}{5} \cdot 2|g(p) - g(u)|, p \geq u, q \leq v. \tag{4.8}$$

Adding (4.7) and (4.8), we get the inequality (4.6) with $k = \frac{9}{10}$. Other cases can be similarity shown. Here (0, 0) is a coupled coincidence point.

## Application to integral equations

Here give an application of a result in the previous section to a problem of an integral equation. Similar applications to integral equation problem results have been made in [2, 4, 21]. We consider the integral equation

$$p(t) = \int_a^t (m_1(t,s) + m_2(t,s))(f_1(s,p(s)) + f_2(s,p(s)))ds + h(t), t \in [a,\infty), \tag{5.1}$$

where $h \in L[a,\infty)$, $m_1(t,s), m_2(t,s), f_1(s,p(s)), f_2(s,p(s))$ are real-valued functions that are measurable both in t and s.

**Assumption 5.1** The functions $m_1(t,s), m_2(t,s), f_1(s,p(s)), f_2(s,p(s))$ satisfy the following:

(1) $m_1(t,s) \geq 0$, $t,s \in [a,\infty)$ and $\int_a^t sup_{s\in[0,\infty)}|m_1(t,s)| dt = \frac{M_1}{2} < \infty$,
(2) $m_2(t,s) \leq 0$, $t,s \in [a,\infty)$ and $\int_a^t sup_{s\in[0,\infty)}|m_2(t,s)| dt = \frac{M_2}{2} < \infty$,
(3) $f_1(s,p(s)), f_2(s,p(s) \in L[a,\infty)$ for all $p \in L[a,\infty)$ and there exist $\lambda, \mu > 0$ such that

$$0 \leq f_1(s,p(s)) - f_1(s,q(s)) \leq \lambda(p(s) - q(s)) \text{ and } -\mu(p(s) - q(s)) \leq f_2(s,p(s)) - f_2(s,q(s)) \leq 0$$

for all $p,q \in L[a,\infty)$ with $q(s)b \leq p(s)$.

**Definition 5.2** An element $(c,w) \in L[a,\infty) \times L[a,\infty)$ is called a coupled lower and upper solution of the integral equation (5.1) if $c(t) \leq w(t)$ and

$$c(t) \leq \int_a^t (m_1(t,s)(f_1(s,c(s)) + f_2(s,w(s)))ds + \int_a^t (m_2(t,s)(f_1(s,c(s)) + f_2(s,w(s)))ds + h(t), w(t) \leq \int_a^t (m_1(t,s)(f_1(s,w(s)) + f_2(s,c(s)))ds + \int_a^t (m_2(t,s)(f_1(s,w(s)) + f_2(s,c(s)))ds + h(t), \text{ where } t \in L[a,\infty).$$

**Theorem 5.3** With the Assumption 5.1, if (5.1) has a coupled lower and upper solution and $\frac{(\lambda+\mu)(M_1+M_2)}{2} < 1$, then it has a solution in $L[a,\infty)$.

*Proof* We consider the complete $G$-metric space $(A, G)$ where $A = L[a,\infty)$ and $G(p,q,r) = \int_a^\infty [|p(t) - q(t)| + |q(t) - r(t)| + |r(t) - p(t)|]dt$, for all $p,q,r \in A$. For every $p,q \in A$, let

$$F(p,q)(t) = \int_a^t (m_1(t,s)(f_1(s,p(s)) + f_2(s,q(s)))ds + \int_a^t m_2(t,s)(f_1(s,p(s)) + f_2(s,q(s)))ds + h(t), t \in [a,\infty).$$

It follows from Theorem 3.2 of [15] that $F(p, q)$ has the mixed monotone property. Now for $p,q,u,v \in X$, $p \geq q$, $u \leq v$, we have

$$F(p,q)(t) - F(u,v)(t) = \Big|\int_a^t (m_1(t,s)(f_1(s,p(s)) + f_2(s,q(s)))ds + \int_a^t m_2(t,s)(f_1(s,q(s)) + f_2(s,p(s)))ds - (\int_a^t (m_1(t,s)(f_1(s,u(s)) + f_2(s,v(s)))ds + \int_a^t m_2(t,s)(f_1(s,v(s)) + f_2(s,u(s)))ds)\Big|,$$
$$= \Big|\int_a^t (m_1(t,s)[(f_1(s,p(s)) - (f_1(s,u(s))) - (f_2(s,v(s)) - f_2(s,q(s)))]ds + \int_a^t m_2(t,s)[(f_1(s,v(s)) - f_1(s,q(s))) - (f_2(s,p(s)) - f_2(s,u(s)))]ds\Big|, \leq \Big|\int_a^t (m_1(t,s)[\lambda(p(s) - u(s)) + \mu(q(s) - v(s))]ds + \int_a^t m_2(t,s)[(\lambda(v(s) - q(s)) + \mu(p(s) - u(s))]ds\Big|.$$

Since $m_1(t,s) \geq 0$, $m_2(t,s) \leq 0$, $p(s) - u(s) \geq 0$, $v(s) - q(s) \geq 0$ and $\lambda, \mu \geq 0$, we have

$$2|F(p,q)(t) - F(u,v)(t)| \leq 2(\int_a^t (m_1(t,s)[\lambda(p(s) - u(s)) + \mu(q(s) - v(s))]ds + \int_a^t m_2(t,s)[(\lambda(v(s) - q(s)) + \mu(p(s) - u(s))]ds) \leq sup_{s\in[0,\infty)}|m_1(t,s)| \int_a^t [\lambda 2|(p(s) - u(s))| + \mu 2|(q(s) - v(s))|]ds + sup_{s\in[0,\infty)}|m_2(t,s) \int_a^t m_2(t,s)[(\lambda 2|(v(s) - q(s))| + \mu 2|(p(s) - u(s))|]ds \leq sup_{s\in[0,\infty)}|m_1(t,s)| \int_a^\infty [\lambda|(p(s) - u(s))| + \mu|(q(s) - v(s))|]ds + sup_{s\in[0,\infty)}|m_2(t,s) \int_a^\infty m_2(t,s)[(\lambda 2|(v(s) - q(s))| + \mu 2|(p(s) - u(s))|]ds.$$

Therefore,

$$2|F(p,q)(t) - F(u,v)(t)| \leq sup_{s\in[0,\infty)}|m_1(t,s)| \int_a^\infty [\lambda 2|(p(s) - u(s))| + \mu 2|(q(s) - v(s))|]ds + sup_{s\in[0,\infty)}|m_2(t,s) \int_a^\infty m_2(t,s)[(\lambda 2|(v(s) - q(s))| + \mu 2|(p(s) - u(s))|]ds \leq sup_{s\in[0,\infty)}|m_1(t,s)| \int_a^\infty [\lambda G(p,p,u) + \mu G(q,q,v)]ds + sup_{s\in[0,\infty)}|m_2(t,s) \int_a^\infty m_2(t,s)[(\lambda G(q,q,v) + \mu G(p,p,u)]ds$$

$$2\int_a^\infty |F(p,q)(t) - F(u,v)(t)|dt \leq \int_a^\infty sup_{s\in[0,\infty)}|m_1(t,s)|[\lambda G(p,p,u) + \mu G(q,q,v)]dt + \int_a^\infty sup_{s\in[0,\infty)}|m_2(t,s)|[(\lambda G(q,q,v) + \mu G(p,p,u)]dt \leq \frac{M_1}{2}[\lambda G(p,p,u) + \mu G(q,q,v)] + \frac{M_2}{2}[(\lambda G(q,q,v) + \mu G(p,p,u)]$$

$$G(F(p,q), F(p,q), F(u,v)) \leq \frac{M_1}{2}[\lambda G(p,p,u) + \mu G(q,q,v)] + \frac{M_2}{2}[(\lambda G(q,q,v) + \mu G(p,p,u)] \tag{5.2}$$

Similarly, we have

$$G(F(q,p), F(q,p), F(v,u)) \leq \frac{M_1}{2}[\lambda G(q,q,v) + \mu G(q,q,u)] + \frac{M_2}{2}[(\lambda G(p,p,u) + \mu G(q,q,v)]. \tag{5.3}$$

Adding (5.2) and (5.3), we have

$$G(F(p,q),F(p,q),F(u,v)) + G(F(q,p),F(q,p),F(v,u)) \leq \frac{(M_1+M_2)(\lambda+\mu)}{2}(G(p,p,u)+G(q,q,v))ds.$$

Again if $(c, w)$ be a coupled lower and upper solution of the integral equation (5.1), then $c \leq F(c,w)$ and $w \geq F(w,c)$, which show that every condition of Corollary 4.2 are satisfied by taking $g = I$. By an application of Corollary 4.2 it follows that (5.1) has a solution in $L[a,\infty)$. □

**Acknowledgments** The authors are sincerely grateful to the referee for suggestions which have been found of great value in the improvement of results in this paper.

## References

1. Abbas, M., Rhoades, B.E.: Common fixed point results for non commuting mappings without continuity in generalised metric spaces. Appl. Math. Comput. **215**, 262–269 (2009)
2. Berzig, M., Samet, B.: An extension of coupled fixed point's concept in higher dimension and applications. Comput. Math. Appl. **63**, 1319–1334 (2012)
3. Bhaskar, T.G., Lakshmikantham, V.: Fixed point theorems in partially ordered metric spaces and applications. Nonlinear Anal. **65**, 1379–1393 (2006)
4. Cho, Y.J., Rhoades, B.E., Saadati, R., Samet, B., Shantawi, W.: Nonlinear coupled fixed point theorems in ordered generalized metric spaces with integral type. Fixed Point Theory Appl. **2012**, 8 (2012)
5. Choudhury, B.S., Kumar, S., Asha, Das, K.P.: Some fixed point theorems in *G*- metric spaces. Math. Sci. Lett. **1**, 25–31 (2012)
6. Choudhury, B.S., Das, K.P., Das, P.: Extensions of Banach's and Kannan's results in fuzzy metric spaces. Commun. Korean Math. Soc. **27**, 265–277 (2012)
7. Choudhury, B.S., Das, K.P., Das, P.: Coupled coincidence point results for compatible mappings in partially ordered fuzzy metric spaces. Fuzzy Sets Syst. **222**, 84–97 (2013)
8. Chugh, R., Kadian, T., Rani, A., Rhoades, B.E.: Property P in *G*-metric spaces. Fixed Point Theory Appl., Volume 2010, Article ID 401684. doi:10.1155/2010/401684
9. George, A., Veeramani, P.: On some result in fuzzy metric space. Fuzzy Sets Syst. **64**, 395–399 (1994)
10. Gholizadeh, L.: A fixed point theorem in generalized ordered metric spaces with application. J. Nonlinear Sci. Appl. **6**, 244–251 (2013)
11. Gregori, V., Sapena, A.: On fixed-point theorems in fuzzy metric spaces. Fuzzy Sets Syst. **125**, 245–252 (2002)
12. Hadžić, O., Pap, E.: Fixed Point Theory in Probabilistic Metric Space. Kluwer Academic Publishers, Dordrecht (2001)
13. Hu, X.Q., Luo, Q.: Coupled coincidence point theorems for contractions in generalized fuzzy metric spaces. Fixed Point Theory Appl. **2012**, 196 (2012)
14. Lakshmikantham, V., Ćirić, L.: Coupled fixed point theorems for nonlinear contractions in partially ordered metric spaces. Nonlinear Anal. **70**, 4341–4349 (2009)
15. Luong, N.V., Thung, N.X.: Coupled fixed point theorems for mixed monotone mappings and an application to integral equations. Comput. Math. Appl. **62**, 4238–4248 (2011)
16. Mustafa, Z., Sims, B.: Some remarks concerning *D*-metric spaces. Int. Conf. Fixed Point Theory Appl. **22**, 189–198 (2004)
17. Mustafa, Z., Sims, B.: A new approach to generalized metric spaces. J. Nonlinear Convex. Anal. **7**, 289–297 (2006)
18. Rao, K.P.R., Altun, I., Hima Bindu, S.: Coupled fixed point theorems in generalized fuzzy metric spaces. Adv. Fuzzy Syst., Volume 2011, Article ID 986748, 6 p
19. Saadati, R., Vaezpour, S.M., Vetro, P., Rhoades, B.E.: Fixed point theorems in generalized partially ordered *G*-metric spaces. Math. Comput. Model. **52**, 797–801 (2010)
20. Sihag, V., Vats, R.K.: Coupled coincidence point result in partially ordered generalized metric spaces. Tankang J. Math. **43**, 609–619 (2012)
21. Sintunavarat, W., Kumam, P.: Coupled fixed point results for non linear integral equations. J. Egypt. Math. Soc. **21**, 266–272 (2013)
22. Sun, G., Yang, K.: Generalized fuzzy metric spaces with properties. Res. J. Appl. Sci. Eng. Technol. **2**, 673–678 (2010)
23. Zadeh, L.: Fuzzy sets. Inf. Control. **8**, 338–353 (1965)

# 25

# Numerical solution for solving special eighth-order linear boundary value problems using Legendre Galerkin method

Zaffer Elahi[1] · Ghazala Akram[1] · Shahid Saeed Siddiqi[2]

**Abstract** In this paper, Galerkin method has been introduced using Legendre polynomials as basis functions over the interval $[-1, 1]$ to solve the eighth-order linear boundary value problems with two-point boundary conditions. Legendre Galerkin method is an effective tool in numerically solving such problems. The performance and applicability of the method is illustrated through some examples that reveal the method presents much better results. The obtained numerical results are convincing and very close to the analytical ones.

**Keywords** Galerkin method · Legendre polynomials · Eighth order · Numerical solutions

## Introduction

Consider the general eighth-order linear differential equation of the form

$$Lu(x) = u^{(8)}(x) + \sum_{i=0}^{7} a_i(x)u^{(i)}(x) = f(x), \quad x \in [-1, 1], \tag{1}$$

subject to the following boundary conditions

✉ Ghazala Akram
toghazala2003@yahoo.com

Zaffer Elahi
zafferelahi@gmail.com

Shahid Saeed Siddiqi
shahidsiddiqiprof@yahoo.co.uk

1 Department of Mathematics, University of the Punjab, Quaid-e-Azam Campus, Lahore 54590, Pakistan

2 Abdus Salam School of Mathematical Sciences (ASSMS), Government College University, Lahore 54600, Pakistan

$$u^{(j)}(-1) = u^{(j)}(1) = 0, \quad j = 0, 1, 2, 3, \tag{2}$$

where $u(x)$, $f(x)$ are continuous functions in the space $\ell^2]-1,1[$ and $a_i(x) = x^i$.

The boundary value problems of higher order have been investigated due to their mathematical importance and the potential for the applications in hydrodynamic and hydromagnetic stability [1, 2]. Higher-order boundary value problems arise in many fields. For instance, sixth- and eighth-order differential equations are modelled by thermal instability as ordinary convection and overstability in horizontal layer of fluid heated from below subject to the action of rotation [2, 3]. Generally, such problems are known to arise in astrophysics. The narrow convecting layers bounded by stable layers which are believed to surround A-type stars may be modelled by sixth-order boundary value problems [4]. Dynamo actions in some stars may be modelled by such equations [5]. Shen [6] derived an eighth-order differential equation by governing bending and axial vibrations. Equations for the equilibrium in terms of components for an orthotropic thin circular cylindrical shell subjected to a load that is symmetric about the shell result in an eighth-order differential equations as shown by Paliwal and Pande [7]. Bishop et al. [8] showed that an eighth-order differential equation arises in torsional vibration of uniform beams. Existence and uniqueness of solutions of *2n-th* order boundary value problems are discussed by Agarwal [9, 10]. The analytical solutions of such problems cannot be found easily. Therefore, the authors suggested different approximate algorithms using Legendre, Hermite and Laguerre [11, 12] polynomials. Among them, Legendre polynomials have extensively been particularly used in the area of physics and engineering. For instance, Legendre and assocaited Legendre polynomials are also widely used in [13–15], to solve the fractional problems. Spectral methods also have gained a good reputation among numerical analysts as a robust numerical tool for a wide variety of problems in applied mathematics and scientific computing. Many researchers used the spectral approach [16, 17] to solve the ordinary and partial differential equations, respectively. A selective review for getting the numerical solution of the eighth-order boundary value problems is presented here. Boutayeb and Twizell [18] used finite difference methods, Akram and Siddiqi [19, 20] used nonic and non-polynomial spline functions, respectively, Akram and Rehman [21] developed reproducing kernel space, Viswanadham and Ballem [22] used Galerkin method with quintic B-spline, Inc and Evans [23] constructed Adomian decomposition method, Wazwaz [24] developed modified Adomian decomposition method, Siddiqi and Iftikhar [25] used homotopy analysis method, and Ballem and Viswanadham [26] presented the Galerkin method with septic B-splines, whereas Abbasbandy and Shirzadi [27] developed variational iteration method.

In this paper, the Legendre Galerkin method has been elaborated for the solution of linear eighth-order boundary value problem with two-point boundary conditions defined in Eq. (1) with Eq. (2).

In "Preliminaries", some important definitions, lemmas and theorems regarding Legendre polynomials are discussed. Legendre Galerkin method is explained in "Description of the method". Convergence and error analysis of the method are discussed in "Convergence and error analysis". The transformation of nonhomogeneous boundary conditions and change of interval are discussed in "Handling of boundary conditions and solution domain". The practical usefulness and applicability of the method have been discussed via examples in "Numerical examples".

## Preliminaries

Legendre polynomials are widely used as a mathematical tool in applied sciences as well as in engineering field. These polynomials are defined precisely and easily differentiated and integrated as well.

Legendre polynomials of degree $n$ over the interval $[-1, 1]$ is defined as

$$L_n(x) = \frac{1}{2^n}\sum_{k=0}^{N}(-1)^k \frac{(2n-2k)!}{k!(n-k)!(n-2k)!}x^{n-2k},$$

where

$$N = \begin{cases} \dfrac{n}{2}, & \text{if } n = 0, 2, 4, \ldots \\ \dfrac{(n-1)}{2}, & \text{if } n = 1, 3, 5, \ldots \end{cases}$$

and satisfy the following recurrence relations

$$(2n+1)L_n(x) = L'_{n+1}(x) - L'_{n-1}(x), \tag{3}$$

$$nL_n(x) = xL'_n(x) - L'_{n-1}(x). \tag{4}$$

Legendre polynomials are orthogonal on $[-1, 1]$ with respect to the weight function 1, i.e.

$$\int_{-1}^{1} L_m(x)L_n(x)\mathrm{d}x = \begin{cases} \dfrac{2}{2n+1}, & \text{if } m = n \\ 0, & \text{if } m \neq n \end{cases} \tag{5}$$

and

$$\int_{-1}^{1} L_n(x)\mathrm{d}x = \begin{cases} 2, & \text{if } n = 0 \\ 0, & \text{if } n > 0. \end{cases} \tag{6}$$

**Lemma 2.1** *Let $n$ and $m$ be any two integers such that $n - m \leq N$ and $m > 0$, then*

$$\int_{-1}^{1} L_n(x)L''_{n-m}(x)\mathrm{d}x = 0.$$

*Proof* Integrating the left hand side by parts and using Eq. (6) yield the result.

**Lemma 2.2** *Let $n$ and $m$ be any two integers such that $n \geq m$, then*

$$\int_{-1}^{1} L_n(x)L'_m(x)\mathrm{d}x = 0.$$

*Proof* The proof is divided into two parts.

**Case I** For $n = m$, we have

$$\int_{-1}^{1} L_n(x)L'_n(x)\mathrm{d}x = \left[\frac{1}{2}\{L_n(x)\}^2\right]_{-1}^{1} = 0.$$

**Case II** For $n > m$, the integral on the left, using Eqs. (3) and (6), can be written as

$$\begin{aligned}\int_{-1}^{1} L_n(x)L'_m(x)\mathrm{d}x &= \int_{-1}^{1}\left[(2m-1)L_{m-1}(x) + L'_{m-2}(x)\right]L_n(x)\mathrm{d}x \\ &= \int_{-1}^{1} L_n(x)L'_{m-2}(x)\mathrm{d}x \\ &= \int_{-1}^{1} L_n(x)L'_{m-4}(x)\mathrm{d}x \\ &\;\;\vdots \\ &= \begin{cases} \int_{-1}^{1} L_n(x)L'_0(x)\mathrm{d}x, & \text{if } m = even \\ \int_{-1}^{1} L_n(x)L'_1(x)\mathrm{d}x, & \text{if } m = odd \end{cases} \\ &= 0.\end{aligned}$$

**Theorem 2.1** *Let $n$ and $m$ be any two integers such that $n, m \leq N$ , then*

(1) $\int_{-1}^{1} L_n'(x)L_m(x)\mathrm{d}x = \begin{cases} 2, & \text{if } n = m+i \\ 0, & \text{if } n \neq m+i \text{ or } n \leq m, \end{cases}$

(2) $\int_{-1}^{1} L_n''(x)L_m(x)\mathrm{d}x$

$= \begin{cases} n(n+1) - m(m+1), & \text{if } n \neq m+i \\ 0, & \text{if } n = m+i \text{ or } n \leq m, \end{cases}$

*where* $i = 1, 3, 5, \ldots, 2k+1 \leq N - m.$

*Proof*

(1) Integrating $\int_{-1}^{1} L_n'(x)L_m(x)\mathrm{d}x$ by parts gives

$$\int_{-1}^{1} L_n'(x)L_m(x)\mathrm{d}x = [L_n(x)L_m(x)]_{-1}^{1} - \int_{-1}^{1} L_n(x)L_m'(x)\mathrm{d}x$$
$$= \left[1 + (-1)^{n+m+1}\right] - \int_{-1}^{1} L_n(x)L_m'(x)\mathrm{d}x. \tag{7}$$

For $n = m+i, i = 1, 3, 5, \ldots, 2k+1 \leq N - m$ and using Lemma 2.2 lead to

$$\int_{-1}^{1} L_n'(x)L_m(x)\mathrm{d}x = 2.$$

For $n = m+i, i = 0, 2, 4, \ldots, 2k \leq N - m$, Eq. (7) yields

$$\int_{-1}^{1} L_n'(x)L_m(x)\mathrm{d}x = 0.$$

For $n \leq m$ and considering the previous cases with Lemma 2.2 yield $\int_{-1}^{1} L_n'(x)L_m(x)\mathrm{d}x = 0$.

(2) The proof is divided into four parts.

(a) For $n = m+i, \quad i = 2, 4, 6, \ldots, 2k \leq N - m.$

$$\int_{-1}^{1} L_n''(x)L_m(x)\mathrm{d}x = \left[L_n'(x)L_{n-i}(x)\right]_{-1}^{1} - \int_{-1}^{1} L_n'(x)L_{n-i}'(x)\mathrm{d}x$$
$$= n(n+1) - \left[L_n(x)L_{n-i}'(x)\right]_{-1}^{1}$$
$$+ \int_{-1}^{1} L_n(x)L_{n-i}''(x)\mathrm{d}x$$
$$= n(n+1) - \left[L_n(x)L_{n-i}'(x)\right]_{-1}^{1},$$
[using Lemma 2.1]
$$= n(n+1) - m(m+1).$$

(b) For $n = m+i, i = 1, 3, 5, \ldots, 2k+1 \leq N - m$, then $\int_{-1}^{1} L_n''(x)L_m(x)\mathrm{d}x = 0$.

(c) For $n = m$, then

$$\int_{-1}^{1} L_n''(x)L_m(x)\mathrm{d}x = \left[L_n'(x)L_{n-i}(x)\right]_{-1}^{1} - \int_{-1}^{1} L_n'(x)L_{n-i}'(x)\mathrm{d}x$$
$$= n(n+1) - m(m+1) = 0.$$

(d) For $n < m$, then integrating $\int_{-1}^{1} L_n''(x)L_m(x)\mathrm{d}x$ by parts and using Eq. (6) leads to $\int_{-1}^{1} L_n''(x)L_m(x)\mathrm{d}x = 0$.

## Description of the method

To solve the linear eighth-order boundary value problem (1) by the Galerkin method along with Legendre basis, $u(x)$ is approximated as

$$u(x) = \sum_{j=0}^{n} \alpha_j L_j(x), \tag{8}$$

where $\alpha_j's$, $j = 0, 1, 2, \ldots, n$ are the Legendre coefficients. To determine these coefficients $\alpha_j$, orthogonalizing the residual with respect to the basis functions, i.e.

$$\langle u^{(8)}(x), L_r(x)\rangle + \sum_{i=0}^{7} \langle a_i(x)u^{(i)}(x), L_r(x)\rangle - \langle f(x), L_r(x)\rangle = 0, \tag{9}$$

where

$$\langle \phi, \psi \rangle = \int_{-1}^{1} \phi(x)\psi(x)\mathrm{d}x.$$

We approximate the integrals in Eq. (9) by integrating by parts such that all derivatives transfer from $u$ to $L_r$. For convenience, few of the inner products of Eq. (9) can be calculated, as

$$\langle a_3(x)u^{(3)}(x), L_r(x)\rangle = -\int_{-1}^{1} u(x)[a_3(x)L_r(x)]^{(3)}\mathrm{d}x, \tag{10}$$

$$\langle a_2(x)u^{(2)}(x), L_r(x)\rangle = \int_{-1}^{1} u(x)[a_2(x)L_r(x)]^{(2)}\mathrm{d}x, \tag{11}$$

$$\langle a_1(x)u'(x), L_r(x)\rangle = -\int_{-1}^{1} u(x)[a_1(x)L_r(x)]'\mathrm{d}x, \tag{12}$$

$$\langle a_0(x)u(x), L_r(x)\rangle = \int_{-1}^{1} a_0(x)u(x)L_r(x)\mathrm{d}x, \tag{13}$$

and $\langle f(x), L_r(x)\rangle \simeq \sum_{k=0}^{m} \dfrac{2f(x_k)L_r(x_k)}{\left[(1 - x_k^2)(L_m'(x_k))^2\right]}.$ (14)

**Lemma 3.1** *The following relations hold*:

1. $$\langle u^{(8)}(x), L_r(x)\rangle = \sum_{k=4}^{7} (-1)^{k+1}\left[u^{(k)}(x)L_r^{(7-k)}(x)\right]_{-1}^{1} + \int_{-1}^{1} u(x)L_r^{(8)}(x)\mathrm{d}x, \tag{15}$$

2. $$\langle a_7(x)u^{(7)}(x), L_r(x)\rangle = \sum_{k=4}^{6}(-1)^k\left[u^{(k)}(x)\{a_7(x)L_r(x)\}^{(6-k)}\right]_{-1}^{1} - \int_{-1}^{1}u(x)\{a_7(x)L_r(x)\}^{(7)}\mathrm{d}x, \tag{16}$$

3. $$\langle a_6(x)u^{(6)}(x), L_r(x)\rangle = \sum_{k=4}^{5}(-1)^{k+1}\left[u^{(k)}(x)\{a_6(x)L_r(x)\}^{(5-k)}\right]_{-1}^{1} + \int_{-1}^{1}u(x)\{a_6(x)L_r(x)\}^{(6)}\mathrm{d}x, \tag{17}$$

4. $$\langle a_5(x)u^{(5)}(x), L_r(x)\rangle = \left[u^{(4)}(x)a_5(x)L_r(x)\right]_{-1}^{1} - \int_{-1}^{1}u(x)\{a_5(x)L_r(x)\}^{(5)}\mathrm{d}x, \tag{18}$$

5. $$\langle a_4(x)u^{(4)}(x), L_r(x)\rangle = \int_{-1}^{1}u(x)\{a_4(x)L_r(x)\}^{(4)}\mathrm{d}x. \tag{19}$$

*Proof*

1. As

$$\langle u^{(8)}(x), L_r(x)\rangle = \int_{-1}^{1}u^{(8)}(x)L_r(x)\mathrm{d}x.$$

Integrating the right hand terms of the above equation by parts leads to

$$\langle u^{(8)}(x), L_r(x)\rangle = B_{T,8} + \sum_{k=4}^{7}(-1)^{k+1}\left[u^{(k)}(x)L_r^{(7-k)}(x)\right]_{-1}^{1} + \int_{-1}^{1}u(x)L_r^{(8)}(x)\mathrm{d}x,$$

where the boundary term

$$B_{T,8} = \sum_{k=0}^{3}(-1)^{k+1}\left[u^{(k)}(x)L_r^{(7-k)}(x)\right]_{-1}^{1}$$

is zero using the boundary conditions defined in Eq. (2) yielding the relation.

2. The inner product of $\{a_7(x)u^{(7)}(x)\}$ with $L_r(x)$ is obtained, as

$$\langle a_7(x)u^{(7)}(x), L_r(x)\rangle = B_{T,7} + \sum_{k=4}^{6}(-1)^k\left[u^{(k)}(x)\{a_7(x)L_r(x)\}^{(6-k)}\right]_{-1}^{1} - \int_{-1}^{1}u(x)\{a_7(x)L_r(x)\}^{(7)}\mathrm{d}x,$$

where the boundary term

$$B_{T,7} = \sum_{k=0}^{3}(-1)^k\left[u^{(k)}(x)\{a_7(x)L_r(x)\}^{(6-k)}\right]_{-1}^{1} = 0$$

gives the relation. The other relations can be obtained similarly.

**Theorem 3.1** *If Eq. (8) is the assumed approximate solution of the boundary value problem (1)–(2), then the discrete system for determining the coefficients $\{\alpha_j\}_{j=0}^{n}$ is given by*

$$\sum_{j=0}^{n}\Bigg[\sum_{k=4}^{7}(-1)^{k+1}\left[L_j^{(k)}(x)L_r^{(7-k)}(x)\right]_{-1}^{1} + \sum_{k=4}^{6}(-1)^k\left[L_j^{(k)}(x)\{a_7(x)L_r(x)\}^{(6-k)}\right]_{-1}^{1} + \sum_{k=4}^{5}(-1)^{k+1}\left[L_j^{(k)}(x)\{a_6(x)L_r(x)\}^{(5-k)}\right]_{-1}^{1} + \left[L_j^{(4)}(x)a_5(x)L_r(x)\right]_{-1}^{1} + \sum_{q=0}^{8}(-1)^q\int_{-1}^{1}L_j(x)\{a_q(x)L_r(x)\}^{(q)}\mathrm{d}x\Bigg]\alpha_j = \sum_{k=0}^{m}\frac{2f(x_k)L_r(x_k)}{\left[(1-x_k^2)(L_m'(x_k))^2\right]}, \quad 0\leq r\leq n. \tag{20}$$

*It can be written, in matrix form, as*

$$AX = B, \tag{21}$$

*where*

$$A = \begin{pmatrix} \mu_{0,0}+\nu_{0,0} & \mu_{1,0}+\nu_{1,0} & \mu_{2,0}+\nu_{2,0} & \cdots & \mu_{n,0}+\nu_{n,0} \\ \mu_{0,1}+\nu_{0,1} & \mu_{1,1}+\nu_{1,1} & \mu_{2,1}+\nu_{2,1} & \cdots & \mu_{n,1}+\nu_{n,1} \\ \mu_{0,2}+\nu_{0,2} & \mu_{1,2}+\nu_{1,2} & \mu_{2,2}+\nu_{2,2} & \cdots & \mu_{n,2}+\nu_{n,2} \\ \cdot & \cdot & \cdot & \cdot & \cdot \\ \cdot & \cdot & \cdot & \cdot & \cdot \\ \cdot & \cdot & \cdot & \cdot & \cdot \\ \mu_{0,n}+\nu_{0,n} & \mu_{1,n}+\nu_{1,n} & \mu_{2,n}+\nu_{2,n} & \cdots & \mu_{n,n}+\nu_{n,n} \end{pmatrix}$$

*and*

$$\mu_{j,r} = \sum_{q=0}^{8}(-1)^q\int_{-1}^{1}L_j(x)\{a_q(x)L_r(x)\}^{(q)}\mathrm{d}x, \quad a_8(x)=1,$$

$$\nu_{j,r} = \Bigg[\sum_{k=4}^{7}(-1)^{k+1}L_j^{(k)}(x)L_r^{(7-k)}(x) + \sum_{k=4}^{6}(-1)^kL_j^{(k)}(x)\{a_7(x)L_r(x)\}^{(6-k)} + \sum_{k=4}^{5}(-1)^{k+1}L_j^{(k)}(x)\{a_6(x)L_r(x)\}^{(5-k)} + L_j^{(4)}(x)a_5(x)L_r(x)\Bigg]_{-1}^{1}.$$

*The term $\mu_{j,r}$ can be calculated using the results given in Sect. 2, while the boundary term $\nu_{j,r}$ can be calculated as*

$$\left[L_j^{(4)}(x)a_5(x)L_r(x)\right]_{-1}^{1}=\frac{1}{384}\{1-(-1)^{n+r+j}\}\prod_{i=0}^{7}(j-i+4),$$

$$\sum_{k=4}^{5}(-1)^{k+1}\left[L_j^{(k)}(x)\{a_6(x)L_r(x)\}^{(5-k)}\right]_{-1}^{1}$$
$$=\frac{1}{3840}\left\{1-(-1)^{n+r+j-1}\right\}\prod_{i=0}^{9}(j-i+5)$$
$$-\frac{1}{768}\{2n+r(r+1)\}\{1+(-1)^{n+r+j}\}$$
$$\times\prod_{i=0}^{7}(j-i+4),$$

$$\sum_{k=4}^{6}(-1)^{k}\left[L_j^{(k)}(x)\{a_7(x)L_r(x)\}^{(6-k)}\right]_{-1}^{1}$$
$$=\frac{1}{46{,}080}\{1-(-1)^{n+r+j}\}\prod_{i=0}^{11}(j-i+6)$$
$$-\frac{1}{7680}\{2n+r(r+1)\}\left\{1+(-1)^{n+r+j-1}\right\}$$
$$\times\prod_{i=0}^{9}(j-i+5)+\frac{1}{3072}\{8n(n-1)+8nr(r+1)$$
$$+(r+2)(r+1)(r)(r-1)\}\{1-(-1)^{n+r+j}\}$$
$$\times\prod_{i=0}^{7}(j-i+4),$$

$$\sum_{k=4}^{7}(-1)^{k+1}\left[L_j^{(k)}(x)L_r^{(7-k)}(x)\right]_{-1}^{1}$$
$$=\frac{1}{645{,}120}\{1+(-1)^{r+j}\}\prod_{i=0}^{13}(j-i+7)$$
$$-\frac{1}{92{,}160}(r(r+1))\left\{1-(-1)^{r+j-1}\right\}\prod_{i=0}^{11}(j-i+6)$$
$$+\frac{1}{30{,}720}\left\{1-(-1)^{r+j-1}\right\}\prod_{i=0}^{3}(r-i+2)\prod_{i=0}^{9}(j-i+5)$$
$$-\frac{1}{18{,}432}\left\{1-(-1)^{r+j-1}\right\}\prod_{i=0}^{5}(r-i+3)\prod_{i=0}^{7}(j-i+4).$$

*After solving the linear system* (21) *having* $(n+1)$ *equations with* $(n+1)$ *unknowns yield, the column vector* $X=(\alpha_0,\alpha_1,\alpha_2,\ldots,\alpha_n)^{\mathrm{T}}$. *Thus, u(x) can now be approximated by Eq.* (8).

## Convergence and error analysis

In this section, the convergence and error analysis of the Legendre Galerkin method have been studied in detail.

### Convergence of the method

**Lemma 4.1** *Let* $x(t)\in H^k]-1,1[$ *(a Sobolev space) and let* $x_n(t)=\sum_{i=0}^{n}c_iL_n(t)$ *be the best approximation polynomial of* $x(t)$ in the $\ell^2$*-norm, then*

$$\|x(t)-x_n(t)\|_{\ell^2[-1,1]}\le c_0n^{-k}\|x(t)\|_{H^k]-1,1[},$$

*and* $c_0$ *is a non-negative constant which depends on the selected norm and is free from* $x(t)$ *and* $n$.

*Proof* [28–30].

**Theorem 4.1** *Assume* $\kappa:X\to X$ *is bounded, with* $X$ *a Banach space, and* $\lambda-\kappa:X\to X$ *is bijective. Further, assume*

$$\|\kappa-\kappa L_n\|\to 0\quad\text{as } n\to\infty,$$

*then for all sufficiently large n, say* $n\ge N$, *the operator* $(\lambda-\kappa L_n)^{-1}$ *exists as a bounded operator from X to X. Moreover, it is uniformly bounded such that*

$$\sup_{n\ge N}\|(\lambda-\kappa L_n)^{-1}\|<\infty.$$

*For the solution of* $(\lambda-\kappa L_n)x_m=L_ny$, $x_m\in X$ *and* $(\lambda-\kappa)x=y$,

$$x-x_m=\lambda(\lambda-L_n\kappa)^{-1}(x-L_n(x)),$$

$$\frac{|\lambda|}{\|\lambda-\kappa L_n\|}\|x-L_n(x)\|\le\|x-x_n\|\le|\lambda|\|(\lambda-\kappa L_n)^{-1}\|\|x-L_n(x)\|.$$

*Proof* [31].

Consequently, the approximation rate of Legendre polynomials is $n^{-k}$ with respect to Lemma 4.1, and also from Theorem 4.1, $\|x-x_n\|$ converge to zero as soon as $\|x-L_n\|$.

### Error analysis of the method

In this subsection, an error estimator for eighth-order boundary value problems using Legendre Galerkin approximation has been discussed.

Consider $e_n(x)=u(x)-u_n(x)$ as the error function of Legendre approximation $u_n(x)$ to $u(x)$, where $u(x)$ is the exact solution of Eq. (1) with boundary conditions defined in Eq. (2). So, $u_n(x)$ satisfies the following problem:

$$u_n^{(8)}(x)+\sum_{i=0}^{7}a_i(x)u_n^{(i)}(x)=f(x)+P_n(x),\quad x\in[-1,1],\tag{22}$$

with boundary conditions

$$u_n^{(i)}(-1)=u_n^{(i)}(1)=0,\quad i=0,1,2,3,\tag{23}$$

where $P_n(x)$ is a perturbation term linked with $u_n(x)$ obtained as follows

$$P_n(x) = u_n^{(8)}(x) + \sum_{i=0}^{7} a_i(x)u_n^{(i)}(x) - f(x), \quad i = 0, 1, 2, 3. \tag{24}$$

We find an approximation $e_{n,N}(x)$ to $e_n(x)$ in the same way as in description of the method, for the solution of Eq. (1) with Eq. (2). Subtracting Eqs. (22) and (23) from Eqs. (1) and (2), respectively, yields the error function of the form

$$P_n(x) = -e_n^{(8)}(x) - \sum_{i=0}^{7} a_i(x)e_n^{(i)}(x) \tag{25}$$

and

$$e_n^{(i)}(-1) = e_n^{(i)}(1) = 0, \quad i = 0, 1, 2, 3. \tag{26}$$

We solve this problem using the Legendre Galerkin method to get the approximation $e_{n,N}(x)$.

## Handling of boundary conditions and solution domain

If the boundary conditions are nonhomogeneous or the solution domain is $[a, b]$, then these conditions are converted to homogeneous conditions and the domain of the solution must be converted to $[-1, 1]$. Consider

$$Lu(t) = u^{(8)}(t) + \sum_{i=0}^{7} a_i(t)u^{(i)}(t) = f(t), \quad t \in [a, b], \tag{27}$$

subject to the following boundary conditions

$$u^{(j)}(a) = \theta_j, \quad u^{(j)}(b) = \phi_j, \quad j = 0, 1, 2, 3. \tag{28}$$

Using the linear transformation $t = \frac{b-a}{2}x + \frac{b+a}{2}$, then Eq. (27) takes the form

$$Lu(x) = \left(\frac{2}{b-a}\right)^8 u^{(8)}(x) + \sum_{i=0}^{7} a_i(\chi)\left(\frac{2}{b-a}\right)^i u^{(i)}(x) = f(\chi), \quad x \in [-1, 1], \tag{29}$$

where

$$\chi = \frac{b-a}{2}x + \frac{b+a}{2},$$

subject to the following boundary conditions

$$u^{(j)}(-1) = \left(\frac{2}{b-a}\right)^j \theta_j = \Theta_j, \quad u^{(j)}(1) = \left(\frac{2}{b-a}\right)^j \phi_j = \Phi_j, \quad j = 0, 1, 2, 3. \tag{30}$$

To transform the nonhomogeneous boundary conditions in Eq. (30) to homogeneous boundary conditions, we replace

$$u(x) = \Psi(x) + \Omega(x), \tag{31}$$

where $\Psi(x)$ is the interpolating polynomial such that $\Psi^{(j)}(-1) = \Theta_j$ and $\Psi^{(j)}(1) = \Phi_j$, $j = 0, 1, 2, 3$. Also,

$$\Omega(x) = \sum_{j=0}^{7} \eta_j x^j$$

and

$$\eta_0 = \frac{1}{96}(48\Theta_0 + 33\Theta_1 + 9\Theta_2 + \Theta_3 + 48\Phi_0 - 33\Phi_1 + 9\Phi_2 - \Phi_3),$$
$$\eta_1 = \frac{1}{96}(-105\Theta_0 - 57\Theta_1 - 12\Theta_2 - \Theta_3 + 105\Phi_0 - 57\Phi_1 + 12\Phi_2 - \Phi_3),$$
$$\eta_2 = \frac{1}{32}(-15\Theta_1 - 7\Theta_2 - \Theta_3 + 15\Phi_1 - 7\Phi_2 + \Phi_3),$$
$$\eta_3 = \frac{1}{32}(35\Theta_0 + 35\Theta_1 + 10\Theta_2 + \Theta_3 - 35\Phi_0 + 35\Phi_1 - 10\Phi_2 + \Phi_3),$$
$$\eta_4 = \frac{1}{32}(5\Theta_1 + 5\Theta_2 + \Theta_3 - 5\Phi_1 + 5\Phi_2 - \Phi_3),$$
$$\eta_5 = \frac{1}{32}(-21\Theta_0 - 21\Theta_1 - 8\Theta_2 - \Theta_3 + 21\Phi_0 - 21\Phi_1 + 8\Phi_2 - \Phi_3),$$
$$\eta_6 = \frac{1}{96}(-3\Theta_1 - 3\Theta_2 - \Theta_3 + 3\Phi_1 - 3\Phi_2 + \Phi_3),$$
$$\eta_7 = \frac{1}{96}(15\Theta_0 + 15\Theta_1 + 6\Theta_2 + \Theta_3 - 15\Phi_0 + 15\Phi_1 - 6\Phi_2 + \Phi_3).$$

The problem takes the form:

$$L\Psi(x) = \Psi^{(8)}(x) + \sum_{i=0}^{7} a_i(x)\Psi^{(i)}(x) = f^*(x), \quad x \in [-1, 1], \tag{32}$$

subject to the following boundary conditions

$$\Psi^{(j)}(-1) = 0, \quad \Psi^{(j)}(1) = 0, \quad j = 0, 1, 2, 3, \tag{33}$$

where

$$f^*(x) = f(x) - L\Omega(x) = f(x) - \sum_{i=0}^{7} a_i(x)\Omega^{(i)}(x).$$

Let

$$\Psi(x) = \sum_{j=0}^{n} \alpha_j L_j(x) \tag{34}$$

be an approximate solution of Eq. (32). Then,

$$u(x) = \sum_{j=0}^{n} \alpha_j L_j(x) + \Omega(x) \tag{35}$$

be the approximate solution of Eq. (31). Using the inverse linear transformation $x = \frac{2}{b-a}t - \frac{b+a}{b-a}$ in Eq. (35) yields the approximate solution $u(t)$ of Eq. (27).

## Numerical examples

Some examples have been constructed to measure the accuracy of the proposed method. Numerical results obtained by the method show the betterment of the method also.

*Example 1* Consider the following differential equation:

$$u^{(8)}(x) + xu(x) = -(48 + 15x + x^3)e^x, \quad x \in [0, 1], \tag{36}$$

subject to the boundary conditions

$$\begin{aligned} &u(0) = 0, \quad u(1) = 0, \quad u'(0) = 1 \quad u'(1) = -e, \\ &u''(0) = 0, \quad u''(1) = -4e, \quad u'''(0) = -3, \\ &u'''(1) = -9e. \end{aligned} \tag{37}$$

The exact solution of the problem is $u(x) = x(1-x)e^x$.

The proposed method is implemented to the problem for $n = 10$. The comparison between the absolute errors of the proposed method and that developed by Viswanadham and Ballem [22] is shown in Table 1 and Fig. 1, respectively.

**Table 1** Numerical results for Example 1

| $x$ | Proposed error | Viswanadham and Ballem [22] |
|---|---|---|
| 0.1 | $1.44002 \times 10^{-7}$ | $5.215406 \times 10^{-8}$ |
| 0.2 | $1.45828 \times 10^{-6}$ | $2.220273 \times 10^{-6}$ |
| 0.3 | $4.38794 \times 10^{-6}$ | $7.003546 \times 10^{-6}$ |
| 0.4 | $7.59131 \times 10^{-6}$ | $1.114607 \times 10^{-5}$ |
| 0.5 | $9.06535 \times 10^{-6}$ | $1.227856 \times 10^{-5}$ |
| 0.6 | $7.81068 \times 10^{-6}$ | $8.881092 \times 10^{-6}$ |
| 0.7 | $4.64523 \times 10^{-6}$ | $2.533197 \times 10^{-6}$ |
| 0.8 | $1.58842 \times 10^{-6}$ | $1.817942 \times 10^{-6}$ |
| 0.9 | $1.61391 \times 10^{-7}$ | $2.041459 \times 10^{-6}$ |

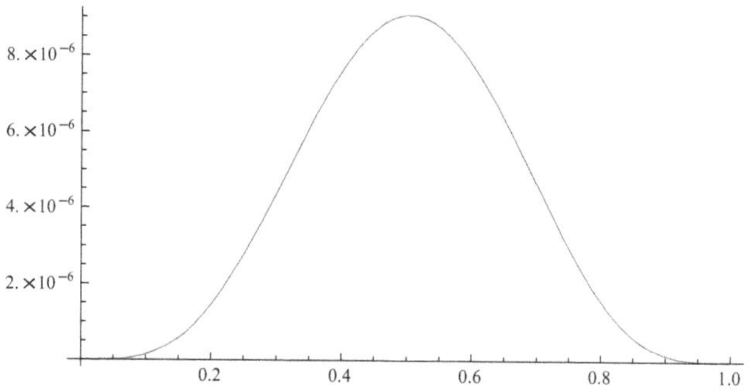


**Fig. 1** Absolute errors for Example 1

*Example 2* Consider the following differential equation:

$$\begin{aligned} &u^{(8)}(x) + u^{(7)}(x) + 2u^{(6)}(x) + 2u^{(5)}(x) + 2u^{(4)}(x) + 2xu^{(3)}(x) \\ &\quad + 2u^{(2)}(x) + x^2u^{(1)}(x) \\ &\quad + xu(x) = -(x^4 - 2x^3 + 14x - 27)\cos x \\ &\quad - (3x^3 - 13x^2 + 11x + 17)\sin x, \quad x \in [0, 1], \end{aligned} \tag{38}$$

subject to the boundary conditions

$$\begin{aligned} &u(0) = 0, \quad u(1) = 0, \quad u'(0) = -1, \quad u'(1) = 2\sin 1, \\ &u''(0) = 0, \quad u''(1) = 4\cos 1 + 2\sin 1, \\ &u'''(0) = 7, \quad u'''(1) = 6\cos 1 - 6\sin 1. \end{aligned} \tag{39}$$

The exact solution of the problem is $u(x) = (x^2 - 1)\sin x$.

The proposed method is implemented to the problem for $n = 10$. The comparison between the absolute errors of the proposed method and that developed by Ballem and Viswanadham [26] is shown in Table 2 and Fig. 2, respectively.

**Table 2** Comparison of numerical results for Example 2

| $x$ | Proposed error | Ballem and Viswanadham [26] |
|---|---|---|
| 0.1 | $5.03731 \times 10^{-8}$ | $4.239380 \times 10^{-6}$ |
| 0.2 | $5.1436 \times 10^{-7}$ | $9.983778 \times 10^{-6}$ |
| 0.3 | $1.55915 \times 10^{-6}$ | $5.096197 \times 10^{-6}$ |
| 0.4 | $2.71487 \times 10^{-6}$ | $7.629395 \times 10^{-6}$ |
| 0.5 | $3.26015 \times 10^{-6}$ | $1.493096 \times 10^{-5}$ |
| 0.6 | $2.82218 \times 10^{-6}$ | $2.288818 \times 10^{-5}$ |
| 0.7 | $1.68491 \times 10^{-6}$ | $2.276897 \times 10^{-5}$ |
| 0.8 | $5.77885 \times 10^{-7}$ | $1.943111 \times 10^{-5}$ |
| 0.9 | $5.88442 \times 10^{-8}$ | $1.323223 \times 10^{-5}$ |

*Example 3* Consider the following differential equation:

$$\begin{aligned} &u^{(8)}(x) + u^{(7)}(x) + 2u^{(6)}(x) + 2u^{(5)}(x) + 2u^{(4)}(x) + 2u^{(3)}(x) \\ &\quad + 2u^{(2)}(x) + u^{(1)}(x) \\ &\quad + u(x) = 14\cos x - 16\sin x - 4x\sin x, \\ &x \in [0, 1], \end{aligned} \tag{40}$$

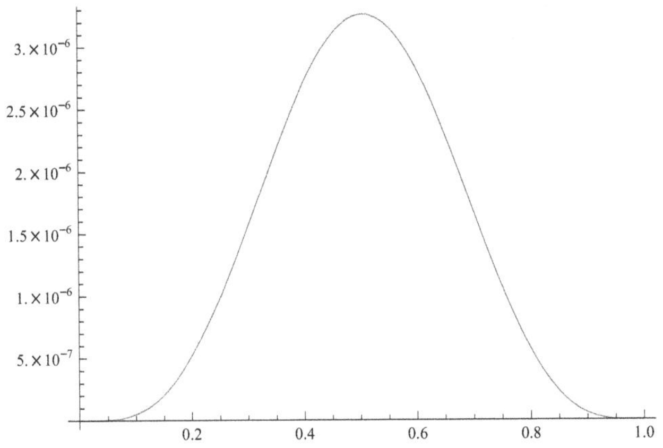


**Fig. 2** The absolute errors for Example 2

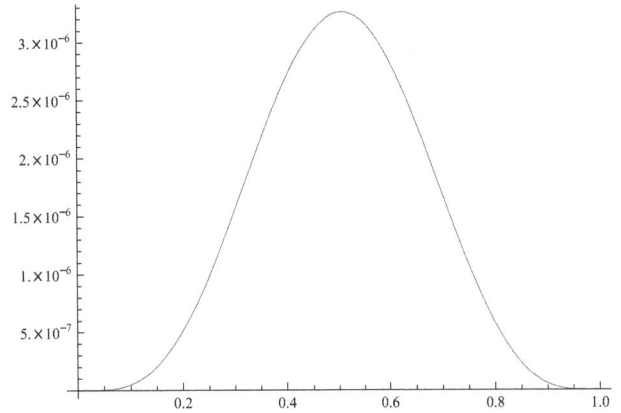

**Table 3** Comparison of numerical results for Example 3

| $x$ | Proposed error | Viswanadham and Ballem [22] |
|---|---|---|
| 0.1 | $5.03731 \times 10^{-8}$ | $3.799796 \times 10^{-7}$ |
| 0.2 | $5.1436 \times 10^{-7}$ | $2.145767 \times 10^{-6}$ |
| 0.3 | $1.55915 \times 10^{-6}$ | $5.632639 \times 10^{-6}$ |
| 0.4 | $2.71487 \times 10^{-6}$ | $9.745359 \times 10^{-6}$ |
| 0.5 | $3.26015 \times 10^{-6}$ | $1.138449 \times 10^{-5}$ |
| 0.6 | $2.82218 \times 10^{-6}$ | $1.013279 \times 10^{-5}$ |
| 0.7 | $1.68491 \times 10^{-6}$ | $7.271767 \times 10^{-6}$ |
| 0.8 | $5.77885 \times 10^{-7}$ | $3.874302 \times 10^{-6}$ |
| 0.9 | $5.88442 \times 10^{-8}$ | $1.430511 \times 10^{-6}$ |

subject to the boundary conditions

$$\begin{aligned} &u(0)=0, \quad u(1)=0, \quad u'(0)=-1, \quad u'(1)=2sin1, \\ &u''(0)=0, \quad u''(1)=4\cos 1+2\sin 1, \\ &u'''(0)=7, \quad u'''(1)=6\cos 1-6\sin 1. \end{aligned} \tag{41}$$

The exact solution of the problem is $u(x)=(x^2-1)\sin x$.

The proposed method is implemented to the problem for $n=10$. The comparison between the absolute errors of the proposed method and that developed by Viswanadham and Ballem [22] is shown in Table 3 and Fig. 3, respectively.

## Conclusion

In this paper, Galerkin method using Legendre polynomials as basis function has been developed to approximate the linear eighth-order boundary value problems. In this method, the nonhomogeneous boundary conditions are transformed to the homogeneous boundary conditions and the solution domain is converted to the interval $[-1,1]$. By

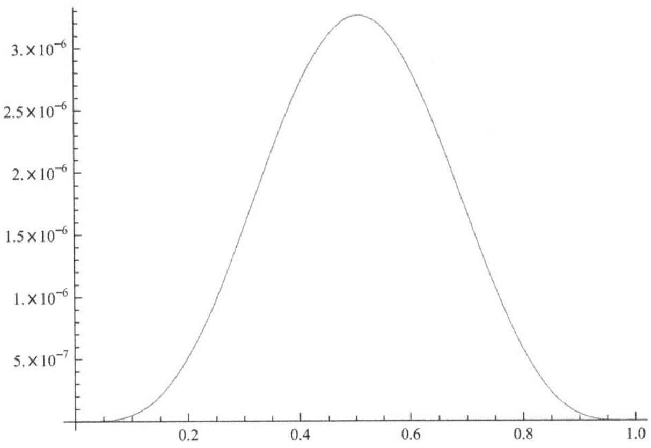


**Fig. 3** Absolute errors for Example 3

comparing the results of the proposed method with other existing methods, it is found that the results are improved and become remarkable. Consequently, the solution may converge efficiently to the analytical one by increasing the order of the problem.

## References

1. Karageorghis, A., Phillips, T.N., Davies, A.R.: Spectral collocation methods for the primary two-point boundary values problem in modelling viscoelastic flows. Int. J. Numer. Methods Eng. **26**(4), 805–813 (1988)
2. Chandrasekhar, S.: Hydrodynamic and Hydromagnetic Stability. Calrendon Press, Oxford (1961) (reprinted: Dover Books, New York, 1981)
3. Boutayeb, A.: Numerical methods for high-order ordinary differential equations with application to eigenvalue problems. Ph.D. thesis, Brunel University, chapter 1, p 112 (1990)
4. Toomre, J., Zahn, J.R., Labour, J., Spiegel, E.A.: Stellar convection theory II: single-mode study of the second convection zone in A-type stars. Astrophys. J. **207**, 545–563 (1976)
5. Glatzmaier, G.A.: Numerical simulations of stellar convection dynamics III: at the base of the convection zone. Geophys. Astrophys. Fluid Dyn. **31**, 137–150 (1985)
6. Shen, I.Y.: Hybrid damping through intelligent constrained layer layer treatments. ASME J. Vib. Acoust. **116**, 341–349 (1994)
7. Paliwal, D.N., Pande, A.: Orthotropic cylindrical pressure vessels under line load. Int. J. Press. Vessels Pip. **76**, 455–459 (1999)
8. Bishop, R.E.D., Cannon, S.M., Miao, S.: On coupled bending and torsional vibration of uniform beams. J. Sound Vib. **131**, 457–464 (1989)
9. Agarwal, R.P.: Boundary Value Problems for Higher Order Differential Equations. World Scientific, Singapore (1986)
10. Agarwal, R.P., Akrivis, G.: Boundary value problems occurring in plate deflection theory. J. Comput. Appl. Math. **8**, 145–154 (1982)
11. Bell, W.W.: Special Function for Scientist and Engineer. D. Van Nostrand Company Ltd., London (1967)
12. Arfken, G.: Mathematical Methods for Physics, 2nd edn. Academic Press Inc, New York (1970)
13. Ezz-Eldien, S.S.: New quadrature approach based on operational matrix for solving a class of fractional variational problems. J. Comput. Phys. **317**, 362–381 (2016)
14. Bhrawy, A.H., Ezz-Eldien, S.S.: A new Legendre operational

technique for delay fractional optimal control problems. Calcolo (2015). doi:10.1007/s10092-015-0160-1
15. Ezz-Eldien, S.S., Doha, E.H., Baleanu, D., Bhrawy, A.H.: A numerical approach based on Legendre orthonormal polynomials for numerical solutions of fractional optimal control problems. J. Vib. Control (2015). doi:10.1177/1077546315573916
16. Bhrawy, A.H., Doha, E.H., Ezz-Eldien, S.S., Gorder, R.A.V.: A new Jacobi spectral collocation method for solving 1 + 1 fractional Schrodinger equations and fractional coupled Schrodinger systems. Eur. Phys. J. Plus **129**(12), 1–21 (2014)
17. Doha, E.H., Bhrawy, A.H., Ezz-Eldien, S.S.: An efficient Legendre spectral tau matrix formulation for solving fractional sub diffusion and reaction sub diffusion equations. J. Comput. Nonlinear Dyn. **10**(2), 021019-8 (2015). doi:10.1115/1.4027944
18. Boutayeb, A., Twizell, E.H.: Finite difference methods for the solution of eighth order boundary value problems. Int. J. Comput. Math. **48**, 63–75 (1993)
19. Akram, Ghazala, Siddiqi, Shahid S.: Nonic spline solutions of eighth order boundary value problems. Appl. Math. Comput. **182**, 829–845 (2006)
20. Siddiqi, S.S., Akram, G.: Solution of eighth order boundary value problems using non polynomial spline technique. Int. J. Comput. Math. **84**, 347–368 (2007)
21. Akram, G., Rehman, H.U.: Numerical solution of eighth order boundary value problems in reproducing kernel space. Numer. Algorithms **62**(3), 527–540 (2013)
22. Viswanadham, K.N.S., Ballem, S.: Numerical solution of eighth order boundary value problems by Galerkin method with quintic B-splines. Int. J. Comput. Appl. **89**(15), 7–13 (2014)
23. Inc, M., Evan, D.J.: An efficient approach to approximate solution of eighth order boundary value problems. Int. J. Comput. Math. **81**, 685–692 (2004)
24. Wazwaz, A.M.: The numerical solution of special eighth order boundary value problems by the modified decomposition method. Neural Parallel Sci. Comput. **8**(2), 133–146 (2000)
25. Siddiqi, S.S., Iftikhar, M.: Numerical solution of higher order boundary value problems. Abstr. Appl. Anal. (2013). doi:10.1155/2013/427521
26. Ballem, S., Viswanadham, K.N.S.: Numerical solution of eighth order boundary value problems by Galerkin method with septic B-splines. Proc. Eng. **127**, 1370–1377 (2015)
27. Abbasbandy, S., Shirzadi, A.: The Variational Iteration method for a class of eighth order boundary value differential equations. Z. Naturforsch. **63a**, 745–751 (2008)
28. Fathy, M., El-Gamel, M., El-Azab, M.: Legendre–Galerkin method for the linear Fredholm integrodifferential equations. Appl. Math. Comput. **243**, 789–800 (2014)
29. Maleknejad, K., Nouri, K., Yousefi, M.: Discussion on convergence of legendre polynomial for numerical solution of integral equations. Appl. Math. Comput. **193**, 335–339 (2007)
30. Canuto, C., Hussaini, M.Y., Quarteroni, A., Zang, T.A.: Spectral Methods on Fluid Dynamics. Springer, Berlin (1988)
31. Atkinson, K.E.: The Numerical Solution of Integral Equations of the Second Kind. Cambridge University Press, Cambridge (1997)

# Permissions

All chapters in this book were first published in MS, by Springer International Publishing AG.; hereby published with permission under the Creative Commons Attribution License or equivalent. Every chapter published in this book has been scrutinized by our experts. Their significance has been extensively debated. The topics covered herein carry significant findings which will fuel the growth of the discipline. They may even be implemented as practical applications or may be referred to as a beginning point for another development.

The contributors of this book come from diverse backgrounds, making this book a truly international effort. This book will bring forth new frontiers with its revolutionizing research information and detailed analysis of the nascent developments around the world.

We would like to thank all the contributing authors for lending their expertise to make the book truly unique. They have played a crucial role in the development of this book. Without their invaluable contributions this book wouldn't have been possible. They have made vital efforts to compile up to date information on the varied aspects of this subject to make this book a valuable addition to the collection of many professionals and students.

This book was conceptualized with the vision of imparting up-to-date information and advanced data in this field. To ensure the same, a matchless editorial board was set up. Every individual on the board went through rigorous rounds of assessment to prove their worth. After which they invested a large part of their time researching and compiling the most relevant data for our readers.

The editorial board has been involved in producing this book since its inception. They have spent rigorous hours researching and exploring the diverse topics which have resulted in the successful publishing of this book. They have passed on their knowledge of decades through this book. To expedite this challenging task, the publisher supported the team at every step. A small team of assistant editors was also appointed to further simplify the editing procedure and attain best results for the readers.

Apart from the editorial board, the designing team has also invested a significant amount of their time in understanding the subject and creating the most relevant covers. They scrutinized every image to scout for the most suitable representation of the subject and create an appropriate cover for the book.

The publishing team has been an ardent support to the editorial, designing and production team. Their endless efforts to recruit the best for this project, has resulted in the accomplishment of this book. They are a veteran in the field of academics and their pool of knowledge is as vast as their experience in printing. Their expertise and guidance has proved useful at every step. Their uncompromising quality standards have made this book an exceptional effort. Their encouragement from time to time has been an inspiration for everyone.

The publisher and the editorial board hope that this book will prove to be a valuable piece of knowledge for researchers, students, practitioners and scholars across the globe.

# List of Contributors

**Maryam Momeni**
Department of Mathematics, Ahvaz Branch, Islamic Azad University (IAU), Ahvaz, Iran

**Taher Yazdanpanah**
Department of Mathematics, Persian Gulf University, 75169 Bushehr, Iran

**Mohsen Rabbani and Reza Arab**
Department of Mathematics, Sari Branch, Islamic Azad University, Sari, Iran

**Rais Ahmad, Mohd. Ishtyak, Mijanur Rahaman and Iqbal Ahmad**
Department of Mathematics, Aligarh Muslim University, Aligarh 202002, India

**Monireh Nosrati Sahlan and Hadi Feyzollahzadeh**
Department of Mathematics and Computer Science, Technical Faculty, University of Bonab, Bonab, Iran

**K. Kaladhar**
Department of Mathematics, National Institute of Technology Puducherry, Karaikal 609605, India

**Ch. RamReddy, D. Srinivasacharya and T. Pradeepa**
Department of Mathematics, National Institute of Technology Warangal, Warangal, India

**Mehmet Zeki Sarikaya and Hatice Yaldiz**
Department of Mathematics, Faculty of Science and Arts, Düzce University, Konuralp Campus, Düzce, Turkey

**Samet Erden**
Department of Mathematics, Faculty of Science, Bartın University, Konuralp Campus, Bartin, Turkey

**Randhir Singh**
Department of Applied Mathematics, Birla Institute of Technology, Mesra, Ranchi 835215, India

**Abdul-Majid Wazwaz**
Department of Mathematics, Saint Xavier University, Chicago IL 60655, USA

**Asyieh Ebrahimzadeh**
Young Researchers and Elite Club, Najafabad Branch, Islamic Azad University, Najafabad, Iran

**Raheleh Khanduzi**
Department of Mathematics, Gonbad-e-Kavous University, Gonbad-e-Kavous, Iran

**Gumrah Uysal**
Department of Computer Technologies, Division of Technology of Information Security, Karabuk University, Karabuk 78050, Turkey

**Ertan Ibikli**
Faculty of Science, Department of Mathematics, Ankara University, Anadolu, Ankara 06100, Turkey

**Rituparna Chutia**
Department of Chemistry, Indian Institute of Technology, Guwahati 781039, Assam, India

**Rekhamoni Gogoi**
North Lakhimpur Girls' Higher Secondary School, Lakhimpur 787001, Assam, India

**D. Datta**
Health Physics Division, Bhabha Atomic Research Division, Mumbai 400085, India

**B. Asady, F. Hakimzadegan and R. Nazarlue**
Department of Mathematics, Islamic Azad University, Science and Branch, Arāk, Iran

**H. Nouriani and R. Ezzati**
Department of Mathematics, Karaj Branch, Islamic Azad University, Karaj, Iran

**Pooja Khandelwal and Arshad Khan**
Department of Mathematics, Jamia Millia Islamia, New Delhi 110025, India

**Anupam Sharma, Mohammad Imdad and Aftab Alam**
Department of Mathematics, Aligarh Muslim University, Aligarh 202 002, India

**M. Mustafa Bahşı**
Department of Mechanical Engineering, Celal Bayar University, Muradiye Campus, Manisa, Turkey

**Ayşe Kurt Bahşı and Mehmet Sezer**
Department of Mathematics, Celal Bayar University, Muradiye Campus, Manisa, Turkey

**Mehmet Çevik**
Department of Mechanical Engineering, Izmir Katip Çelebi University, Çiğli Main Campus, Izmir, Turkey

**Le Anh Minh**
Department of Mathematical Analysis, Hong Duc University, Thanh Hóa, Vietnam

**Dang Dinh Chau**
Department of Mathematics, Hanoi University of Science, VNU, Hanoi, Vietnam

**K. Kaladhar and E. Komuraiah**
Department of Mathematics, National Institute of Technology Puducherry, Karaikal 609605, India

**M. Emin Özdemir**
Department of Mathematics, K.K. Education Faculty, Atatürk University, Kampus, 25240 Erzurum, Turkey

**Saeid Abbasbandy and Elyas Shivanian**
Department of Mathematics, Imam Khomeini International University, Ghazvin 34149-16818, Iran

**Güvenc¸ Arslan and Sevgi Yurt Oncel**
Department of Statistics, Kırıkkale University, Yahs¸ihan, 71450 Kırıkkale, Turkey

**K. Arun Prakash, M. Suresh and S. Vengataasalam**
Department of Mathematics, Kongu Engineering College, Erode, Tamil Nadu, India

**Mahmood Otadi and Maryam Mosleh**
Department of Mathematics, Firoozkooh Branch, Islamic Azad University, Firoozkooh, Iran

**Laurent Hoeltgen**
Chair for Applied Mathematics, Brandenburg University of Technology Cottbus-Senftenberg, Platz der Deutschen Einheit 1, 03046 Cottbus, Germany

**Binayak S. Choudhury and Pradyut Das**
Department of Mathematics, Indian Institute of Engineering Science and Technology, Shibpur, Howrah, West Bengal, India

**Zaffer Elahi and Ghazala Akram**
Department of Mathematics, University of the Punjab, Quaide-Azam Campus, Lahore 54590, Pakistan

**Shahid Saeed Siddiqi**
Abdus Salam School of Mathematical Sciences (ASSMS), Government College University, Lahore 54600, Pakistan

# Index

www.ingramcontent.com/pod-product-compliance
Lightning Source LLC
Chambersburg PA
CBHW082025190326
41458CB00010B/3274
*9781682855713*